Hands-On PLC Programming with RSLogix™ 500 and LogixPro®

The study of PLCs and computer control requires extensive hands-on experimentation and the use of hardware with possible physical access to high-voltage supply and devices. Users must take full precautions, be aware of safety requirements, and must not touch any of the power supply or hardware wiring at all time.

About the Authors

Dr. Eman Kamel holds a BS in electrical engineering from Cairo University, MS in electrical and computer engineering from the University of Cincinnati, and a Ph.D. in industrial engineering from University of Louisville. Dr. Eman had more than 20 years of experience in process automation at several companies, including Dow Chemical, GE Jet Engine, Philip Morris Co., VITOK Engineers, Evana Tools, and PLC Automation. She designed and implemented PLC-based automation projects in several application areas, including tobacco manufacturing, chemical process control, wastewater treatment, plastic sheets processing, and irrigation water level control. She has extensive experience with Siemens and Allen-Bradley PLC programming, instrumentation, communication, and user interfaces. She has also used and developed good knowledge and experience with other types of PLCs, including GE, TI, Modicon, Telemechanic, Furnace, and Reliance. Eman developed customized and interactive PLC/computer control training for several applications. She has extensive computer-aided instruction experience and implementation in the areas of PLCs, computer control, and automation at several universities.

Dr. Khaled Kamel is currently a professor of computer science at Texas Southern University (TSU). He worked as full-time faculty and administrator at the University of Louisville Engineering School, where he was a professor and the chair of the Computer Engineering and Computer Science department. He also worked as instrumentation engineer at GE Jet Engine. He served as the founding dean of the College of Information Technology at the United Arab Emirates University and the College of Computer Science & Information Technology at the Abu Dhabi University. Dr. Kamel received a BS in electrical engineering from Cairo University, a BS in mathematics from Ain Shams University, MS in computer science from Waterloo University, and a Ph.D. in electronics and communication engineering from the University of Cincinnati.Dr. Kamel worked as principle investigator on several government and industry grants, which included the supervision of over 100 graduate research Master and Doctoral students over the past 25 years. His current research interest is more interdisciplinary in nature but focused in the areas of Industrial Control, Sensory Fusion, and Distributed Computing.

Hands-On PLC Programming with RSLogix™ 500 and LogixPro®

Eman Kamel, Ph.D.

Khaled Kamel, Ph.D.

Mc
Graw
Hill
Education

New York Chicago San Francisco Athens
London Madrid Mexico City Milan
New Delhi Singapore Sydney Toronto

Cataloging-in-Publication Data is on file with the Library of Congress.

McGraw-Hill Education books are available at special quantity discounts to use as premiums and sales promotions or for use in corporate training programs. To contact a representative, please visit the Contact Us page at www.mhprofessional.com.

Hands-On PLC Programming with RSLogix™ 500 and LogixPro®

1 2 3 4 5 6 7 8 9   QVS   21 20 19 18 17 16

ISBN 978-1-25-964434-4
MHID 1-25-964434-0

Sponsoring Editor Michael McCabe	**Proofreader** Hardik Popli
Editorial Supervisor Donna M. Martone	**Indexer** Robert Swanson
Acquisitions Coordinator Lauren Rogers	**Production Supervisor** Pamela A. Pelton
Project Manager Moumita Majumdar, Cenveo® Publisher Services	**Composition** Cenveo Publisher Services
Copy Editor Ragini Pandey	**Art Director, Cover** Jeff Weeks

Contents

Preface

This book offers readers an introduction to PLC programming with focus on real industrial process automation applications. Rockwell Allen-Bradley SLC-500 PLC hardware configuration and the LogixPro 500 simulation software are described and used throughout the book. A small and inexpensive training setup with a power supply, processor, inputs/outputs discrete modules, analog inputs/outputs module, ON/OFF switches, push button switches, LEDs/light indicators, processor-integrated multiports communication ports, and a programing laptop was used to illustrate all programming concepts and the implementation of parts of completed automation projects by the authors in the past 20 years. All programming and project implementation in the first five chapters are described using the LogixPro 500 SLC-500 PLC simulation software. PLC hardware setup or software is not needed for the LogixPro use, but you must have the LogixPro 500 simulation software installed on your computer/laptop. If you have access to a training unit or SLC-500 hardware, you can use the AB RSLogix software.

Each chapter contains a set of homework questions and small laboratory design, programming, debugging, or maintenance projects. Two comprehensive capstone design projects are detailed at the end of this book in Chap. 9. All programs and system configurations described in chapters or included in the end of chapters homework assignments are fully implemented and tested. Complete solutions to all end-of-chapter laboratory assignments are available for instructors on request. Odd-numbered homework problem solutions are included in the book appendix. Concepts of process control and automation are introduced in Chap. 1. Chapter 2 details the fundamental of relay logic programming. It also covers the architecture and operation of PLCs. Configuration, operation, and the programming of timers and counters are the focus of our coverage in Chap. 3.

Chapter 4 is dedicated to the coverage of mathematical, logic, and commonly used commands operations with an emphasis on their use in real-time industrial applications. Ladder programming, for both PLC ladder logic and HMI user interface, are discussed in detail in Chap. 5. Modular structured programming design is used with emphasis on industrial standers and safety. Coverage is specific to the Allen-Bradley SLC-500 processor and the LogixPro 500 software, but the concepts are applicable to other systems.

System checkouts and troubleshooting are typically the most challenging and time-consuming tasks in industrial automation/process control applications. Chapter 6 contains common design and troubleshooting techniques. It also addresses critical issues of validation, hazards, safety standards, and protection against hardware/software failures or malfunction. Analog programming and associated instrumentation is covered in Chap. 7. Configuration, interface, scaling, calibration, and associated user interface are briefly covered.

Chapter 8 presents a comprehensive introduction to open- and closed-loop digital process control. Topics covered include sensors, actuators, on/off control, feedback control, PID tuning, and measures of good control. This chapter is intended to provide users with the understanding of the big picture of a control system in terms of system tasks, requirements, and overall expectations. It can best serve advanced engineering/technology, CS, or IT students as a prerequisite to the fundamentals and hands-on activities covered in the first seven chapters of the book. It can also serve other readers as a recap to the skills learned in previous chapters.

The book concludes with a comprehensive case study in Chap. 9. The case details the specifications of an irrigation canal downstream water level control. Coverage proceeds from the specification level to the final system design/implementation with associated documentation. The project is a small part of a much larger project implemented by the authors in Egypt more than 10 years ago. All implementations are redone using the AB SLC-500 PLC system. A second case study commonly used in waste water treatment facilities PLC control, wet wells pumping station, is briefly covered.

Recent advances in industrial process control have produced more intelligent and compact PLC hardware as the one we adopted in this book, AB SLC-500 system. It has also made available an extremely user-friendly development software for structured ladder programming, communication, easier configuration, modular design, documentation, and overall system troubleshooting. These advancements have created many opportunities for challenging and rewarding careers in the areas of PLC technology and process automation. This book is intended for a senior-level, one-semester course in an academic setting with the expectation of weekly hands-on laboratory work outside the class. Chapters 1 through 5 can serve as the content for a one-quarter course with adequate laboratory time. The book can also be used for a two full-week's industrial training in a small group setting with adequate training setup for each user. Successful career opportunity in the demanding field of PLC control and automation requires acquisition of the skills in this book along with adequate hands-on experience.

Eman Kamel, Ph.D.
Senior Control Engineer
PLC Automation

Khaled Kamel, Ph.D.
Professor, CS Department
Texas Southern University

CHAPTER 1

Introduction to PLC Control Systems and Automation

This chapter is an introduction to the world of programmable logic controllers (PLCs) and their evolution over the past fifty years as the top choice and the most dominant among all systems available for process control and automation applications.

AG 170 kW hydropower generator built and installed in 1912.

Chapter Objectives

- Understand concepts of process control.
- Realize the history of PLC and relay logic.
- Understand PLC hardware architecture.
- Understand the characteristics of hardwired and PLC systems.

A programmable logic controller is a microprocessor-based computer unit that can perform control functions of many types and varying levels of complexity. The first commercial PLC system was developed in the early 1970s to replace hardwired relay controls used in large manufacturing assembly plants. The initial use of PLCs covered automotive, jet engines, and large chemical plants. PLCs are used today in many tasks, including robotics, conveyor systems, manufacturing controls, process controls, electric power plants, wastewater treatment, and security applications. This chapter is an introduction to the world of PLCs and their evolution over the past fifty years as the top choice and the most dominant among all systems available for process control and automation applications.

1.1 Control System Overview

A control system is a device or set of structures designed to manage, command, direct, or regulate the behavior of other devices or system. The entire control system can be viewed as a multivariable process having a number of inputs and outputs, which can affect the behavior of the process. Figure 1.1 shows this functional view of control systems. This section is intended as a brief introduction, and will be covered in more detail in Chapter 7.

1.1.1 Process Overview

In the industrial world, the word *process* refers to an interacting set of operations that lead to the manufacture or development of a product. In the chemical industry, it refers to the operations necessary to take an assemblage of raw materials and cause them to react in some prescribed fashion to produce a desired end product, such as gasoline. In the food industry, it means to take raw materials and operate on them in such a manner that an edible high-quality product results. In each use, and in all other cases in the process industries, the end product must have certain specified properties, which depend on the conditions of the reactions and operations that produce them. The word *control* is used to describe the steps necessary to ensure that the regulated conditions produce the correct properties in the product.

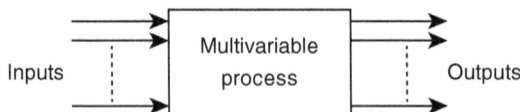

FIGURE 1.1 Control systems—functional view.

A process can be described by an equation. Suppose we let a product be defined by a set of properties: P_1, P_2, ..., and P_n. Each of these properties must have a certain value for the product to be correct. Examples of properties are color, density, chemical composition, and size. The process can be assumed to have m variables characterizing its unique behavior. Some of these variables can also be categorized as input, output, process property, and internal or external system parameters. The following equations express a process property and a variable as a function of process variables and time.

$$P_i = F(v_1, v_2, \ldots v_m, t)$$
$$v_i = G(v_1, v_2, \ldots v_m, t)$$

where P_i = the ith process property
v_i = the ith process variable
t = time

To produce a product with the specified properties, some or all the m process variables must be maintained at specific values in real time. Figure 1.2 shows free water flow through a tank, similar to rain flow in a home gutter system or a small creek. The tank acts in a way to slow the flow rate through the piping structure.

The output flow rate is proportional to the water head in the tank. Water level inside the tank will rise as the input flow rate increases. At the same time output flow rate will increase with noticeable increase in the tank water level. Assuming a large enough tank, level stability will be reached when flow in is equal to flow out. This simple process has three primary variables: flow in, flow out, and tank level. All three variables can be measured and, if desired, can also be controlled. The tank level is said to be a self-regulated variable.

Some of the variables in a process may exhibit the property of self-regulation, whereby they will naturally maintain a certain value under normal conditions. Small disturbances will not affect the tank level stability due to its self-regulating characteristic. A small increase in tank flow in will cause a slight increase in the water level. An increase in water level will cause an increase in the flow out, which will eventually produce a new stable tank level. Large disturbances in the tank input flow may force

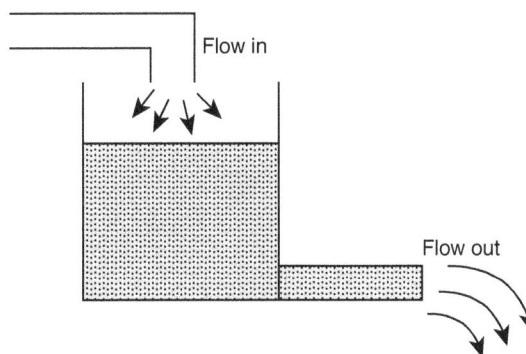

FIGURE 1.2 Water flow tank process.

undesired changes in the tank level. Control of variables is necessary to maintain the properties of the product, the tank level in our example, within specification. In general, the value of a variable v actually depends on many other variables in the process and also on time.

1.1.2   Manual Control Operation

In a manual control system, humans are involved in monitoring the process and making the decisions necessary to bring about desired changes in the process. Computers and advanced digital technologies may be used to automate a wide variety of process operation, status, command, and decision support functions. Sensors and measurement instruments are used to produce different process variables status, while final control elements or actuators are used to force changes in the process. As shown in Figure 1.3, humans close the control loop and establish the connection between measured values, desired conditions, and the needed activation of the final control elements.

Manual control is widely available and can be effective for simple and small applications. The initial cost of such systems might be relatively smaller than automated ones, but the long-term cost is typically much higher. It is difficult for operators to achieve the same control and quality due to various factors, such as different levels of domain expertise and unexpected changes in the process. The costs of operation and training can also become a burden unless certain functions are automated. Most systems start by using manual control or existed previously through manual operation. As the system owners acquire and accumulate process control experience over time, they use this knowledge to make process improvements and eventually automate the control system.

The introduction of digital computers in the control loop has allowed the development of more flexible control systems, including higher-level functions and advanced algorithms. Furthermore, most current complex control systems can not be implemented without the application of digital hardware. However, the simple sequence of sensing, control, and actuation for the classic feedback control becomes more complex as well. A real-time system is one in which the correctness of a result depends not only on the logical correctness of the calculation but also on the time at which the different

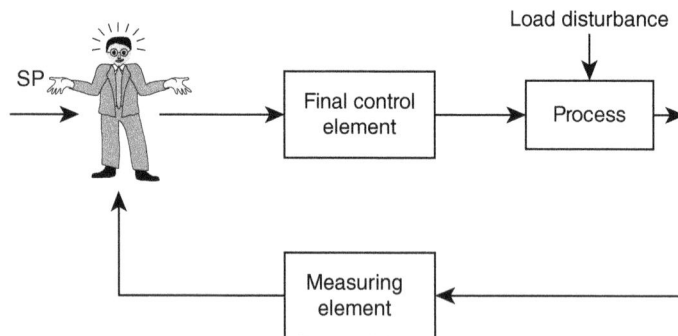

FIGURE 1.3 Manual control systems.

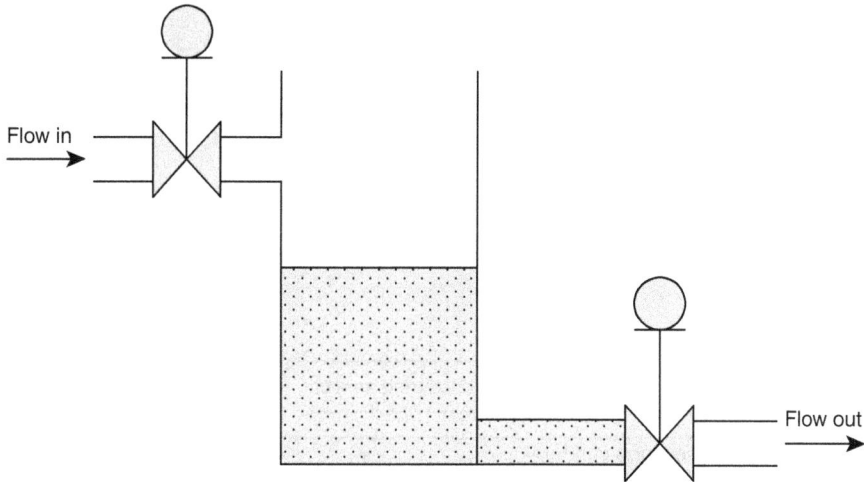

FIGURE 1.4   Tank level manual controls.

tasks are executed. Time is one of the most important entities of the system, and there are timing constraints associated with systems tasks. Such tasks normally have to control or react to events that take place in the outside world, which are happening in "real time." Thus, a real-time task must be able to keep up with the external events with which it is concerned.

Figure 1.4 shows a simple manual control system. The level in the tank varies as a function of the flow rate through the input valve and the flow rate through the output valve. The level is the control or controlled variable, which can be measured and regulated through valve control and adjustment at the input or the output flow or both. The two valves can be motorized and activated from an easy-to-use operator interface. Valve position variations are achieved through an operator input based on observed process real-time conditions. We will see next that the operator can easily be eliminated.

1.1.3   Automated System Building Blocks
The closed control loop shown in Figure 1.5 consists of the following five blocks:

- Process
- Measurement
- Error detector
- Controller
- Control element

In manual control, the operator is expected to perform the task of error detection and control. Observations and actions taken by operators can lack both consistency and reliability. The limitation of manual control can be eliminated through the implementation of closed-loop systems and the associated process control strategies. Details of such

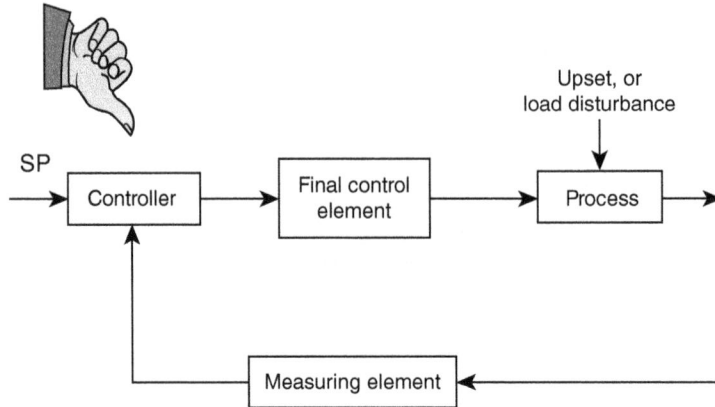

FIGURE **1.5** Closed-loop control.

strategies will be detailed in Chapter 7. Figure 1.5 shows a block diagram of a single-variable closed-loop control. The controller can be implemented using various technologies, including hardwired relay circuits, digital computers, and more often the PLC systems.

It is impossible to achieve perfect control, but in the real world it is not needed. We can always live with small errors within our acceptable quality range. An oven with a desired temperature of 500°F can achieve the same results at 499.99°F. In most cases we are limited by the precision and cost of the actual sensors. There is no good justification for spending more money to achieve unwanted/unnecessary gains in precision.

Errors in real time are used to judge the quality of the system design and its associated controller. The errors can be measured in three ways as explained below using the following definitions:

Absolute error = set point − measured value

Error as percentage of set point = absolute error/set point ∗ 100

Error as percentage of range = absolute error/range ∗ 100

Range = maximum value − minimum value

Errors are commonly expressed as percentage of range and occasionally as percent of set point but rarely as an absolute value. Also, most process variables are commonly also quantified as percentage of the defined range. This quantification allows for universal input/output (I/O) PLC computer interfaces regardless of the physical nature of the sensory and actuating devices. A PLC analog input module having several input slots can accommodate and process temperature, pressure, motor speed, viscosity, and many other measurements in exactly the same way. Later chapters will detail the PLC hardware and software as applied to real-world industrial control applications. Even though the implementation focus will be on the AB SLC 500, the concepts covered will apply to other PLCs with no or very little modifications. International standards and the success of open system architectures are the main reasons for the universal nature of today's PLC technology and its compatibility.

1.1.4 Direct/Reverse Acting Controller

The controller can be designed to provide an output that is either directly proportional to the amount of error in the process or inversely proportional to the error. This type of controller behavior is labeled as direct and reverse action, respectively. We will demonstrate this concept using the liquid tank of Figure 1.4 for a level control process. The error in level is expressed as the difference between the set point and the measured value. The following are the two possible control strategies based on the error value:

- *Direct acting:* In this control strategy, we regulate the tank level by adjusting the position of the inlet valve while keeping the outlet valve position fixed. If the error is positive (set point larger than the measured value), then the controller output (inlet valve position) will increase. This will allow more flow into the tank and will cause the level to increase, which in turn will reduce the process error. The behavior of the controller in this case is known as direct acting.

- *Reverse acting:* In this control strategy, the level is regulated by adjusting the position of the outlet valve while keeping the inlet valve position fixed. If the error is positive (set point greater than the measured value), then the controller output (outlet valve position) will decrease. This will allow less flow out of the tank and will cause the level to increase, which in turn will reduce the process error. The behavior of the controller in this case is known as reverse acting.

1.2 Hardwired Systems Overview

Prior to the widespread use of PLCs in process control and automation, hardwired relay control systems or analog single-loop controllers were used. This section will briefly introduce relay systems and the logic used in process control. It is important that the reader understands the fundamentals of relays in order to fully appreciate the role of PLCs in replacing relays, simplifying process control design/implementation, and enhancing process quality at a much lower overall system cost. Coverage in this section is limited to functionality and application without much detail of either electrical or mechanical characteristics.

1.2.1 Conventional Relays

In this section we will learn how a relay actually works. A relay is an electromagnetic switch having a coil and a set of associated contacts of a typical relay, as shown in Figure 1.6. Contacts can be either normally open or normally closed. An electromagnetic field is generated once voltage is applied to the coil. This electromagnetic field generates a force that pulls the contacts of the relay, causing them to make or break the controlled external circuit connection. These electrically actuated devices are used in automobiles and industrial applications to control whether a high-power device is switched ON or OFF. While it is possible to have a device, such as a large industrial motor or ignition system, directly powered by an electrical circuit without the use of a relay, such choice is neither safe nor practical. For example, in a factory, a motor control may be placed far away from the high-voltage electrical motor and its power source for safety reasons. In this case, it is more practical to have a low-power electrical relay coil circuit control a high-power relay contacts than to directly wire a

FIGURE **1.6** Typical industrial relays.

high-power electrical switch from the control area to the motor and its independent power supply.

Figure 1.7 shows a control relay (CR1) with two contacts normally open (CR1-1) and normally closed (CR1-2). On the left side of the figure, power is not applied to the coil (CR1) and the two contacts are in the normal state. On the right side of the same figure, power is applied to the coil and the two contacts switch state; the normally open contact closes and the normally closed contact opens.

Figure 1.8 shows a simple relay circuit for controlling a bell using a single pole single through (SPST) switch; pressing the switch causes the bell to sound. A relay is typically used to control a device that requires high voltage or draws large current.

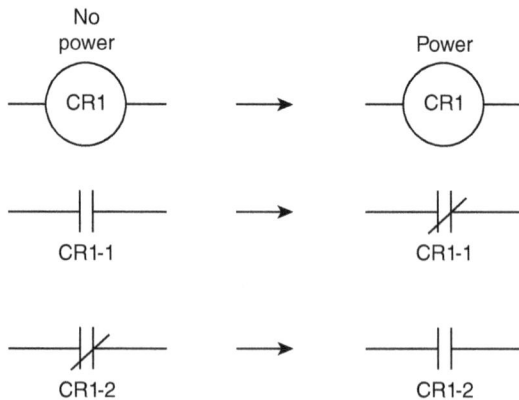

FIGURE **1.7** Relay with two contacts normally open and normally closed.

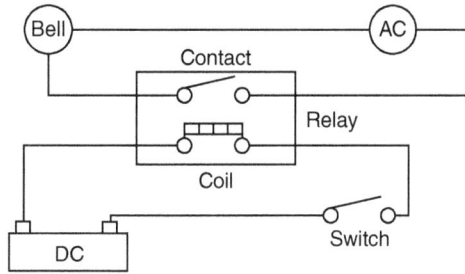

FIGURE 1.8 Simple relay circuit.

The relay allows full power to the device without needing a mechanical switch that can carry the high current. A switch is normally used to control the low-power side, the relay coil side. Notice that we have two separate circuits: the bottom uses the dc low power, while the top utilizes the ac high power. The two circuits are only connected through electromagnetic field coupling. The low-power dc side is connected to the coil while the high-power ac side in this example is located in the field away from the control room. The two sides are normally powered from two independent sources in a typical industrial facility automated application. Of course it is not cost effective to replace the relay in this example with a PLC, but it does for a real application with hundreds or thousands of I/O devices.

1.2.2 Relay Logic System

Relay logic systems are control structures appropriate for both industrial and municipal applications. The operations/processes that will be controlled by relay logic systems are hardwired, unlike programmable logic control systems. These systems are inflexible and can be difficult to modify after deployment. Since the operation of relay logic controllers is built directly into the device, it is easy to troubleshoot the system should any problems arise. Such control systems are developed with fixed features for specific applications. Typically, large pumps and motors will be equipped with hardwired relay control to protect them against damage under overloads and other undesired working conditions. Programmable logic control systems provide much—needed flexibility and allow for future continuous quality improvements in the process.

Figure 1.9 shows two relay circuits for implementing two inputs "AND" and "OR" logic functions, respectively. Each relay has two magnetic coils and associated normally closed (NC) contacts. The two inputs are connected to one side of each of the two coils and the other end of the coil is connected to the ground. The contacts are connected in a predefined manner to produce the desired output as a function of the two inputs. Input A and Input B can be at either the Ground level (0/Low logic/False logic) or the +V level (1/High logic/True logic). The AND arrangement produces the +V logic (High logic) only when the two inputs are high while the OR configuration produces the Ground logic (Low logic) only when the two inputs are low. Notice that the relay operation involves electrical (coils and power supply) and mechanical (moving contacts) components.

Schematic diagrams for relay logic circuits are often called logic diagrams. A relay logic circuit is an electrical diagram consisting of lines/networks/rungs in which each

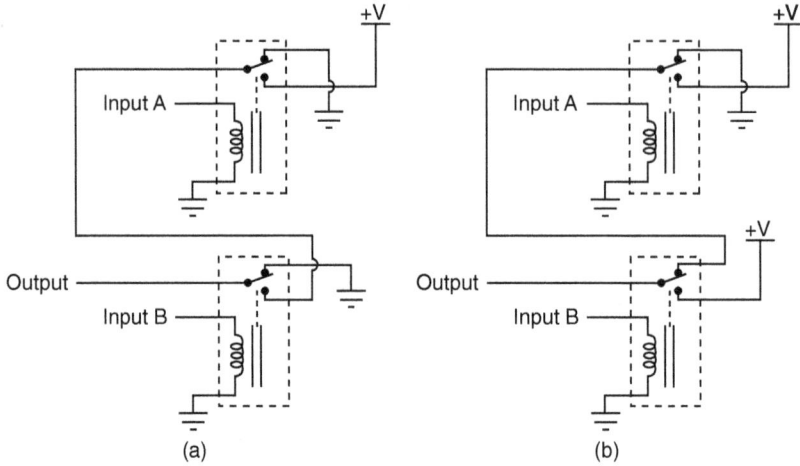

Figure 1.9   (a) AND logic function; (b) OR logic function.

must have continuity to enable the intended output device. A typical circuit consists of a number of rungs, with each controlling an output. This output is controlled through a combination of input or output conditions (such as switches and control relays) connected in series, parallel, or series-parallel to obtain the desired logic to drive the output. Relay logic diagrams represent the physical interconnection of devices. It is possible to design a relay logic diagram directly from the narrative description of a process control event sequence. In ladder logic diagrams, an electromechanical relay coil is shown as a circle and the contacts actuated by the coil as two parallel lines. Given this notation, the relay line logic diagrams for AND and OR logic functions are shown in Figure 1.10.

The "L_1" and "L_2" designations in the logic diagram refer to the two poles of a 120 ac voltage power supply. L_1 is the hot side of the supply and L_2 is the ground/neutral side. Output devices are always connected to L_2. Any device overloads that are to be

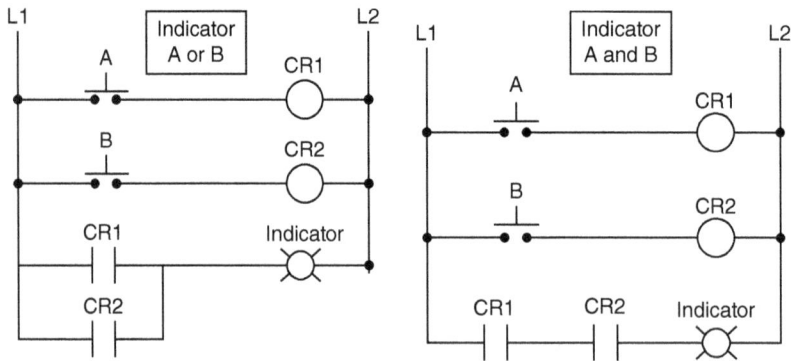

Figure 1.10   Relay line logic diagrams.

included must be shown between the output device and L_2; otherwise the output device must be the last component before L_2. Input devices are always shown between L_1 and the output device. Relay contacts control devices may be connected either in series, parallel, or a combination of both.

1.2.3 Control Relay Application

Relays are widely used in process control and automation applications. PLCs have gained much acceptance in the last thirty years and have gradually replaced most of the old, hardwired relay-based control systems. It is important that we understand the old relay control systems in order to appreciate and make the transition to the more powerful, easier to implement, cheaper to maintain, and reliable PLC control. This section documents two simple relay control applications.

Figure 1.11 shows the line diagram for a common application of an electromechanical relay DC motor control circuitry. A momentary normally open (NO) push button switch starts the motor and another normally closed (NC) push button switch-stops the motor. The control relay contact is used to latch the start push button after it is released. Another contact associated with the same relay is used to start the motor. Pressing the stop push button at any time will interrupt the flow of electricity supply to the motor and cause it to stop.

Another application is shown in Figure 1.12. The line diagram illustrates how a three-contact relay is used to control two pilot lights. The desired control is accomplished using two push button switches; PB1 starts the operation and PB2 terminates it at any time.

Below are the critical steps for this example:

- With no power applied to the control relay the contacts are in normal state. The normally open is open and the normally close is close. The green pilot light (G) receives power and turns ON as indicated by the green fill light. The red pilot light (R) is OFF, as shown.

- Rung 1: Once PB1 is pressed, CR1 coil becomes energized; this in turn makes contact. CR1-1 closes and maintains power to CR1 through the normally closed push button PB2.

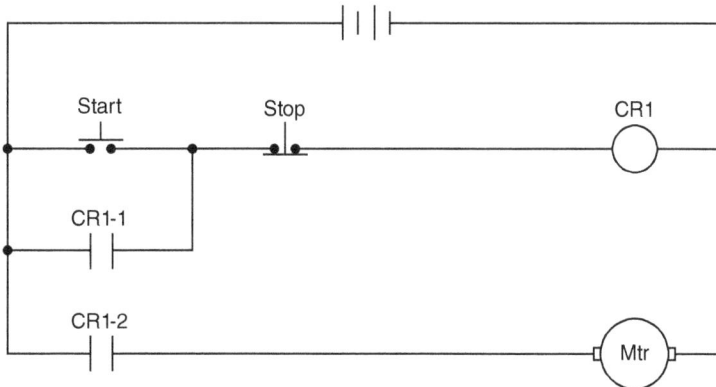

FIGURE 1.11 DC motor controls.

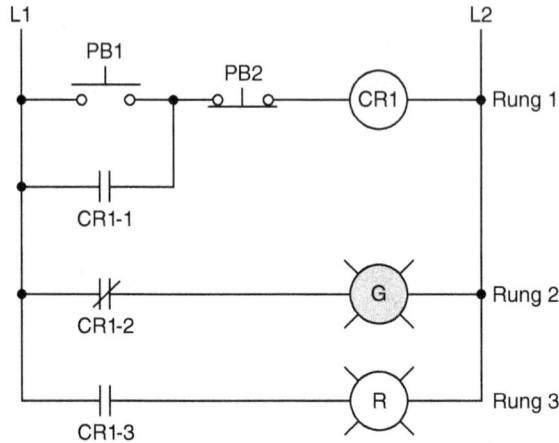

FIGURE **1.12**   Relay controlling two pilot lights.

- When CR1 energizes the contacts switch state, the normally open closes and the normally close opens. This will turn OFF the green light in rung 2 and turn ON the red light on rung 3.
- When the PB2 push button is pressed, the control relay loses power and the contacts switch to the normal state. This results in turning the green light ON and the red light OFF.

1.2.4   Motor Magnetic Starters

A magnetic starter is used to control high power to a motor, as shown in Figure 1.13. Three of the motor magnetic starter contacts are used to connect the three phases of the high voltage supply. In addition, overload relays are physically attached in series with the three-phase supply voltage (L1, L2, and L3) for the motor's protection. Figure 1.14 shows a low-power motor starter circuit at the Motor Control Center (MCC). START and STOP PB switches start and stop the motor through the control of its magnetic starter. The magnetic starter contact M-4 is used to latch the motor start action.

Figure 1.15 illustrates a line diagram of a magnetic reversing motor starter controlled by forward and reverse push buttons. Pressing the Forward push button completes the forward coil circuit from L1 to L2. Energizing coil F in turn energizes two

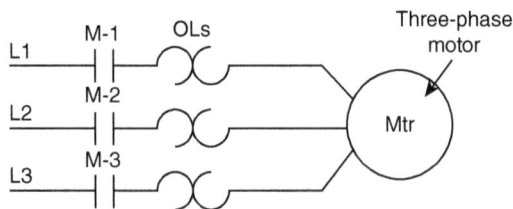

FIGURE **1.13**   High-power motor circuit.

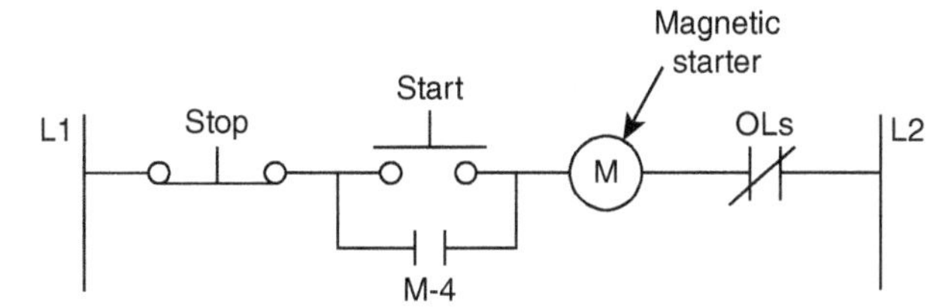

Figure 1.14 Low-power MCC starter circuit.

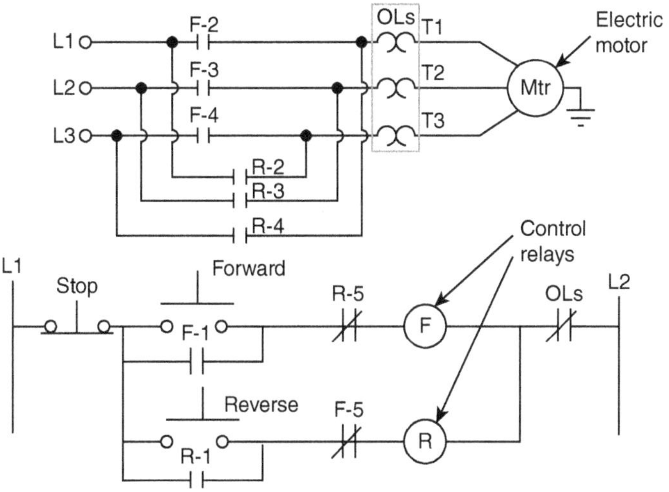

Figure 1.15 Control of reversing motor starter.

auxiliary contacts, F-1 and F-5. F-1 provides a latch around the forward push button maintaining coil F energized. The normally closed contact F-5 will prevent the motor from running in the reverse direction if the reverse PB is pressed before the stop PB while the motor is running in the forward direction. The lower part of Figure 1.15 illustrates a line diagram of the magnetic reversing starter controlled by forward and reverse push buttons.

Pressing the Reverse push button completes the reverse coil circuit from L1 to L2, energizing coil R, which in turn energizes two auxiliary contacts, R-1 and R-5. R-1 provides a latch around the reverse push button, maintaining coil R energized. The normally closed contact R-5 will prevent the motor from running in the forward direction if the reverse button is pressed before the stop while the motor is running in the reverse direction.

Reversing the motor running direction is accomplished by switching two of the motor input voltage phases, phase 1 and phase 3 in this case. When coil R energizes R-2, R-3 and R-4 are closed; L1 connects to T3, L3 to T1, and L2 to T2 causing the motor to run in the reverse direction.

Vertical gate control for downstream water level regulation is one such application, which makes use of this motor-running-direction reversal. A desired increase in downstream water level requires running the motor in certain direction, which causes the gate to move upward. Running the motor in the opposite direction will cause the downstream water level to decrease. Movements in both directions are accomplished by using a single motor. These motors are heavy-load, high-power devices/actuators with widespread use in industrial process control and automation applications. Typical cost for each such motor is high, and they come ready equipped with a magnetic starter with all needed instrumentation and protective gear, such as overloads relay contacts.

1.2.5   Latch and Unlatch Control Relay

Latch and unlatch control relay work exactly like the Set Reset flip flop used in digital logic design. Set is the latch coil and Reset is the unlatch coil. It is designed to maintain the contact status when power is removed from the coil, as shown in Figure 1.16. Figure 1.17 shows the line logic diagram for the latch and unlatch control relay.

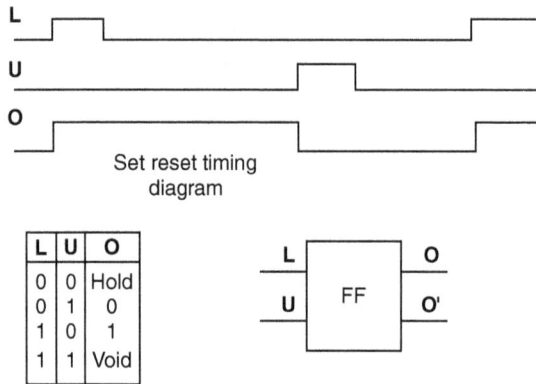

Set reset timing diagram

L	U	O
0	0	Hold
0	1	0
1	0	1
1	1	Void

FIGURE 1.16   Latch and unlatch operation.

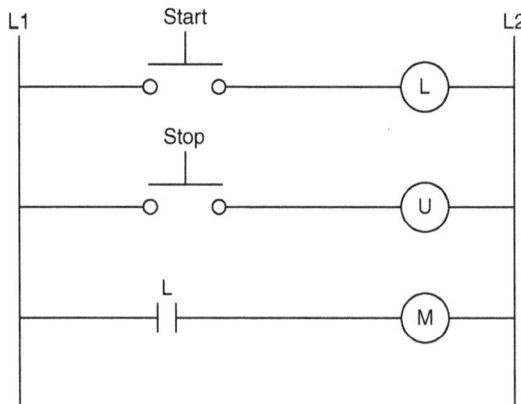

FIGURE 1.17   Latch/unlatch control line diagram.

Once the Start push button is pressed, coil L receives power and energizes. After Start PB is released, coil L does not receive power but maintains the energized status. The L contact will close and cause motor M to run. To stop motor M, the Stop push button must be pressed to switch the status of the latch unlatch relay to the unlatch status.

The Start and the Stop PB switches are interlocked through hard wiring. Either action can be activated at any time but never both at the same time. The Start (latch) and the Stop (unlatch) can be generated through program logic events instead of the two PB switches shown, such as when the temperature in a chemical reactor exceeding certain range or the level in a boiler drum is below certain threshold.

1.3 PLC's Overview

This section is intended as a brief introduction to PLC, its history/evolution, hardware/software architectures, and the advantages expected from its use relative to other available choices for process control and automation.

1.3.1 What Is a PLC?

A PLC is an industrial computer that receives inputs from input devices, then evaluates these inputs in relation to stored program logic and generates outputs to control peripheral output devices. The I/O modules and a PLC functional block diagram are shown in Figure 1.18. Input devices are sampled and the corresponding PLC image table is updated in real time. The user's program, loaded in the PLC memory through the programming device, resolves the predefined application logic and updates the output internal logic table. Output devices are driven in real time according to the output table updated values.

Standard interfaces for both input and output devices are available for the automation of any existing or new application. These interfaces are workable with all types of PLCs regardless of the selected vendor. Sensors and actuators allow the PLC to interface to all kinds of analog and ON/OFF devices through the use of digital I/O modules, analog-to-digital converters, digital-to-analog converters, and adequate isolation

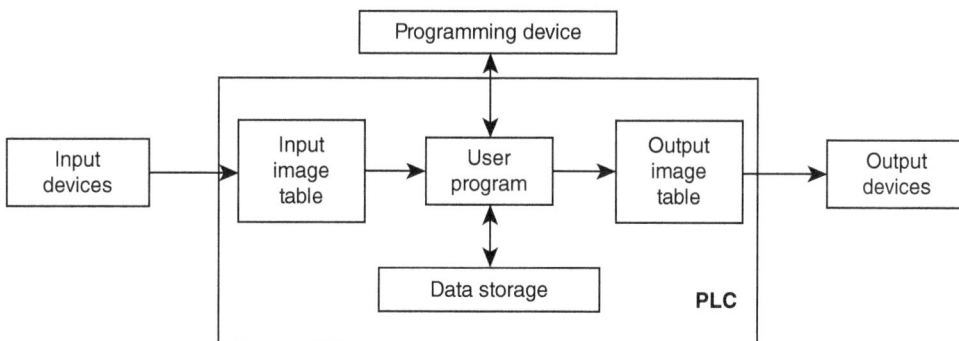

FIGURE 1.18 Inputs/outputs PLC architecture.

circuits. Apart the power supply input and the I/O interfaces, all signals inside the PLC are digital and low voltage. Details of PLC hardware and interfaces will be discussed later in this book.

Since the first deployment of PLCs five decades ago, old and new vendors have competed to produce more advanced and easier to use systems with associated user-friendly development and communication tools. Figure 1.19 shows a variety of popular PLCs used in the industry. You should notice the diversity in size and, obviously, associated capabilities, which not only allow cost accommodation but also enable the design and implementation of complex distributed control systems. The vast majority of available vendors allow the integration of other PLCs as part of a networked distributed control system. It is also possible to implement extremely large system control on one PLC system with a large number of interconnected chassis and modules.

Wikipedia, the free encyclopedia, states that "a PLC or programmable controller is a digital computer used for automation of electromechanical processes, such as control of machinery on factory assembly lines, amusement rides, or light fixtures." PLCs are used in many industries and machines. Unlike general-purpose computers, a PLC is designed for multiple inputs and output arrangements, extended temperature ranges, immunity to electrical noise, and resistance to vibration and impact. Programs to control machine operation are typically stored in battery-backed-up or nonvolatile memory. A PLC is an example of a hardwired real-time system since output results must be produced in response to input conditions within a bounded time, otherwise unintended operation will result. Most of the electromechanical components needed for hardwired control relay systems are completely eliminated resulting in great reduction in space, power consumption, and maintenance requirements.

A PLC is a device that can replace the necessary sequential relay circuits needed for process control. The PLC works by sampling its inputs and depending upon their state, actuating its outputs to bring about desired changes in the controlled system. The user

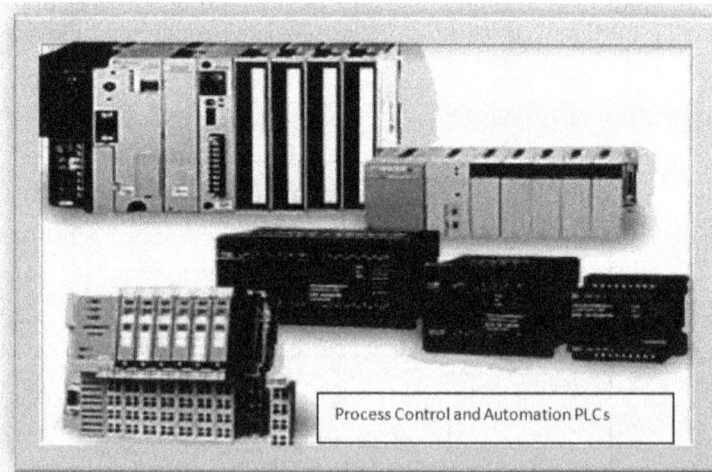

Process Control and Automation PLCs

Figure 1.19 Typical industrial PLCs.

enters a program, usually via software that allows control systems to achieve the desired results. Programs are typically written in ladder logic but higher-level development environments are also available. The IEC 1131-3 standard (International Electrotechnical Commission global standard for industrial control Programming) has tried to merge PLC programming languages under one international standard. We now have PLCs that are programmable in function block diagrams, instruction lists, C, and structured text—all at the same time! Personal computers (PCs) are also being used to replace PLCs in some applications.

PLCs are used in a vast majority of real-world applications. The evolution of the globally competitive economy has mandated industries and organizations to commit investments in digital process control and automation using PLCs. Wastewater treatment, machining, packaging, robotics, material handling, automated assembly, or countless other industries are extensively using PLCs. Those who are not using this technology are wasting money, time, quality, and competitiveness. Almost all applications that use electric, mechanical, or hydraulic devices control have a need for a PLC.

For example, let's assume that when a switch turns on we want to turn a solenoid on for 15 seconds and then turn it off regardless of the duration of the switch on position. We can accomplish this task with a simple external timer. What if our process included 100 switches and solenoids? We would need 100 external timers to handle the new requirements. What if the process also needed to count how many times the switches turned ON individually? We have to employ a large number of external counters along with external timers. All this would require extensive wiring, energy, space, and expensive maintenance requirements. As you can see, the bigger the process, the more the need for a PLC. We can simply program the PLC to count its inputs and turn the solenoids ON for the specified time.

1.3.2 History of PLCs

Prior to the introduction of PLCs, all production and process control tasks were implemented using relay-based systems. Industrialists were dealing with this inflexible and expensive control systems issues for decades. Upgrading a relay-based machine control production system means that the whole production system changes, which is very expensive and time consuming. In 1960s, General Motors (GM) issued a proposal for the replacement of relay-based machines. The *PLC history* was all started with an industrialist named Richard E. Morley, who was also one of the founders of Modicon Corporation, in response to GM's proposal. Morley finally created the first PLC in 1969. It was sold in 1977 to Gould Electronics and was presented to GM. This first PLC is now kept safely at the company headquarters.

plcdev.com lists the timeline shown in Figure 1.20 of the development of the PLC by different manufacturers. It spans the period from 1968 to 2005. The new SLC500 was introduced by Rockwell Automation/Allen-Bradley in 1994. It was designed to provide an easy-to-use and scalable infrastructure for small and large distributed control applications. Details of the SLC500 and associated interfaces—including hardware, software, human machine interface (HMI)'s, communication, and networking, along with Industrial Control application implementation using this Allen Bradley infrastructure—will be the focus of this book. Reduction in size, lower cost, larger capabilities, standard interfaces, open communication protocols, user-friendly development environment, and human machine interface tools are the trend in the evolvement of PLC, as shown in the history chart.

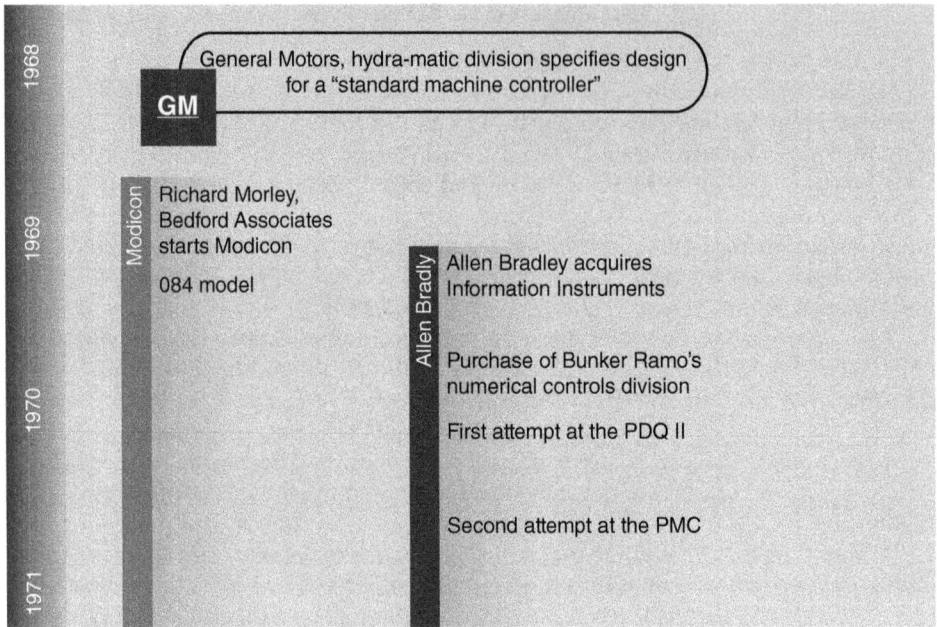

FIGURE 1.20 1968 to 1971 early PLC systems**.

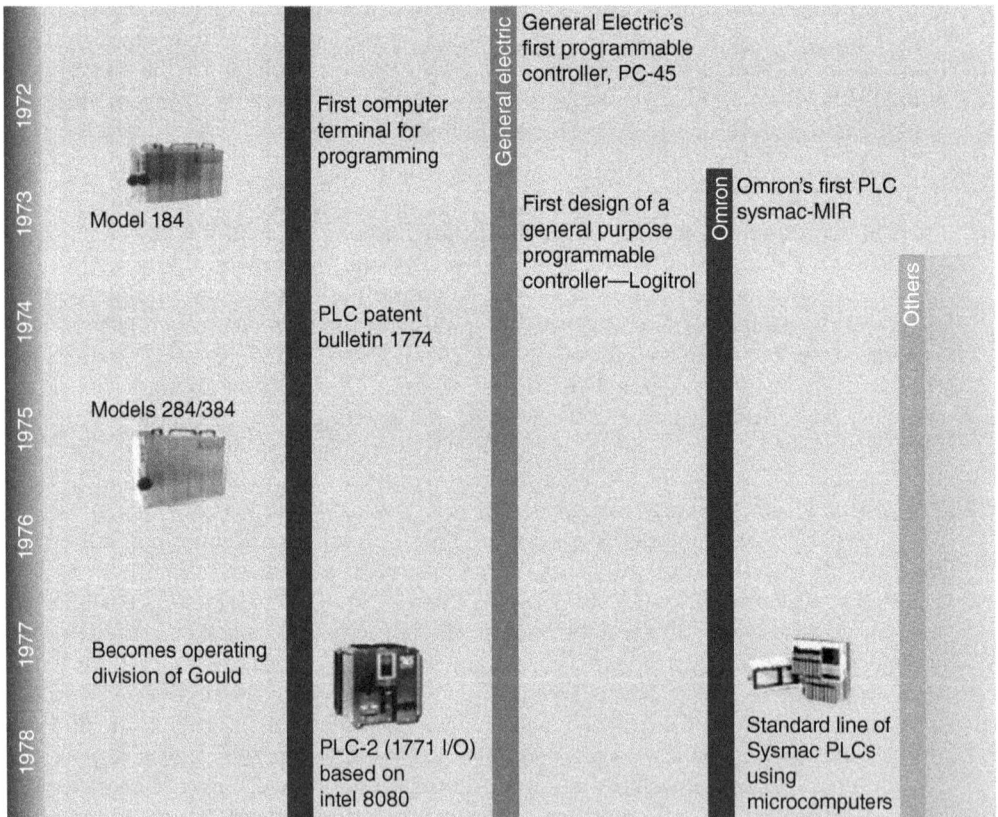

FIGURE 1.20 1972 to 1978 first-generation PLC systems**. *(Continued)*

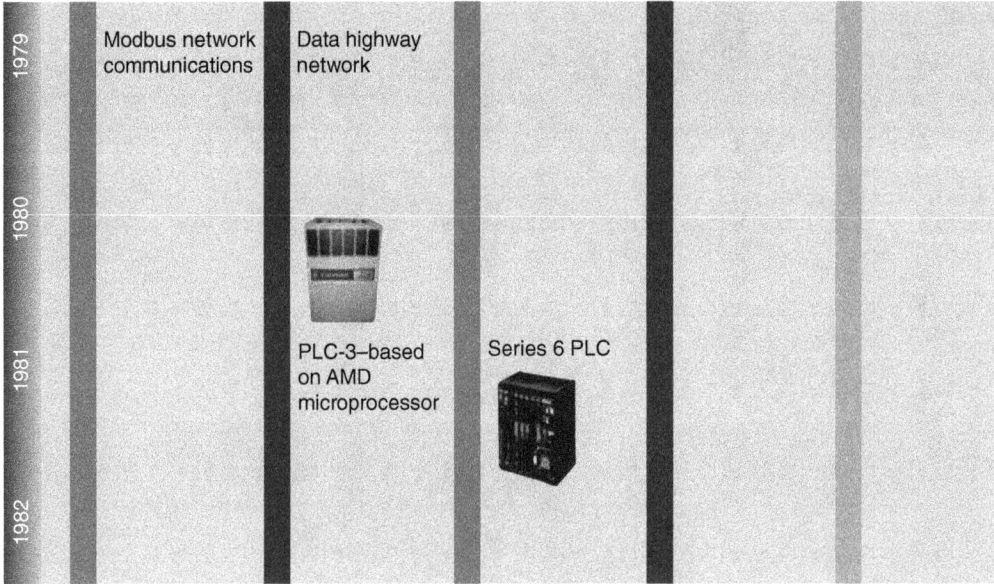

FIGURE 1.20 1979 to 1982 early second-generation PLCs**. (*Continued*)

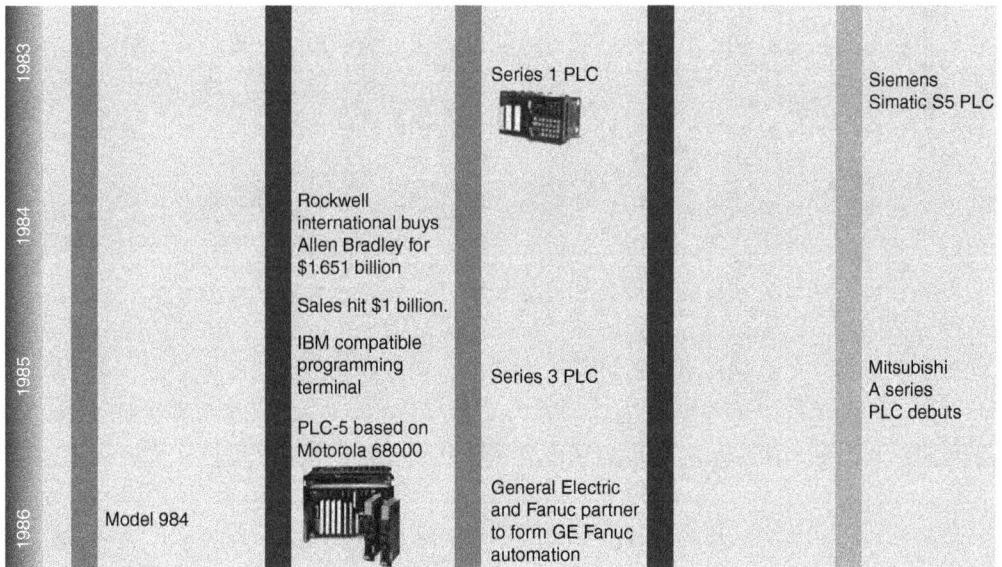

FIGURE 1.20 1983 to 1986 second-generation PLCs**. (*Continued*)

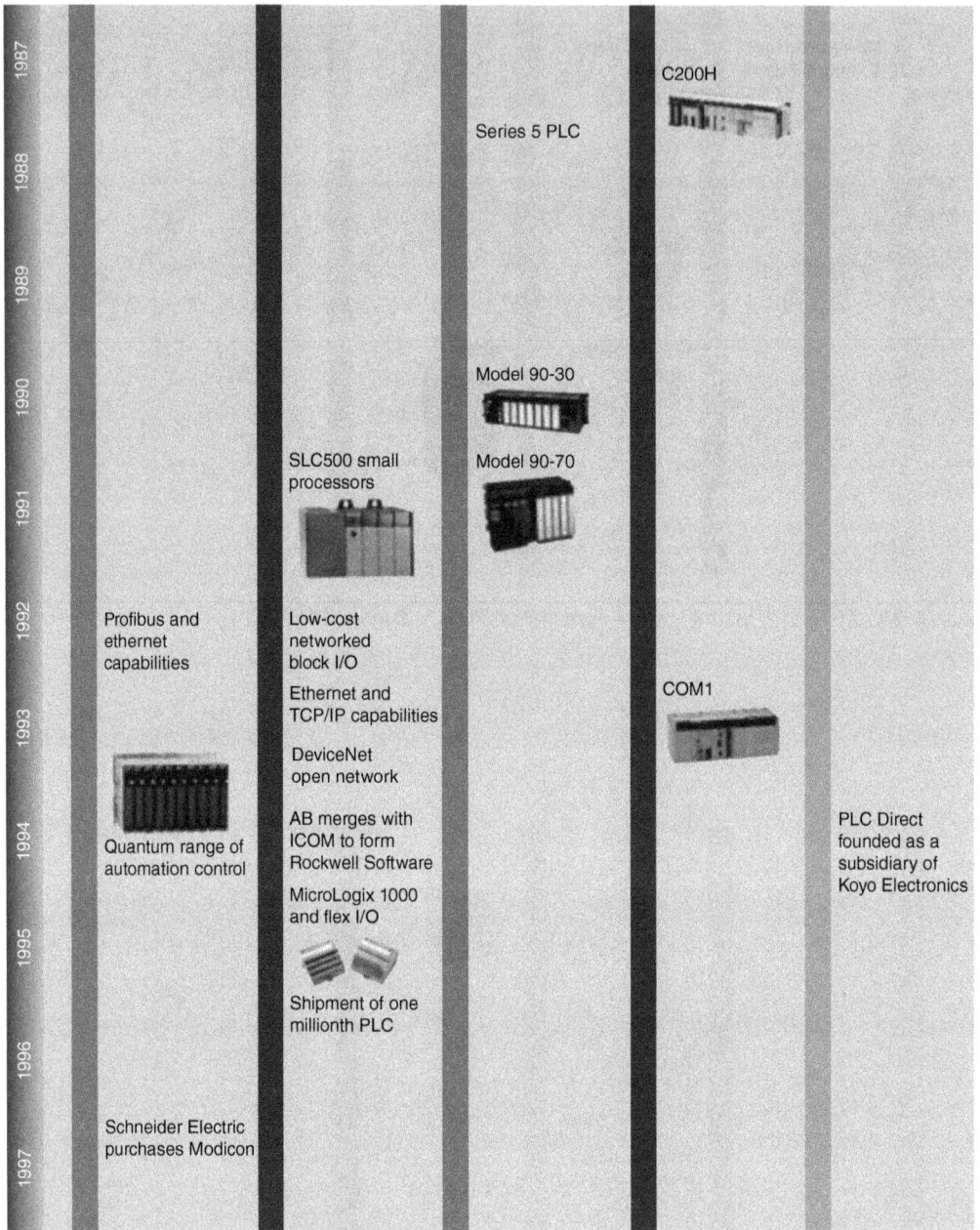

FIGURE 1.20 1987 to 1997 early third-generation PLCs**. (*Continued*)

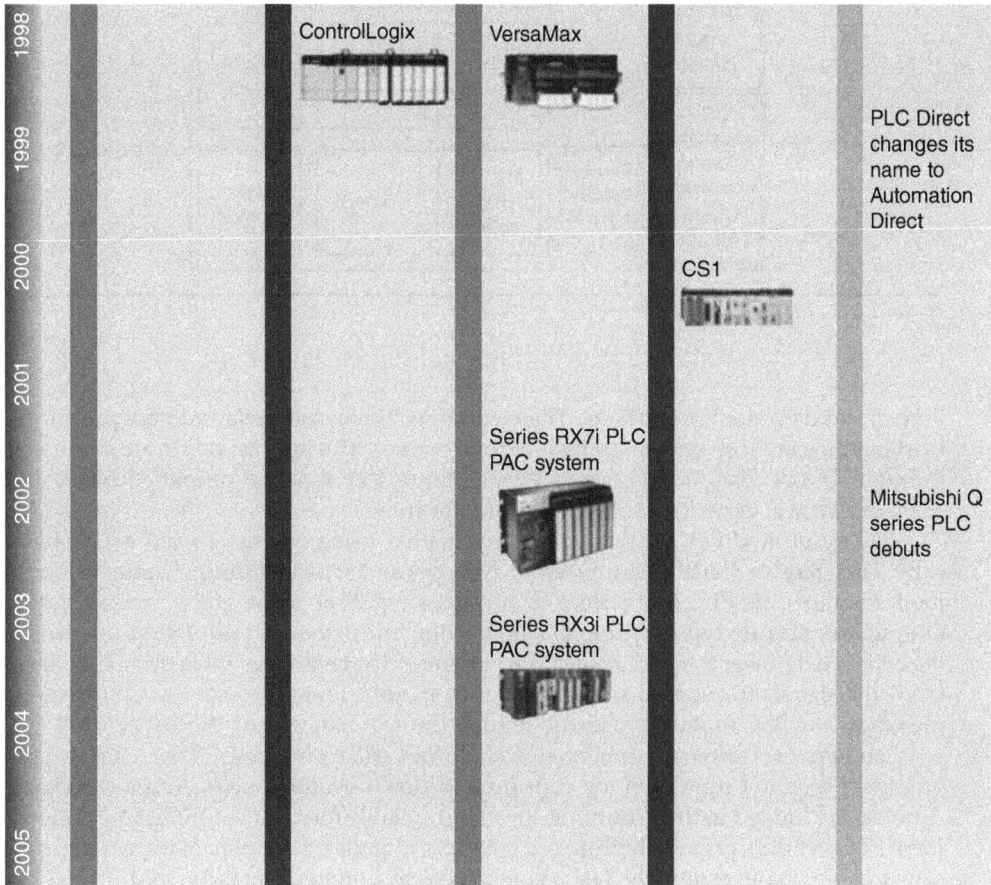

FIGURE 1.20   1998 to 2005 third-generation PLCs**. (*Continued*)
**PLC history chart (R. Morley, the father of PLC)

PLC's history is displayed in time categories starting from the early systems introduced from 1968 to 1971. This is followed by a span of six years, labeled as the first PLC generation. The second generation started in 1979 and covered a period of seven years, ending in 1986. This period showed greater number of vendors, mostly from existing U.S. companies in addition to German and Japanese firms. The early third-generation started 1987 and lasted for ten years, followed by a lasting period of continued growth and advancement in both hardware and software tools, which led to a wide deployment of PLCs in most manufacturing automation and process control activities.

1.3.3   PLC Architecture

A typical PLC mainly consists of a CPU (central processing unit), power supply, memory, communication module, and appropriate circuits to handle I/O data. A PLC can be viewed as an intelligent box having hundreds or thousands of separate relays, counters,

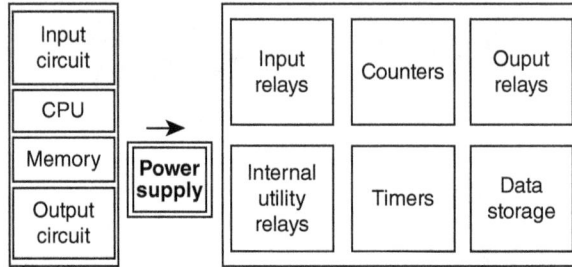

Figure 1.21   PLC architecture.

timers, and data storage locations. These counters, timers, and relays do not physically exist but they are software-simulated internal entities. The internal relays are simulated through bit locations in memory registers. Figure 1.21 shows a simplified block diagram of a typical generic PLC hardware architecture.

PLC input modules are typically implemented using transistors and exist physically. They receive signals from external switches and sensors through contacts. These modules allow the PLC to interface to and get a real-time sense of the process status. Output modules are typically implemented using transistors and use TRIACs to switch the connected power to the output coil when the output reference bit is true. They send ON/OFF signals to external solenoids, lights, motors, and other devices. These modules allow the PLC to interface to and regulate, in real time, the controlled process.

Counters are software simulated and do not exist physically. They can be programmed to count up, down, or both up and down events/pulses. These simulated counters are limited in their counting speed but suitable for most real-time applications. Most PLC vendors provide high-speed counters modules that are hardware based and can accommodate extremely fast events. Typical counters include UP-COUNTER, DOWN-COUNTER, and UP/DOWN-COUNTERS. Timers are also software simulated and do not exist physically. The most common types are the ON-DELAY, OFF-DELAY, and RETENTIVE timers. Timing increments vary but are typically larger than one thousands of a second. The vast majority of process control applications make extensive use of timers and counters in a variety of ways and applications, which will be detailed in Chapter 3.

Data storage is a high-speed memory/registers assigned to simply store data. They are usually used in math or data manipulation as temporary storage. They also used to store values associated with timers, counters, I/O signals, and user interface parameters. Communication buffers and related networking and user interface tasks also make use of high-speed storage. Typically, they can also be used to store data and programs when power is removed from the PLC. Upon power-up the same contents, which existed before power was removed, will still be available.

1.3.4   Hardwired System Replacement

As stated in the previous section, PLCs were introduced to replace hardwired relays. In this section we will introduce the process of replacing the relay logic control by a PLC. The example we will use to demonstrate this replacement process may not be very cost effective for the use of a PLC but it will demonstrate the fundamental concepts. As shown earlier, the first step is to create the process ladder logic diagram/flow chart. PLCs do not understand these schematic diagrams but most vendors provide software to convert ladder logic diagrams into machine code, which shields users from actually having to learn the PLC processor's specific code. Still, we have to translate all process logic into the standard symbols that the PLC recognizes. Terms like switch, solenoid, relay, bell, motor, and other physical devices are not recognized by PLCs. Instead input, output, coil, contact, timer, counter, and other terms are utilized.

Ladder logic diagrams use standard symbols and associated addresses to uniquely represent different elements and events. Two vertical bars, representing L_1 and L_2, span the entire diagram and are called the power/voltage bus bars. All networks/rungs start at the far left, L_1, and proceed to the right ending at L_2. Power flows from left to right through available closed circuits. Inputs like switches are assigned the contact symbol of a relay, as shown in Figure 1.22. Output like the bell is assigned the coil symbol of a relay as shown in Figure 1.22. The ac/dc supply is an external power source and is thus not shown in the ladder logic diagram. The PLC executes the logic and turns an output ON or OFF using TRIAC switching interface without any regard to the physical device connected to that output.

The PLC must know the location of each input, output, or other elements used in our application. For example, where are the switch and the bell going to be physically connected to the PLC? The PLC has pre-specified I/O addresses in a wide variety of signal forms and sizes to interface with all types of devices. For now assume that our input (a push button switch) will be labeled 0000 and the output (a bell) will be called 0500. The final step converts the schematic into a logical sequence of events telling the PLC what to do when certain real-time events or conditions are satisfied. In our example we obviously want the bell to sound while the push button switch is being pressed. Electric power connection to the bell is made while the push button switch is being pressed. Once the push button is released, electric power connection to the bell is removed. The only requirement for this small system to work is to have the push button connect to the PLC input module and for the bell to be wired to the PLC output module, as will be shown later. Figure 1.23 shows the logic diagram for our simple example. More real, comprehensive industrial control examples and extensive coverage will illustrate this concept in Chapter 2.

Two vertical bars, representing L_1 (the hot phase) and L_2 (the neutral phase), span the entire diagram and are called the power/voltage bus bars. All rungs start at the far left, L_1, and proceed to the right ending at L_2. Power flows from left to right through available closed circuits. Figure 1.24 and Figure 1.25 show the results of converting a hardwired control relay to a PLC ladder logic control. The first example implements a simple motor control using momentary START and STOP push buttons used to initiate

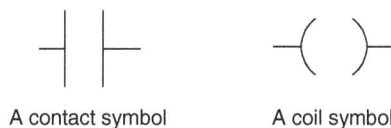

A contact symbol   A coil symbol

FIGURE 1.22 Contact and coil symbols.

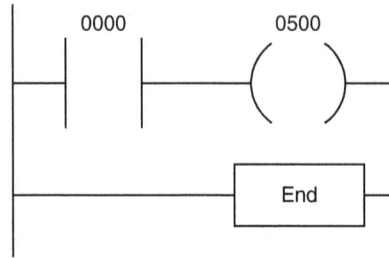

Figure 1.23   Bell logic diagram.

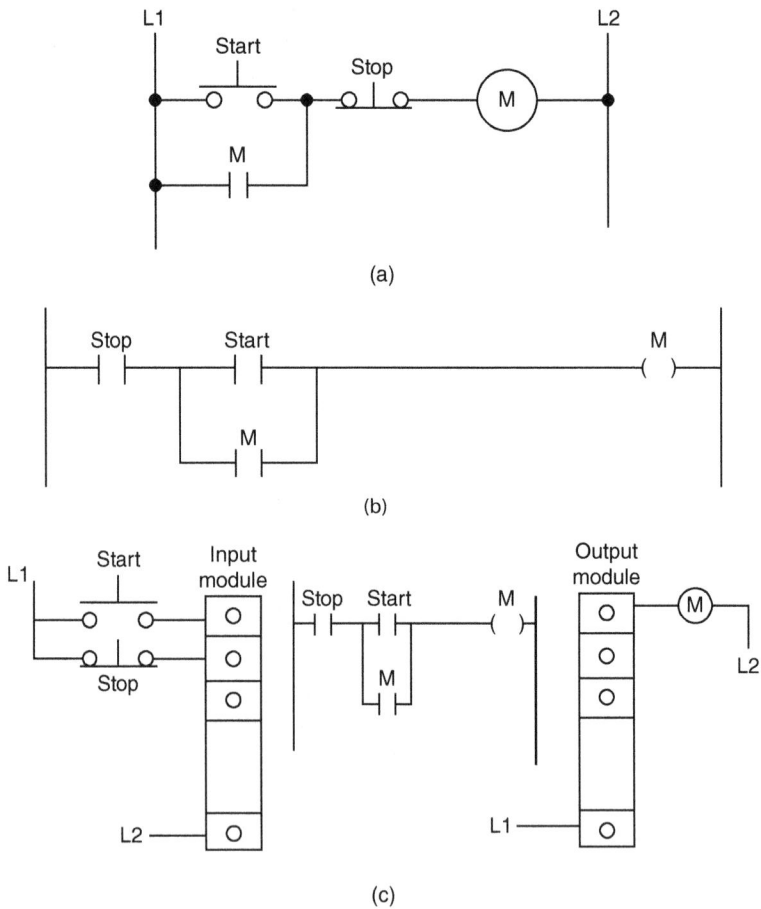

Figure 1.24   (a) Hardwired motor start/stop control relay; (b) motor start/stop PLC ladder logic control; (c) motor start/stop PLC ladder logic in relation to I/O modules.

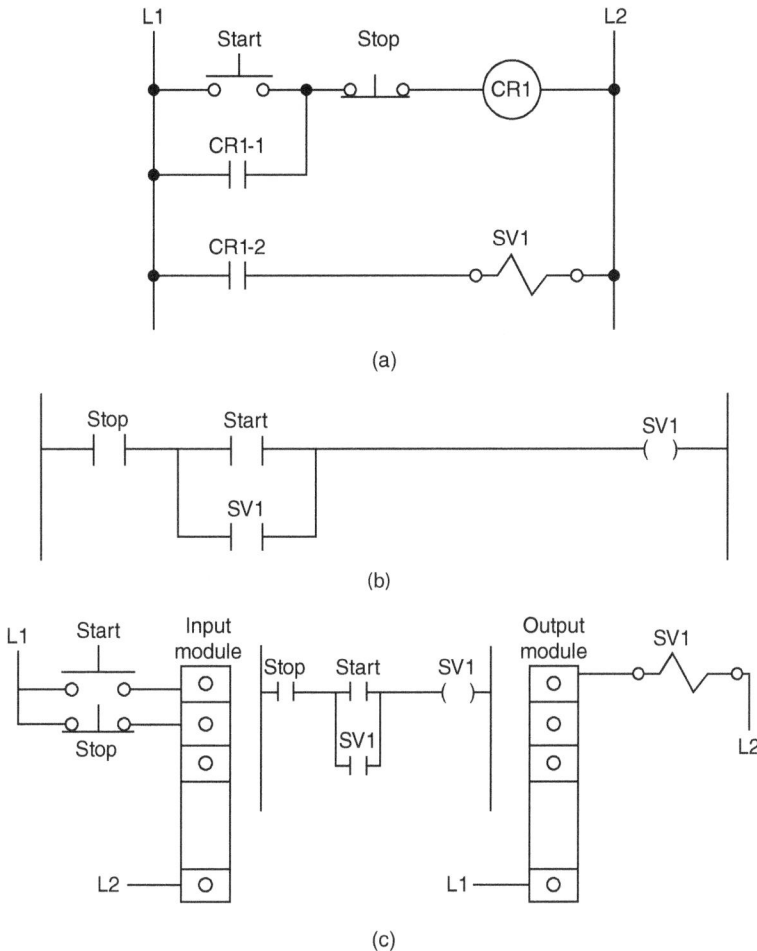

Figure 1.25 (a) Hardwired solenoid valve relay control; (b) solenoid valve PLC ladder logic control; (c) solenoid valve PLC ladder logic control in relation to I/O modules.

or stop the motor—only in the Auto mode. The Start PB is a normally open contact, which closes while the switch is pressed and opens when released. The Stop PB is a normally closed contact, which opens while the switch is pressed and closes once released. The second example shows a simple solenoid valve control using Start and Stop momentary push buttons. The solenoid valve is activated once the Start PB is pressed and deactivated through the Stop switch action.

1.3.5 PLC Ladder Logic

The PLCs use a ladder logic program, which is similar to the line diagram used in hardwired relay control system. Figure 1.26 describes the control circuit for a ladder logic program rung, which is composed of three basic sections: the signal, the decision,

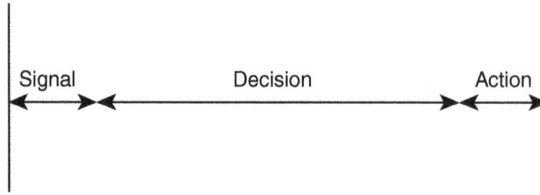

FIGURE **1.26**   Ladder rung/network.

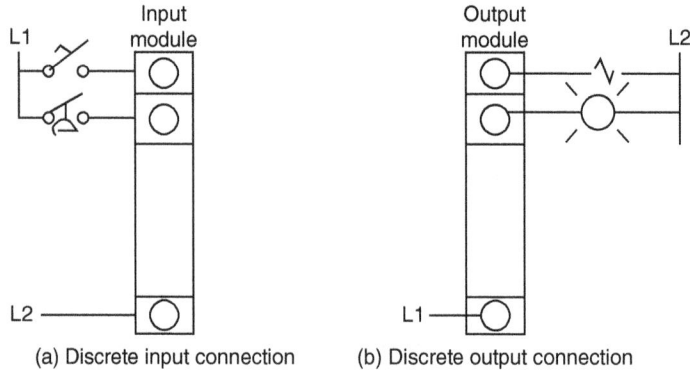

(a) Discrete input connection   (b) Discrete output connection

FIGURE **1.27**   Input/output terminal connection.

and the action. The PLC input modules scan the input signals; the CPU executes the ladder logic program in relation to the input status and makes a decision. The output modules update and drive all output devices. The following sections show the I/O terminal connection and describe the digital I/O addressing format.

As shown in Figure 1.27a, the input devices are connected to the input module through the hot L1, while neutral is connected directly to the input module. Figure 1.27b shows the outputs wired to the output terminal module, the outputs are wired to the output terminal module, and the neutral L2 connected to the output devices. The figure shows two digital inputs, a foot and pressure switches and two outputs, a solenoid and a pilot light.

1.4   Manual/Auto Motor Control Operation

Figure 1.28 shows a manual/auto (M/A) control of a three-phase induction motor. While the M/A switch held in the Manual position, pressing the Start push button energizes the magnetic motor starter, M. This Start PB is only for manual operation. Since the Start push button is a normally open momentary switch, the power to the magnetic starter is maintained through the latch with the auxiliary contact M1-1 around the Start push button. When the M/A switch is placed in Auto position, the digital output module receives the hot L1 through the Auto switch. When rung logic in the software for the

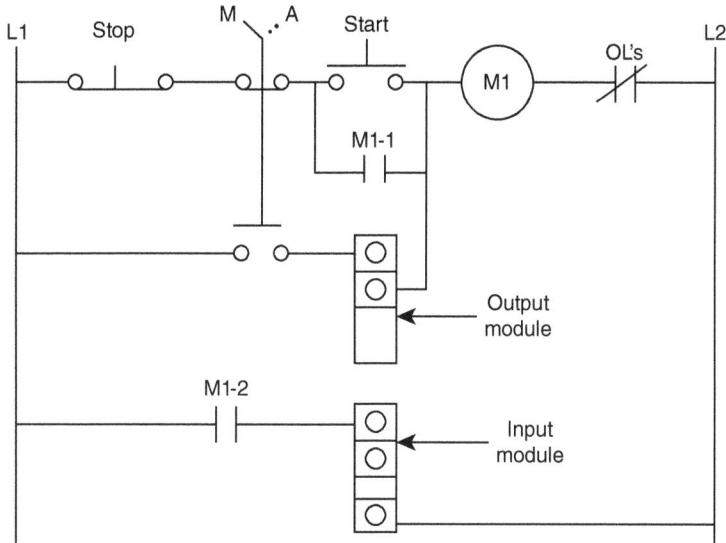

Figure 1.28 Manual/auto motor control PLC connection.

output is true, switching of L1 occurs by the TRIAC switch inside the output module and the magnetic starter is energized causing the motor to run. Notice that the manual Start and Stop PBs have no effect on the automatic operation of the motor, they neither can stop the motor nor jog it. Motor status can be monitored with normally open contact M1-2 wired between the hot L1 and the digital input module in either of the two modes of operation. The neutral is directly connected to the input module. The motor overload conditions, which are typically deployed in the motor for protection and safety operation, are combined and shown in the PLC wiring connection in Figure 1.28. This safety and protection is part of the standard safety requirement for most industrial motors, as defined in the Electrical National Code. Overloads can be caused by excessive heat, current, or load among other factors.

1.5 SLC-500 LogixPro Simulator Setup

LogixPro 500 is an interactive Allen Bradley SLC 500 PLC educational tool based on the ProSim-II simulation software. ProSim was developed to assist students in the acquisition of programming skills used in the control of process-oriented equipment and systems. Although designed primarily for use with PLCs, the flexible interface of the ProSim package readily allows it to be used with any computer programming language that supports ActiveX objects in a MS Windows environment.

1.5.1 The LogixPro 500 Screen

Typical industrial processes, such as material transfer operations using a conveyor or batch mixing tanks complete with pumps and metering, are graphically displayed on the screen. The displayed processes are fully animated and when used with a PLC or

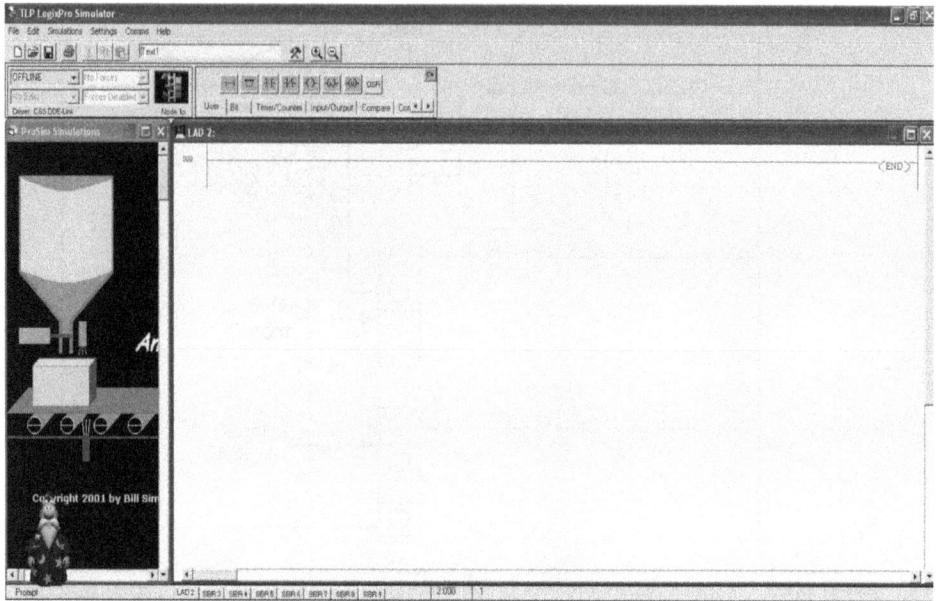

FIGURE 1.29   LogixPro 500 main screen.

PLC emulator, will respond to the signals of the PLC in the same manner as actual process equipment and sensors would respond. For example, a properly addressed PLC output can be used to start a pump on a mixing tank. Once started, visual indication of the pump's rotation and fluid flow will appear on the computer screen. In addition, a digital signal representative of fluid flow would be transferred back as an input to the PLC program. Figure 1.29 shows the LogixPro 500 main screen.

1.5.2   Editing Your Program

If you are familiar with Windows and know how to use a mouse, then you are going to find LogixPro 500 editing a breeze. Both instructions and rungs are selected simply by clicking on them with the left mouse button. Deleting is then just a matter of hitting the Del key on your keyboard. Double clicking (two quick clicks) with the left mouse button allows you to edit an instruction's address, while right clicking (right mouse button) displays a pop-up menu of related editing commands. Online help is also available for quick reference.

1.5.3   Debugging Your Program

If you take a look at the PLC panel, you'll notice an adjustable Speed Control. This is not a component or a feature in normal PLCs, but is provided with LogixPro 500 so that you may adjust the speed of the simulations to suit your particular computer. The simulator's adjustable speed panel does not show till you load and run the program. When the simulation is slowed, so is the PLC scanning. You can use this effectively when trying to debug your program. Set the scan slow enough and you can easily monitor how your program's instructions are responding. This capability may not be typical of real PLCs but for training purposes, you will find that it is an invaluable debugging tool.

1.5.4   RSLogix Documentation

Be sure to check out the entries listed under "RSLogix/LogixPro 500 Reference Documents & Links" on the lower half of the LogixPro 500 Index page. Also, if you have the space on your hard drive, then seriously consider installing the "AB SLC 500® Instruction Set Reference Manual."

1.5.5   I/O Simulator Screen

The I/O simulator screen shown in Figure 1.30 provides two discrete 16-bit inputs at address I:1 and I:3, one 16-bit four-digit BCD input at address I:5 , two discrete 16-bit outputs/coils at address O:2 and O:4, and one 16-bit four-digit BCD output display at address O:6. Only discrete I/Os are accommodated under LogixPro 500 simulation. Analog programming, debugging, and simulation will be discussed later using a simple hardware trainer and the AB SLC 500 software. The same trainer will be used to perform HMI and PLC communication/networking tasks. Note: You can configure inputs, by right-clicking on the switch, as push buttons being normally close/normally open or Single Pole Single Through (SPST) being OPEN/CLOSE. The output light color can be configured by right-clicking (red, green, and yellow). The LogixPro simulator does not support real analog I/O programming. It also does not support floating point

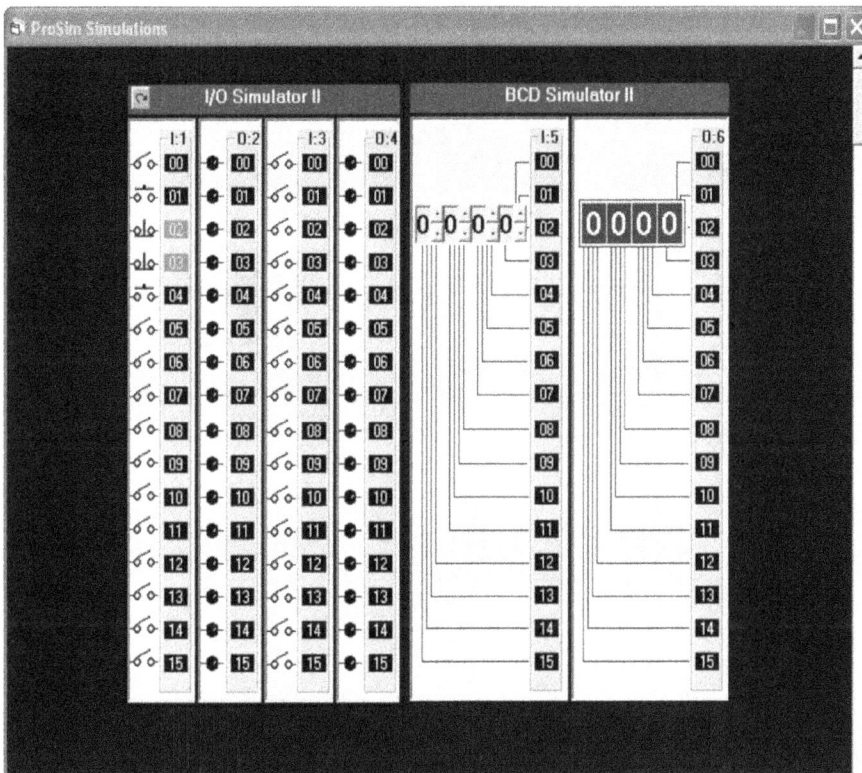

FIGURE **1.30**   Discrete I/O simulation screen.

arithmetic, which is not supported by the phased out unsupported SLC5/01 and SLC5/02 processors.

Every I/O is assigned a unique address according to its terminal position within the I/O modules, which are assumed to be 16 points per module for the two discrete inputs and the two discrete outputs modules. These addresses are fixed and cannot be changed. Its configuration can be changed off line, while its status can be altered during program execution in the online mode. The four-digit BCD simulator allows manipulation of the input in BCD format. The display is also in BCD format. This often requires internal PLC conversion of inputs from BCD to decimal binary and outputs from decimal binary to BCD. Additional I/Os can be used in the program up to the limit defined for the target processor used, but those I/O elements will not be accessible from the I/O simulation panel. All I/O elements, along with other data used in your program, are accessible from the project tree. Closing the I/O simulator will show the project tree and allow examination and manipulation of different program and data files as will be described next.

1.5.6 LogixPro 500 PLC Software

The LogixPro 500 software's main screen has three parts: the project tree, the PLC instructions panel, and the PLC control panel, as shown in Figure 1.31. You need to close an open simulator in order to be able to access the project tree.

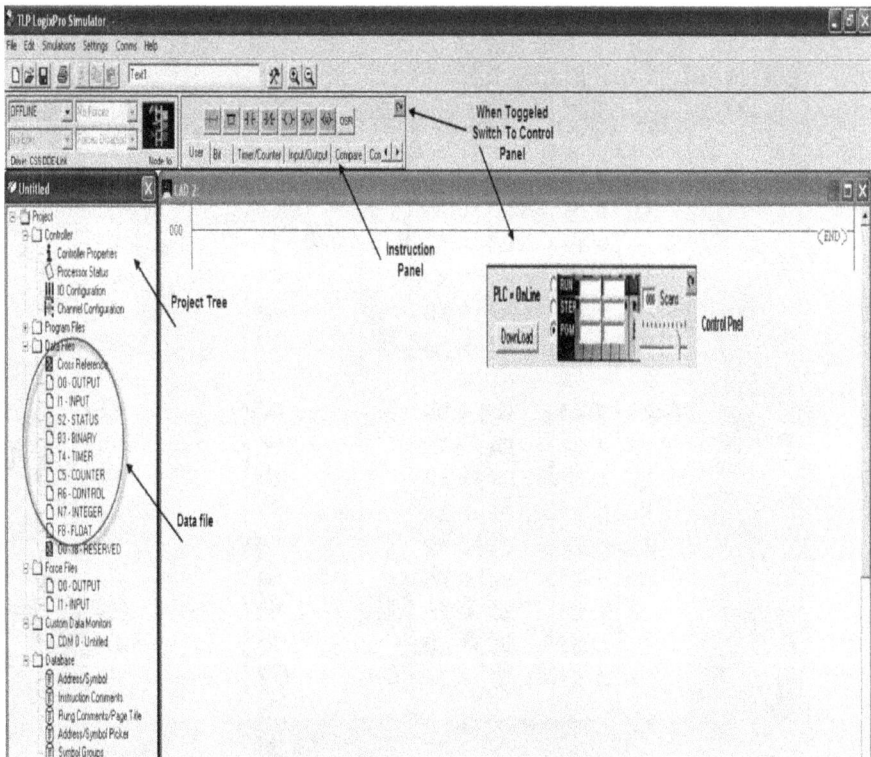

FIGURE 1.31 LogixPro 500 simulator main screen.

The project tree functions look and behave like Windows Explorer, with the following main features:

- As with other Windows programs, a folder with plus (+) sign can be expanded to show its contents.
- A folder with minus (−) sign can be collapsed to hide its contents.
- Using the window toolbar, you can perform the following:
  - Open files
  - Delete files
  - Copy files
  - Rename files
  - Create new file

The size of the data files for the LogixPro 500 is less than the real RSLogix500 software, which will be detailed in Chapter 2. The LogixPro 500 instruction panel also gives limited coverage than the RSLogix 500; for example, advance instructions are not included.

1.6 Process Control Choices

PLCs are not the only devices available for controlling a process or automating a system. Control relays and PCs can be used to implement the same control. Each choice may be of a benefit depending upon the control application. This debate has been going on for a long time while the mix of technologies advanced at an incredible rate. With continuing trend of PLC prices going down, size shrinking, and performance improving, the choice in favor of PLCs has become less of a debate. Still, system owners and designers have to ask themselves if using a PLC is really an overkill for an intended process control or automation application. Table 1.1 summarizes a brief comparison between PLCs and control relays with important issues to be considered.

A dedicated controller is a single instrument that is dedicated to controlling one process variable such as temperature for a heating control. They typically use PID (proportional integral derivative) control and have the advantage of an all-in-one package, typically with displays and buttons. These controllers can be an excellent tool to use in simple applications. PLCs can compete functionally and financially with these controllers, especially when several controllers are needed. PLCs offer a greater degree of flexibility and can be programmed to handle existing and future scenarios.

PCs can also be fitted with special hardware and software for use in process control applications. PCs can provide advantage in certain control tasks relative to PLCs, but their use is not as widespread as PLCs. A hybrid networked system of PLCs and PCs is in wide use in large, distributed control applications. Table 1.2 shows a brief comparison between PLCs and PCs with important issues to be considered.

Issue of PLC and Control Relays Comparison	PLCs	Control Relays
Control Logic Changes	Changes in logic can easily be implemented in software.	Changes require more complex hardware modifications.
Deployment on Different Systems	Easier to customize and download software.	Requires construction of new control panels.
Future Expansion	New I/O modules, expansion chassis, HMI's, and software patches can be added. Networked control systems can be utilized.	Expansion is possible but at higher cost.
Reliability	PLCs are more robust and redundancy is available.	Less reliable because of the use of individual components.
Down time	Troubleshooting/changes can be made online with no downtime.	Changes or troubleshooting often requires the system to go offline.
Space Requirement	Space requirement rapidly decreases as the number of relays increase.	Huge space requirement for a system with large number of relays.
Data Acquisition and Communication	PLCs support data collection, analysis, and communication.	Not directly or easily possible.
Maintenance and Speed of Control	Less maintenance and faster speed of control.	Mechanical parts require more maintenance and reduce speed of control.
Cost	Effective cost and performance for a wide range of process control applications.	Can be cost effective for very small systems.

TABLE 1.1   PLC and control relay comparison.

Chapter 1: Home Work Problems and Laboratory Projects

1) Define the following:
   a. Set point variable.
   b. Controlled variable.
   c. Manipulated variable.
   d. Direct acting control.
   e. Reverse acting control.

2) What is the meaning of the word "process" in a chemical industry?

3) Define What is an open-loop controller?

4) What is the difference between the following:
   a. Open- and closed-loop control
   b. Manual and automated control
   c. Direct acting and reverse acting control

5) List at least three advantages of PLC control over hardwired relay control.

Comparison Issues	PLCs	PCs
Environment	PLCs are specifically designed for harsh conditions with electrical noise, magnetic fields, vibration, extreme temperatures, or humidity.	Common PCs are not designed for harsh environments. Industrial PCs are available but at much higher cost.
Ease of Use	By design, PLCs are friendlier to technicians since they are programmed in ladder logic and have easy connections.	Operating systems like Windows, UNIX, and Linux are common. Connecting I/Os to the PC is not always as easy.
Flexibility	PLCs in rack format are easy to exchange and expand. They are designed for modularity and expansion.	Typical PCs are limited by the number of special cards they can accommodate and are not easily expandable.
Speed	PLCs execute a single program in sequential order and have better ability to handle real-time events.	PCs are designed to handle multiple tasks. Real-time operating systems can handle real-time events.
Reliability	A PLC rarely crashes over long periods of time.	Chances of a PC locking up and crashing are more.
Programming Languages	PLCs languages used are typically ladder logic, function block, or structured text.	PCs are very flexible and powerful in providing a wide variety of programming tools.
Data Management	Memory is limited in its ability to store and analyze large data.	PCs excel in long-term data storage, modeling, simulation, and trending.
Cost	Hard to compare pricing due to many variables, like I/O counts, hardware needed, programming software, etc.	

TABLE **1.2**   PLC and PC control system comparison.

6) Explain the advantages of using a logic diagram or flow chart in programming.

7) Explain the steps used in implementing a single-variable, closed-loop control.

8) Define the following:
   a. Absolute error
   b. Error as a percent of set point
   c. Error as a percent of range

9) If an oven set point = 220°C, Measured value = 200°C, and Range = 200–250°C, answer the following:
   a. What is the absolute error?
   b. What is the error as a percent of set point?
   c. What is the error as a percent of range?
   d. Repeat the above parts for a measured temperature value of 230°C assuming the same set point and range.

10) Explain why the Electrical National Code demands users to control a motor's start/stop using normally open/normally close momentary push button switches instead of maintain switches.

11) Explain the following:

   a. The function of a process controller.
   b. The function of the final control element.
   c. The main objectives of process control.

12) Study the circuit in Figure 1.32 and answer the following questions:

   a. What logic gate type does the indicator represents?
   b. What is the status (ON/OFF) of the indicator if push buttons A and B are pressed and released one time?
   c. What is the status of the indicator if push buttons A and B are pressed and maintained closed all the time?
   d. What is the status of the indicator if push button A or B is pressed at any one time?
   e. Show how you can modify the circuit to maintain the indicator status ON if push button A or B is pressed and maintained closed.
   f. Modify the circuit in Figure 1.32 to maintain the indicator ON once the two push buttons are activated.
   g. Add a STOP push button to turn the indicator OFF and restart the process at any time.

13) Figure 1.33 shows a line diagram for an auto/manual motor control circuit. The start/stop push buttons should start and stop the motor only if the Auto/Manual switch is in Manual position. As shown in Figure 1.33, the circuit has error(s). Perform the following:

   a. Define the error(s).
   b. Redraw the circuit to correct the error.

14) Find and explain the status of CR1, M1, and SV1 in Figure 1.34 under the following conditions:

   a. PB1 is not pushed, and LS1 is open.
   b. PB1 is pushed, and LS1 is open.
   c. PB1 is pushed, and LS1 is close.

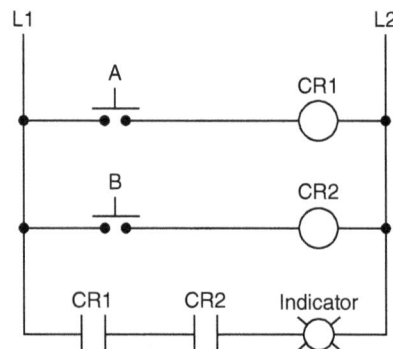

FIGURE 1.32   AND logic gate indicator.

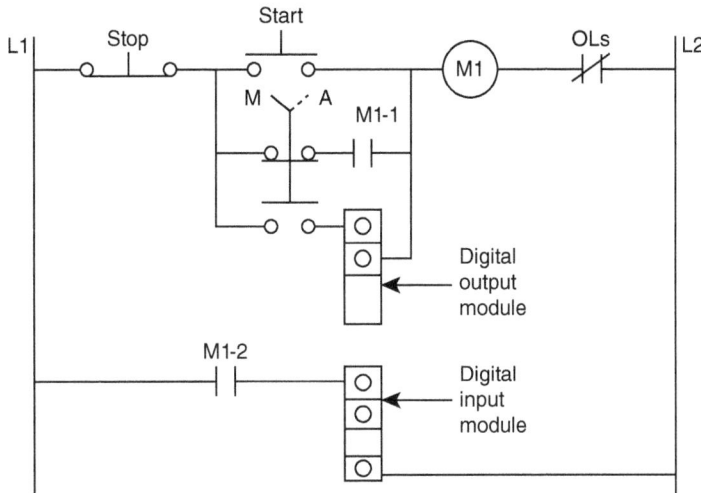

FIGURE **1.33** Problem 13 incorrect auto/manual control.

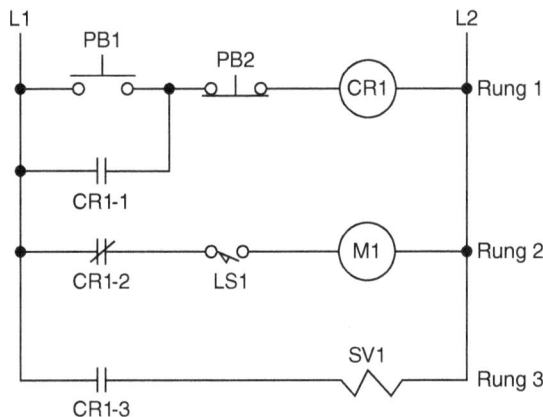

FIGURE **1.34** Problem 14 wired control relay for motor and solenoid valve activation.

15) In some applications, such as motion control, machine tooling, and material handling, the operator should be able to turn the motor on forward/reverse a few seconds to move the load slightly in the forward or reverse direction. This type of motor control is called jogging. Modify Figure 1.14 to include a run/jog switch.

16) Using internet resources, write a two pages report summarizing the history and evolution of PLC's.

17) Figure 1.17 shows a latch/unlatch logic line diagram controlling a motor. Redraw the line diagram to include interlock contacts in order to prevent the operator from pushing the two push buttons (Start/Stop) simultaneously.

18) Process control logic can be implemented using relays, PLCs, or PCs. Construct a comparison between the three options based on cost, scalability, and historical developments / deployments. Use the internet to construct your research.

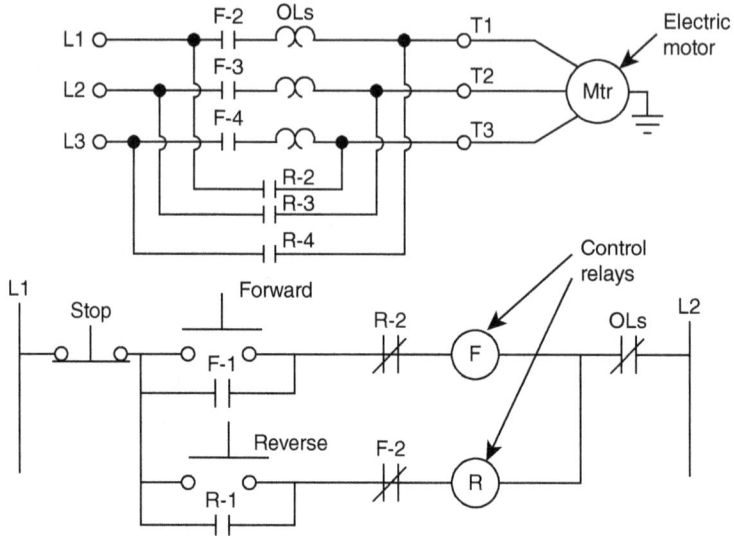

FIGURE 1.35 Wiring and relay logic for problem 19.

19) Figure 1.35 shows a three-phase reversible (runs forward or reverse direction) induction motor wiring and the associated relay logic diagram. Identify and correct the wrong motor wiring and explain the reason for the rewiring.

Laboratory 1.1—LogixPro 500 Program Creation

The objective of this laboratory is to get users familiar with the LogixPro500 simulator software. Use the following steps to create a ladder logic program in LogixPro 500:

- From the Simulation menu bar close the existing simulation and open the I/O simulator.
- Open a new file and from select processor enter the processor type, as shown in Figure 1.36.

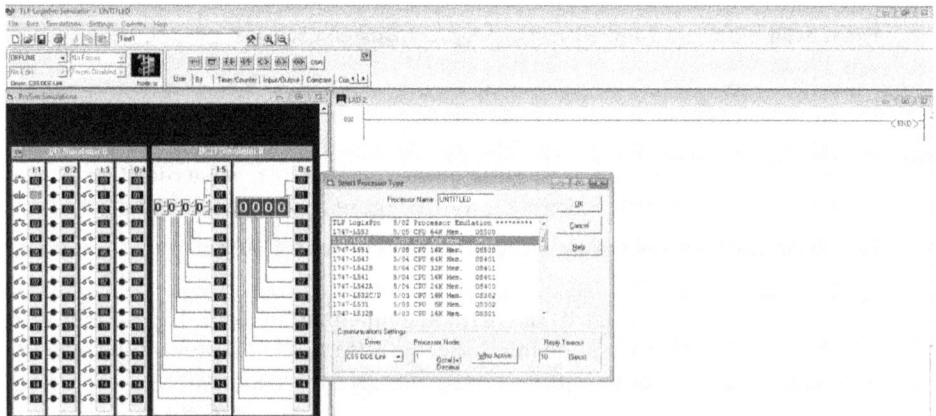

FIGURE 1.36 I/O simulation panel in LogixPro 500.

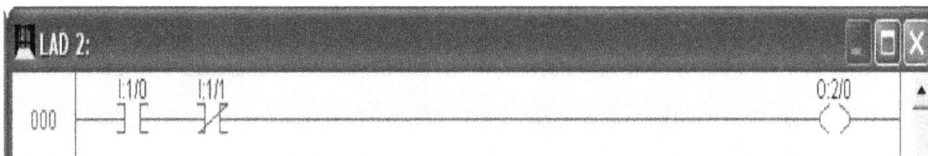

FIGURE 1.37   One rung program in LogixPro 500.

- Collapse the I/O simulation screen back to its normal size by clicking on the same (center) button you used to maximize the simulation's window. You should now be able to see both the simulation and program windows again. If you wish, you can adjust the relative size of these windows by dragging the bar that divides them with your mouse.
- Enter the single rung program shown in Figure 1.37, which consists of two input instructions (XIC [Examine If Closed]), Examine If Open, and a single output instruction (OTE [Output Energize]).
- Click on the XIC instruction with your left mouse button (left click), then drag and drop to the input section of rung 0 and enter the address (I:1/0) or drag and drop the address from the I/O simulator. Repeat for XIO address (I: 1/1) from the I/O simulator as shown in Figure 1.38.
- Click on the OTE instruction with your left mouse button (left click), then drag and drop to the output section rung 0 and enter the address (O:2/0) or drag and drop the address as shown in the Figure 1.38.
- You can drag and drop addresses between ladder rungs.
- Right click on the XIC instruction and select "Edit Symbol" from the drop-down menu that appears. Another textbox will appear where you can type in a name (SS1), (SS2) to associate with these addresses, as shown in Figure 1.39. As before, a click anywhere else will close the box.

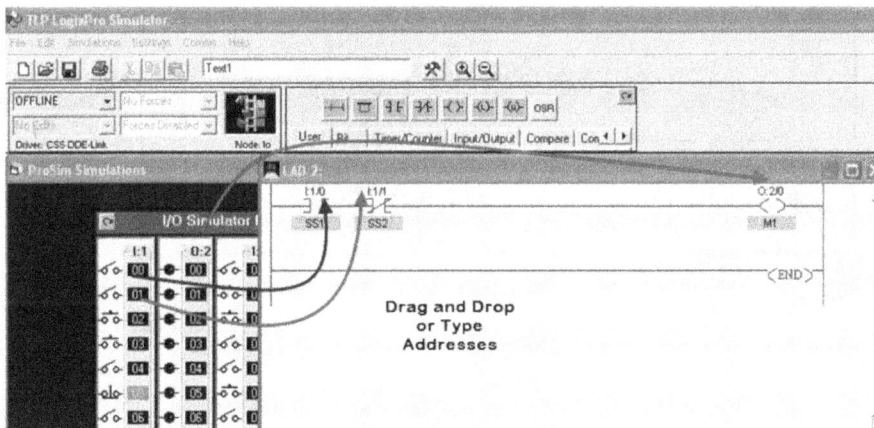

FIGURE 1.38   Program creation in LogixPro 500.

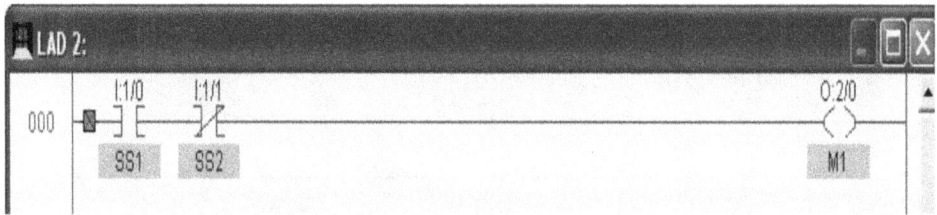

FIGURE 1.39   Editing tag symbols in LogixPro 500.

- Enter the address and symbol for the OTE instruction and your first RSLogix program will now be complete. Before continuing however, double check that the addresses of your instructions are correct.

Laboratory 1.2—Program Testing

It is now time to "Download" your program to the PLC. First click on the "Toggle" button at the top right corner of the Edit panel, which will bring the PLC panel into view, as shown in Figure 1.40.

Click on the "Download" button to initiate the downloading of your program to the PLC. Once download is complete, click inside the "RUN" option selection circle to start PLC scanning. The simulator screen shown in Figure 1.41 should now be in view. For this laboratory we will be using the I/O simulator. Use your mouse to click on I: 1/0 switch and note the change in the status color (yellow) of the terminal to which the switch is connected to and the ladder rung inputs: Switch1, Switch2, and output M1.

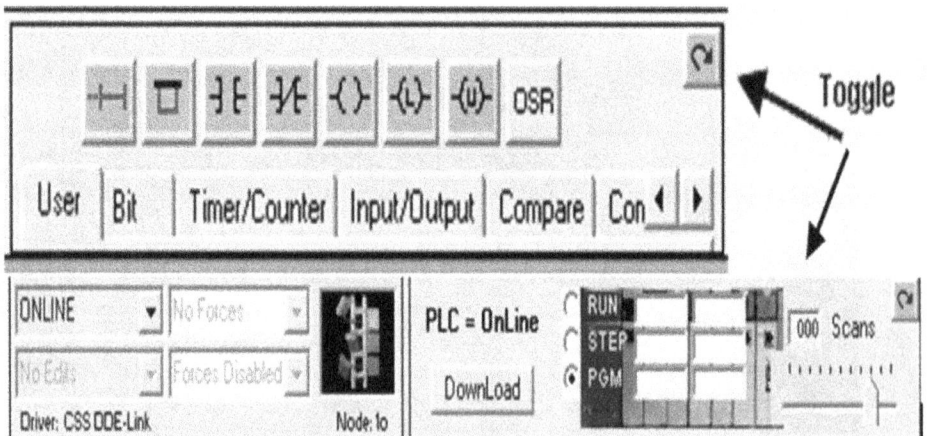

FIGURE 1.40   PLC panel.

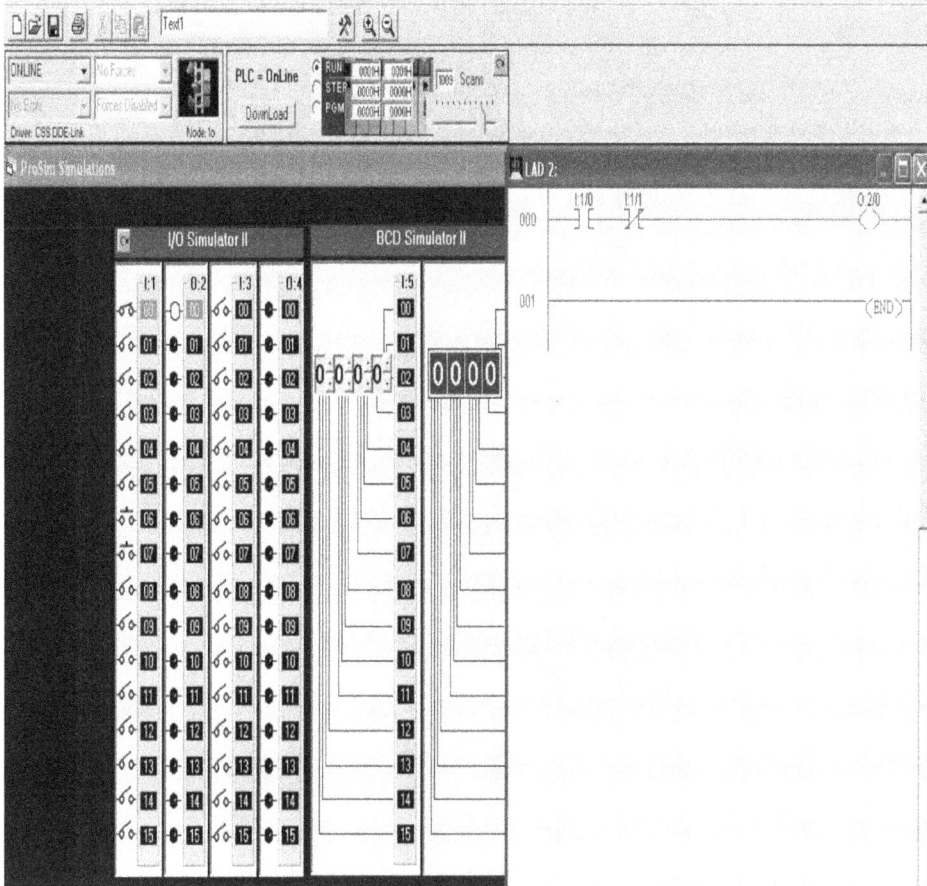

FIGURE **1.41**   LogixPro 500 simulator screen.

Laboratory 1.3—Converting Hardwired Control Relay to PLC Ladder Logic

Refer to Figure 1.34 and use the following I/O addresses to convert the hardwired control relay to a PLC ladder program:

Input	Tag Name	Output	Tag Name
PB1	SS3	CR1	PL2
PB2	SS4	M1	PL3
LS1	SS5	SV1	PL4

Requirements:

- Assign the three required physical discrete input addresses from the Simulator panel.

- Assign the three required physical discrete output addresses from the Simulator panel.

- Change the selector switches to the desired type and status.
- Change the discrete output indicators to three different colors.
- Program the ladder logics required.
- Download the program and perform the check out.
- Submit your reports with your comments on the laboratory.

CHAPTER 2

Fundamentals of PLC Logic Programming

T his chapter will focus on PLC hardware fundamentals and logic programming for the Allen-Bradley (AB) SLC-500. The concepts covered are also applicable to other types of PLCs.

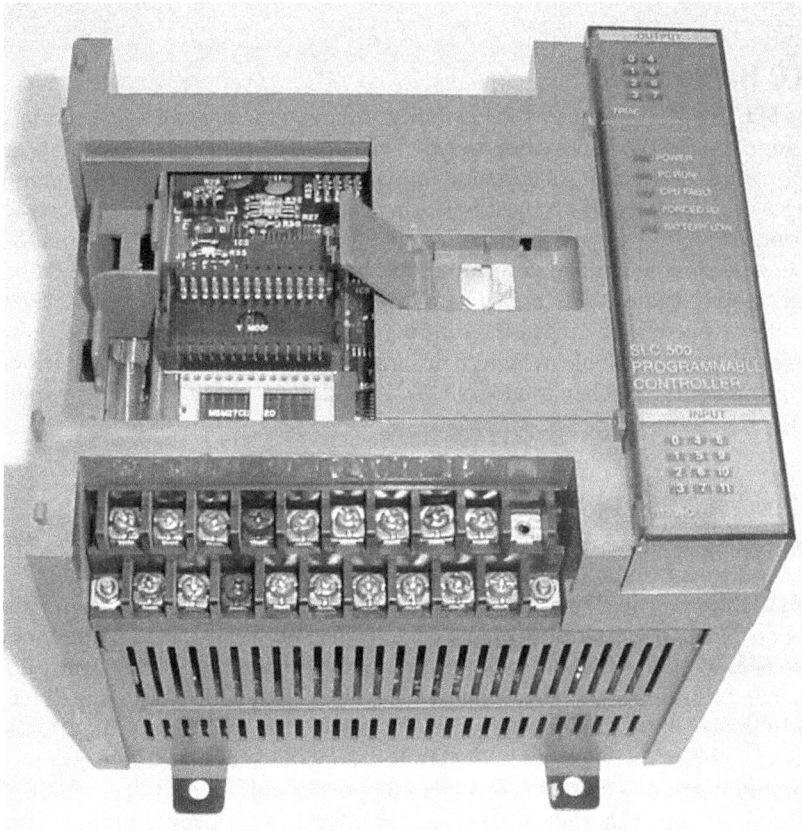

Hands-on PLC programming with RSLogix™ 500 and LogixPro®

Chapter Objectives

- Understand PLC hardware and memory organization.
- Understand ladder logic diagrams and programming.
- Understand combinational and sequential logic instructions.
- Use ladder programming in industrial process control.

A programmable logic controller is a special digital computer used for industrial and process automation; such as manufacturing assembly lines, chemical plants, waste water treatment, or amusement rides. PLC programming is a major task as, in addition to requiring knowledge of the specific ladder logic development environment and associated utilities, it assumes familiarity with the control application domain. Understanding PLC hardware details, Human Machine Interfaces (HMIs), and communication fundamentals is a must. This chapter focuses on PLC hardware fundamentals and logic programming for the Allen-Bradley SLC-500 family of PLCs. HMIs and communication will be addressed in later chapters.

2.1 PLC Hardware

The SLC-500 PLC provides the flexibility and power to control a wide variety of devices in support of most automation needs. The compact design, flexible configuration, and a powerful instruction set combine to make the SLC-500 a perfect solution for controlling a wide variety of applications. The system combines a microprocessor, a power supply, input circuits, and output circuits in a compact housing to create a powerful PLC. After you download your program, the central processing unit (CPU) contains the logic required to monitor and control the devices in your application. The CPU monitors the inputs and changes the outputs according to the logic of your user program, which can include Boolean logic, counting, timing, complex math operations, and communications with other intelligent devices. A typical SLC-500 PLC system building block is shown in Fig. 2.1.

2.1.1 SLC-500 Processor

The SLC processors offer a wide range of choices in memory size, I/O capacity, instruction set, and communication ports to allow the designer to tailor a control system to the exact application requirements. The processor can configure modular controllers of up to 4096 inputs plus 4096 outputs and an instruction memory capacity of 1 to 64K words. Different versions are available for faster communication port and shorter execution time; this includes the SLC 5/01, SLC 5/02, SLC 5/03, SLC 5/04, and the SLC 5/05 shown in Fig. 2.2. Factors for the selection of the processor should include pricing, performance, scalability, and the exact requirements of the project. Communication is available via DH-485 port, a built-in DH+ port, a 10/100BASE-T (10 or 100 Mbps Ethernet port), operator interfaces, or programming terminals interface. The processor module has a set of LEDs used to indicate important PLC hardware status information.

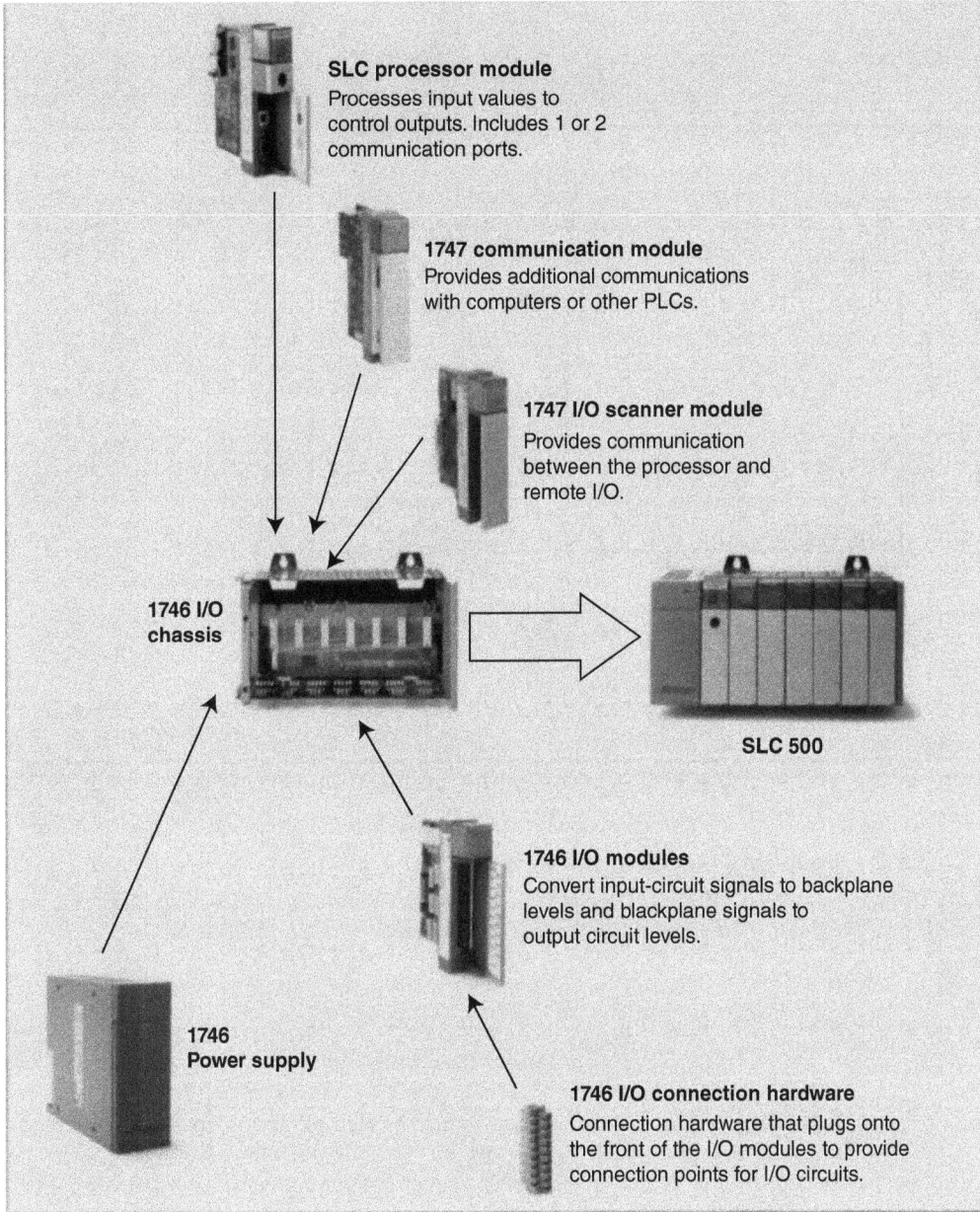

SLC processor module
Processes input values to
control outputs. Includes 1 or 2
communication ports.

1747 communication module
Provides additional communications
with computers or other PLCs.

1747 I/O scanner module
Provides communication
between the processor and
remote I/O.

**1746 I/O
chassis**

SLC 500

1746 I/O modules
Convert input-circuit signals to backplane
levels and blackplane signals to
output circuit levels.

**1746
Power supply**

1746 I/O connection hardware
Connection hardware that plugs onto
the front of the I/O modules to provide
connection points for I/O circuits.

FIGURE 2.1   Typical SLC-500 configuration. (*Image courtesy of Allen-Bradley, a Rockwell Automation business.*)

Battery (provides
backup power for the
CMOS RAM)

Memory
module

SLC 5/05 CPU

Run ☐☐ Force
FLT ☐☐ ENET
Batt ☐☐ RS232
Run Rem Prog

Key switch

Channel 1
ethernet
(10base-T)

Channel 0
RS232
(DH485, D51,
or ASCII)

IP address

Hardware
address

xxxxxx

Write-on
area for
IP address

Operating system memory
module download protection
jumper

Location of serial and
catalog numbers

Left side view

Front view

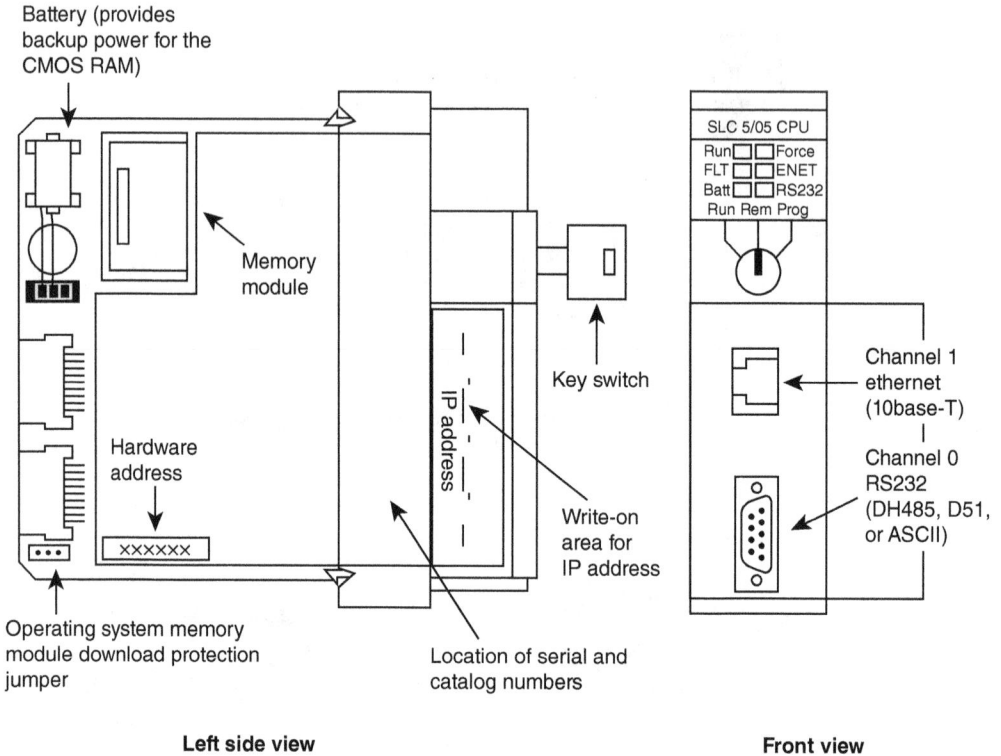

FIGURE 2.2 SLC 5/05 modular processor. (*Image courtesy of Allen-Bradley, a Rockwell Automation business.*)

2.1.2 Operating Modes of the CPU

The CPU has three modes of operation: Program, Remote, and Run. The following are the characteristics of each of the three CPU modes and two additional states:

- In the Stop state, the CPU is not executing the program. Projects cannot be executed in this mode.

- In the Program mode, the processor is accepting and compiling new instructions, either as a new program or as changes to an existing program.

- In the Test state, which is available from the Remote mode, the processor reads inputs and solves ladder program, but it does not allow field devices to be energized. Test mode is used to test a program during installation, maintenance, or troubleshooting.

- In the Remote mode, changing the operating mode is done from a personal or industrial computer through remote communication. Some processors have Run-Rem-Prog key switch on the processor module to change operating modes. Changing the operating mode to Run or Prog using the key switch is called local mode.

- In Run mode, the scan cycle is executed repeatedly in the processor memory and outputs are activated according to the implemented program logic. Once the creating or editing of a program is complete, the processor is put into the Run mode to execute the stored code. Programs cannot be downloaded in this mode.

2.1.3 Communication Modules

The SLC-500 family provides communication modules (CMs), which provide critical additional functionality to the system. There are three communication modules: RS232/RS485, Allen Bradley Data Highway (DH), and Ethernet. The CPU supports up to three communication modules. The following communication interfaces are available with the SLC family processors:

- MicroLogix 1000—DH 485 and Ethernet
- 5/01 processor—DH-485
- 5/02 processor—DH-485
- 5/03 processor—Serial and DH-485
- 5/04 processor—Serial and DH+
- 5/05 processor—Serial and Ethernet

2.1.4 Input/Output Modules

The input/output modules are of three types: digital, analog, and special. The digital I/Os are discrete ON/OFF voltage type signals, analog I/Os are variable (minimum to maximum value) voltage or current signals, and special. Special modules accommodate other types of signal interfaces as in high speed counters, ASCII, BCD, and RTD modules. All modules are designed according to universal standards independent of the application interface devices used. Analog modules can accommodate all kinds of sensors and actuators typically used in process control and automation. An example of a special module is the high-speed pulse (HSP) counter or the ASCII module.

2.1.4.1 Digital Input Modules

As shown in Fig. 2.3, the input module performs four main tasks: it senses the presence of an input signal, maps the input signal, which is typically a 120-V ac or 24-V dc, to a low dc voltage, isolates the input signal from the mapped output signal, and outputs a dc signal to be sensed by the PLC processor (CPU) during the input scan cycle.

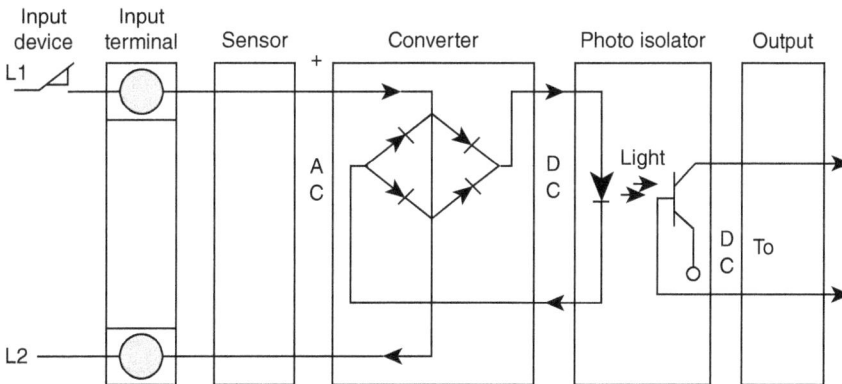

FIGURE 2.3 Digital input modules.

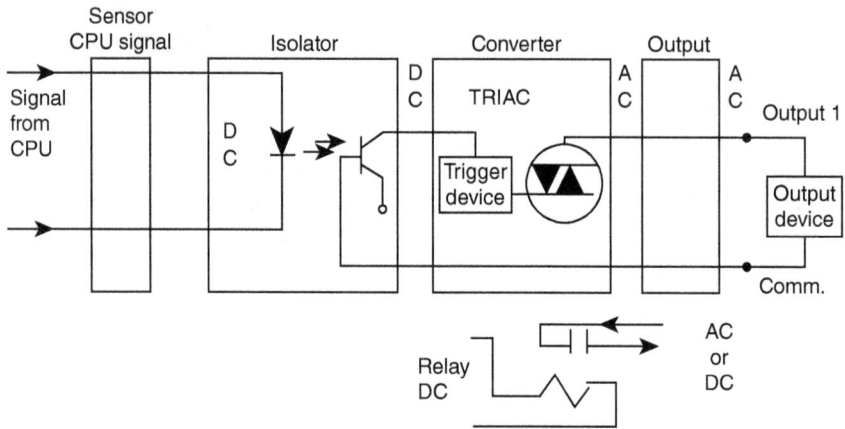

Figure 2.4   Digital output module.

2.1.4.2   Digital Output Modules

The output module operates in the opposite manner to the input module. It acts as a TRIAC switch connecting any selected input to the module's ac or dc voltage as shown in Fig. 2.4. The output module having the matched terminal address receives an output command by the CPU. The output module performs the voltage switching to the selected terminal output point.

TRIAC, from TRIode for Alternating Current, is the trade name for an electronic component that can conduct current in either direction when it is triggered (turned on), and is formally called a bidirectional triode thyristor or bilateral triode. The bidirectional property makes TRIACs very convenient switches for ac circuits, also allowing them to control very large power flows with milliampere-scale gate currents.

2.1.5   Power Supply

The power supply's main function is to convert the 120/220 ac voltage input to various dc voltages required for the PLC components and the CPU operation. The power supply has three main components: line conditioner, rectifier, and a voltage regulator. The line conditioner purifies the input ac voltage waveform to a smoothed sine wave. The rectifier converts the stepped down input ac voltage to the required dc voltage level as shown in Fig. 2.5. The voltage regulator maintains a constant dc output voltage level by filtering and reducing existing ripples.

2.1.6   SLC-500 Memory Organization/Specifications

The SLC-500 processor module contains the CPU, which is the brain of the PLC. The processor interfaces with the memory used to store programs and data needed to perform desired functions or control tasks. Memory is organized into three sections: executive, scratch pad, and processor file. The executive section uses a read-only memory (ROM) and stores the management program that enables the CPU to interpret the data and actions entered by the PLC operators/users. The scratch pad area is used as temporary storage and thus utilizes a random access memory (RAM).

The processor file area is used to store and allow the manipulation of user's programs and associated data—program files and data files. Ladder logic programs and

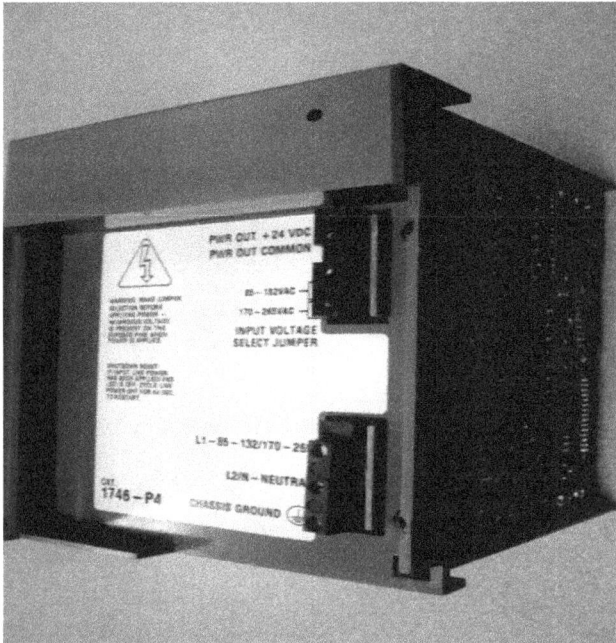

FIGURE **2.5**   SLC-500 power supply.

system configuration information are stored in this section. The program files use data files to perform their functions. Each program file contains up to 256 folders addressed 0 to 255; system functions (password, processor type, and I/O modules slots in the PLC rack), reserved storage, user program, and subroutines folders. The data files, up to 256 files, contain the information used by the program files addressed from 0 to 255. Each data file consists of 256 elements; some are one 16-bit word (as in bit file), while others are three 16-bit words (as in counters files).

2.1.7   Processor Memory Map and Program Organization

This section briefly introduces the structure for the AB SLC-500 processor memory organization. It also covers the concept of structured programming through the use of subroutines and advanced programming techniques.

2.1.7.1   Memory Areas

The processor memory area is divided into three sections. Each memory area stores the user program, user data, and configuration. The following is a brief description of each section:

- Load memory is a nonvolatile storage for user program, data, and configuration.
- Work memory is a volatile storage work area for some elements of the user project while the user program is executing.
- Retentive memory is a nonvolatile storage used for a limited quantity of work memory values.

File type	File number	Words per element
Output image	0	1
Input image	1	1
Status	2	1
Bit	3	1
Timer	4	3
Counter	5	3
Control	6	3
Integer	7	1
Floating point	8	2
Assign file type as needed		

FIGURE 2.6   RSLogix 500 file structure.

2.1.7.2   Memory Map

The processor memory map is divided into several data files as shown in Fig. 2.6, where each data file consists of an operand and elements tags such as inputs, outputs, and bit memory. The CPU identifies these operands based on a numerical absolute address.

Fig. 2.7 shows the file structure and associated addressing format for the RSLogix 500 PLC system. The maximum number of elements per file depends on the PLC series selected.

General addressing format is expressed in symbols and letters as follows:

$$\# \, x \, f: e. \, s/b$$

Element designator   Bit designator

O:00 2/04

Output   Rack   Group   Terminal

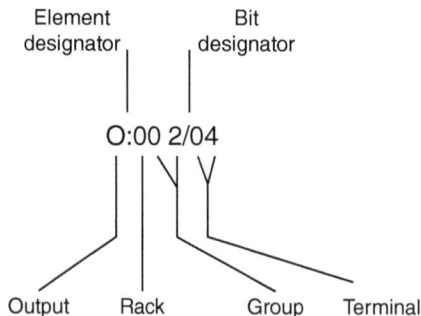

FIGURE 2.7   Addressing format.

where # = file address identifier used in file arithmetic and logic instructions, which perform on several registers at the same time. This field is optional.

x = file type identifier used as follows:

O = Output	R = Control
I = Input	N = Integer
S = Status	F = Floating
T = Timer	B = Bit
C = Counter	

f = file numbers that are defined as follows:

0 = output	5 = counter
1 = input	6 = control
2 = status	7 = integer
3 = bit	8 = floating
4 = timer	9 to 999 = assigned as needed

e = element number is processor dependent, which can take on a value in the following ranges:

0–30	I/O files
0–31	status files
0–999	all other files

s = subelement number or mnemonic
b = bit number within the word

The following are examples of valid addressing under the LogixPro 500:

- O: 002/03 refers to output word 2 bit #3. The leading digit in the element number (002) designates the rack number 0 group number 2, and (/03) designates terminal number 3.

- B3: 6/2 refers to bit file number 3, element 6, and bit #2.

- C5:0. PRE refers to counter file 5, element 0 preset word. The counter has three words: preset, accumulated, and control words.

2.2   Ladder Logic Diagrams

The PLC uses a language called *ladder logic program,* which is similar to the line diagram used in hardwired relay control system. Fig. 2.8 describes the control circuit for a ladder logic rung, which is composed of three basic sections: the signal, the decision, and the action.

The PLC input modules scan the input signals; the CPU executes the ladder logic program in relation to the input status and makes a decision. The output modules update and drive all output devices. The program scan process or what is referred to as PLC events in the operating cycle is summarized in Tab. 2.1. Description of each scan

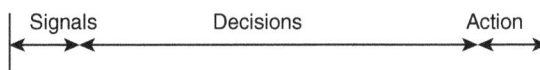

Figure 2.8   Ladder diagram rung.

Event in Operating Cycle	Description
Input scan	The status of each input module is read, and the input image table in the processor is updated with the information
Program scan	The ladder program is executed
Output scan	The output image table information is transferred to the output module
Communication	Communication with computer and other devices takes place
Processor overhead	Internal housekeeping in the processor takes place, including updates of the status file and internal time base

TABLE 2.1 PLC Program Scan Cycle

operation is also given. The following sections show the I/O terminal connection and describe the digital I/O addressing format.

2.2.1 PLC Input/Output Terminal Connection

As shown in Fig. 2.9a, input devices are connected to the input module through the hot (L1), while neutral is connected directly to the input module. Figure 2.9b shows the output devices wired to the output terminal module through the neutral (L2), while hot is directly connected to the output module.

Figure 2.10 shows a ladder logic rung. This is very similar to the line diagram used in hardwired relay. Every instruction is examined if true (true means the bit in the PLC

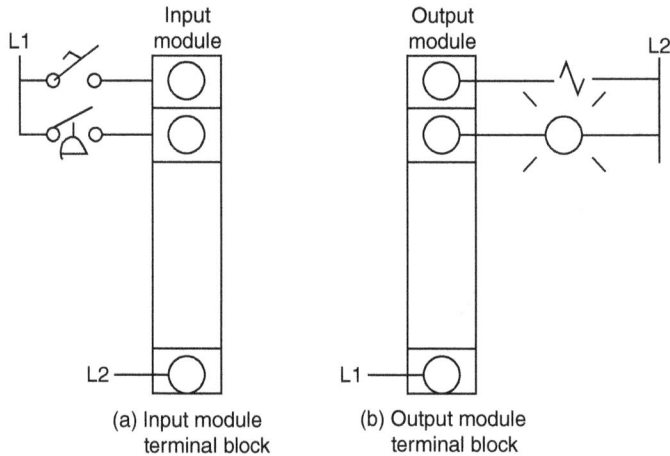

FIGURE 2.9 I/O PLC discrete I/O connections.

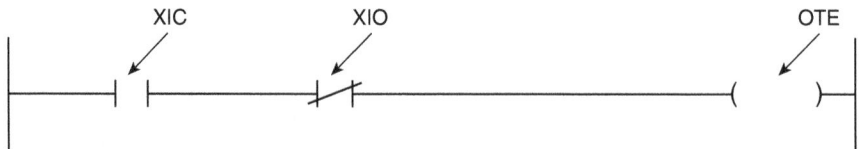

FIGURE 2.10 Typical ladder logic rung.

memory is assigned a value of 1). If true, continuity of the rung is maintained or power flow is maintained. When all input instructions are true, the output will be set to 1 (ON). The rung shown has two input elements (representing decision) and a coil (representing action) as previously shown in Fig. 2.8. A brief description of the three commonly used instructions is discussed in the next section. The AB SLC-500 system notation, hardware, and software development tools will be assumed throughout this book and in all implemented examples and projects.

2.2.2 PLC Boolean Instructions

Three common instructions are used during the scanning of a typical ladder rung: examine if closed (XIC), examine if open (XIO), and output energize (OTE). Two are associated with inputs and the third is associated with output. These inputs and outputs can be either critical or program-related and intermediate conditions. Each of the three instructions is detailed as follows.

Examine If Closed (XIC): As shown in Fig. 2.11, when the switch is closed, the XIC instruction is true and continuity of power is maintained, as indicated by a bold solid line.

As shown in Fig. 2.11*a*, the input device is connected to the input module through the hot (L1), while the neutral connects directly to the module. When switch SS1 is open, the instruction (XIC) in the PLC rung is false (assigned image table memory bit is 0). Figure 2.11*b* shows the switch in the close position, the XIC instruction in the PLC rung is true (assigned image table memory bit is 1) and power flows through the rung, as indicated by the solid line around the instruction.

Examine If Open (XIO): As shown in Fig. 2.12: When SS1 is open, the XIO instruction is true and continuity of power is maintained, as indicated by a bold solid line.

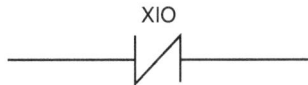

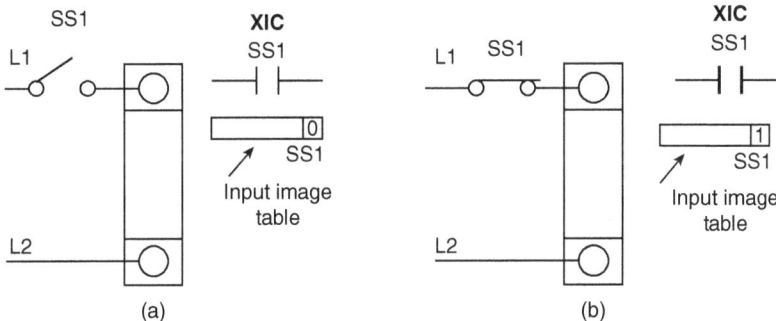

FIGURE 2.11 Hardwired connections and the associated instruction status.

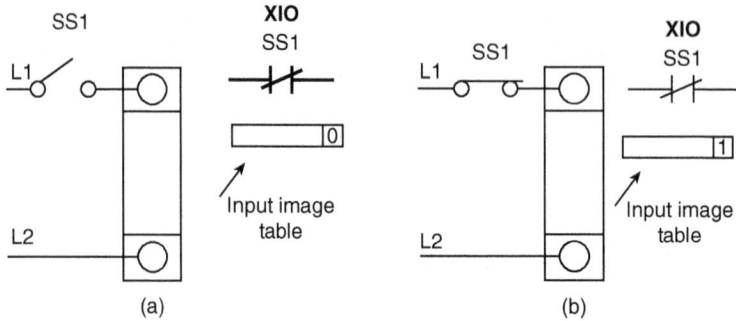

FIGURE **2.12** PLC discrete input connections and associated instruction status.

As shown in Fig. 2.12*a*, the input device is connected to the input module through the hot (L1), while neutral connects directly to the module. When switch SS1 is open the instruction XIO in the PLC rung is true and power flow through the rung as indicated by the solid line around the instruction. Figure 2.12*b* shows the switch in the close position, the XIO instruction in the PLC rung is false and power does not flow through the rung.

Output Energize (OTE): The output energize instruction writes a value for an output bit in the PLC image table memory based on the power flow status preceding the instruction; if all preceding conditions are true, the output instruction becomes true. The output signal for the associated control actuator is wired to the output coil terminal. Coils are assigned unique memory bit addresses. No two coils can have the same address except for the latch/unlatch (L/U) instruction, which has been discussed in the next chapter. Hardwired connection associated with the OTE instruction is illustrated in Fig. 2.13.

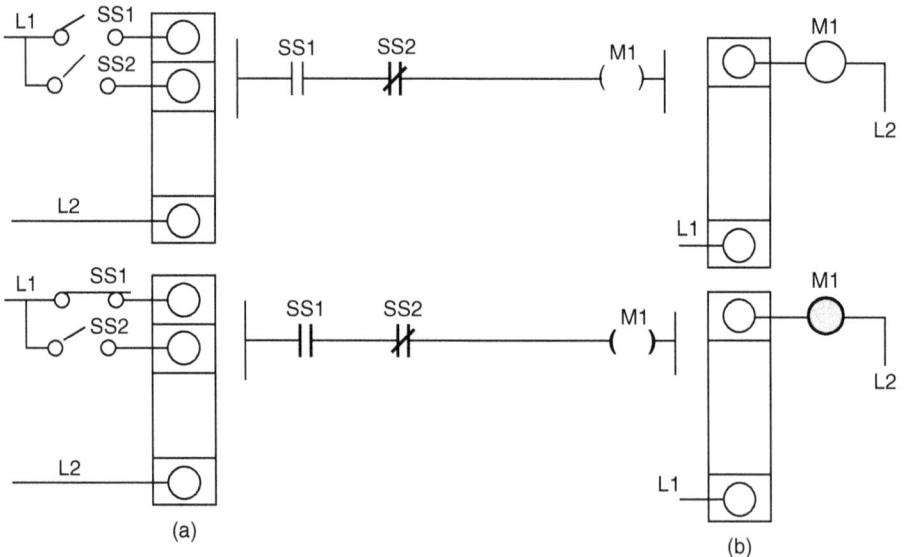

FIGURE **2.13** PLC I/O connections and the associated instructions status.

The memory bit associated with the coil instruction is updated every scan of the ladder in the following manner:

If the output energize memory bit is set to 1, power flows through that output coil.

If the output energize memory bit is set to 0, power will not flow through that output coil.

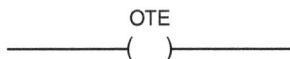

$$\text{OTE}$$
$$———( )———$$

Figure 2.13 shows the three instructions: examine if closed (XIC), examine if opened (XIO), and output energize (OTE) hardwired connections of I/O in relation to the PLC instructions. Notice that when an instruction is true power flows through the instruction.

2.2.3 LogixPro 500 Data Files

LogixPro is a low-cost PLC simulator designed for the Rockwell Allen Bradley RSLogix 500 software. There is no need for any PLC hardware or actual I/O devices when using the simulator. The commercial edition of LogixPro, which is part of the book materials, provides animated process simulations, including an I/O simulator with configurable switch types/status, color-selectable output indicators, four-digit thumbwheel switches, and a four-digit BCD digital readout as shown in Fig. 2.14. Several process animation simulators are also included and have been used extensively in Chap. 5 and in several homework and small projects throughout the book. User-defined data files are not available in LogixPro and there is no need to configure the processor, the I/O modules, or the communications. The close button above the shown I/O simulator will toggle the open window between the

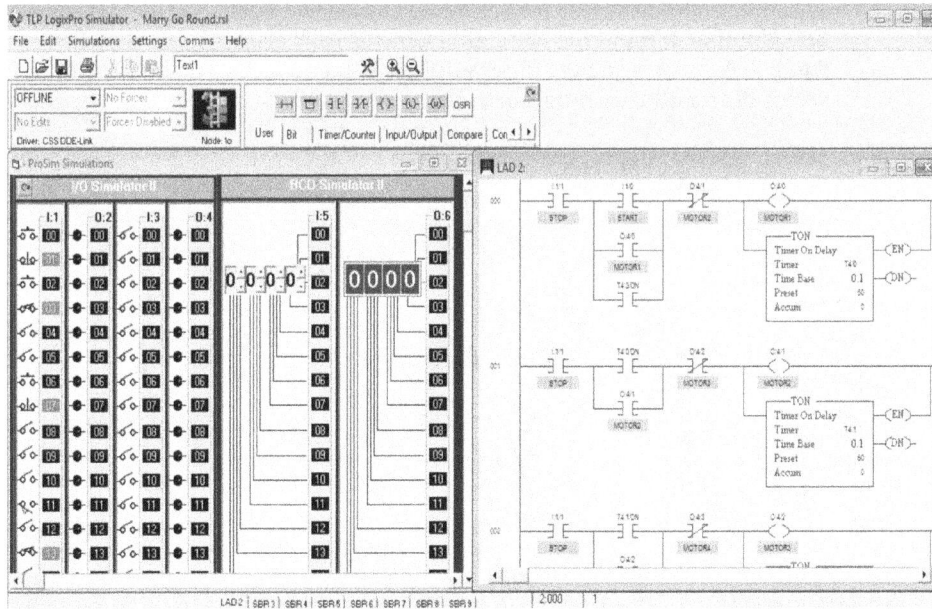

Figure 2.14 LogixPro main screen.

project view and the simulator view. The simulator provides an excellent environment for process control application programming implementation and debugging before transitioning to the real hardware and the process/system final checkout.

The LogixPro's screens and looks are very similar to the RSLogix 500 interface. The instruction toolbar is the same except that some instructions, such as for analog programming, are not available in LogixPro. There is an additional button in the upper right corner, the toggle button, which will toggle the PLC simulator from offline to online mode. The simulations will only function when you write PLC ladder logic and download to the processor to control them. The switches wired to the input modules can be closed by moving the mouse pointer over the switch until the pointer changes to a pointing finger hand. Click the mouse button and the status of the switch will change from open to close or close to open. The type and status of the switch can be changed by right-clicking on the switch. The switch type can be changed to a normally open PB, a normally closed PB, a normally open limit switch, a normally closed limit switch, or a normally open toggle switch. The color of the output indicators can also be changed by right-clicking on an indicator and selecting a color from the pop-up menu.

LogixPro 500 data files are accessible from the project tree. The LogixPro 500 I/O addresses are expressed in decimals. The following is a listing of the file definition in the LogixPro 500 simulation software.

File Number 0: From the project tree, click OUTPUT. The output file has eight elements each element consists of one word, O: 0 to O: 7. A word contains 16 bits, for example, O: 3/0 to O: 3/15. The first two words; O: 0 and O: 1, are not accessible for all user programs. The output file screen is shown in Fig. 2.15. Other files can be selected and displayed in the same manner. Unlike the RSLogix 500 software, only one file window can be displayed/monitored at any time. You can toggle the display between different file windows. You can also toggle between the project tree and the selected simulator view.

File Number 1: Input file has eight elements each element consists of one word, I: 0 to I: 7. A word is 16 bits, for example, I: 2/0 to I: 2/15. The first word, I: 0, is not accessible for all user programs.

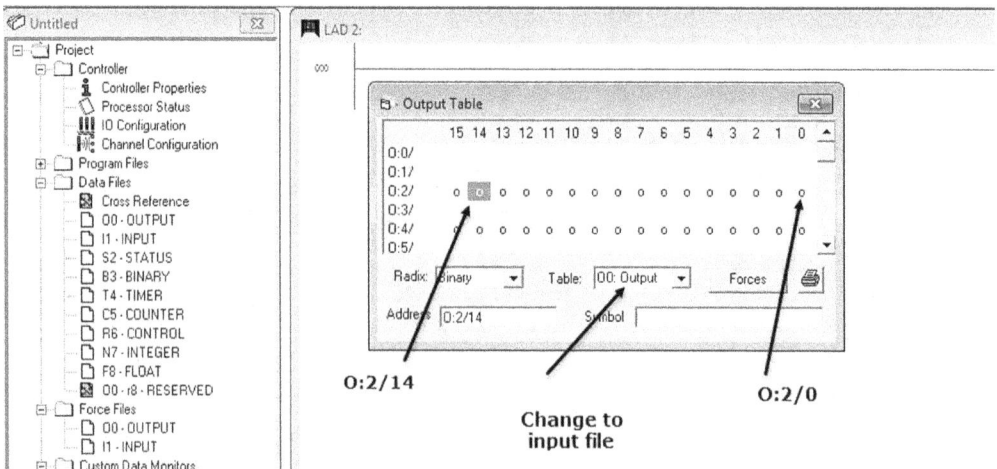

FIGURE 2.15   Data file selection/display.

File Number 2: Status file has fifty elements, S2:0 to S2:49. Each element consists of one word, for example, S2:0/0 to S2:0/15.

File Number 3: Bit file has one hundred elements, B3:0 to B3:99. Each element consists of one word, for example, B3:0/0 to B3:0/15.

File Number 4: Timer file has one hundred elements, T4:0 to T4:99. Each element has three words; the timer file could be used as ON-Delay (TON), OFF-Delay (TOF), or Retentive (RTO). Three words are reserved under each timer element; the preset, the accumulated, and the status word. For T4:0 the address words are T4:0.PRE, T4:0.ACC, and three bits out of the status words; T4:0/EN (timer enable bit), T4:0/DN (timer done bit), and T4:0/TT (timer timing bit).

File Number 5: Counter file has one hundred elements, C5:0 to C5:99. Each element has three words: preset, accumulated, and status. C5:0 address words are C5:0. PRE, C5:0.ACC, and five bits out of the status words; C5:0/UN, C5:0/OV, C5:0/ DN, C5:0/CD, and C5:0/CU.

File Number 6: Control file has one hundred elements, R6:0 to R6:99. Control file is used for advance instructions, which are not part of the laboratory projects in this book.

File Number 7: Integer file has one hundred elements, N7:0 to N7:99. Each element consists of one word; for example, N7:0 goes from N7:0/0 to N7:0/15. There are four different Radix for each type word file: binary, octal, decimal, and HEX-BCD.

File Number 8: Floating file has one hundred elements, F8:0 to F8:99. Each element is one word in LogixPro 500. It does not support floating point arithmetic's and use "F" values as integers. In RSLogix500 each element has two words for floating points decimal.

2.3 Combinational Logic

Digital systems logic elements are classified as combinational or sequential. Combinational logic includes AND, OR, NOT, NAND, NOR, XOR, and XNOR gates. The basic combinational logic operations are described next followed by RSLogix500 implementation using LogixPro 500.

2.3.1 Logic Gate Operations

AND logic gate: If all inputs are true then the output becomes true, otherwise the output is false.

The truth table in Fig. 2.16*b* shows the output logic for all possible input conditions for the two inputs AND gate. The product sign (.) is used as the Boolean operator for the AND gate operation. Figure 2.16*c* shows the ladder logic for an AND operation.

OR logic gate: If all inputs are false then the output is false, otherwise the output is true. A sum sign (+) is used as the Boolean operator for OR logic operation as shown in Fig. 2.17.

NOT logic gate: The output of the NOT (inverter) gate is the inverse of the input logic. If the input variable to the inverter is labeled "A" then the inverted output is known as "A NOT." This is also shown as "A'," or A with a bar over the top as shown in Fig. 2.18.

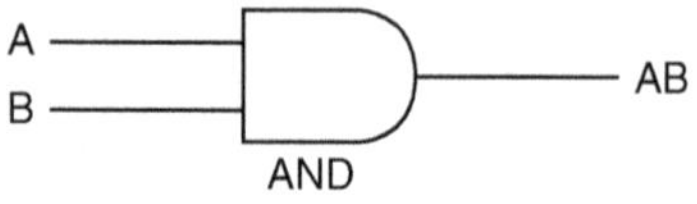

(a)

AND gate logic		
A	B	A.B
0	0	0
0	1	0
1	0	0
1	1	1

(b)

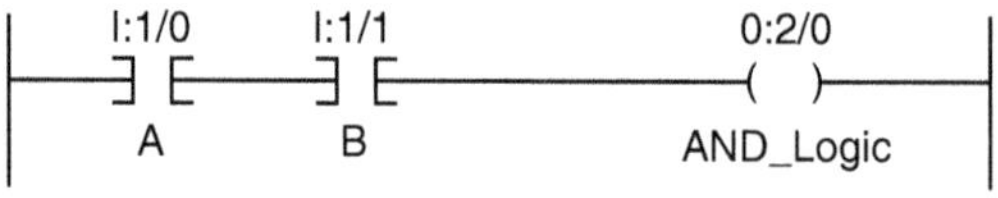

(c)

FIGURE 2.16 (*a*) AND operation symbol; (*b*) AND operation rules; (*c*) AND operation ladder logic.

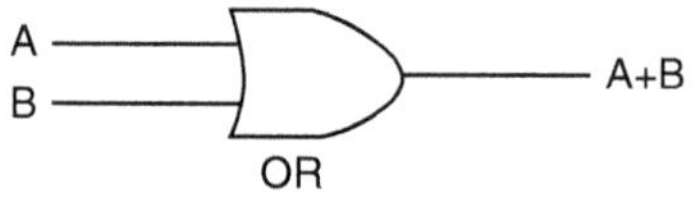

(a)

OR gate logic		
A	B	A+B
0	0	0
0	1	1
1	0	1
1	1	1

(b)

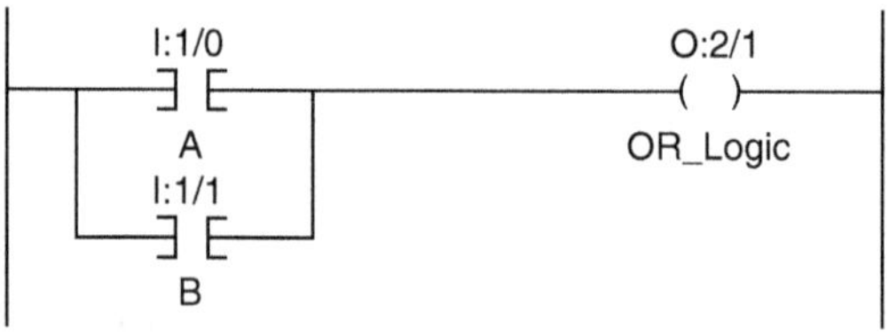

(c)

FIGURE 2.17 (*a*) OR operation symbol; (*b*) OR operation rules; (*c*) OR operation ladder logic.

NAND logic gate: If all inputs are true then the NAND gate output turns false, otherwise the output is true as shown in Fig. 2.19. This is a NOT-AND gate, which is equal to an AND gate followed by a NOT gate.

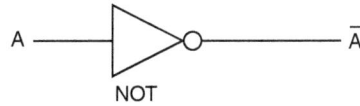

A ———▷o— $\overline{A}$

NOT

(a)

NOT gate logic	
A	$\overline{A}$
0	1
1	0

(b)

Figure 2.18 (a) NOT logic gate; (b) NOT logic rules.

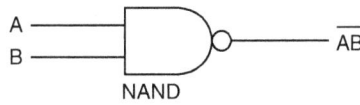

A ——⟩o— $\overline{AB}$
B ——

NAND

(a)

NAND gate logic		
A	B	$\overline{A.B}$
0	0	1
0	1	1
1	0	1
1	1	0

(b)

Figure 2.19 (a) NAND logic gate; (b) NAND logic rules.

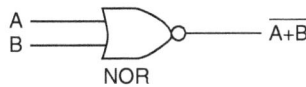

A ——⟩o— $\overline{A+B}$
B ——

NOR

(a)

NOR gate logic		
A	B	$\overline{A + B}$
0	0	1
0	1	0
1	0	0
1	1	0

(b)

Figure 2.20 (a) NOR logic gate; (b) NOR logic rules.

NOR logic gate: If all inputs are false then the NOR gate output becomes true, otherwise the gate output is false as shown in Fig. 2.20. This is a NOT-OR gate, which is equal to an OR gate followed by a NOT gate.

(a)

XOR gate logic		
A	B	A⊕B
0	0	0
0	1	1
1	0	1
1	1	0

(b)

(c)

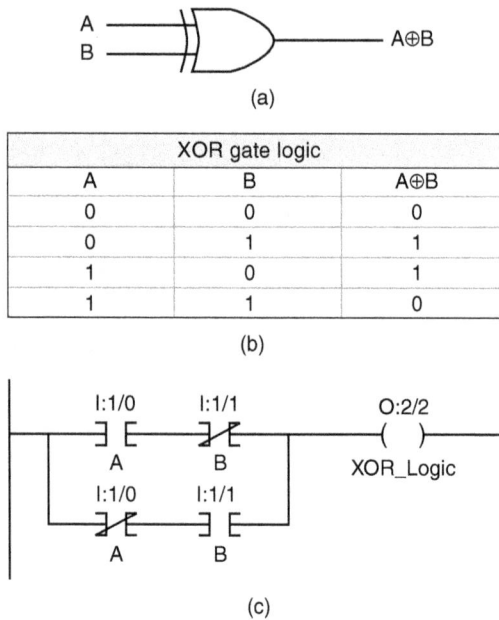

FIGURE 2.21 (a) XOR logic gate; (b) XOR logic rules; (c) XOR ladder logic.

XOR logic gate: If the two inputs are different logic level then the gate output becomes true, otherwise the gate output is false as shown in Fig. 2.21. An inside circle plus sign ⊕ (Boolean operator) is used to show the XOR operation.

XNOR logic gate: If the two inputs are the same logic level then the gate logic output is true, otherwise the gate output is false as shown in Fig. 2.22. A negated ⊕ sign or bar over the top of the XOR output is used to show the XNOR operation.

Table 2.2 shows the symbolic representation FOR THE seven logic gates that are used to document PLC program logic elements.

Table 2.3 is a summary truth table of the input/output combinations for the NOT gate together with all possible input/output combinations for the other gate functions.

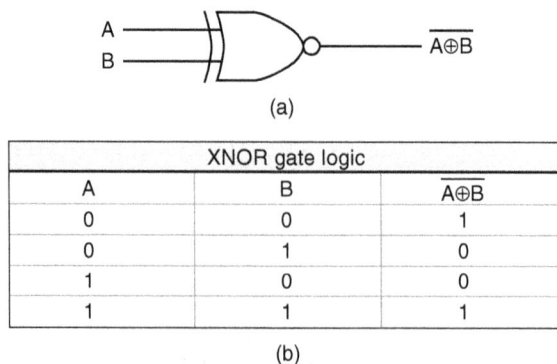

(a)

XNOR gate logic		
A	B	$\overline{A \oplus B}$
0	0	1
0	1	0
1	0	0
1	1	1

(b)

FIGURE 2.22 (a) XNOR logic gate; (b) XNOR logic rules.

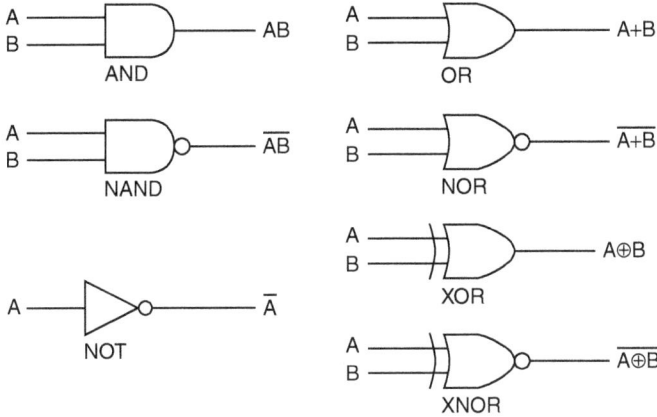

TABLE 2.2 Logic Gate Symbols

		Inputs		Outputs					
		A	B	AND	NAND	OR	NOR	EXOR	EXNOR
NOT gate		0	0	0	1	0	1	0	1
A	Ā	0	1	0	1	1	0	1	0
0	1	1	0	0	1	1	0	1	0
1	0	1	1	1	0	1	0	0	1

TABLE 2.3 Logic Gates Representation Using Truth Table

Note that an "n bit" inputs have 2^n rows. The names for the XOR/EXOR and XNOR/EXNOR are interchangeably used in the literature.

2.3.2 LogixPro 500 Implementation Examples

Example 2.3.2.1 This example uses the OR logic to set the most significant bit of a 16-bit memory register B3:0. The following is a brief description of the ladder instruction operation:

When Tag_IN is off, the OR instruction is not executed. The MSB of B3:0 is reset (0) and B3:1 MSB is set (1) as shown in Fig. 2.23.

When Tag_IN is ON, the OR instruction is executed. B3:0 is ORed bit by bit with B3:1 and the result is placed in B3:0. The operation results in setting the MSB of B3:0 as shown in Fig. 2.24.

Example 2.3.2.2 The use of XIC instructions to START/STOP motor from two different locations (A or B) is shown in Fig. 2.25. Figure 2.26 shows a one rung ladder program to start/stop a motor from two different locations (A or B). There are two normally open push button switches, START_A and START_B, to start the motor from either of the two locations and two normally closed push-button switches, STOP_A and STOP_B, to stop the motor from either location. Figure 2.26 shows the motor running after one of the start push buttons is pressed, which causes power to flow to the motor. This will activate the XIC contact latching around START_A/START_B. If either of the two stop PBs is pressed; the motor will lose power and the XIC around the start push buttons will be off. Notice that the two start PBs are implemented as an OR condition, while the two stop PBs are shown as an AND condition.

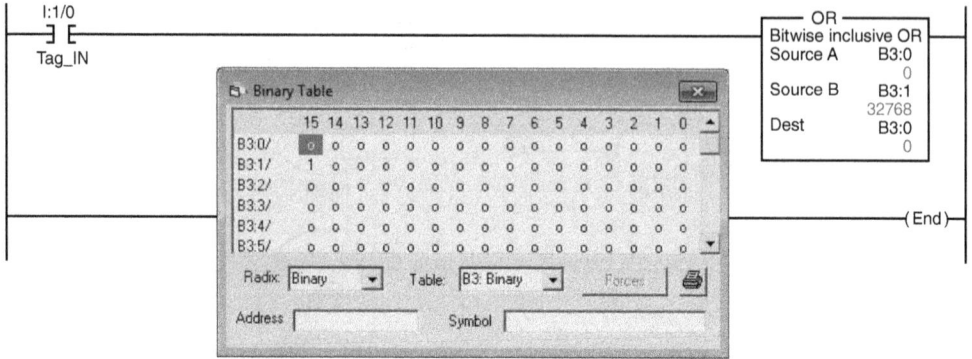

FIGURE 2.23   Status before the OR activation.

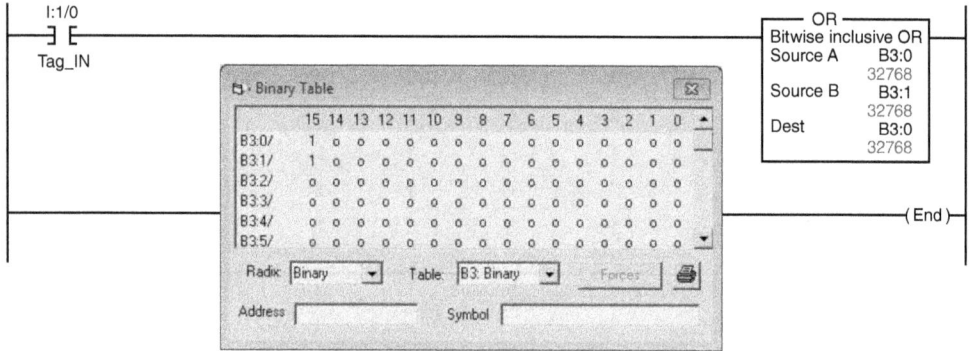

FIGURE 2.24   Status during the OR activation.

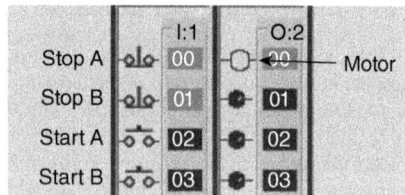

FIGURE 2.25   LogixPro 500 panel configuration.

Example 2.3.2.3   Figure 2.27 shows the implementation of a flashing pilot light (PL) when two toggle switches are set at opposite conditions (XOR condition). The FLASH_BIT tag with address B3:0/0 is initially OFF (0 value). After the first scan the same bit will be set assuming a valid XOR condition, which will turn the pilot light ON. The flashing bit gets reset after the second processor scan causing the pilot light to go OFF. The flashing is done at the scan rate, which is extremely fast for a human

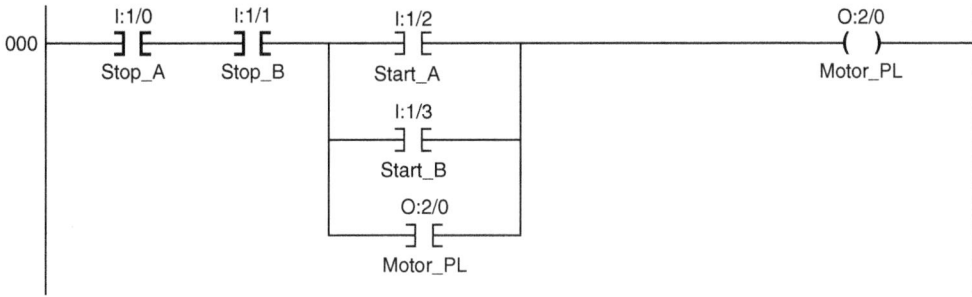

FIGURE 2.26   Two start/stop motor rung.

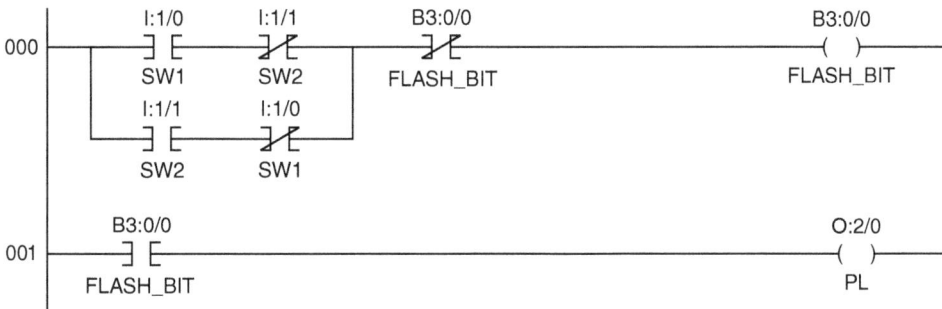

FIGURE 2.27   Two toggle switches flashing pilot light.

display. In Chap. 3 we will learn how to use timers to control an event rate. We will also identify available status bits called flashing bits, which can be used to provide an ON/OFF action at different preprogrammed rates.

2.3.3   I/O Testing Using the Force Function

The first step in the checkout process includes testing of all input and output device wiring to the PLC modules. This process is done without the execution of the implemented PLC logic and is a must in order to insure both the accuracy of all the wiring and the desired functioning of all I/O devices/interfaces.

2.3.3.1   Testing Output Using Force ON/OFF Function

This section covers output forcing in the Allen-Bradley LogixPro 500 system. Switch the selected "I: 1/0" switch on the simulator to open position. From the project tree click on output file then click "Forces", Fig. 2.28. Enter "1" in the bit you will force (O: 2/0) and then enable the force, Fig. 2.29. Monitor the PL (O: 2/0), which represents motor M1 on the I/O simulator, as you force the output ON while the "I: 1/0" (SS1) and "I: 1/1" (SS2) switches are in the OPEN position. The program must be online in order to use the force functions.

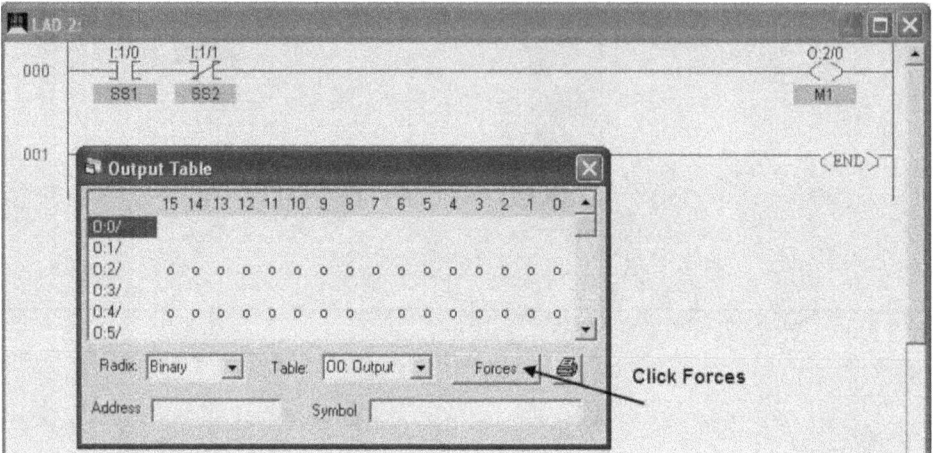

FIGURE 2.28   Forcing action.

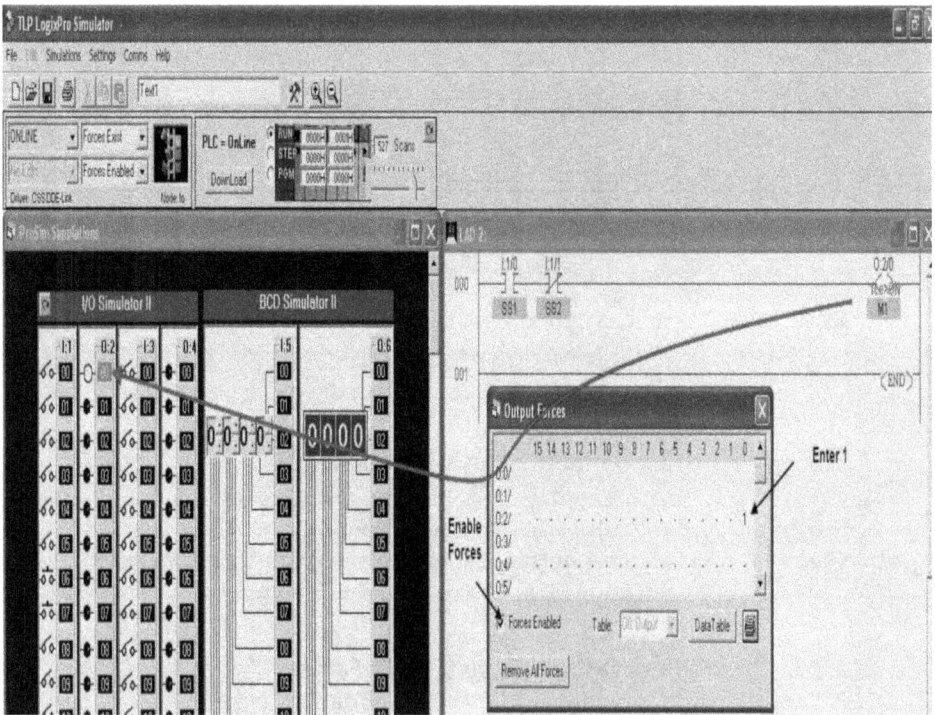

FIGURE 2.29   Enabling output 0: 2/0 force action.

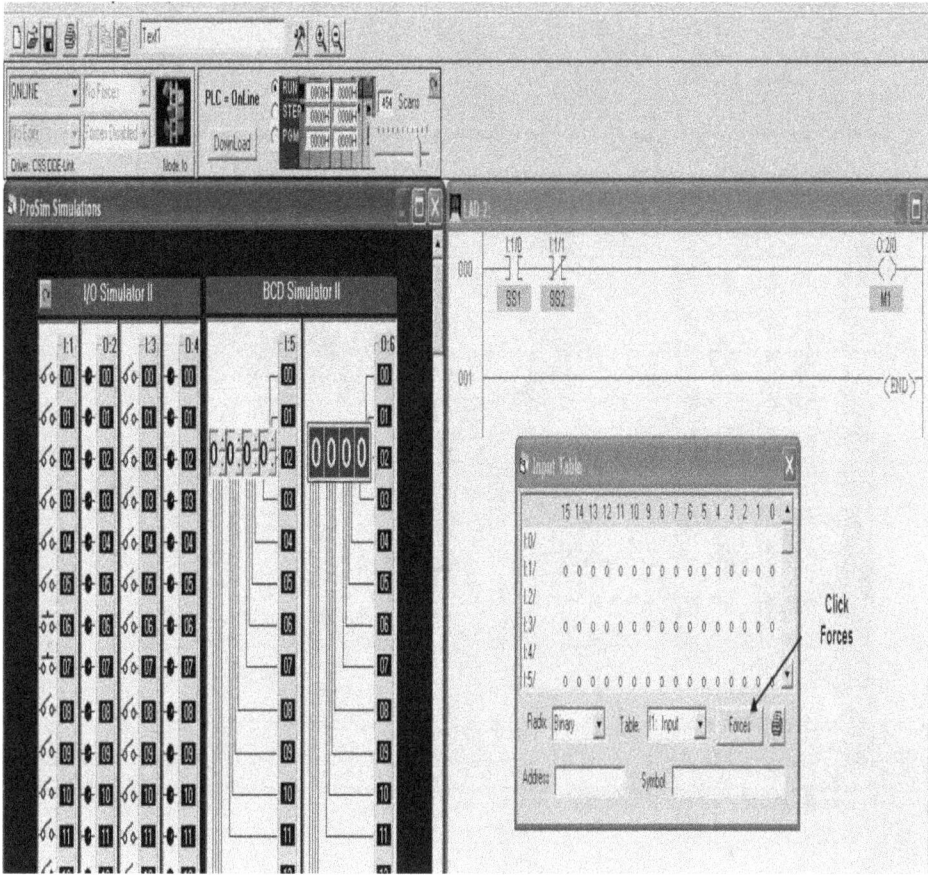

FIGURE 2.30   Project tree force actions.

2.3.3.2   Testing Input Using Force ON/OFF Function

The first step in the checkout process includes testing of all inputs and outputs device wiring to or from the PLC modules. This section covers input forcing the in Allen Bradley LogixPro 500 system. As shown in Fig. 2.30, from the project tree menu under force file, click Forces table then enter the address you need to force and enter Force value (True or False) as shown in Fig. 2.31.

- Switch I: 1/0 on the simulator to open position.
- From project tree click on input file then click forces.
- Enter 1 in the bit you will force (I: 1/0) then enable the forces.

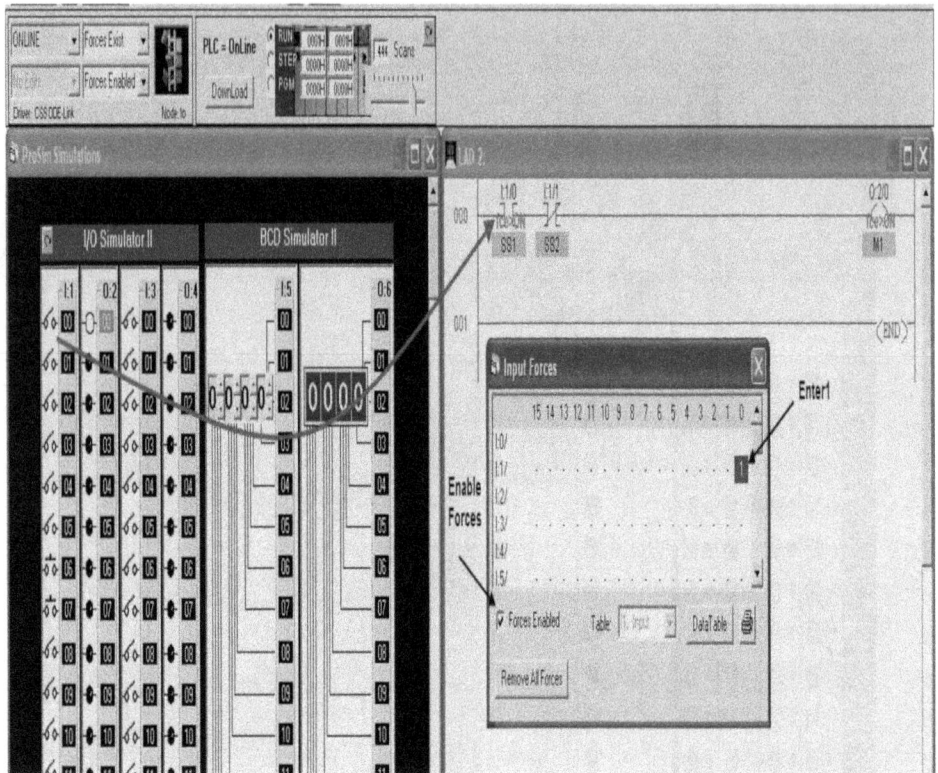

FIGURE **2.31**   Enabling input force action.

2.4   Combinational Word Logic Operations

The word logic instructions perform the combinational operation on word size instead of individual bits. All logic instructions applicable to bit operands are available for word operands with each word defined as 16 consecutive bits, which must start at a valid word address as defined in the SLC 500 PLC system. Words used for the input operands can be a physical discrete input word or any of the valid file words. The destination can also be a physical discrete output word or any of the possible valid file words.

2.4.1   AND Word Logic Operation

The AND logic operation ANDs two words at source A (N7:0) and source B (N7:1) inputs bit by bit. Figure 2.32a shows the AND instruction. If "Tag_IN" is true, the "AND logic operation" instruction is executed as shown in Fig. 2.32b. The value of source A and the value of source B are ANDed. The result is mapped bit-for-bit and the output is placed in the destination (N7:2).

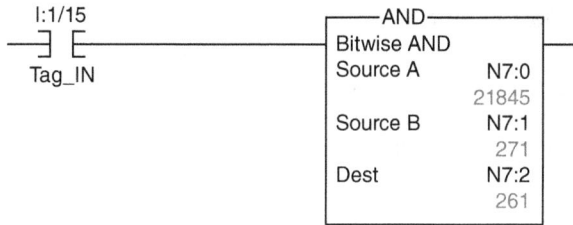

I:1/15
Tag_IN

AND
Bitwise AND
Source A   N7:0
                    21845
Source B   N7:1
                    271
Dest   N7:2
                    261

(a)

Parameter	Address name	Value
Source A	N7:0	0101 0101 0101 0101
Source B	N7:1	0000 0001 0000 1111
Dest	N7:2	0000 0001 0000 0101

(b)

FIGURE 2.32   (a) AND word instruction; (b) AND word instruction execution result.

2.4.2   OR Word Logic Operation

The OR logic operation ORs two words at source A (N7:0) and source B (N7:1) inputs bit by bit. Figure 2.33*a* shows the OR instruction. If "Tag_IN" is true, the "OR logic operation" instruction is executed as shown in Fig. 2.33*b*. The value of source A and the value of source B are ORed. The result is mapped bit-for-bit and the output is placed in the destination (N7:2).

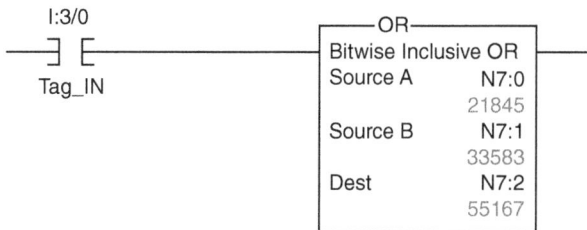

I:3/0
Tag_IN

OR
Bitwise Inclusive OR
Source A   N7:0
                    21845
Source B   N7:1
                    33583
Dest   N7:2
                    55167

(a)

Parameter	Address Name	Value
Source A	N7:0	0101 0101 0 101 0101
Source B	N7:1	1000 001 1 0 010 1111
Dest	N7:2	1101 0111 0 111 1111

(b)

FIGURE 2.33   (a) OR word instruction; (b) OR word instruction execution result.

2.4.3 XOR Word Logic Operation

The XOR logic operation XORs two words at source A (N7:0) and source B (N7:1) inputs bit by bit. Figure 2.34a shows the XOR instruction. If "Tag_IN" is true, the "XOR logic operation" instruction is executed as shown in Fig. 2.34b. The value of source A and the value of source B are XORed. The result is mapped bit-for-bit and the output is placed in the destination (N7:2).

2.4.4 NOT Word Instruction

Figure 2.35 shows the LogixPro 500 implementation of the NOT word instruction. If (TAG_IN) is true, the NOT instruction is executed. The NOT instruction inverts all bits of the source register (N7:0) and the output is placed in the destination register (N7:1).

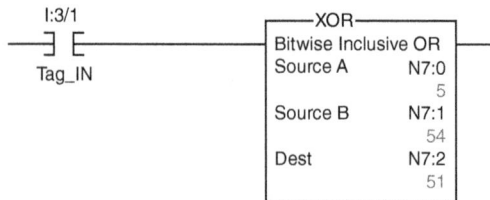

(a)

Parameter	Address name	Value			
Source A	N7:0	0000	0 000	0 000	0101
Source B	N7:1	0000	0 000	0 011	0110
Dest	N7:2	0000	0 000	0 011	0011

(b)

Figure 2.34 (a) XOR word instruction; (b) XOR word instruction execution result.

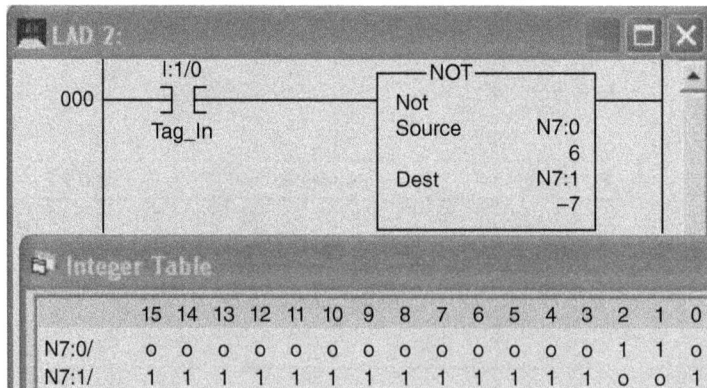

Figure 2.35 NOT word instruction.

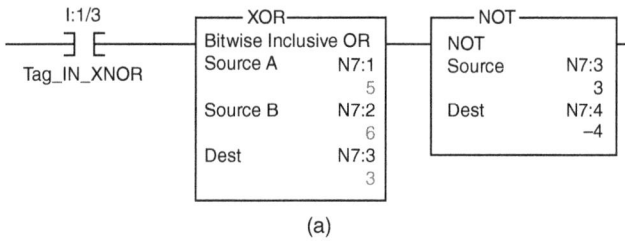

(a)

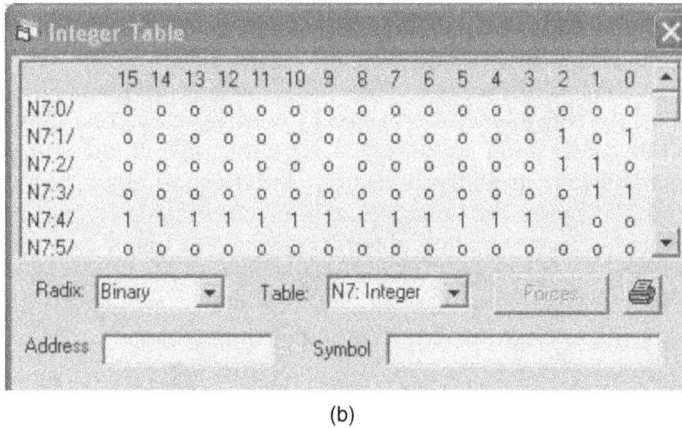

(b)

Figure 2.36   (a) XNOR word instruction; (b) XNOR word execution results.

The source and destination parameters must be word addresses. Notice that the result of "−7" in the destination is the one's complement of "6" in the source (MSB is 1 indicating negative number). Two's complement of 6 is its one's complement "−7" + 1, which is "−6."

Example 2.4.4.1   Figure 2.36a shows the ladder for forming XNOR instruction out of an XOR and NOT instructions. Figure 2.36b shows the execution results.

If TAG_IN_XNOR is true, the XOR and NOT instructions are executed. The value of N7:1 is exclusive or with the value of N7:2 bit by bit, and the result is placed in N7:3. N7:4 MSB is 1, indicating that the number is negative so the two's complement of the value of N7:4 is evaluated and placed in N7:4.

2.5   Latch, Unlatch, Subroutine, and One-Shot Instructions

The latch and unlatch instructions are associated with one output coil. The latch will force the associated coil ON and maintains it to be ON until the unlatch instruction is executed regardless of the scanned program code between the two actions. One-shot instructions are triggered once during a single scan. These instructions are typically used in initialization/configuration tasks and to capture or issue edge triggered events.

2.5.1 Latch and Unlatch Instructions

This section covers the commonly used instructions, latch (L) and unlatch (U). The two actions are associated with the same physical output and are independent of the rest of the ladder logic.

2.5.1.1 Output Latch Instruction (L)—Set 1 Bit

The instruction is only executed if the preceding logic for the same rung is true (power flows to the L coil), then (L) is activated. When preceding rung input is false, then L maintains the active status. L remains active until unlatch action is executed.

2.5.1.2 Output Unlatch Instruction (U)—Reset 1 Bit

The instruction is only executed if the preceding logic for the same rung is true (power flows to the (U) coil, then (U) is activated which resets the (L) coil. When the preceding rung input (power flows to the coil) is false, then L maintain the inactive status. The latch (L) and the unlatch (U) must be assigned the same bit address.

Figure 2.37 shows the timing diagram for latch/unlatch instructions. Notice that the two signals cannot be active at the same time. The figure assumes that O represents the logic status of the physical output associated with the L or U coil rungs.

2.5.2 Positive/Negative Edge One-Shot Instruction

Figure 2.38 demonstrates the one-shot rising (OSR) operation for positive edge transitions. Transitions are defined over selected discrete input signals, output signals, or events. Notice that B3/0 is the same as B3:0/0. Also B3/5:2 is the same as B3/82.

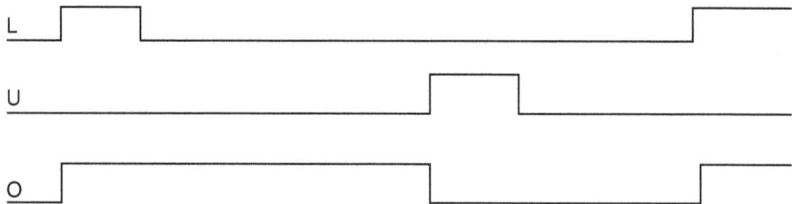

FIGURE **2.37** Latch/unlatch timing diagram.

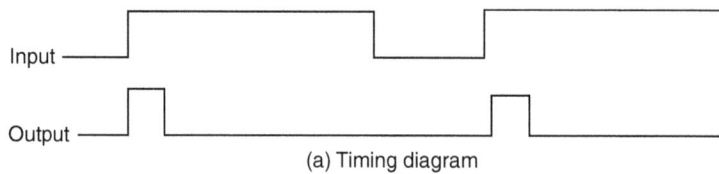

(a) Timing diagram

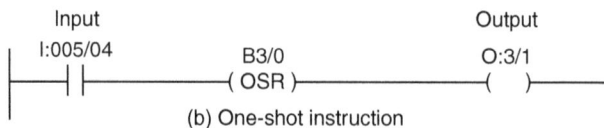

(b) One-shot instruction

FIGURE **2.38** Positive edge one-shot logic.

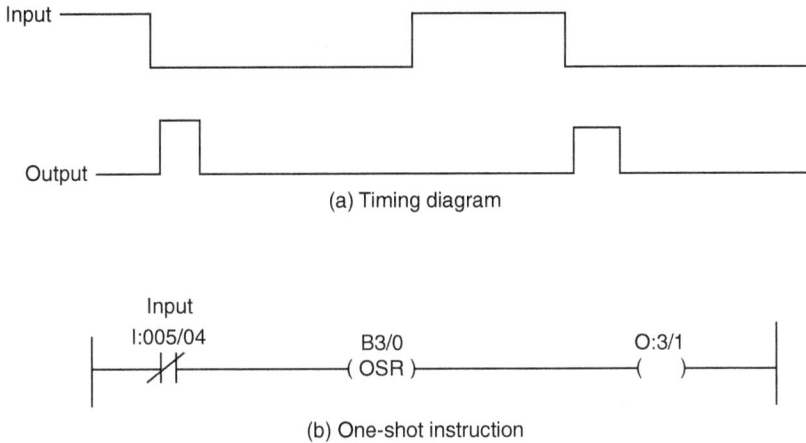

(a) Timing diagram

(b) One-shot instruction

Figure 2.39 Negative edge one-shot logic.

Figure 2.39 illustrates the one-shot operation for the negative edge transition. Each one shot must be assigned a unique bit address.

Notice that the one shot shown for the positive and the negative transitions in Figs. 2.38 and 2.39 use the same bit address B3/0, which is not allowed in the same program. Every one-shot, positive or negative, must be assigned a unique bit address. Output coils also must be assigned unique addresses even though some PLCs allow repeated coil address in the same program.

Example 2.5.2.1 Positive edge instruction is used in an initialization rung, as shown in Fig. 2.40. A positive edge trigger instruction causes this rung to execute once when Auto/Manual switch is in Auto, which clears all timers/counters and accumulated values.

MOV instruction is covered in more detail in Chap. 3. Another way of causing an initialization block to execute only once during the first scan after powering the PLC is to use available status bit S:1/15, which goes ON/True only during the first scan. The bit is used as an open contact condition for the execution of the initialization block or the JSR instruction rung, which calls the subroutine. JSR instruction and the nesting of subroutines are covered in Sec. 2.5.4.

Example 2.5.2.2 The use of latch/unlatch instructions to control a conveyor belt start/stop from two different locations. Figure 2.41 shows a conveyor belt that can be activated electrically. There are two push button switches at the beginning of the belt (Location A): S1 for start and S2 for stop. There are also two push button switches at the end of the belt (Location B): S3 for start and S4 for stop. It is possible to start or stop the belt from either of the two locations. The LogixPro 500 simulator and associated two rungs implementation are shown in Fig. 2.42. Rung 1: the conveyor built is switched on when START_A or START_B is pressed. The motor run action is maintained by using the latch instruction (L). Rung 2: the conveyor built is switched off when STOP_A or STOP_B is pressed. The motor stop action is achieved by using the unlatch instruction (U) on the same output coil. Notice that output coils do not repeat in the ladder program except for the latch and the corresponding unlatch coils.

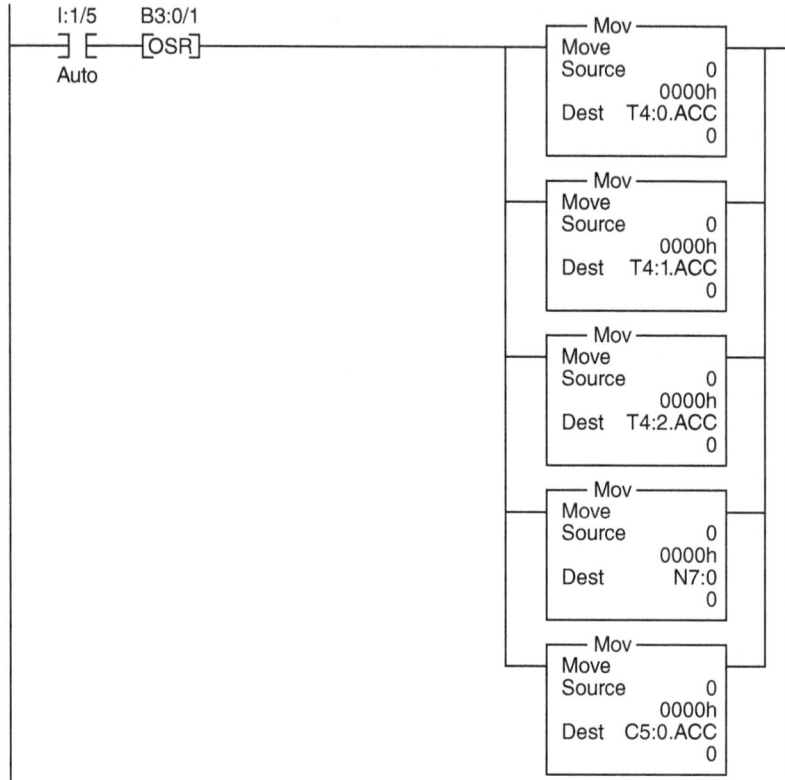

FIGURE 2.40 One-shot initialization rung.

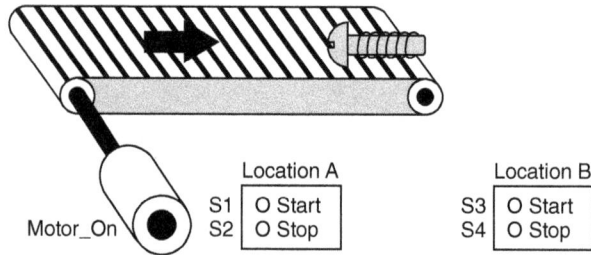

FIGURE 2.41 Conveyor system start/stop.

Example 2.5.2.3 A failed to start signal is received from a motor when output is issued and motor running signal is not received in 5 seconds. The power to the magnetic starter must be removed once the fail to start condition is detected. Figure 2.43 implements the motor failed to start alarm using the latch/unlatch instruction. The following are the logic details:

Rung0: If START PB is pressed while failed to start alarm (FTS_ALARM) is inactive, motor M1 will start.

Rung1: When motor M1 is active and the motor running input signal (M1_RUNNING_INPUT) OFF, timer 0 will start.

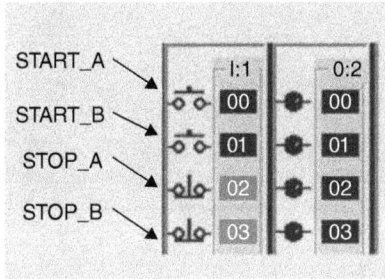

(a)

(b)

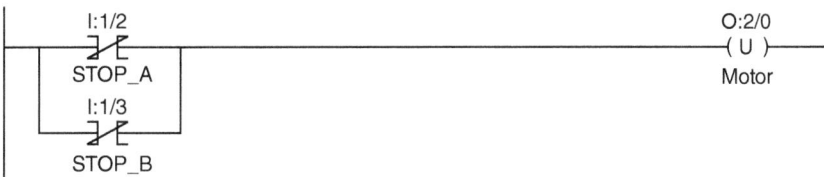

(c)

FIGURE 2.42 (a) Conveyor system I/O panel configuration; (b) Conveyor system start/stop rung 1; (c) Conveyor system start/stop rung 2.

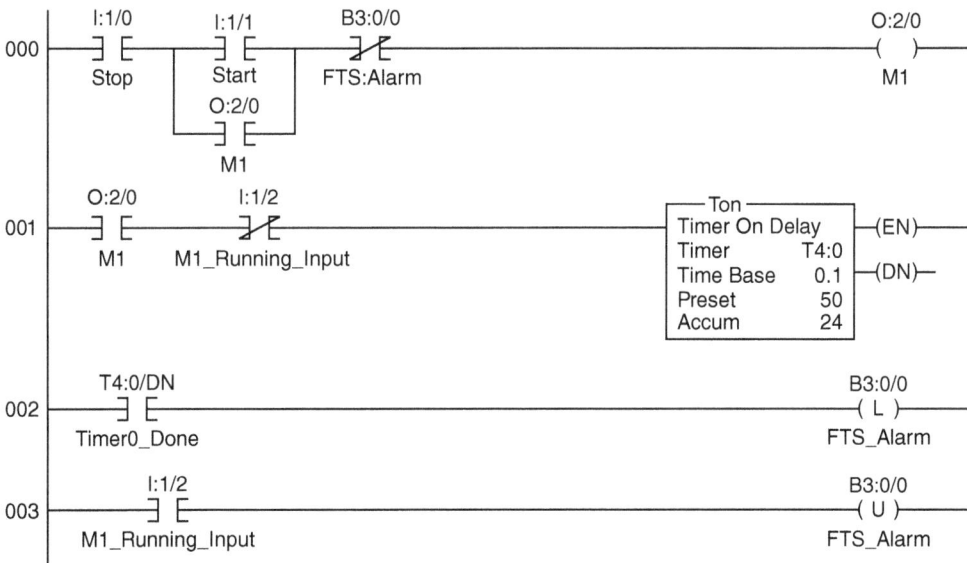

FIGURE 2.43 Motor fail to start alarm logic.

Rung2: If motor running input signal is not received within 5 seconds; timer 0 done bit (T4:0/DN) will be set and a fail to start alarm will be latched on.

Rung3: When motor running input signals (M1_RUNNING_INPUT) is set; the fail to start alarm will clear. This will indicate that motor M1 problem is fixed otherwise the alarm will remain and power will not be allowed to the motor starter.

2.5.3   File shift Instructions

This section examines two of the commonly used memory register shift instructions, the Bit Shift Right (**BSR**) and the Bit Shift Left (**BSL**). Each of the two instructions has a file, control, bit address, length, enable output (EN) connection, and done output (DN) connection. The following are the specification for the BSL and BSR instruction parameters:

File is the address of the bit array you want to shift. You must use the file indicator (#) in the bit array address. You must start the array at a 16-bit element boundary, for example, use bit 0 of element #1, 2, 3, etc.

Control is the unique address of the control structure (48 bits, three 16-bit words) in the control area of memory that stores the instruction's status bits, the size of the array (in number of bits), and the bit pointer (currently not used).

Bit address is the location of the bit that will be added to the array.

Length is the total number of bits to be shifted by the BSR. Bits located to the left of the last bit in the array, up to the next word boundary are not altered by the BSL instruction. Unloaded bit continue shifting left up to the end of the current word. The length field cannot exceed the bit file boundary or else it may cause a major syntax error in the program.

2.5.3.1   BSL Instruction

As shown in Fig. 2.44a; If "TAG_IN" transitions from false-to-true, the BSL instruction shifts the pattern of data one bit through the array to the left, loads (I:1/1) bit into (B3:2/0), and unloads (B3:2/4) into (R6:10/UL). Figure 2.44b shows the execution steps of the BSL instruction. The unload bit of the control register R6:10 and the bit address I:1/1 are shown and will be used in the next example.

Example 2.5.3.1   This example simulates an assembly line parts reject alarm process. A photoelectric sensor (simulated by I:1/0) provides a positive pulse for each rejected part. A reject alarm (simulated by O:2/0) is activated once five rejected parts are accumulated. Figure 2.45a shows the LogixPro 500 I/O panel configuration, and Fig. 4.45b documents the implemented two rungs ladder program. I:1/1 is maintained TRUE during the assemble operation. B3:2 element's word is assumed to be cleared during initialization at the start of the program run. Notice that the shifting to the left continues through the entire word regardless of the defined length parameter. The unloading in this example will always come from B3:2/4 bit.

2.5.3.2   Bit Shift Right Instruction (BSR)

As shown in Fig. 2.46a, If TAG_IN transitions from false-to-true, the BSR instruction shifts the pattern of data one bit to the right, loads (I:1/1) bit of data into bit B3/37 (B3:2/5), and unloads B3/32 (B3:2/0) into (R6:15/UL). Figure 2.46b shows the execution steps of the BSR instruction. The unload bit of the control register R6:15 (R6:15/UL) and the PB input bit address I:1/1 are highlighted.

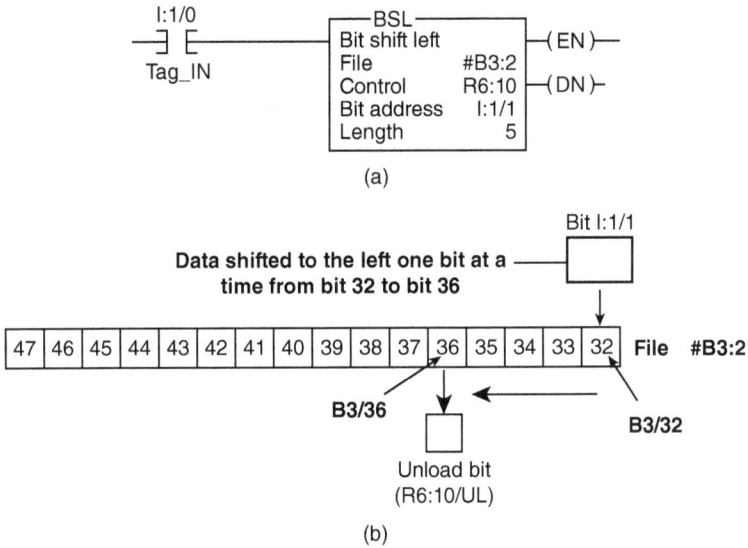

```
        I:1/0                ┌──BSL──────────┐
       ─┤ ├─                 │ Bit shift left │──( EN )──
        Tag_IN               │ File      #B3:2│
                             │ Control   R6:10│──( DN )─
                             │ Bit address I:1/1│
                             │ Length        5│
                             └───────────────┘
                              (a)
```

```
                                           Bit I:1/1
                                          ┌─────┐
 Data shifted to the left one bit at a ───│     │
     time from bit 32 to bit 36           └─────┘
                                             │
 ┌──┬──┬──┬──┬──┬──┬──┬──┬──┬──┬──┬──┬──┬──┬──┬──┐   ▼
 │47│46│45│44│43│42│41│40│39│38│37│36│35│34│33│32│  File  #B3:2
 └──┴──┴──┴──┴──┴──┴──┴──┴──┴──┴──┴╱─┴──┴──┴──┴╲─┘
                              ╱     │      ◄──────  ╲
               B3/36       ┌─────┐                  B3/32
                           │     │
                           └─────┘
                          Unload bit
                          (R6:10/UL)

                              (b)
```

FIGURE 2.44 (a) BSL instruction block; (b) BSL instruction execution steps.

Example 2.5.3.2 Figure 2.47 simulates an assembly line parts reject alarm process but using the BSR instruction. A photo electric sensor (simulated by I:1/0) provides a positive pulse for each rejected part. A reject alarm (simulated by O:2/0) is activated once six rejected parts are accumulated. The input bit address used I:1/1 is assumed to be 1 (ON) simulating a flag for each rejected part. B3:2 is cleared at the start of the assembly operation. The upload bit in the control word R6:15/UL will become true once six parts are rejected. This will initiate the reject alarm O:2/0, tag name REJECT_ALARM. B3:2 assumed to be initialized at zero value.

Example 2.5.3.3 For Fig. 2.48's BSL configuration; if B3:9 = 0000 0000 0000 0100 and B3:10 = 0000 0000 0000 0000, then after the BSL instruction execution of one time the value of B3:10 = 0000 0000 0000 0001 with a "0" is uploaded into R6:10/UL. After the BSL instruction execution of two or more times, the value of B3:10 = 0000 0000 0000 0001 with "1" is uploaded into R6:10/UL.

2.5.4 JSR Instructions and Subroutine Nesting

Structured programming is a common good practice in designing software in general and for large control system implementations. The overall task is divided into logical sections/blocks, each defined for a specified function. Each block can be defined using a subroutine file. The main program coordinates the execution/scanning sequence by calling available subroutines conditionally or unconditionally on every scan. Figure 2.49 shows a simple subroutine calling structure. From the main file (File #2) Jump To Subroutine (JSR) will transfer the program scan to File #3 (SBR 3:), which will execute rung 1 in the subroutine to start/stop motor 1 (M1). Then program execution will return back to the main file to execute rung 1 in the main file (LAD 2:). SBR can be the only instruction in the subroutine first rung. Coil B3:0/0 shown in the subroutine first rung is optional.

(a)

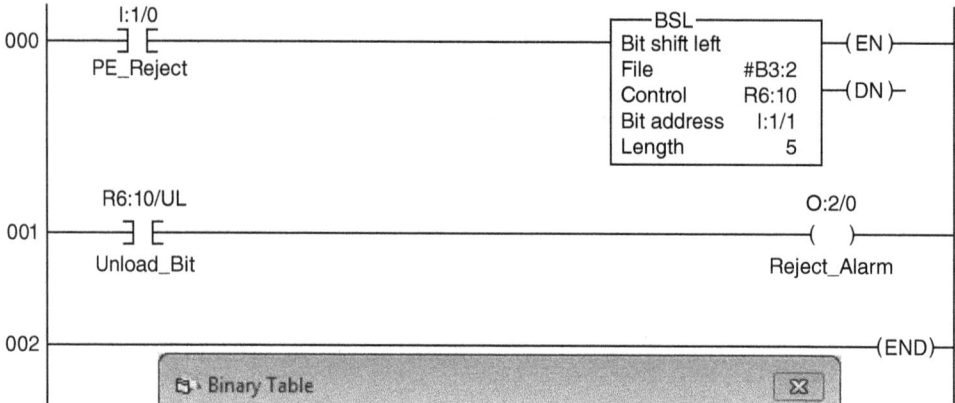

(b)

FIGURE 2.45   (a) BSL instruction LogixPro 500 panel configuration; (b) BSL instruction ladder program.

To enter a parameter into the JSR command, highlight the command and type a file number between 3 and 255. Use the following steps to create a subroutine folder:

1. In the left split screen, left click on the Program Files folder to highlight.

2. Right click on the file folder to open menu box. Select New.

3. Type the same number that was entered into the JMP command.

4. Optional, name the subroutine folder in space provided.

5. Use the tab on the bottom of the ladder logic program to switch between the main logic program and the subroutine program.

(a)

(b)

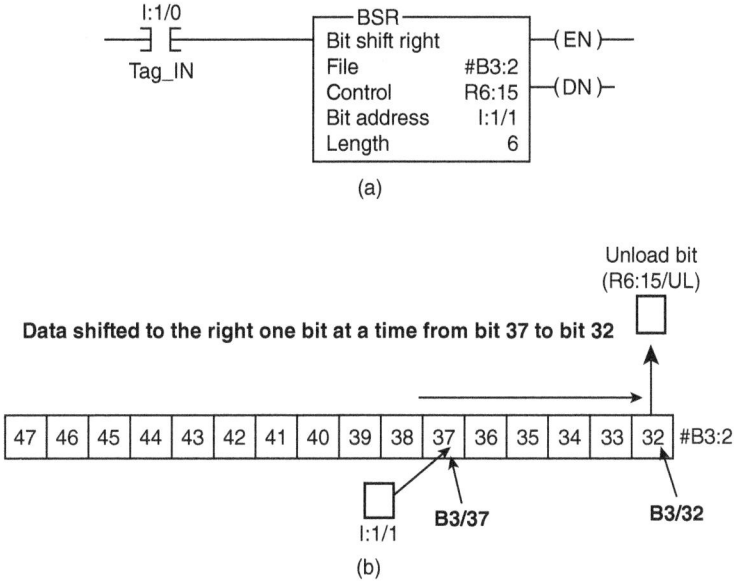

Figure 2.46   (a) BSR instruction block; (b) BSR instruction execution steps.

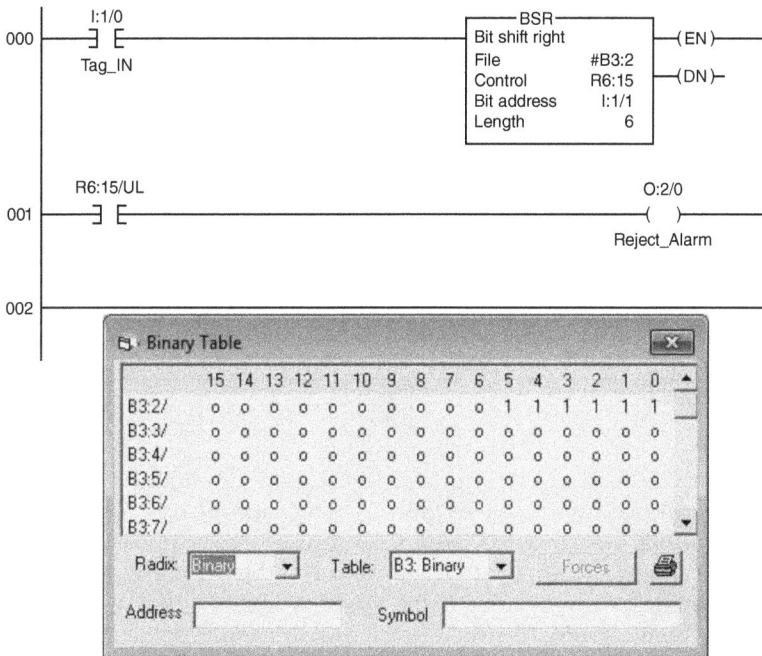

Figure 2.47   BSR instruction LogixPro 500 ladder program.

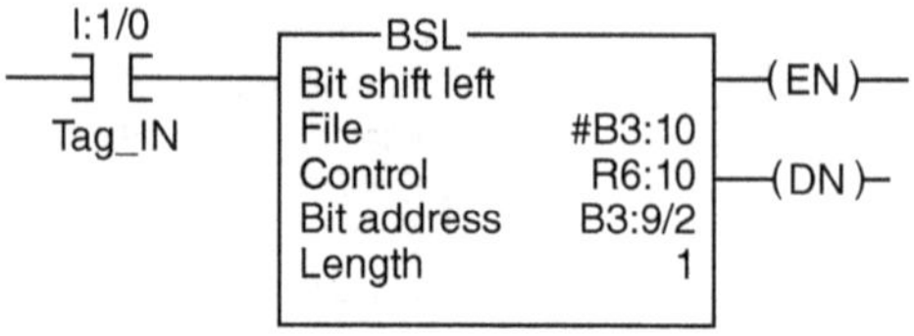

FIGURE 2.48   BSL for Example 2.5.3.3.

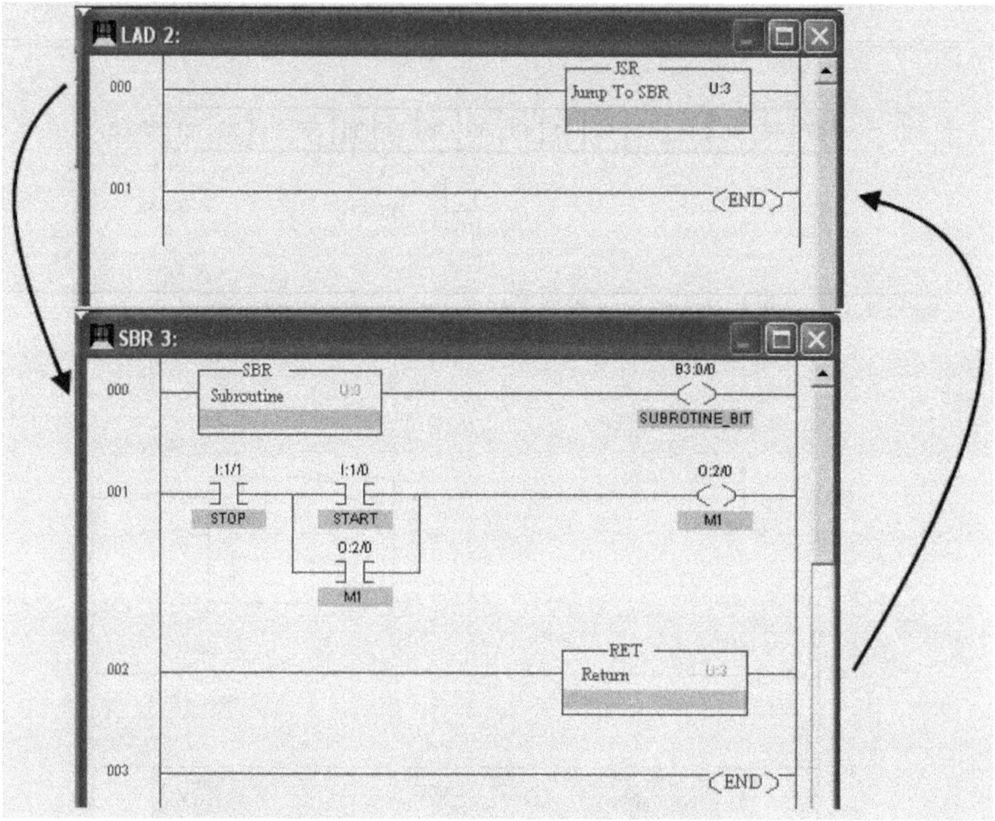

FIGURE 2.49   JSR Instruction and subroutine calling.

Figure 2.50 shows the structure for nesting subroutine calls. The main file #2 (LAD 2:) calls subroutine file #3 (SBR 3:), which in turn calls file #4 (SBR 4:). Once the RET from SBR 4 is executed (scanned), control will return to SBR 3. Once the RET from SBR 3 is executed, control will return to the main program (LAD 2:). Nesting of subroutines will consume stack space, which is used to store needed return addressing information for the calling program/subroutine. Care must be taken in restricting the number of nesting layers in order to minimize the scan cycle time and the stack size.

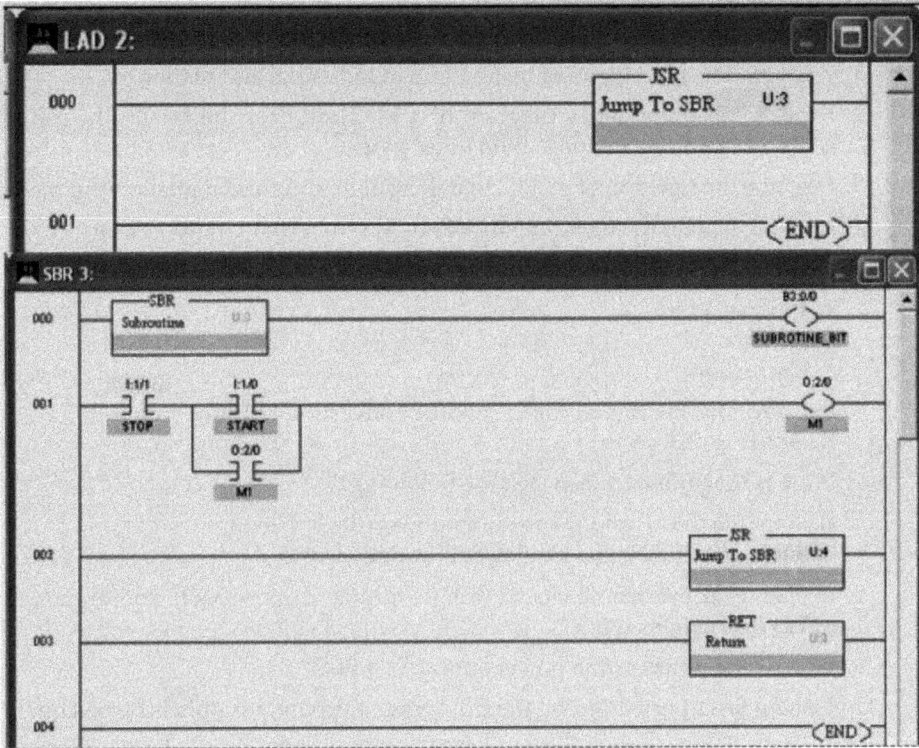

(a)

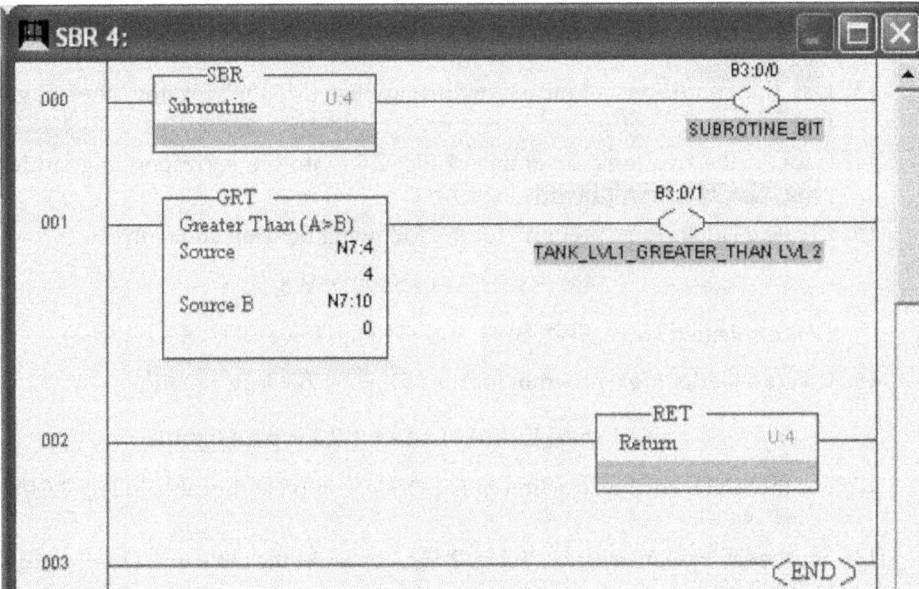

(b)

FIGURE 2.50 (a) Subroutine nesting; (b) subroutine nesting.

Chapter 2: Home Work Problems and Laboratory Projects

1. What are the advantages of using PLCs in industrial automation?

2. Draw a functional block diagram of a PLC and describe briefly the role of each component and its interface with other parts.

3. Describe the function of a PLC digital input module and digital output module.

4. List and explain the operating modes of PLC 5/03 (CPU/processor).

5. Define the following terms:
   a. Program scan
   b. Address
   c. Instruction

6. Explain how the digital input module converts a 120 V ac to a low TTL voltage. Refer to Fig. 2.3.

7. What is the difference between the following?
   a. Main file (file 2) and the remaining files (file 3, 4, ext.).
   b. Subroutine (SBR) and nesting subroutine files.

8. Explain how examine if closed (XIC), examine if open (XIO), and output energize (OTE) instructions work.

9. What is the function of a power supply in a PLC?

10. List the tasks performed by the CPU when the power supply is turned on?

11. Explain the difference between OTE (output energize) and OTL (output latch) instructions. What reverses the status of the output in each case?

12. Write the Boolean equations for the logic diagrams shown in the line diagram in Fig. 2.51a and b.

13. List the conditions required to turn motor (M) ON for the line diagram in Fig. 2.52.

14. Convert the two logic diagrams of Fig. 2.53 into the corresponding ladder logic programs. Hint: Simplify the logic first.

15. Create a ladder logic program for the following Boolean equation:

$$SV = (SW1 + SW2)\,(SW3)$$

   SV is a solenoid valve. SW1, SW2, and SW3 are three ON/OFF switches.

16. Create a ladder logic program for the following Boolean equation:

$$(\overline{SW1} \cdot \overline{SW2}) + (SW3) = PL1,\ (PL1\ \text{is pilot light 1})$$

17. For the AND word instruction in Fig. 2.54a, complete the table in Fig. 2.54b for the "Dest" word.

18. For the OR byte instruction in Fig. 2.55a, complete the table for "Dest" word shown in Fig. 2.55b.

19. For Fig. 2.56a XOR Word instruction, complete the table for the "Dest" word shown in Fig. 2.56b.

20. For the Fig. 2.57a BSR Word instruction, complete the bit table for B3:12 shown in Fig. 2.57b under the following conditions:

a. When I: 1/4 is true and Tag_IN goes from OFF to ON one time.
b. Repeat part (a) if Tag_IN switch from OFF to ON two times.

21. Repeat problem 20 for the BSL instruction.

22. Show how you can program a rung using a logic operation instruction to clear the most significant byte in a memory word location B3:5.

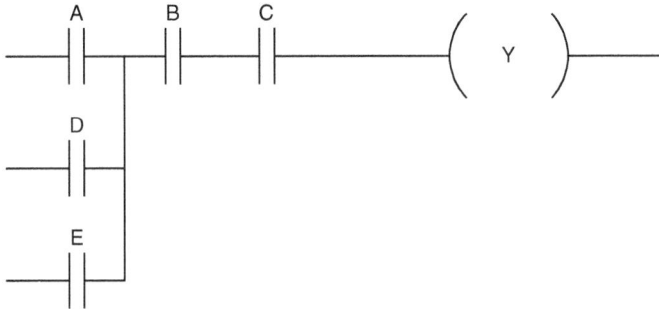

(a)

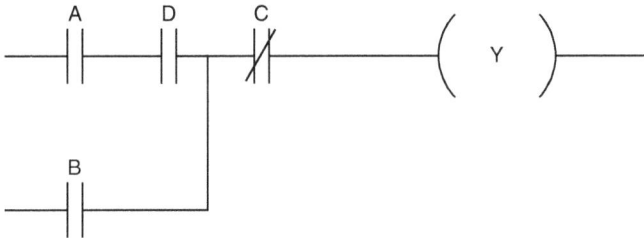

(b)

Figure 2.51 (a) Problem 12 logic diagram; (b) Problem 12 logic diagram.

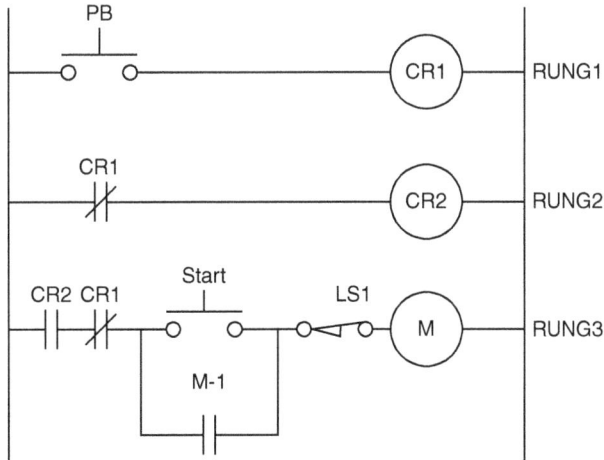

Figure 2.52 Problem 13 line diagram.

(a)

(b)

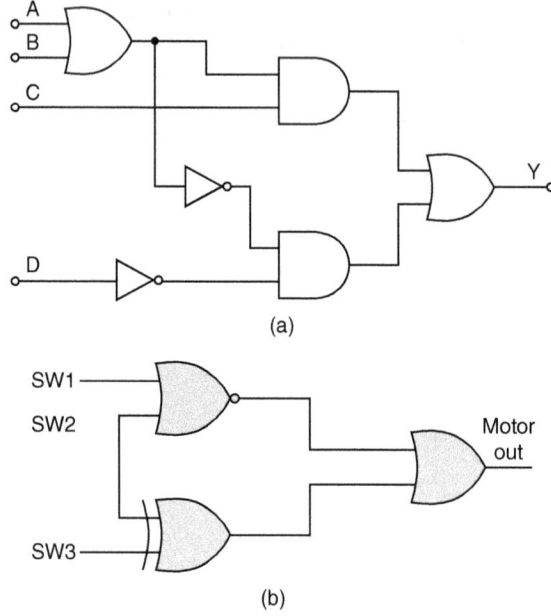

Figure 2.53   (a) Problem 14 logic diagram; (b) Problem 14 logic diagram.

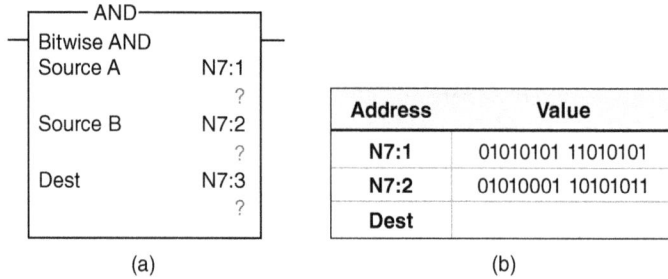

(a)

Address	Value
N7:1	01010101 11010101
N7:2	01010001 10101011
Dest	

(b)

Figure 2.54   (a) Problem 17 AND word instruction; (b) Problem 17 AND word instruction result.

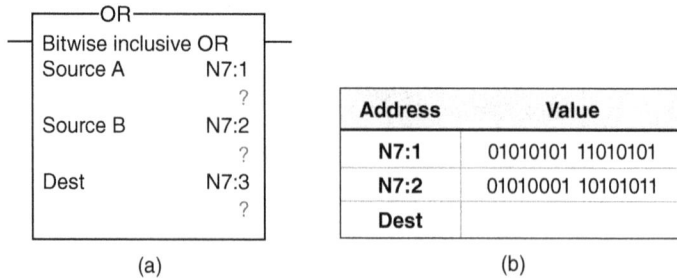

(a)

Address	Value
N7:1	01010101 11010101
N7:2	01010001 10101011
Dest	

(b)

Figure 2.55   (a) Problem 18 OR word instruction; (b) Problem 18 OR word instruction result.

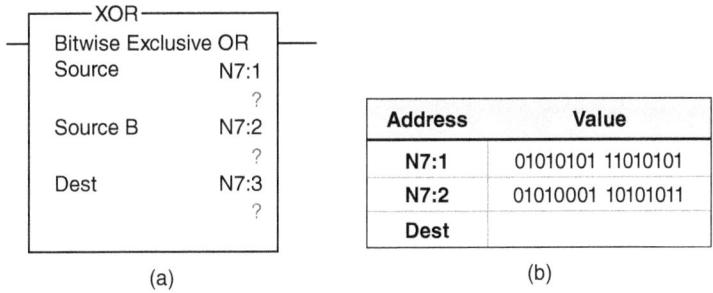

```
        ┌─────XOR──────────┐
      ──┤ Bitwise Exclusive OR ├──
        │ Source        N7:1 │
        │                 ?  │
        │ Source B      N7:2 │
        │                 ?  │
        │ Dest          N7:3 │
        │                 ?  │
        └────────────────────┘
```

Address	Value
N7:1	01010101 11010101
N7:2	01010001 10101011
Dest	

(a)   (b)

FIGURE 2.56   (a) Problem 19 XOR word instruction result; (b) Problem 19 XOR word instruction result.

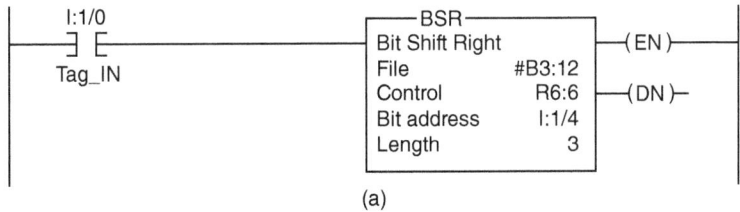

```
   I:1/0                        ┌─────BSR──────────┐
 ──┤↓├──────────────────────────┤ Bit Shift Right      ├──( EN )──
   Tag_IN                       │ File         #B3:12 │
                                │ Control        R6:6 ├──( DN )──
                                │ Bit address   I:1/4 │
                                │ Length           3  │
                                └──────────────────────┘
```

(a)

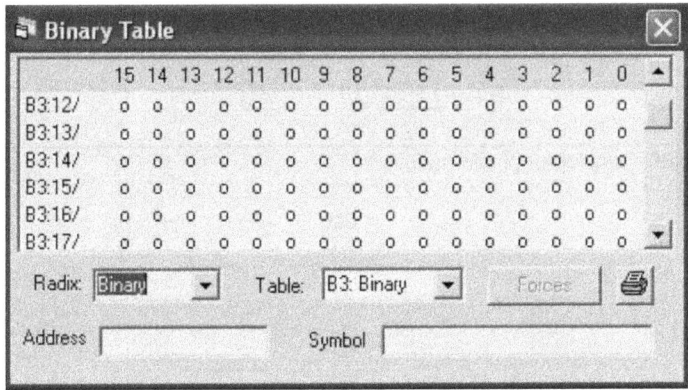

(b)

FIGURE 2.57   (a) Problem 20 SHR word instruction; (b) Problem 20 BSR word instruction result.

23. Explain the negative OSR instruction and complete the timing diagram in Fig. 2.58 assuming output is initially at the low state.

24. Repeat problem 23 for positive going edge input.

25. Write a program to turn on PL1 for one scan when a selector switch SS1 goes from off to on.

26. Repeat problem 25 for negative going input.

27. Explain output latch (L), output unlatch (U) instructions, and complete the timing diagram shown in Fig. 2.59 assuming output (O) is initially low.

28. Show two rungs using bit logic instructions to implement logic NAND and logic XNOR.

29. For the XNOR logic implemented in Fig. 2.60; assume N7:0 has 1 and N7:1 has 5, what is the value of N7:2 and N7:3?

Input ——

Output ——

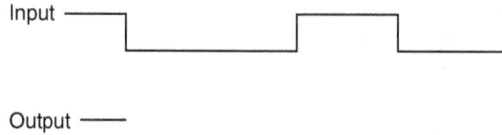

FIGURE **2.58**   Problem 23 negative going OSR instructions.

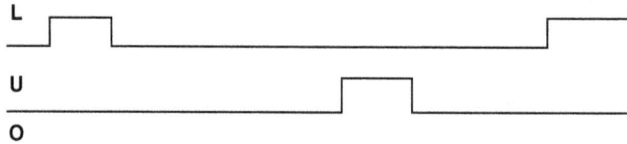

L

U

O

FIGURE **2.59**   Problem 27: latch-unlatch instructions.

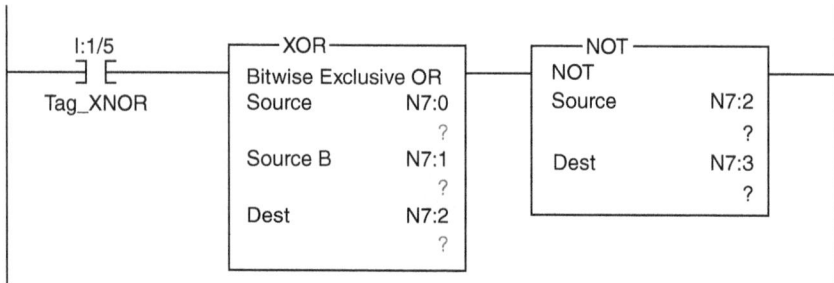

```
   I:1/5          ——XOR——              ——NOT——
   ─┘ └─          Bitwise Exclusive OR  NOT
   Tag_XNOR       Source        N7:0   Source       N7:2
                                  ?                    ?
                  Source B      N7:1   Dest         N7:3
                                  ?                    ?
                  Dest          N7:2
                                  ?
```

FIGURE **2.60**   Ladder for problem 29.

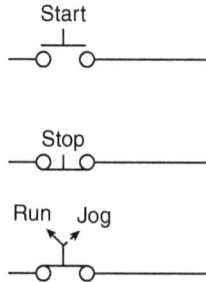

Start
─o o─

Stop
─o‾l‾o─

Run   Jog
─o o─

FIGURE **2.61**   RUN/JOG motor control panel.

30. A system has three input switches; NO START PB, NC STOP PB, and a RUN/JOG selector switch as shown in Fig. 2.61. The RUN/JOG switch status is ON when placed on RUN and OFF when placed on JOG. In order to jog the motor, the START PB must be held down (ON position). Write a ladder logic rung to run a motor M1 if the RUN/JOG switch is placed in RUN, and to jog the motor if placed in JOG.

31. A pump station has four pumps; Pump1, Pump2, Pump3, and Pump4. The running inputs for the pumps are I:1/0, I:1/1, I:1/2, and I:1/3. Write a ladder rung to indicate that the pump station is in use. The pump station is in use (output O:2/0) if any pump is running.

Laboratory 2.1—Programming Combinational Logic

The objective of this laboratory is to get students familiar with the bit file and to step through the logic and the combinational logic programming in the PLC.

Enter the three Rungs shown in Fig. 2.62 to implement the combinational logic operation AND, OR, and XOR. These ladder logic diagrams assumes two input switches (SS0 and SS1) and three output coils (AND_LOGIC, OR_LOGIC, XOR_LOGIC). Add a fourth rung for the XNOP operation with output tag XNOR_LOGIC.

The I/O addresses should be documented as follows:

System Inputs

Tag Name	Address	Comments
SS0	I: 1/0	Simulator Selector Switch 1
SS1	I: 1/1	Simulator Selector Switch 2

System Outputs

Tag Name	Address	Comments
AND_LOGIC	O:2/0	Simulator Pilot Light 1
OR_LOGIC	O:2/1	Simulator Pilot Light 2
XOR_LOGIC	O:2/2	Simulator Pilot Light 3
XNOR_LOGIC	O:2/3	Simulator Pilot Light 4

Requirements:

- From the I/O simulator, toggle SS0, SS1, and monitor the pilot lights.
- Program three more rungs for logic NOT, NAND, and NOR.

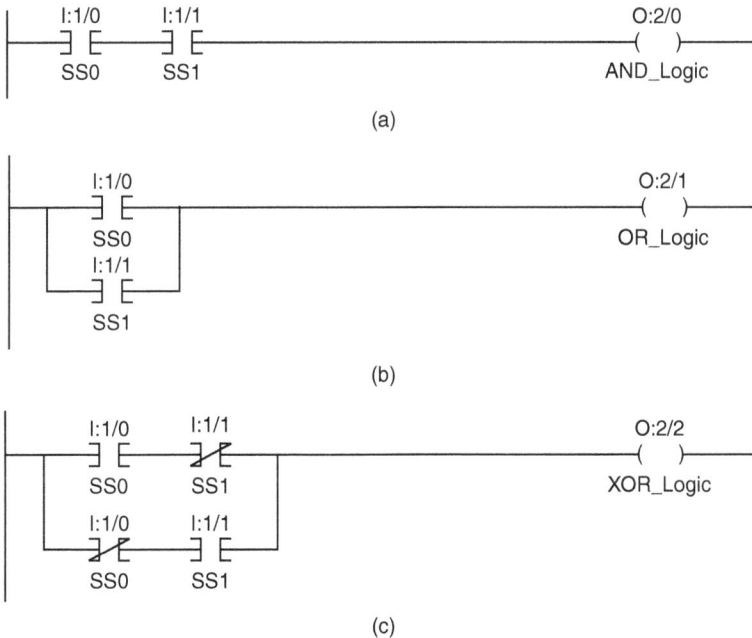

FIGURE 2.62   (a) Laboratory 2.1 rung 1; (b) laboratory 2.1 rung 2; (c) laboratory 2.1 rung 3.

- Reload the program and verify that the new rungs' logic works.
- Perform the check out and document your laboratory report.

Laboratory 2.2—Basic Word Logic Operation Using Structured Programing

The objective of this laboratory is to realize the combinational logic functions used in Laboratory 2.1 in a word logic operations using structured programming.

To create a subroutine (SUB), perform the following steps illustrated in Fig. 2.63:

- In file 2 (main file), lift-click on the program control, and enter jump to subroutine instruction, document AND_LOGIC, and repeats for the remaining subroutines as shown.

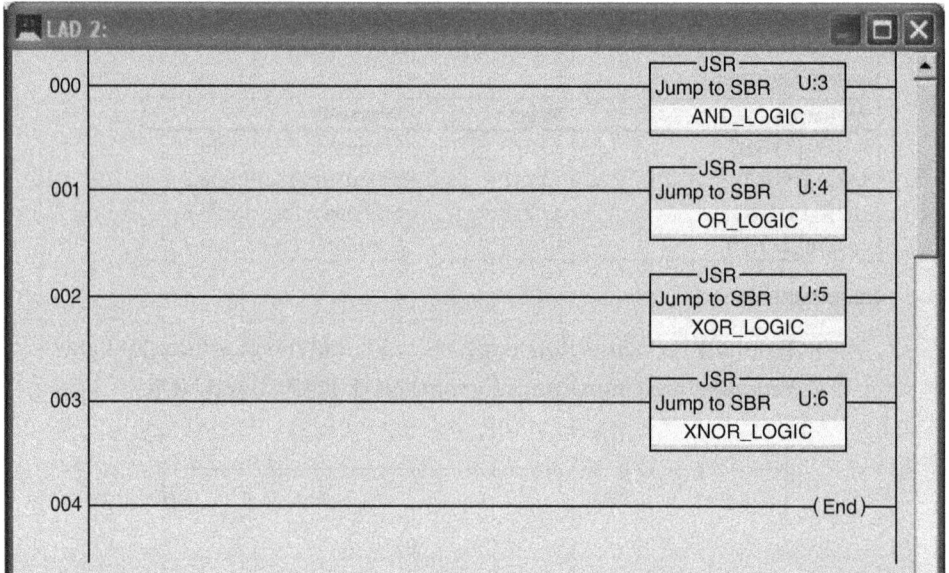

(a)

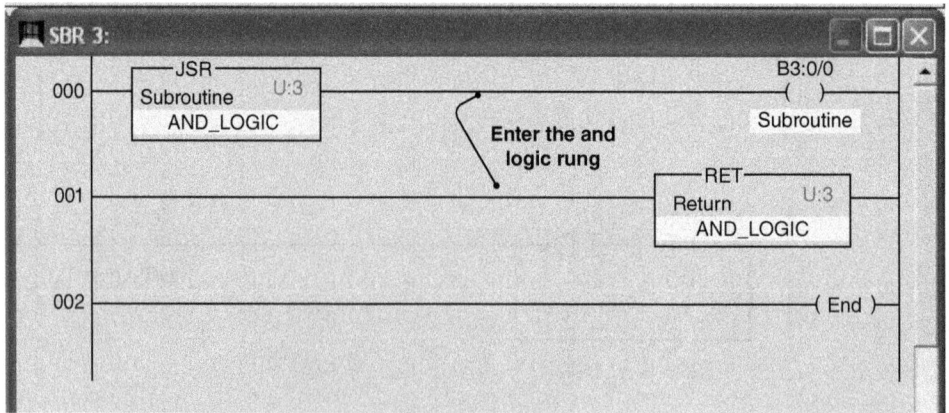

(b)

FIGURE 2.63 (a) Subroutines structure; (b) AND logic subroutine.

- Go to file 3, left-click on program control, and enter subroutine instruction, then enter return instruction, repeats for file 4, 5, and 6 as shown.
- Enter the logic rung for every files between subroutine and return from subroutine.

Requirements:

a. Reprogram the required rungs shown in Fig. 2.64a to c as program structures using three subroutines. The subroutines implement the logics AND, OR, and XOR operations. Using N7:1, N7:2, and N7:3 for TAG_ VALUE1, TAG_VALUE2, and TAG_VALUE RESULT.
b. Download the program to the PLC and go on line.
c. Explain the advantage of the interlock switches between the word logics.
d. Add a fourth subroutine for the logic XNOR using TAG_IN_XNOR.
e. Report your laboratory results in the format used in the Table shown in Fig. 2.65 for the programmed logic operations.

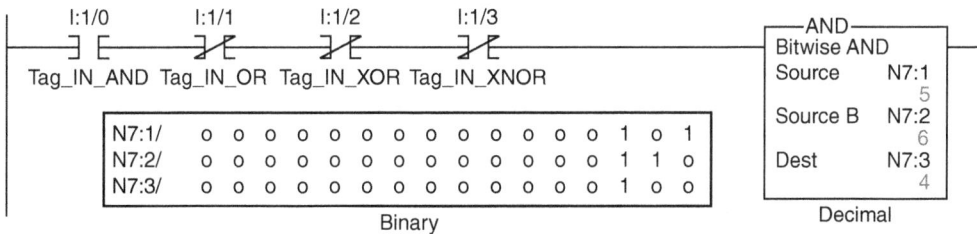

(a)

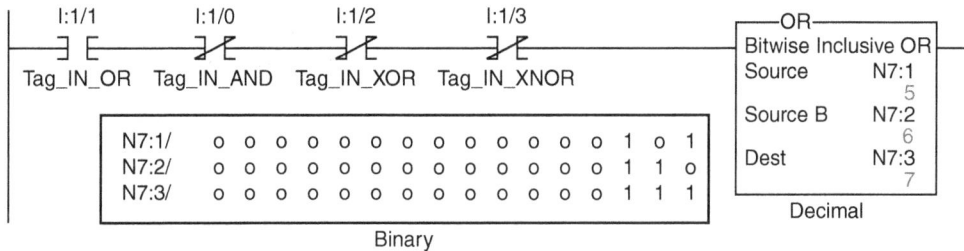

(b)

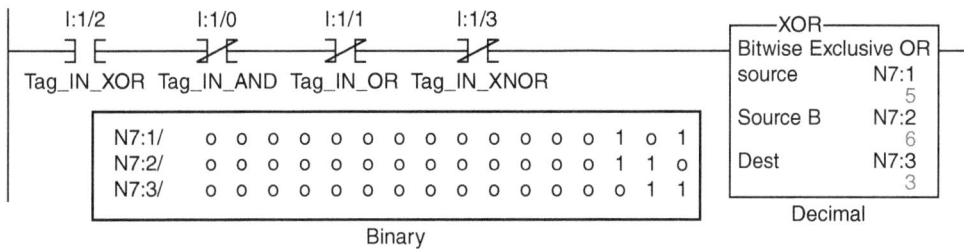

(c)

Figure 2.64   (a) Laboratory 2.2 rung 1; (b) laboratory 2.2 rung 2; (c) laboratory 2.2 rung 3.

N7:1	N7:2	N7:3
15h	30h	
25h	1Ah	
5Bh	11h	

(a)

N7:1	N7:2	N7:3
15h	30h	
25h	1Ah	
5Bh	11h	

(b)

N7:1	N7:2	N7:3
15h	30h	
25h	1Ah	
5Bh	11h	

(c)

N7:1	N7:2	N7:3
15h	30h	
25h	1Ah	
5Bh	11h	

(d)

FIGURE 2.65   (a) Logic AND result; (b) Logic OR result; (c) Logic XOR result; (d) Logic XNOR result.

Laboratory 2.3—Controlling a Conveyor Belt Using Latch and Unlatch Instructions

Figure 2.66 shows a conveyor belt that can be activated electrically. There are two push button switches at the beginning of the belt (Location A): S1 for start and S2 for stop. There are also two push button switches at the end of the belt (Location B): S3 for start and S4 for stop. It is possible to start or stop the belt from either end.

Requirements:

1. Assign and document all I/O addresses.

2. Enter the program shown in Fig. 2.67.

3. Download the program and go online.

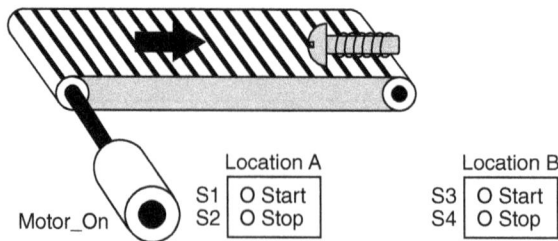

FIGURE 2.66   Conveyor system start/stop.

Rung 1

The conveyor belt motor is switched on when start switch "S1" or "S3" is pressed.

```
    "StartSwitch_Left"      "Motor_ON"
├─────────┤ ├──────────────( L )──────┤
    "StartSwitch_Right"
├─────────┤ ├──────────────────────────
```

Rung 2

The conveyor belt motor is switched off when stop switch "S2" or "S4" is pressed.

```
    "StopSwitch_Left"       "Motor_ON"
├─────────┤/├──────────────( U )──────┤
    "StopSwitch_Right"
├─────────┤/├──────────────────────────
```

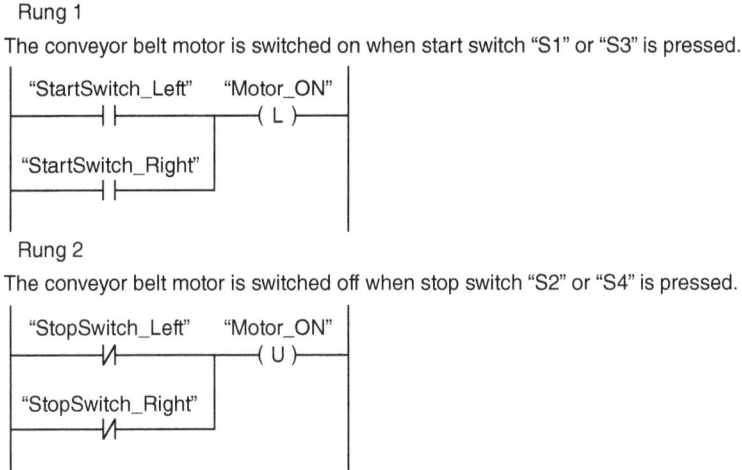

Figure 2.67   Conveyor system ladder program.

4. Provide system check out as detailed in the previous steps.

5. Repeat the laboratory using OTE instruction instead of latch and unlatch.

6. Document your program.

Laboratory 2.4—Conveyor Belt Movement Directions

Figure 2.68 shows a conveyor belt that is equipped with two photoelectric barriers (PEB1 and PEB2). These are designed to detect the direction in which a package is moving on the belt. The barrier provides a true level input pulse of small duration but longer than a one program scan time.

Rung 1 description: If the signal changes from "0" to "1" (positive edge) at "PEB1" and, at the same time, the signal state at "PEB2" is "0", then the package on the belt is moving to the left and pilot light (PL_L) will turn on.

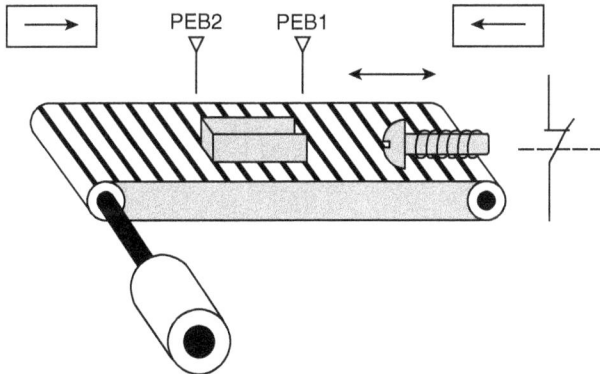

Figure 2.68   Conveyor belt movement's direction detection.

Rung 2 descriptions: If the signal changes from "0" to "1" (positive edge) at "PEB2" and, at the same time, the signal state at "PEB1" is "0", then the package on the belt is moving to the right and Pilot Light (PL_R) will turn ON.

Requirements:

- Two pilot lights (PL_R) and (PL_L) are used to indicate the belt direction status. Using one-shot and output latch/unlatch instructions, write a documented ladder program to detect the conveyor movement direction as indicated by the two pilot lights (right or left).

- Download the program to your PLC and provide the checkout.

CHAPTER 3

Timers and Counters Programming

T his chapter introduces the common types of timers and counters used in PLC programming with an emphasis on the format used in the Allen Bradley (AB) SLC-500 and other families of PLCs. The concepts will be enforced through the use of implemented real industrial application examples.

Velaro high-speed train.

Chapter Objectives

- Understand types, operation, and implementation of timers.
- Understand types, operation, and implementation of counters.
- Understand core and special timing instructions.
- Use timers and counters in industrial process control.

Real-time system behavior is one of the most core issues in control applications. Timers and counters are critical elements used in PLCs, hardwired controllers, and impeded systems to accommodate application events and real-time constrains. This chapter introduces the common types of timers and counters used in PLC programming with an emphasis on the format used in the Allen Bradley (AB) SLC-500 and other compatible families of PLCs. The concepts will be enforced through the use of implemented real industrial application examples. AB SLC-500 timers are available in three different forms: ON delay (TON), OFF delay (TOF), and retentive ON delay (RTO). Other PLC vendors will support the same timers functionality but with slightly different implementation.

3.1 ON-Delay Timers

This timer's main function is the delaying of the rising edge of output (DN) by the predefined period of the preset time (PRE). The timer block is shown in Fig. 3.1. Timer variables are preset (T4:0.PRE), accumulated (T4:0.ACC), and the *status* word that provides events critical information. Only three bits are used from the timer status word: the enable bit (T4:0/EN), the timing bit (T4:0/TT), and the done bit (T4:0/DN). The enable (EN) bit indicates that the timer is enabled, the timing bit indicates that the timer is timing, and timer done bit (DN) indicates that the timer-accumulated value is equal to the preset time. The timer stops timing once DN becomes true. The TON timing diagram is shown in Fig. 3.2. The LogixPro provides one time base (.1 second).

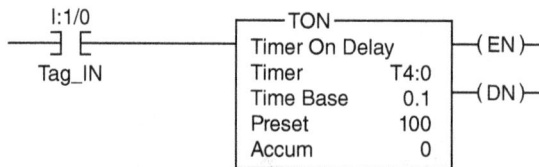

FIGURE 3.1 ON-delay timer instruction.

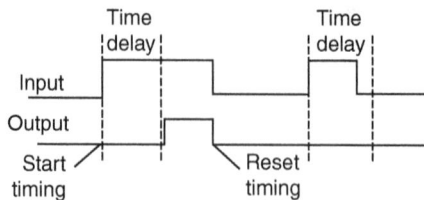

FIGURE 3.2 ON-delay timing diagram.

Preset value in seconds must be multiplied by 10. RSLogix 500 provide two time bases (.1 and 1 second). The Timer output done bit turns ON after a continually enabled timer input for a duration equal or greater than the Timer preset value. If the input resets during the delay, the timer output stays off. Refer to the books supplement website for an *interactive simulator* illustrating the operation of this timer. All three types of timers are available in all PLC systems with minor variations but with the same operational principles.

Example 3.1.1 This example uses an ON-delay timer to generate a pulse every 6 seconds. Figure 3.3 shows the ON-delay timer setting whereas Fig. 3.4 plots the DN bit timing diagram. The following describes the detailed operation of the timer rung:

- Once the program starts to scan, the timer done bit is false (T4:0/DN). This will enable the timer, and T4:0/DN (EN) will be true causing the timer to start timing. The timer timing bit (T4:0/TT) will be set and the accumulated value (T4:0.ACC) will increment.

- When accumulated value equals the preset value, the timer done bit will be set for one scan causing the timer timing bit to reset. The timer DN bit goes off the next scan and the sequence repeats.

- As long as the processor is on, the timer will keep running generating pulse every 6 seconds (timer preset value). Notice that the pulse width is controlled by the scan time. The pulse width can be controlled by latching and unlatching the done bit; an example is provided later in this chapter illustrating the control of the pulse width duration.

Example 3.1.2 This example uses an ON-delay timer to delay motor2's start action by 10 seconds. Figure 3.5 shows another example of ON-delay (TON) application. The timer function is to delay motor 2 from running for 10 seconds after motor 1 runs. When the STOP PB is pressed, the two motors will turn off. Figure 3.5 shows the LogixPro 500 panel configuration and a three-rung implementation. Figure 3.6 plots the resulting timing diagram, whereas Fig. 3.7 shows the two motors' operation relative to the timers' parameters.

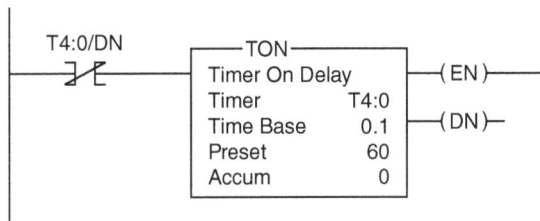

FIGURE 3.3 One scan pulse per 6 seconds.

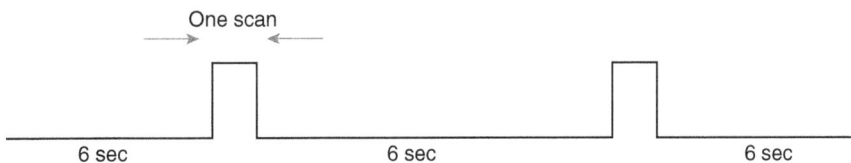

FIGURE 3.4 DN output timing.

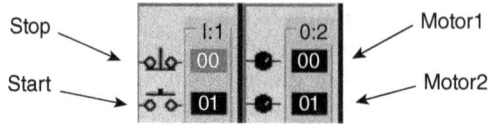

Figure 3.5 Two motors starting sequence.

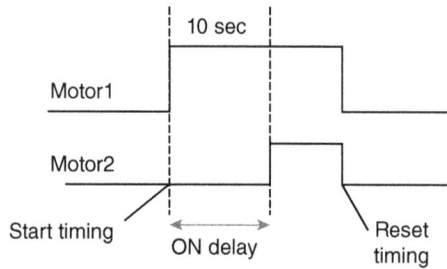

Figure 3.6 Motors' timing diagram.

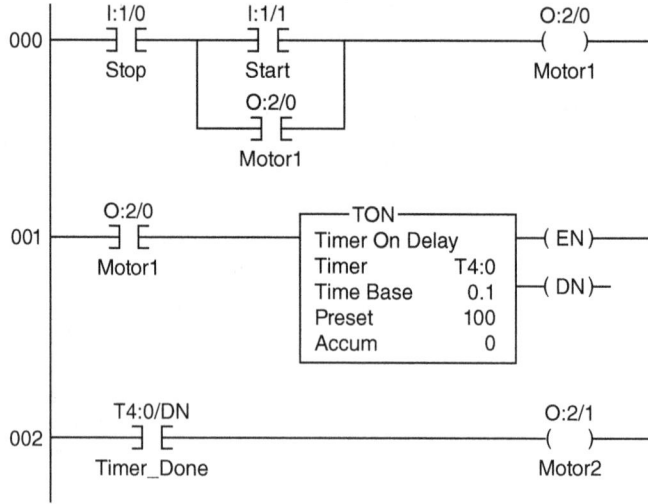

Figure 3.7 Timer's parameters and motor operation.

3.2 Generating a Pulse Timer

The RSLogix 500 does not provide a pulse timer instruction, but it can be generated using the timer timing bit. Figure 3.8 shows a timing diagram for a pulse timer, which generates a pulse with a preset width time. The timer's enable (IN) and the desired timer's done bit (O) are plotted in the timing diagram. A pulse is generated once a positive transition is detected on the timer enable input (IN). The ladder logic needed for the implementation of such pulse timer will be shown in the next example.

Example 3.2.1 This application uses an ON-delay timer to run a vertical gate motor on an irrigation canal for 15 seconds when a pulse signal is issued. Figure 3.9 shows the LogixPro 500 panel configuration and the associated two-rung implementation. Figure 3.10*a* documents the pulse timer timing diagram, whereas Fig. 3.10*b* shows the vertical gate operation relative to the timers' parameters.

When timer VG1_AUTO_START signal is ON (pulse signal), the timers T4:0/EN and T4:0/TT turn on while the T4:0/DN is off. The timer timing bit (T4:0/TT) is latching around the VG1_AUTO_START to keep the timer running for 15 seconds. Once the timer-accumulated value (T4:0.ACC) is equal to

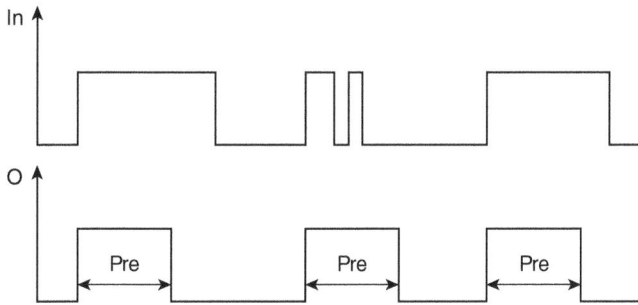

FIGURE 3.8 Pulse timer.

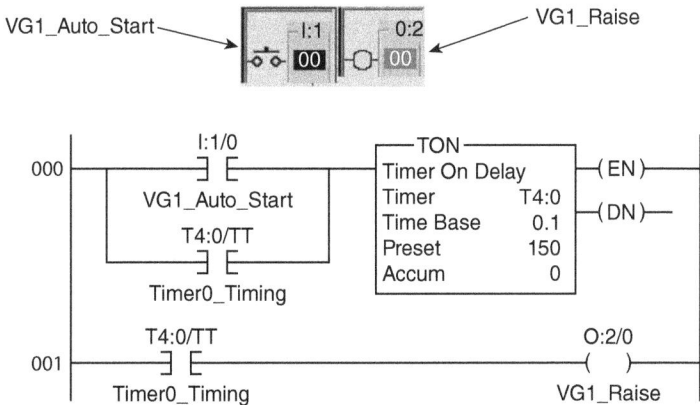

FIGURE 3.9 Pulse timer's implementation.

(a)

(b)

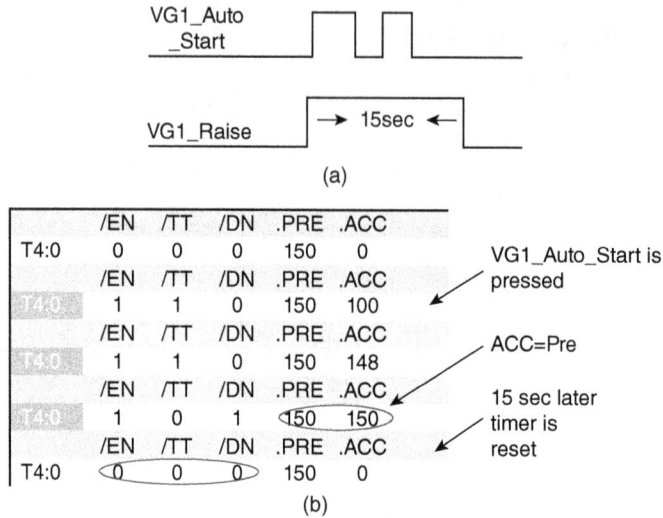

Figure 3.10 (a) Pulse timer timing diagram; (b) pulse timer's parameters and gate operation.

the timer preset value (T4:0.PRE), the timer timing bit (T4:0/TT) will turn off and the timer done bit (T4:0/DN) will be on. This will break the seal around the VG1_AUTO_START and reset the timer as shown in Fig. 3.10 sequence. The VG1_RAISE (motor output) is controlled by the timer timing bit, which is on for only 15 seconds.

Example 3.2.2   This example illustrates the use of ON-delay timers to flash a pilot light. Two ON-delay timers are used to flash a pilot light on/off according to the timing sequence of the ladder program shown in Fig. 3.11a panel configuration, (b) ladder program, and (c) timing diagram. Timer 0 (T4:0/TT) controls the PL ON time and timer 1 (T4:1/TT) controls the PL OFF time. Figure 3.11d shows the timers' parameter values relative to the pilot light operation.

Rung 0: When Man/Auto switch is placed in Auto position, T4:1/TT is OFF (timer 1 is not executed). Timer 0 will run for 5 seconds (PL ON time).

Rung 1: When timer 0 stops running, T4:0/TT is off, which runs timer 1 in rung 1 for 3 seconds (PL off time).

Rung 2: T4:0/TT is controlling the PL, which will flash ON/OFF.

Example 3.2.3   An ON-delay timer is used to debounce a signal simulated by a push button switch. The signal transition from low to high will drive a counter with preset value of 10. Figure 3.12 shows (a) the ladder program, and (b) the timing diagram. The PB switch action is assumed to last less than 1 second, which is the preset value for the TON. The counter instruction is detailed in Sec. 3.5. The following is a brief description of the three rungs:

Rung 0: When PB_SIGNAL transitions from low to high, the TON instruction executes. The timer TT bit will be true for 1 second and latch around PB_SIGNAL to debounce the signal and ignore any signal transition if it happens in less than 1 second.

Rung 1: When the timer TT bit transitions from low to high, the counter increments.

Rung 2: Once the counter-accumulated value is equal to the counter preset value (10), the counter done bit becomes true and PL1 turns on.

Rung 3: Pressing the Reset PB causes the counter to reset.

(a)

(b)

(c)

	/EN	/TT	/DN	.PRE	.ACC	
T4:0	0	0	0	50	0	Auto is off
T4:1	0	0	0	30	0	
	/EN	/TT	/DN	.PRE	.ACC	
T4:0	1	1	0	50	39	Auto is on and T4:0 is running
T4:1	0	0	0	30	0	
	/EN	/TT	/DN	.PRE	.ACC	
T4:0	0	0	0	50	0	Auto is on
	1	1	0	30	15	and T4:1 is running

Flip, flop

(d)

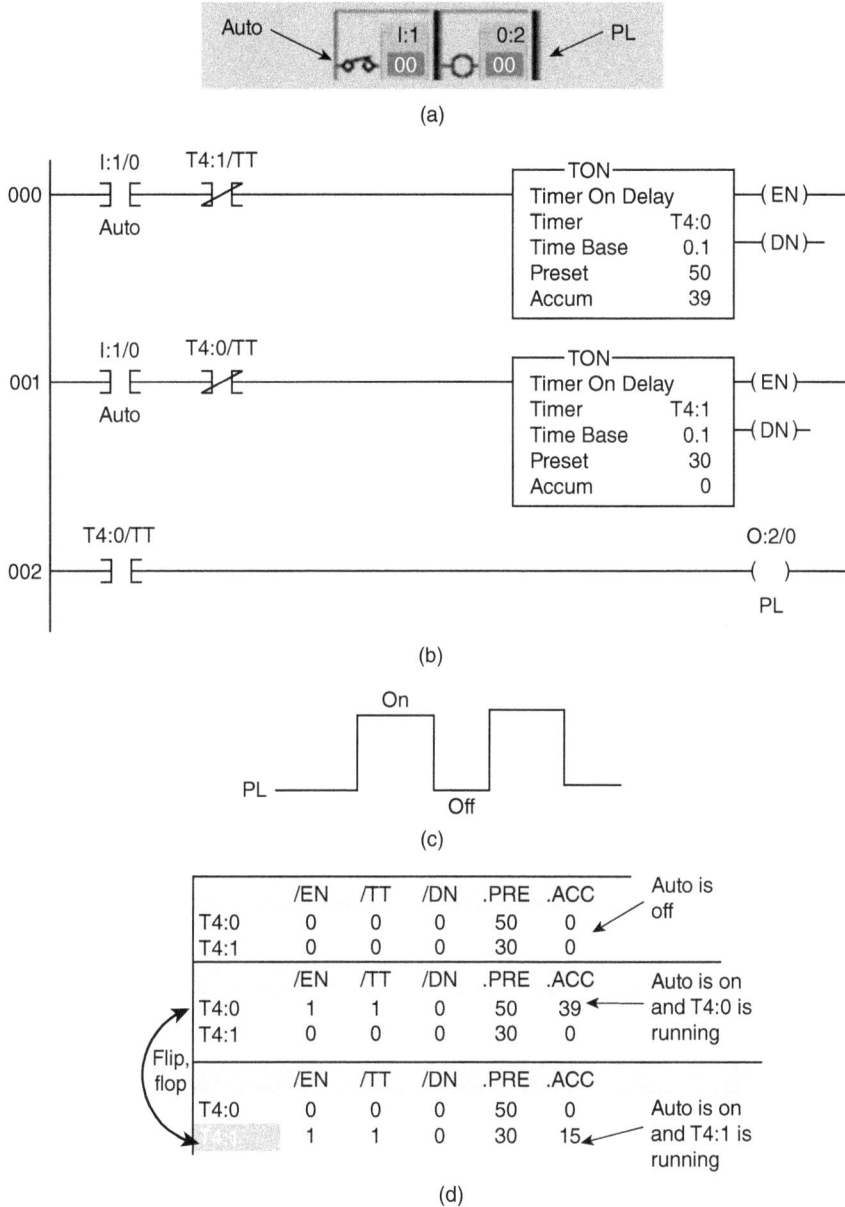

FIGURE 3.11 (a) LogixPro 500 panel configuration; (b) LogixPro 500 ladder program; (c) pilot light timing diagram; (d) Timer's parameters and pilot light operation.

3.3 OFF-Delay Timers

The TOF main function is the delaying of the falling edge of output (DN) by the predefined period of the preset time (PRE). If the input to the instruction (IN) is true, the output is set true. When the input turns off, the timer starts timing. It resets the output

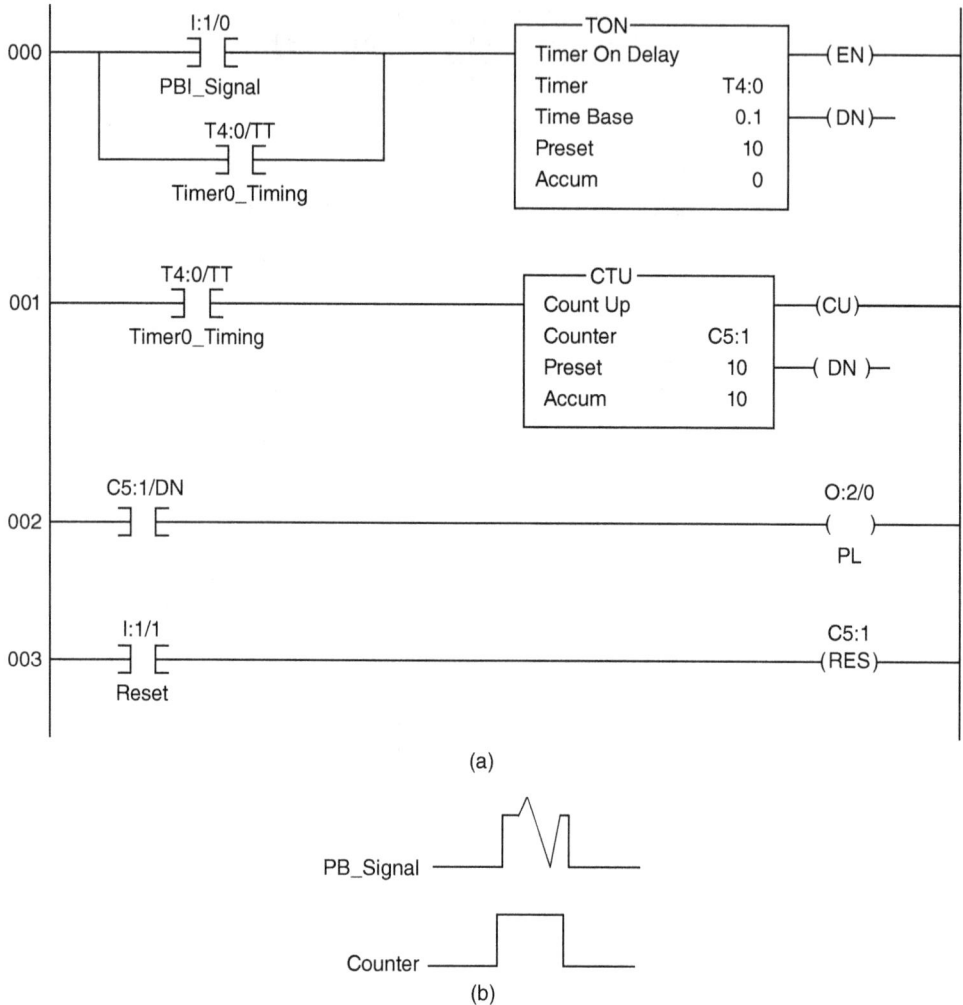

Figure 3.12 (a) Ladder program; (b) timing diagram.

when the timer-accumulated value (ACC) is equal to the timer preset value (PRE). Figure 3.13 illustrates the TOF instruction and Fig. 3.14 shows the timing diagram. Refer to the book's supplement website for an *interactive simulator* illustrating the operation of this timer.

Example 3.3.1   An application uses two motors in a pumping station facility. Both motors run on the activation of a START PB switch. The activation of a STOP PB switch stops motor 1 immediately and then motor 2 after 10 hours. Figure 3.15 shows implementation of two ladder logic rungs using an OFF-delay timer.

Example 3.3.2   This example uses a limit switch and three OFF-delay timers to control three motors. Figure 3.16a shows a simple process-control implementation and Fig. 3.16b shows the LogixPro 500 panel configuration used, which assumes a limit-switch tag name LS1 and three outputs: motor 1 O:2/0, motor 2 O:2/1, and motor 3 O:2/2. Timer preset values of 10, 20, and 30 seconds are used to

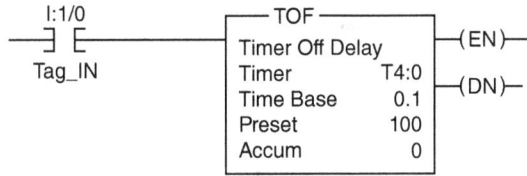

FIGURE **3.13** OFF-delay timer instruction.

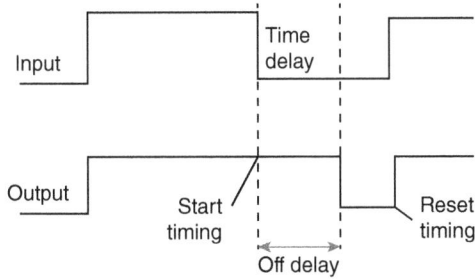

FIGURE **3.14** OFF-delay timing diagram.

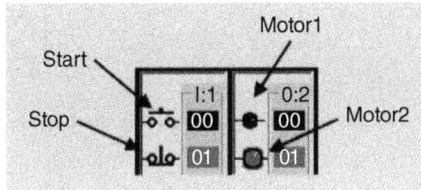

(a)

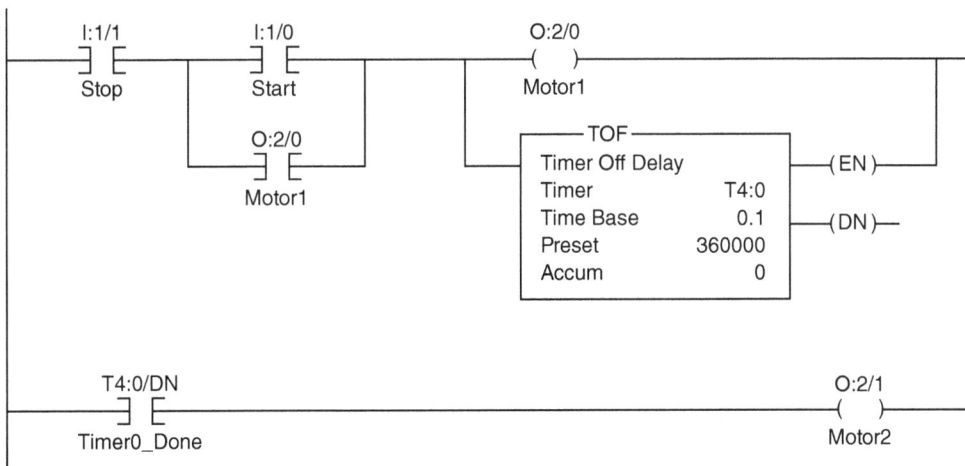

(b)

FIGURE **3.15** (a) Two motors timing panel configuration; (b) two motors timing ladder program.

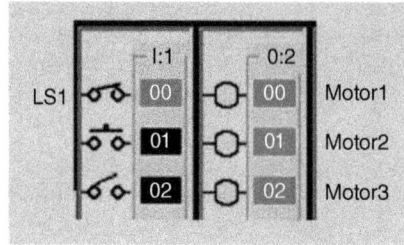

(a)

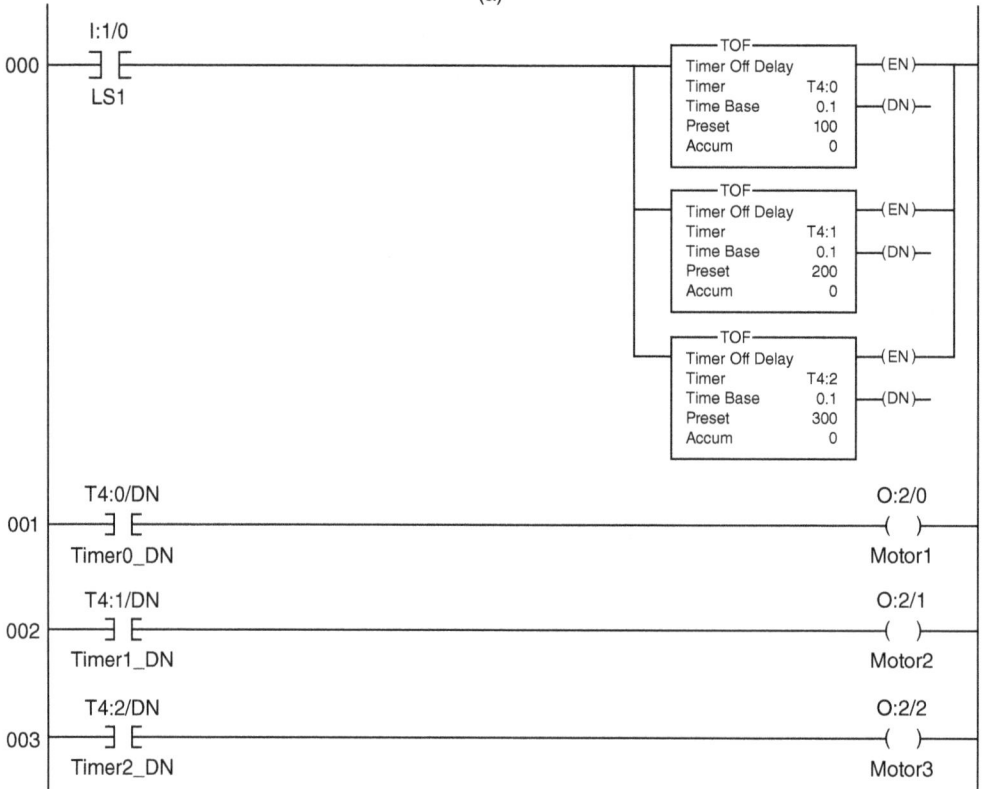

(b)

Figure 3.16   (a) LogixPro 500 panel for Example 3.3.2; (b) LogixPro 500 ladder for Example 3.3.2.

sequence the operation of the three motors. When LS1 is true, all three motors start running. If LS1 opens, motor 1 turns off after 10 seconds, motor 2 turns off after 20 seconds, and motor 3 turns off after 30 seconds. Once LS1 closes, the accumulated values of all three timers are cleared.

3.4   Retentive Timers-On-Delay

Retentive timer-on-delay (RTO) works exactly the same as the ON-delay timer except that the accumulated value is retained while the timer instruction is inactive. To reset the accumulated value to zero, a positive pulse instruction is required to reset the timer (RES). Figure 3.17 shows the RTO instruction and Fig. 3.18 illustrates the RTO timing

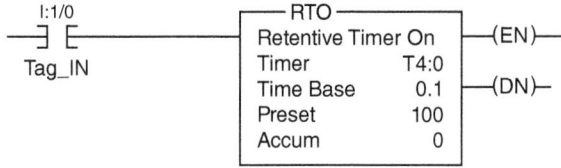

FIGURE **3.17** Retentive timer instructions.

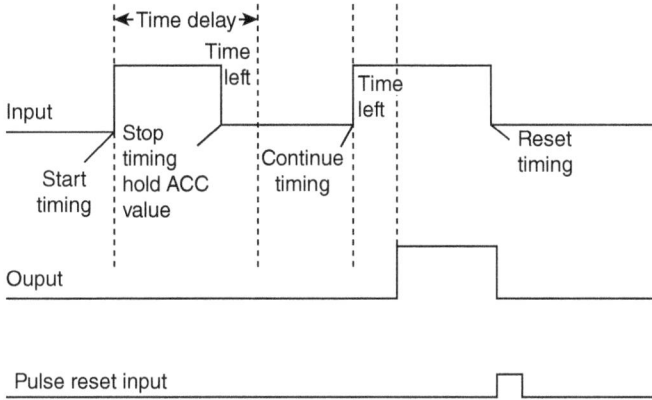

FIGURE **3.18** RTO delay timing diagram.

diagram. Refer to the book's website for an *interactive simulator* illustrating the operation of this timer.

Example 3.4.1 Figure 3.19 shows a ladder logic program using the RTO timer instruction. The RTO ladder logic diagram shown has a Auto/Manual selector switch and the timer preset value is set to 1 hour (when simulated use shorter time). Once the selector switch is placed in Auto, the rung turns true. Figure 3.20 shows the timer reset and timing sequence. The processor scans the ladder and the following events are observed:

- The timer starts timing once the selector switch is placed in Auto. If the selector switch is placed in Manual, the timer stops timing and the timer-accumulated value is retained.

- Once the switch is placed back in Auto, the timer starts timing from where it left.

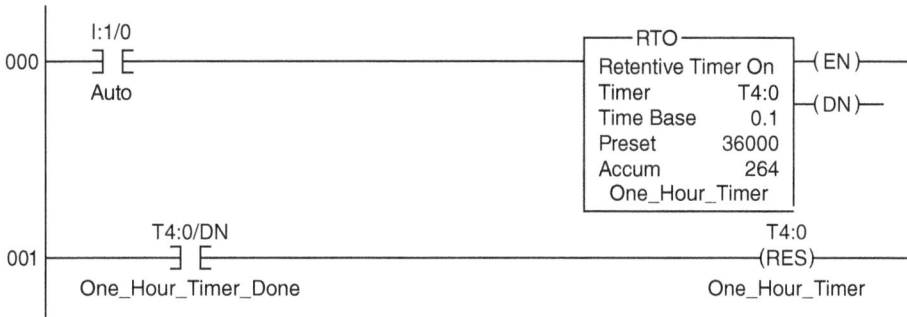

FIGURE **3.19** RTO timer ladder for Example 3.4.1.

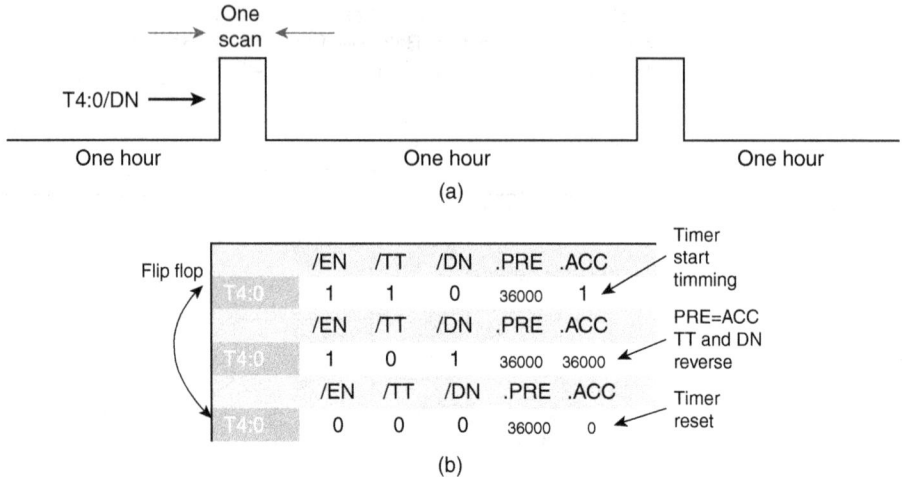

FIGURE **3.20**   (a) RTO timer reset for Example 3.4.1; (b) RTO timer sequence for Example 3.4.1.

- Once the timer-accumulated value T4:0.ACC equals the preset value T4:0.PRE, the timer output (ONE_HOUR_DN) is set to true. The accumulated value T4:0.ACC is reset to zero when the RES (ONE-HOUR-TIMER) instruction is true.
- The RTO in this example continuously recycles on an hourly basis.

3.5 Fundamentals of Counters

Allen Bradley counters are available in two different forms: count-up (CTU) and count-down (CTD). An implemented counter application is analyzed in the next section. The concepts covered in this section follow the SLC-500 notations but can be applied to other PLC brands with minor modification.

3.5.1 Count-Up Counters

The up-counter instruction block is shown in Fig. 3.21. The main function of the up counter is to increment the accumulated value (C5:0.ACC) each time the input (IN) to

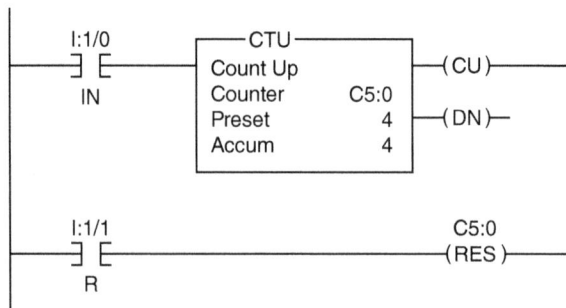

FIGURE **3.21**   Count-up (CTU) instruction format.

the counter transitions from 0 to 1. If the accumulated value is equal to or greater than the preset value (C5:0.PRE), which is set to 4 in Fig. 3.22, (C5:0/DN) output is set. The counter will continue counting up for incoming IN pulses past the preset value until it is reset. When reset input (R) is true, the accumulated value resets to 0.

Example 3.5.1.1 Two normally open contacts IN and R, address I:1/0 and I:1/1, are configured as PB switches to provide the counting pulses and the counter reset (R) action, respectively, as shown in Fig. 3.23. Typically the counting pulses are provided from either a photo sensor or a timing event. The count-up ladder program is shown in Fig. 3.24. Experiment with the two PB inputs and verify the counter operation.

Example 3.5.1.2 A part is moving on a conveyor line crossing a photoelectric cell. The function of the photoelectric cell is to detect passing parts. The conveyor line stops after 100 parts are counted. Figure 3.25 shows implementation of six rungs using the CTU instruction. The first rung assumes examine if closed (XIC) START PB (I: 1/0), examine if closed (XIC) STOP (wired high) push button (I: 1/1), an Auto selector switch (I: 1/2), and motor output (O: 2/0). The second rung uses a count-up counter (CTU) with a preset value of 100, examine if closed (XIC) address I: 1/4 tag name MOTOR_RUN,

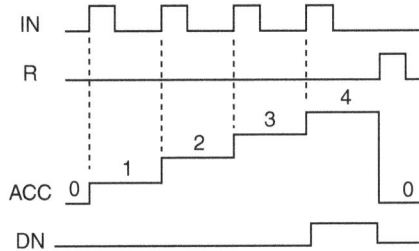

FIGURE 3.22 Count-up (CTU) timing diagram.

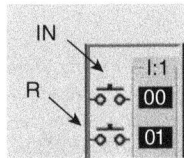

FIGURE 3.23 Panel setting for Example 3.5.1.1.

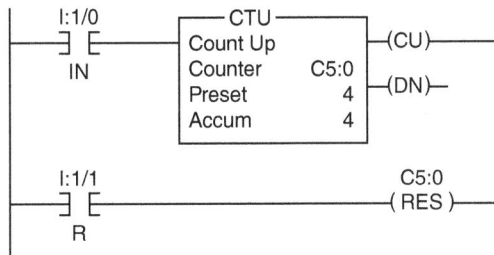

FIGURE 3.24 Count-up (CTU) ladder program with preset = 4.

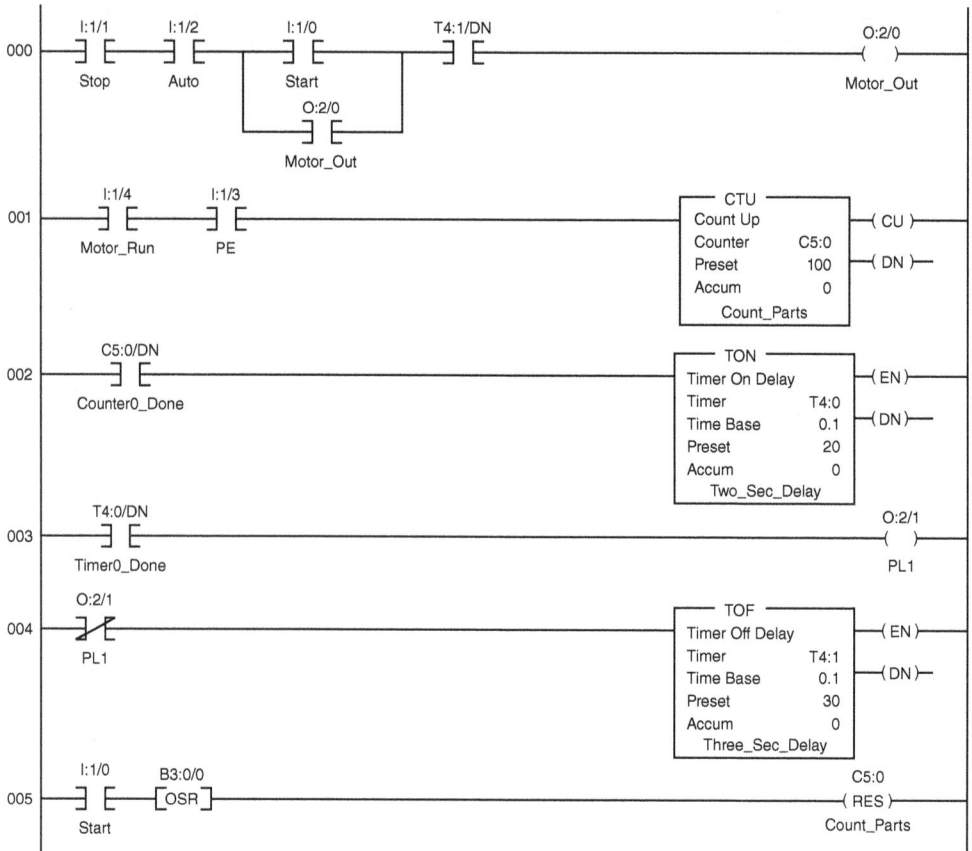

Figure 3.25   Count up Example 3.5.1.2.

and (XIC) address I: 1/3 tag name PE for photoelectric cell. Once selector switch Manual/Auto is placed in Auto position and the START PB is activated, the motor will run immediately and drives a conveyor system. Once the conveyor system runs, an input is received from the motor magnetic starter indicating successful system start (MOTOR_RUN). The motor runs for duration to allow a preset number of parts to be transported over the conveyor system, which is set to 100 in our case. The following are the critical steps for this example.

The first rung initially is false during the first scan because the START PB is false as this switch is not pressed yet (normally open PB) and the Auto mode (I: 1/2) is set. Once START PB is pressed, I: 1/0 becomes true, motor output MOTOR_OUT address O: 2/0 will be true and causes motor to run. During the next scan, O: 2/0 will latch around the START PB and maintains the first rung true.

The second rung has I: 1/4 as the motor running input, which becomes true once the motor starts running and enables the CTU instruction every time a part crosses the PE. When the motor running input (MOTOR_RUN) is set indicating that the conveyor is moving, a photoelectric cell will issue a positive narrow pulse as the part crosses the photoelectric cell, which causes the CTU current count value to increment.

Once the counter preset value equals the current count value, the counter output C5:0/DN becomes true and starts the ON-delay timer T4:0. After a 2-second delay, pilot light (PL1) will be set indicating the end of the cycle. The 2 seconds allow the last part to cross the counting station.

Once PL1 is set, the TOF timer in the fifth rung will lose power and TIMER1 will start timing. Three seconds later, MOTOR_OUT will be false and motor will stop running. The 3 seconds allow the last part to arrive at the end of the conveyor line.

Counter count value will reset to zero in the sixth rung when START PB is pressed.

A solenoid, which is not shown, stops parts from entering the counting station while the conveyor runs during the 5 seconds delay. Notice that only one timer is needed for this implementation. Two different timers were used for the demonstration of TON and TOF instructions.

Example 3.5.1.3   A timer accumulated value can be extended using a counter as is illustrated in this example. Figure 3.26 shows the LogixPro 500 panel setting, using two switches and a light. Figure 3.27 documents the ladder program implementation. Timer T4:0 will recycle every hour and increment the counter C5:0. After 5 hours the counter done bit will be true that turns on the pilot light (PL1).

3.5.2   Count-Down Counters

The main function of count-down counter (CTD) is to decrement the accumulated value (C5:0ACC). Each time the input (IN) to the counter transitions from 0 to 1. If the current accumulated value is greater than or equal to the preset value (C5:0.PRE), the counter output done bit (C5:0/DN) is set; otherwise, the done bit is reset. Figure 3.29 shows the

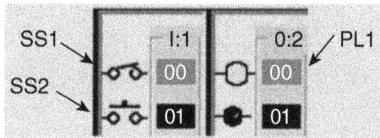

FIGURE 3.26   Cascaded timer panel configuration.

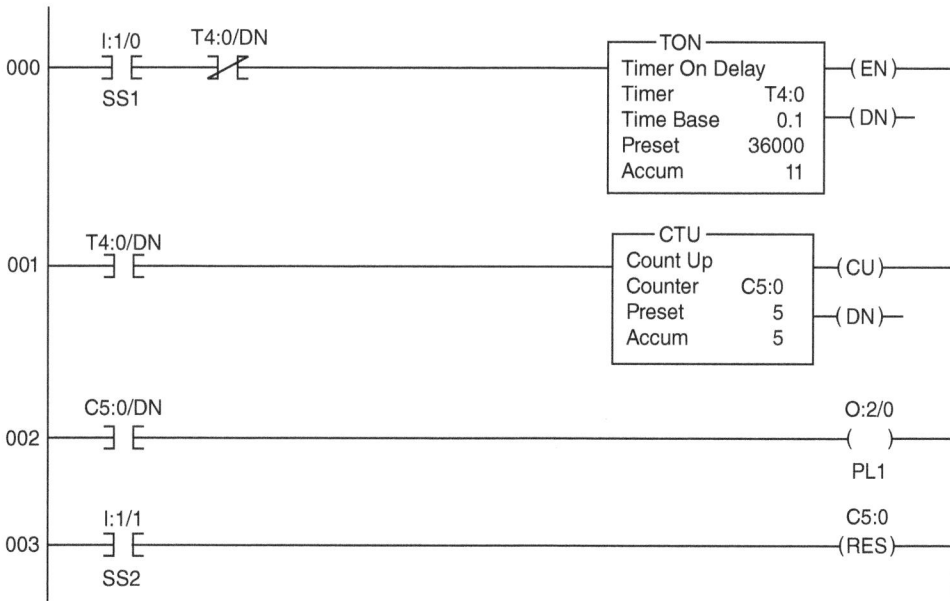

FIGURE 3.27   Cascaded timer ladder program.

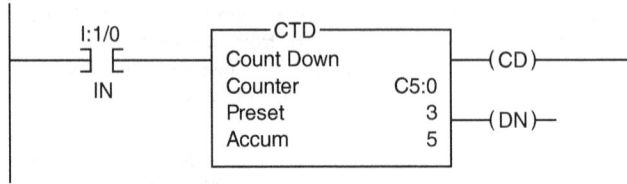

FIGURE **3.28** CTD instruction.

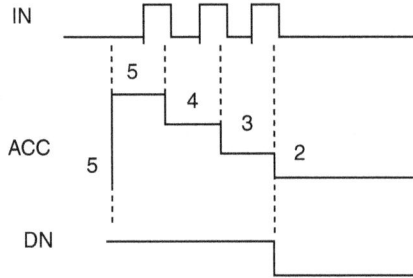

FIGURE **3.29** CTD timing diagram for preset = 3.

CTD instruction, and Fig. 3.29 illustrates the timing diagram using a PB switch for the counter IN.

Example 3.5.2.1 The CTU and CTD counter do not exist as an instruction in the AB SLC-500. They can be constructed using CTU and CTD instructions with the same address as shown in Fig. 3.30. The main function of CTU and CTD is to count up or down the accumulated value based on inputs signifying increase (PE1 pulse) and decrease (PE2 pulse) in an entity, as in cars entering or leaving a parking garage. A RESET pulse will clear the counter-accumulated value and a LOAD pulse moves N7:4 to the counter preset register.

   When PE1 transitions from 0 to 1, the current count value (C5:0.ACC) is incremented. If the signal state at the CTD input triggered by PE2 transitions from 0 to 1, the current count value is decremented. If the signal at the LOAD input changes from 0 to 1 or during the first scan, the counter preset value (C5:0.PRE) is set to the value of the preset parameter (N7:4). Figure 3.31 illustrates a timing sequence for this timer for a preset value of 6. S:1/15 is used to force the load and reset operation during the first scan.

Example 3.5.2.2 The up/down counter can be used to keep track of the good parts count in a conveyor system. A simple conveyor system is configured with two photocells, a sorting station, and an elevated belt for rejected parts. The photoelectric cell PE1 keeps track of the incoming parts count whereas PE2 counts the number of rejected parts. The counting is implemented as shown in Fig. 3.30 with the up-counter enabled through PE1 and the down-counter enabled by PE2. The two counters are assigned the same address, which stores the number of good parts in its accumulated value. Figure 3.31 illustrates a counting sequence from the start to the point of eight total good parts and two rejected bad parts.

Example 3.5.2.3 Figure 3.32 shows a tank with inlet and outlet valves and a level sensor. This application uses counters in the implementation of the tank process control. It assumes two solenoid valves: FILL (SV1) and DRAIN (SV2). The tank level sensor is simulated by a count-up and -down

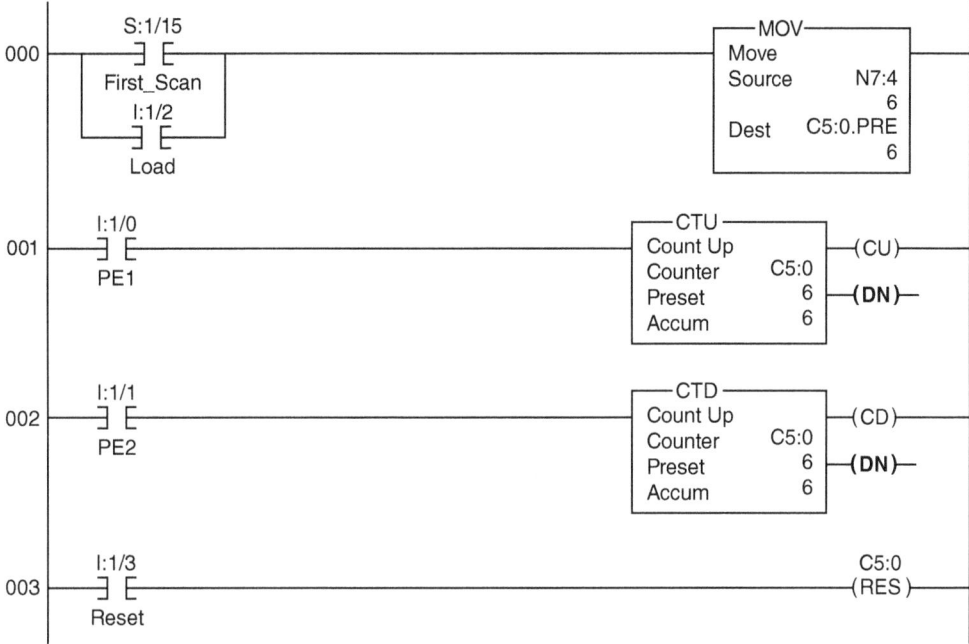

FIGURE 3.30 Count-up/-down ladder program.

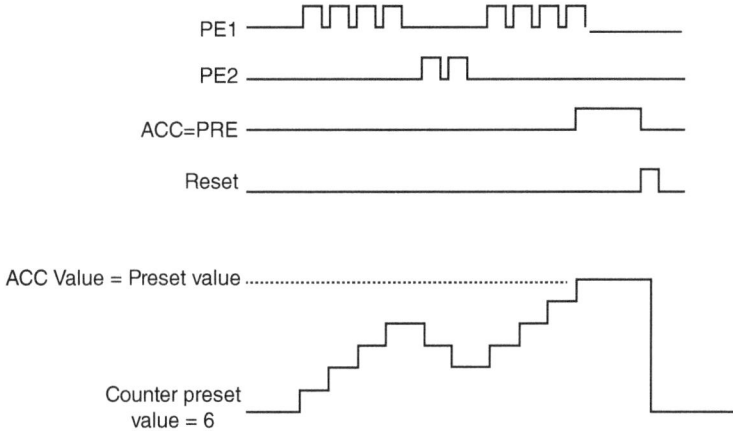

FIGURE 3.31 Count-up/-down timing sequence.

(CTUD) counter. If the tank level is greater than or equal to 50 m, the tank is drained by energizing SV2; otherwise, if tank level is less than or equal to 10 m, fill the tank by activating SV1. The Allen Bradley SLC-500 PLC implementation detailed next uses a processor built-in flash bit S: 4/6 to simulate the tank filling and draining actions. It assumes a 1-m level increase during filling per flash and a 1-m decrease in level during draining per flash. Figure 3.33 shows the panel configuration. The ladder rungs are listed in Fig. 3.34.

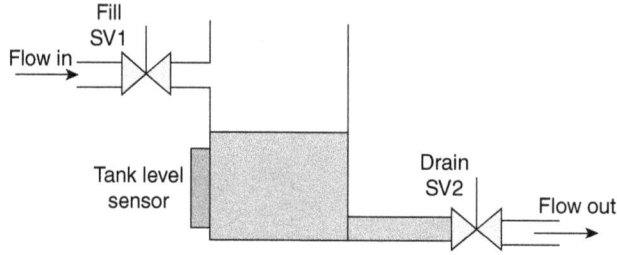

FIGURE 3.32 Fill/drain tank control.

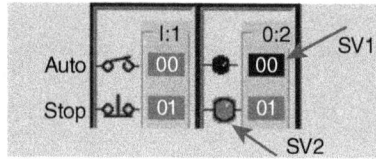

FIGURE 3.33 Tank fill/drain panel configuration.

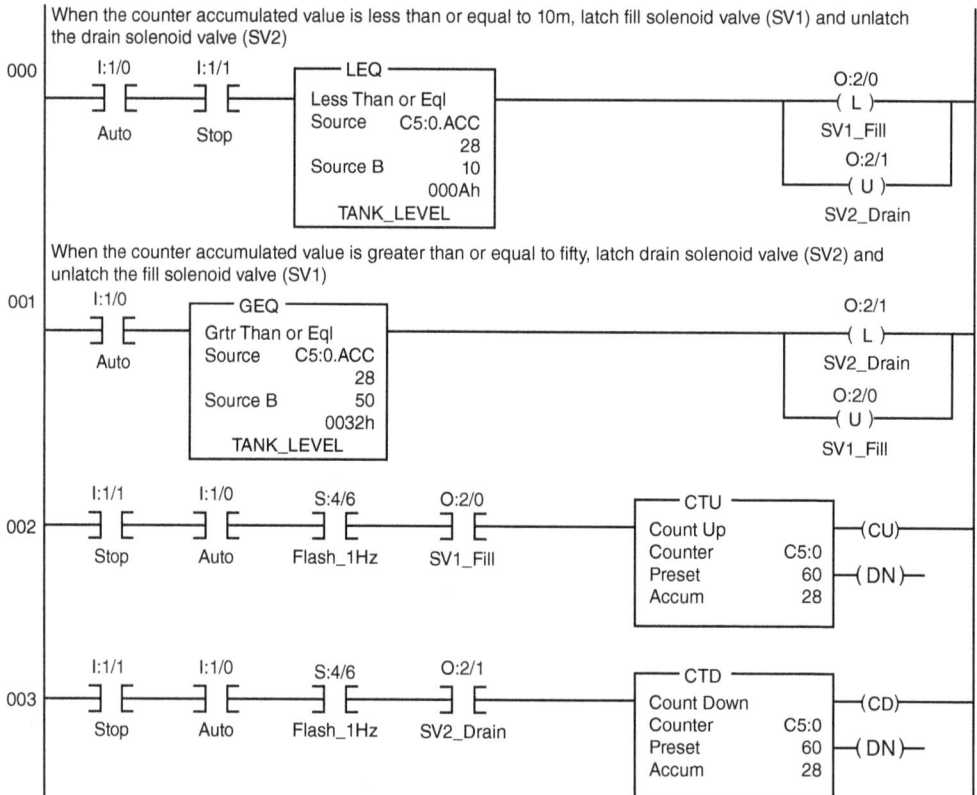

FIGURE 3.34 Tank fill/drain ladder program.

The following is the descriptive documentation for the fill/drain rungs listed in Fig. 3.34:

- Rung 0: Start tank filling and stop draining when the tank level is low; the tank level is less than or equal to 10 m.
- Rung 1: Stop tank filling and start draining when the tank is full; the tank level is greater than or equal to 50 m.
- Rung 2: An up-counter (CTU) updates the tank level during filling, which is given the tag name "C5:0.ACC." The PLC S:4/6 processor bit flashes once every second, which simulates a 1-m change in tank level. Tank control is active when the Auto/Manual selector switch is set to the Auto position.
- Rung 3: A down-counter (CTD) updates the tank level during draining, which is given the tag name "C5:0.ACC." The PLC S:4/6 processor bit flashes once every second, which simulates a 1-m change in tank level. Tank control is active when the Auto/Manual selector switch is set to the Auto position.
- Tank level must be initialized (C5:0.ACC) before the four ladder rungs are scanned. This is typically part of the overall program initialization, which is not shown here.

Chapter 3: Home Work Problems and Laboratory Projects

1. Explain the difference between an ON-delay timer (TON) and a retentive timer (RTO) instructions.

2. Complete the timing diagram for the ON-delay timer shown in Fig. 3.35. The input signal is the timer-enabling input whereas the output is derived by the timer done bit. Indicate when the timer-accumulated value starts and stops timing.

3. The ladder rung shown in Fig. 3.36 is programmed to recycle a timer every 6 minutes (360 seconds). When the program is tested, it does not work correctly, explain why?

4. What is the maximum time you can enter on a preset time for a single timer?

5. What is the status of motor1, motor2, SOL1, and SOL2 in Fig. 3.37 under the following conditions:

   a. When LS1 is ON for more than 20 seconds
   b. When LS1 transition is from ON to OFF after 10 seconds
   c. When LS1 transition is from ON to OFF after 20 seconds

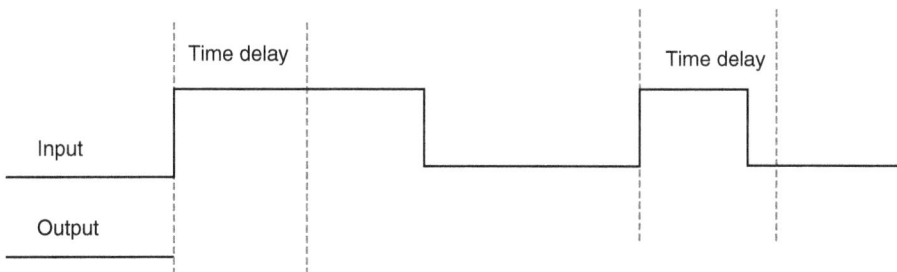

FIGURE 3.35 Problem 2 timing diagram.

```
  I:1/0                                        ┌─RTO──────────────────┐
──┤ ├──────────────────────────────────────────┤ Retentive Timer On    ├─( EN )──
  AUTO                                         │ Timer          T4:0   │
                                               │ Time Base       0.1   ├─( DN )──
                                               │ Preset        36000   │
                                               │ Accum             0   │
                                               │   Reset_Timer0        │
                                               └───────────────────────┘

  T4:0/DN                                                           T4:0
──┤/├──────────────────────────────────────────────────────────────(RES)──────
                                                                  Reset_Timer0
```

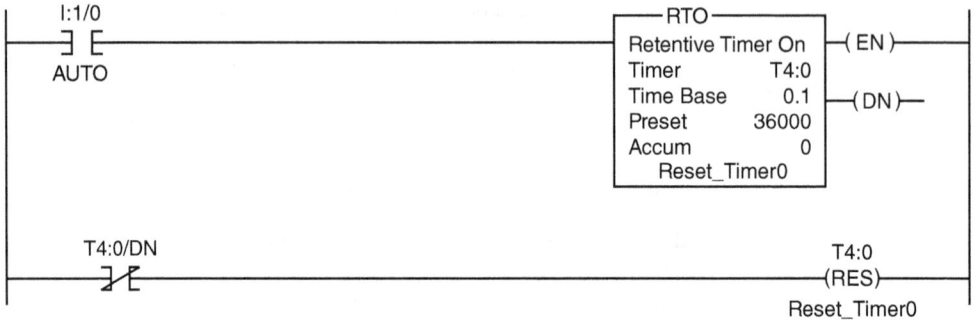

FIGURE 3.36 Problem 3 ladder program.

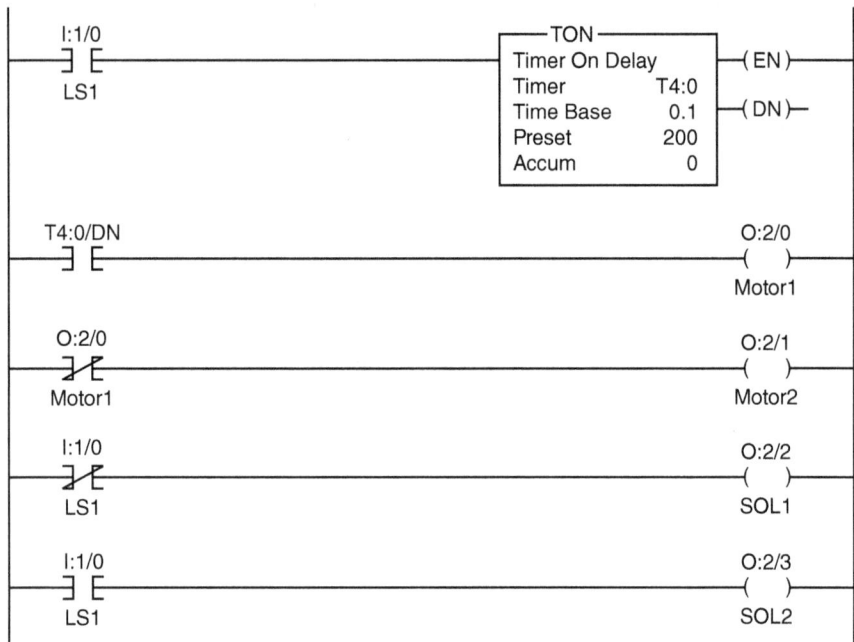

```
  I:1/0                                        ┌─TON──────────────────┐
──┤ ├──────────────────────────────────────────┤ Timer On Delay        ├─( EN )──
  LS1                                          │ Timer          T4:0   │
                                               │ Time Base       0.1   ├─( DN )──
                                               │ Preset          200   │
                                               │ Accum             0   │
                                               └───────────────────────┘

  T4:0/DN                                                           O:2/0
──┤ ├───────────────────────────────────────────────────────────────( )────────
                                                                    Motor1

  O:2/0                                                            O:2/1
──┤/├────────────────────────────────────────────────────────────────( )───────
  Motor1                                                            Motor2

  I:1/0                                                            O:2/2
──┤/├────────────────────────────────────────────────────────────────( )───────
  LS1                                                               SOL1

  I:1/0                                                            O:2/3
──┤ ├─────────────────────────────────────────────────────────────────( )──────
  LS1                                                               SOL2
```

FIGURE 3.37 Problem 5 ladder program.

6. Repeat problem 5 if OFF-delay timer (TOF) is used.

7. If a TON timer rung is active and the ACC value is less than the PRE value, what is the status of timer EN, TT, and DN bits?

8. Repeat problem 7 for a TOF timer.

9. Write a ladder logic program to turn ON motor1 when the START PB is pressed. Ten seconds later motor2 turns on. The STOP PB stops both motors.

10. Motor2 is to be started once motor1 runs. Motor2 will continue running continuously for simulated 15 seconds regardless of motor1's status—running or stopped. If motor1 stops after the 15 seconds and starts again, the same sequence repeats.

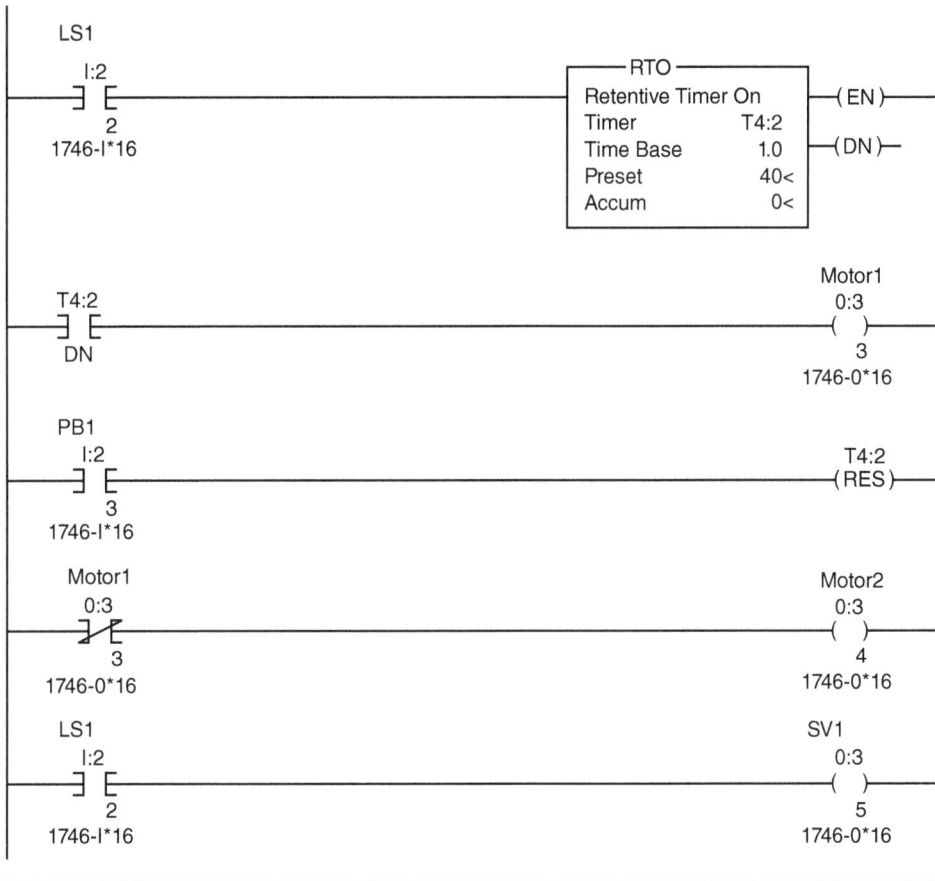

Figure 3.38 Problem 11 rungs.

If motor1 stops during the 15 seconds, it will not affect motor2. Implement the ladder program for motor2 control.

11. Assume that PB1 is a normally open switch and not pressed. Examine the rungs in Fig. 3.38 and answer the following questions:

   a. What is the status of motor1 and motor2, if LS1 is ON for 40 seconds or longer?
   b. What is the status of motor1 after LS1 is ON for 60 seconds and then OFF?
   c. What is the status of motor2 when LS1 is OFF?
   d. What is the status of motor1, motor2, and SV1 if LS1 is maintained ON and PB1 is pressed?

12. A START PB switch is used to start a sequence of pilot lights (simulating real motors), which can be terminated at any time by pressing a STOP PB switch. The sequence starts PL1, then PL 2, then PL 3. The same sequence will repeat until the process is stopped or the cycle is repeated for five times. Use the same or different preset time for each lights.

13. Draw the done output for the up-counter shown in Fig. 3.39.

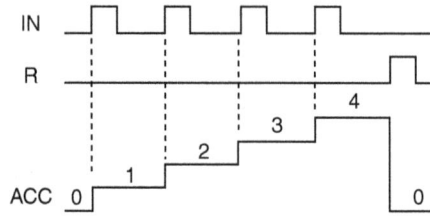

FIGURE **3.39**   Up-counter diagram for problem 13.

14. Repeat problem 11 assuming a TON timer is used in place of the RTO timer.

15. Study the rungs in Fig. 3.40 for the CTU and answer the following questions:
   a. Under what condition will M1 turn ON?
   b. Under what condition will M2 turn ON?
   c. What is the counter-accumulated value (C5:0.ACC) when input I: 2/6 is ON?

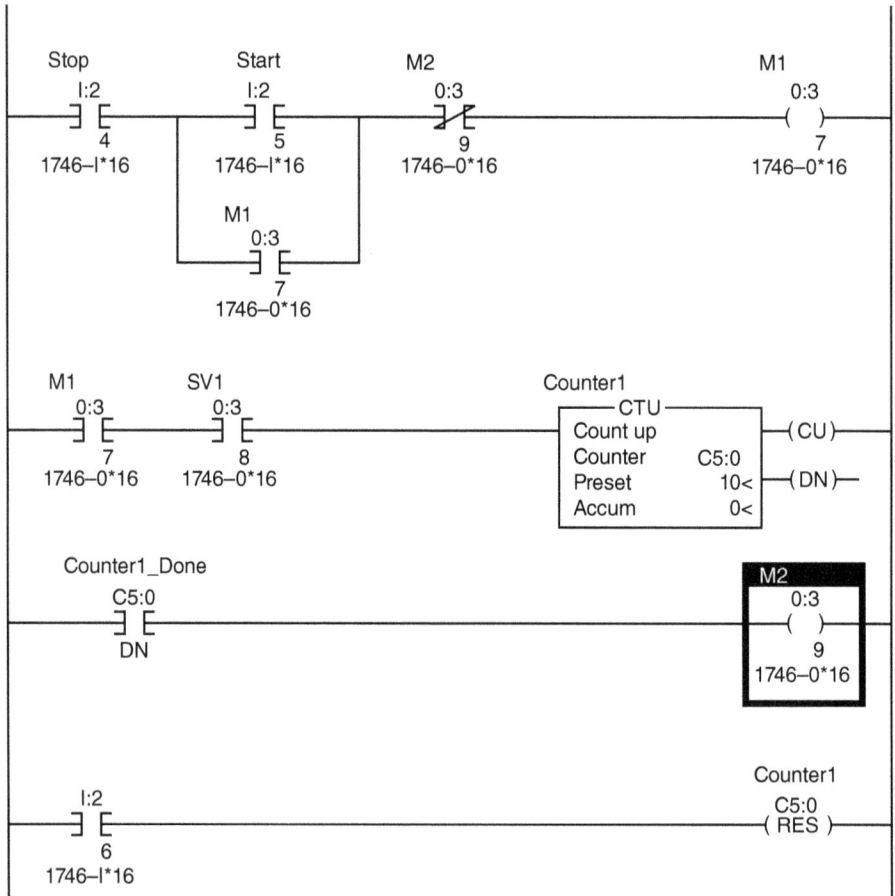

FIGURE **3.40**   Up-counter ladder for problem 15.

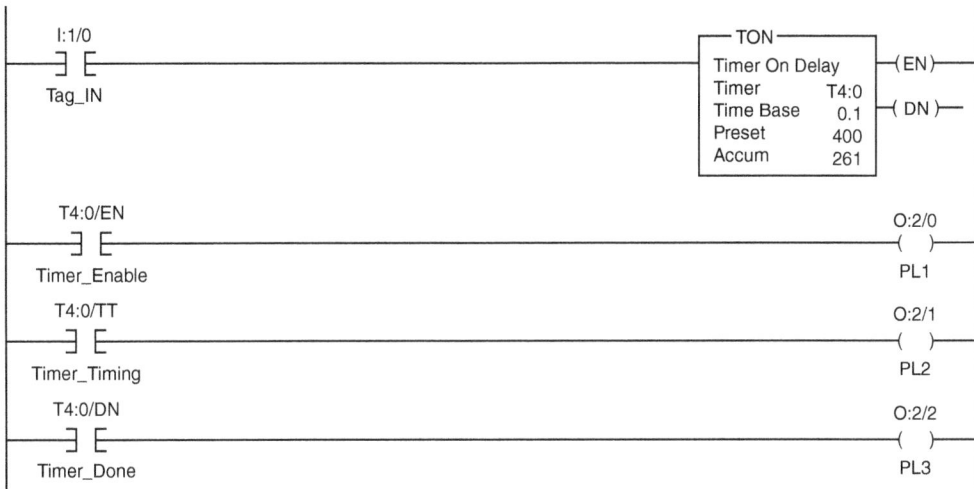

FIGURE 3.41   Ladder program for problem 17.

16. Write a ladder logic program using ON-delay retentive timer and a limit switch to keep track of solenoid valve activation. A PB with tag name "Cycle_Done" is used to reset the timer.

17. What is the status of PL1, PL2, and PL3 (ON/OFF) in Fig. 3.41 for the following conditions:

    a. T4:0.ACC is less than T4:0.PRE and the TAG_IN is ON?
    b. T4:0.ACC is equal to T4:0.PRE and the TAG_IN is ON?

18. Repeat problem 3.17 using TOF instead of TON and answer the following:

    a. What is the status of PL1, PL2, and PL3 if TAG_IN is off?
    b. What is the status of PL1, PL2, and PL3 when TAG_IN transitions from off to on and T4:0.ACC is less than T4:0.PRE?

19. Study the rungs in Fig. 3.42 assuming that the counter preset value (C5:0.pre) holds the value of 10. Answer the following questions:

    a. What is the counter-accumulated value when PB3 is pressed?
    b. What is the counter-accumulated value when PB1 is activated 15 times and PB2 is activated 12 times?
    c. What is the counter-accumulated value when PB4 is pressed?
    d. Under what conditions will PL1 turn on?

20. Write a PLC program to flash a pilot light (PL) ON/OFF. The pilot light (PL) is ON for 5 seconds and OFF for 3 seconds.

    Hint: Use two TON timers.

21. The rungs shown in Figure 3.43 were programmed to indicate Pump fail to start alarm (Output O:2/7) when system is in AUTO, PUMP output is enabled, and the pump running input is not received in 5 seconds. When the rung was simulated, the alarm was OFF under the above conditions. Explain why and fix the problem?

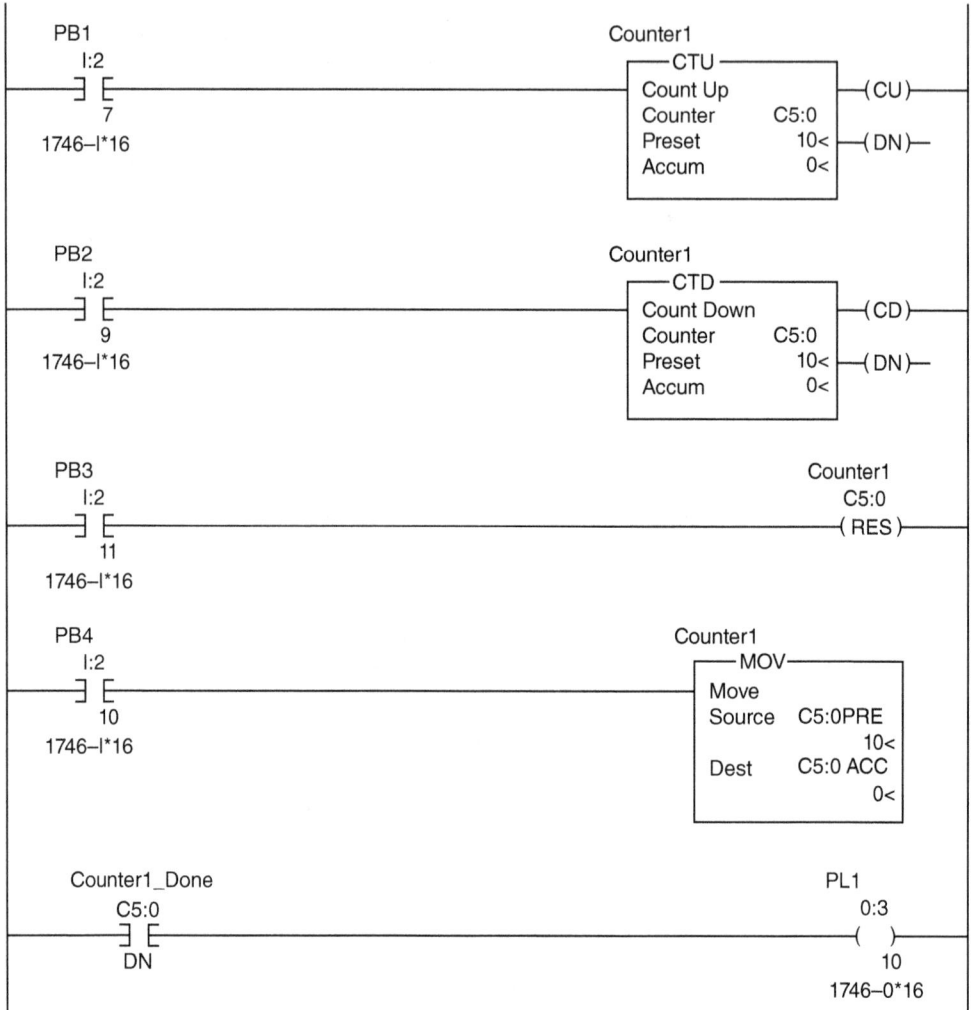

Figure 3.42   Up-/down-counter ladder for problem 19.

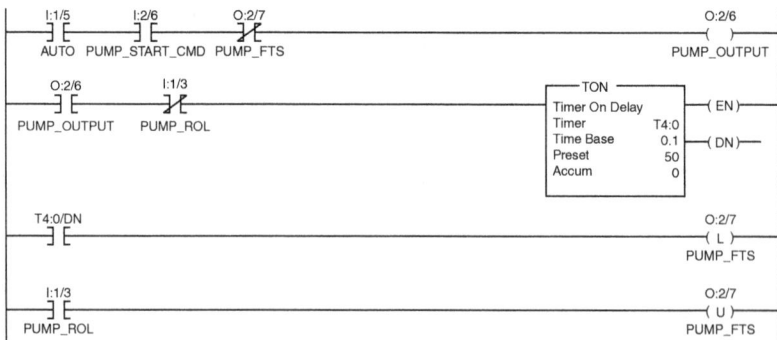

Figure 3.43   Problem 21 cooling system.

22. Write a PLC program to allow the operator to run the conveyor line for luggage transportation by pushing a START PB. The START PB is a normally open momentary switch, and the STOP PB is normally closed and wired high. The following is a brief process description:

    a. When pushing START PB, flash a pilot light (PL) every 2 seconds to warn people that the conveyor belt is about to start.
    b. After 80 seconds, start the conveyor motor and turn OFF the warning light.
    c. The operator stops the conveyor line by pushing the STOP PB.

23. A parking lot has two momentary action sensors to count the number of cars entering or exiting the garage. One sensor is placed in the entrance and the other is placed in the exit. Two messages should be displayed to customers, indicating the status of the parking lot ("Parking Full" or "Parking Not Full"). "Parking Full" is simulated by PL1 and "Parking Not Full" is represented by PL2. Write and document a ladder logic program to implement the above process.

24. Modify problem 19 to count the number of luggage in the conveyor line and stop the motor after 100 luggage items have passed. Add sensors as needed.

25. What is the status of PL1 and PL2 in Fig. 3.44 after PE1 transitions seven times?

26. Study the program in Fig. 3.45 and answer the following:

    a. What is the value of the C5:0.ACC after 10 transitions of the PE1 and 3 transitions of the PE2 instruction?
    b. What is the value of the C5:0.ACC after Reset PB is activated?

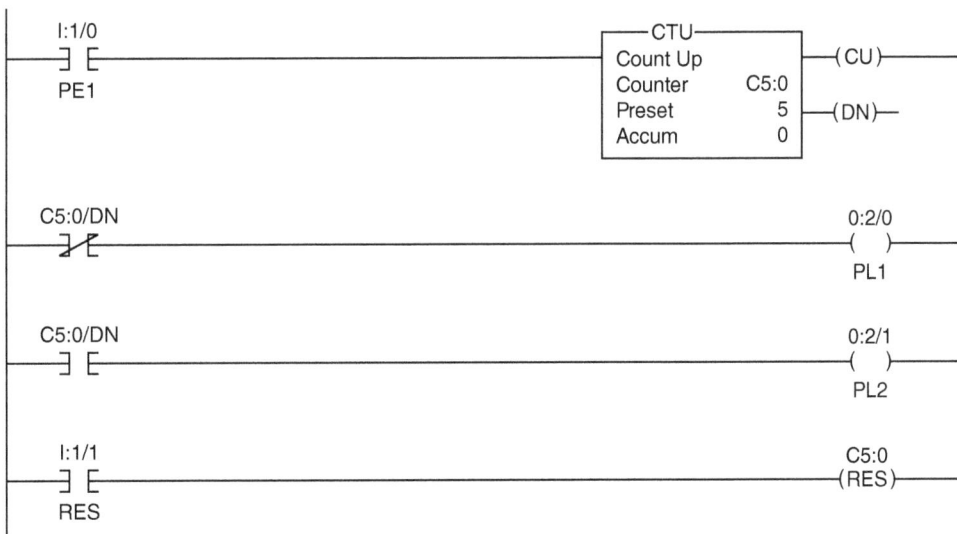

Figure 3.44 Problem 25 ladder program.

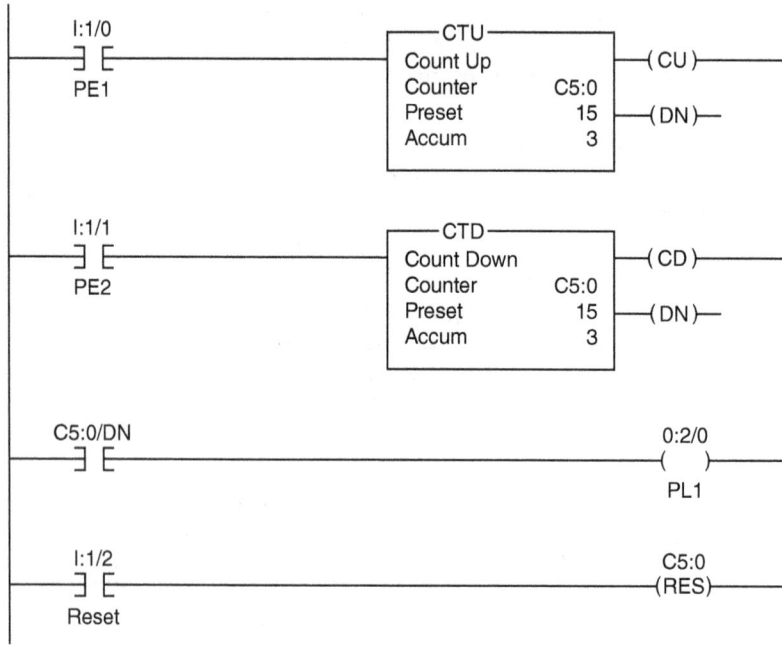

FIGURE 3.45 Problem 26 ladder program.

Laboratory 3.1—Merry-Go-Round

This laboratory will allow students to use timers with interlocking requirements and more complex sequencing.

This laboratory assumes that there are four motors represented by four pilot lights on the LogixPro. A START PB switch is used to start a sequence of pilot lights operation, which can be terminated at any time by pressing a STOP PB switch. The sequence starts by PL1, then PL2, then PL3, and then PL4. The same sequence will repeat until the process is stopped. Each selected PL will turn ON for a period of 5 seconds while the other PLs are OFF.

Requirements:

- Write a documented ladder logic program using ON-delay timers to realize the required timing sequence.
- Assign the I/O addresses and the bit addresses.
- Provide the laboratory checkout.

Laboratory 3.2—Machine Tool Operation

The objective of this laboratory is to get the students familiar with basic timing instructions used to control a machine tool operation.

A machine tool consists of five stations: robot pickup and placement of parts (station 1), pallets feed (station 2), stationary paint (station 3), inspection (station 4), and rejection (station 5) stations. Station 1 operation starts by a robot unloading the finished parts and placing the unfinished parts. Station 2 regulates the feed rate of the pallets, which carry the parts on the conveyor line. Station 3 is a paint work station. Its function is to spray paint the edge of a printed circuit, the unfinished part. Station 4 is the inspection station, which has 10 piano fingers—5 up and 5 down. Its function is to check on the sprayed paint by closing the up and down piano fingers on the edge of the printed circuit and report 1 or 0 based on the paint status. For good paint, a 1 should be reported. The main function of station 5 is to reject the badly painted parts after receiving a signal from station 4. Station 3 is the focus of this laboratory and is described next.

Station 3: Spray Paint Operation

A part-in-place limit switch (LS1) is located at the station to indicate that a part exists at the workstation. A solenoid valve (SV1), when energized, will raise the pallet to the level of the spray paint chamber. A limit switch (LS2) is placed in the upper location, and when closed, it indicates that the part is up at the level of the spray paint chamber. A solenoid valve (SV2), when energized, will spray the part for 5 seconds. When the spray paint cycle is done, SV2 is released to stop the spray painting and the solenoid valve (SV1) should de-energize to lower the pallet to the level of the conveyor belt. Conveyor belt and feed rate control are not the parts of this laboratory.

Requirements:

- Write a documented ladder logic program for station 3 (paint work station) to raise the pallet that carries the printed circuit, maintain in the paint chamber for 5 seconds, and lower the painted part to the conveyor belt.
- Assume and use any sensors required to control the spray paint process.
- Apply the concept of "what if scenarios" to check on and resolve any unusual conditions.
- Provide the checkout.
- Modify this laboratory to do the following:
  - Use Auto/Manual selector switch for station 3.
  - Energize the trap solenoid before raising the part, which prevents the current part from moving and exiting station 3. At the same time, activate the stop solenoid valve, which prevents new parts from entering the station.
  - When the spray paint cycle is done and the solenoid valve (SV1) is de-energized to lower the pallet to the level of the conveyor belt, release the trap solenoid to allow the finished parts to leave the station. At the same time, de-energize the stop solenoid to allow a new part to enter the station.

Discrete Inputs

1. Station 3 Auto/Manual selector switch.

Station 3 clear sensor to indicate that a finished part has exit the station.

Discrete Outputs

1. Station 3 pre-stop solenoid to prevent a new part from entering the station before the finished part exits the station.

Station 3 traps solenoid to trap the part before raising it to the painting chamber.

Laboratory 3.3—Pump Fail to Start Alarm

The objective of this laboratory is to get users familiar with the methods used to trigger an alarm if a motor fails to start using the TON instruction. Four selector switches and two pilot lights are used in this process simulation. SS1 is the pump remote selector switch. SS2 and SS3 are the pumps of high- and low-pressure discrete sensors. SS4 stays off if the pump runs within 5 seconds from the start action. PL1 is a discrete output indicator simulating the Pump Start/Pump Stop.

Requirements:

- List the conditions for successful Pump Start/Pump Stop as shown in Fig. 3.46.
- Add pump does not fail to start as a condition for pump 1 output to continue on. Output to the pump must be turned off once PUMP1_PFS_PL2 changes to the on status. Reprogram the ladder logic to get it to work; explain why.
- Download the program and provide the checkout.

Laboratory 3.4—Vertical Gate Monitoring

A vertical gate is equipped with one motor and two limits switches, LS1 and LS2; the gate is moving in the upper and lower direction. When gate is fully raised, LS1 is closed and LS2 is opened, and when gate is fully lowered, limit switch's, logic is reversed. A running on line input signal is true while the motor is running. This input is false when the motor stops or fail to start.

Requirements: Develop a documented ladder logic program to do the following:

- Turn on green pilot light indicating that gate is fully raised.
- Turn on red pilot light indicating that gate is fully lowered.
- Flash orange pilot light indicating that gate did not reach the fully raised or the fully lowered position in 120 seconds.
- Document your program using logic diagram.

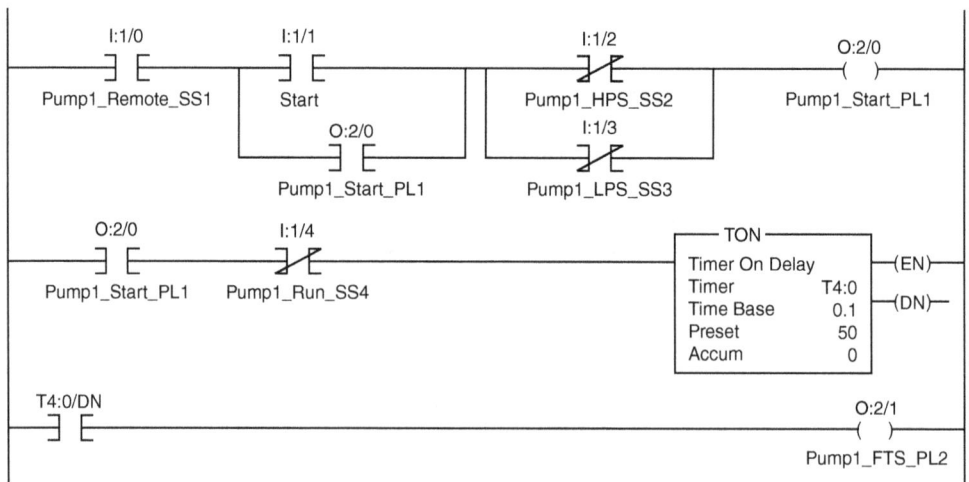

Figure 3.46 Laboratory 3.3 pump fail to start.

Laboratory 3.5—Cooling System Control

The objective of this laboratory is to get student familiar with the use of pulse timers in an industrial application.

In a cold room, Fig. 3.47, the temperature must be maintained below 0°C. Temperature fluctuations are monitored by means of a sensor. If the temperature rises above 0°C, the cooling system is switched on for a predetermined time. The cooling system on lamp is lit during this time. The cooling system and the lamp are turned off if one of the following conditions is fulfilled:

- The sensor reports a temperature below 0°C.
- The preset cooling time has elapsed.
- The STOP PB switch has been pressed.
- If the preset cooling time has expired and the temperature in the cold room is still too high, the cooling system can be restarted by means of the Reset PB switch. Hint: Use TP (generate pulse instruction). Figure 3.48 shows the system addresses for controlling the room temperature.

Requirements:

- Write a documented ladder program using 30 seconds for the cooling system preset time.
- Download the program and provide checkout.
- Flash the cooling system on lamp to indicate that the cooling system is in progress.

Laboratory 3.6—OFF-Delay Control of Three Motors

A simple process control implementation uses three OFF-delay timer (TOF) instructions. A limit switch tag name "LS1" and three outputs, motor 1 "O: 2/0," motor 2 "O: 2/1," and motor 3 "O: 2/2." Timer preset values of 10, 20, and 30 seconds are used to sequence operation of the three motors. When LS1 is true all three motors start running. If LS1 changes from high to low (OFF state), motor 1 turns OFF after 10 seconds, motor 2 turns OFF after 20 seconds, and motor 3 turns OFF after 30 seconds. Once the LS1 returns back to the true state, all three timers' accumulated values are cleared.

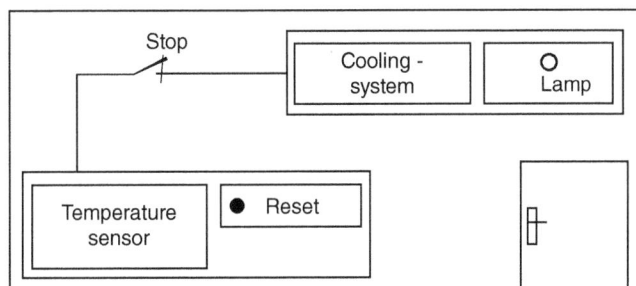

FIGURE 3.47 Laboratory 3.5 cooling system.

Sensor	I:1/0	Temperature sensor signal.
Reset	I:1/1	Restart.
Stop	I:1/2	The cooling system is switched off.
Max cool time	T4:0.PRE	Predetermined cooling time. This tag is defined in the timer preset word.
Curr cool time	T4:0.ACC	Currently elapsed cooling time. This tag is defined in the timer accumulated word.
Lamp	O:2/0	The "cooling system on" lamp is switched ON.
Cool system	O:2/1	The cooling system is switched ON.

FIGURE 3.48   Addresses for cooling temperature (assuming T4:0 timer is used).

Requirements:
- Write a documented ladder logic program using three OFF-delay timers.
- Assign the I/O addresses.
- Provide the laboratory checkout.

Laboratory 3.7—Pump Start/Stop Control for Predefined Calendar

The objective of this laboratory is to familiarize students with RTO timers to keep track of the running time for a pump running on calendar till the calendar expires.

Two float sensors provide indication for water-level (low and high) limits. A pump is to run when water is above high limit and keeps running till the water goes below the low limit (Schmitt trigger logic) or the time to run the pump expires.

Requirements:
- Write a documented ladder logic program.
- Assign the I/O addresses.
- Provide the laboratory checkout.

Laboratory 3.8—Conveyor System Control

The objective of this laboratory is to familiarize users with the basic timing and counting instructions used in conveyor system control.

A part is moving on a conveyor line crossing a photoelectric cell. The function of a photoelectric cell is primarily to count parts. The line stops after 10 parts are counted. The following is the sequence for this control operation:

The STRAT PB starts the conveyor motor after 5 seconds delay. The motor must start only if the Auto/Man switch is placed in the Auto position. The photoelectric cell should count only when the motor is running.

- When the count reaches 10, stop the conveyor after a 2-second delay.
- An ON-pilot light indicates the end of the sequence.
- The STOP PB switch, when activated, takes the system to initial condition.
- The operator can restart the same sequence by activating the START PB switch.

Requirements:

- Assign the system inputs and outputs.
- Enter the ladder rungs shown in Fig. 3.49.
- Add comments to each rung.

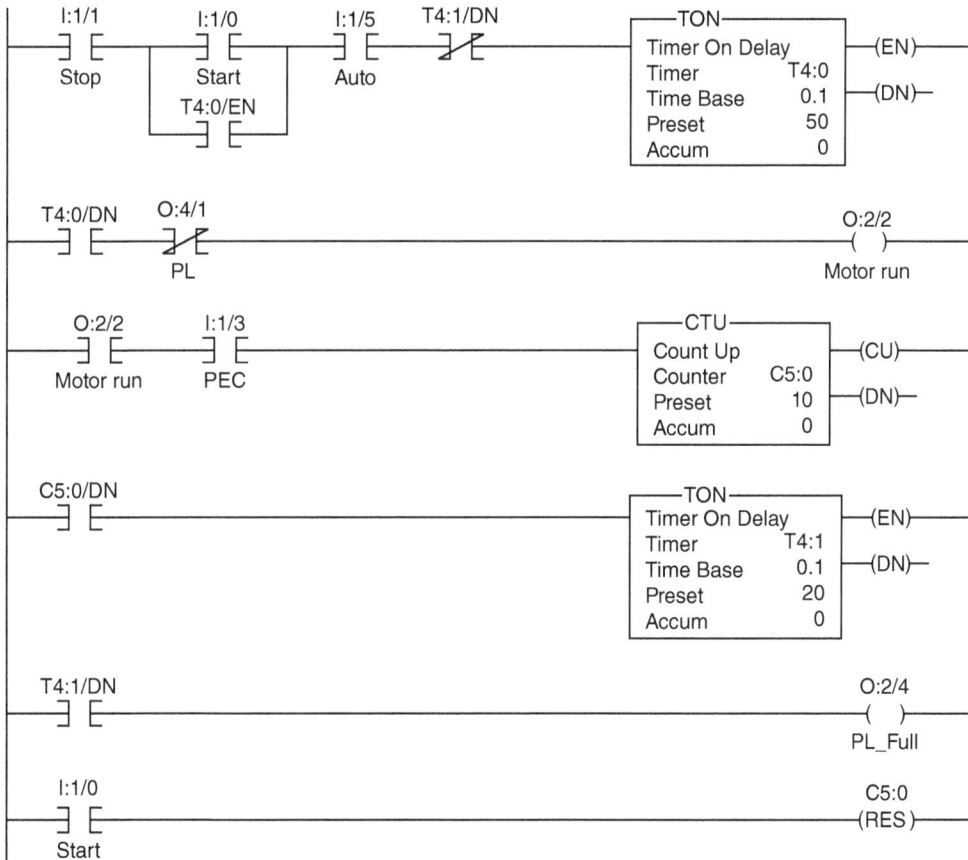

FIGURE 3.49 Laboratory 3.8 ladder program.

- Download the program and go online.
- Simulate the program using the LogixPro I/O simulator and verify whether the program is running according to the process description.

Modify this laboratory such that the part count is retained if the operator activates a new switch, Halt, while the motor is running. The Halt switch stops the motor and halts the counter operation. START PB activation will start the motor and resume the part count. Hint: The MOV (move) instruction covered in Chap. 4 can be used to retain the part count.

Use SS1 for the photoelectric cell input signal, SS2 for the motor running input indication signal, PL1 for the motor starting output signal, and PL2 for the pilot light output signal.

CHAPTER 4

Math, Move, and Comparison Instructions

T his chapter details the three classes of ladder instructions: mathematical, comparison, and move. Numbering systems and representation in the SLC-500 PLC system are covered briefly.

A 1973 electron microscope.

Chapter Objectives

- Understand math instructions.
- Understand move and comparison instructions.
- Be able to implement industrial control applications.
- Be familiar with the SLC-500 PLC development software.

In this chapter, we discuss three classes of PLC ladder instructions: the mathematical, comparison, and move instructions. It complements our coverage of logic and input/output instructions in Chap. 2. It also follows our detailed discussions of timers and counters in Chap. 3. The coverage is not intended to be comprehensive and repetitive, but it lays the ground work for the reader to be able to independently study, comprehend, and use other instructions in any application. The coverage of mathematical instructions is preceded by a short review of numbering systems and numbers representation in the SLC-500 PLC system.

4.1 Math Instructions

Mathematical operations are essential part of digital process control and automation. These operations are performed over input values from analog measurements or user interfaces. It is used to execute control strategies leading to the desired outputs for the regulation of the control process and associated user interface displays. Analog input/output scaling and validation is one of the typical examples of mathematical instruction use.

4.1.1 Numbering Systems

The commonly used decimal number system is based on module 10 representation. The system accommodates both integers and real numbers. Real numbers have an integer part and a fraction part separated by the decimal point. The integer part digits have increasing weight as we move to the left of the decimal point, while the weights decrease for the fraction digits as we move to the right of the decimal point. Consider the decimal number 9623.154, which is equivalent to the following sum:

$$9623.154 = 3*10^0 + 2*10^1 + 6*10^2 + 9*10^3 + 1*10^{-1} + 5*10^{-2} + 4*10^{-3}$$

A normalized scientific representation is normally used for real numbers as follows:

9623.154 = 0.9623154*10^4; the fraction part (0.9623154) is called *mantissa*, while the power of 10 (4) is called *exponent*. This system assumes infinite range and unlimited precession, which can neither be implemented in limited resource systems as in digital computers nor is needed in practical real-time control application.

PLCs as well as digital computers use the binary number system or module 2 arithmetic. Numbers, integers or real, are represented in fixed-size memory word. It is typically 1, 2, 4, or 8 bytes long. The size is either assigned to the default standard or can be altered through available user declaration. We will assume a 1 byte (8-bits) representation to demonstrate this binary system for unsigned and signed integers. Octal (base 8) and hex-decimal (base 16), which are also part of our discussion, are just derivatives of the binary system conveniently used in documentation. The following is a summary of integer's representation:

- Unsigned integer presentation (range from 0 to 255), $(97)_{10} = (0110\ 0001)_2$ and $(1110\ 0001)_2 = (225)_{10}$
- Signed integer presentation (range from −128 to 127), $(97)_{10} = (0110\ 0001)_2$ and $(-97)_{10} = 2$'s complement of $(0110\ 0001)_2 = (1001\ 1110)_2 + 1 = (1001\ 1111)_2$

The following is an example illustrating the requirement of matching operands in arithmetic and move instructions, both in type and size. It is also important when communicating arithmetic data among different communicating nodes:

Example 4.1.1.1 Two signed integers are represented in different formats, the first in a 2-byte word and the second in a 4-byte word. Before we can add up the two integers, the first signed integer needs to be converted into a 4-byte signed integer. The conversion is straight forward if the integer is positive; we just add insignificant 16 zeroes to the integer original representation (hex 0000 0013 for 0013 hex). Notice that the most significant bit in the one-word original integer is zero, indicating a positive integer. What if the integer is negative? In this case we need to represent this negative integer in 32 bits using the two's complement format. Let us assume that the value is −19, which is represented originally as the two's complement of 19 (hex FFED). Representing −19 in 32 bits would produce the hex value FFFF FFED. Notice that the most significant bit is one, signifying a negative integer. So for negative integers, we added an FFFF significant word to the original integer representation.

Normalized real numbers are decomposed into two components: mantissa and exponent. Exponent uses the signed integer representation, mantissa uses binary fraction presentation, and a single bit is used to indicate the real number sign. Four bytes is the default for real numbers presentation for single precision and 8 bytes for double precision. More accuracy in the representation can be achieved through the double-precision declaration. The following is the IEEE 754 Standard for floating point format:

- Single precision: bit 0 to 22 is used for mantissa, 23 to 30 is the exponent, and 31 is the overall number sign (range of -10^{38} to 10^{38}).
- Double precision: bit 0 to 51 is used for mantissa, 52 to 62 is the exponent, and 63 is the overall number sign (range of -10^{308} to 10^{308}).

4.1.2   SLC-500 Data and Numbers Representation

Variables used in the SLC-500 can be declared as real, integer, or Boolean. The smallest memory unit is a bit, which can accommodate a Boolean variable as a discrete input/output. Eight consecutive bits form a memory byte, while 16 consecutive bits (2 bytes) constitute a memory word, which is often used to represent integers. Four consecutive bytes form a double word, which is used to accommodate real and some integer variables. Typical Boolean variables include discrete input/output, comparison flags, computational flags, and other intermediate logic variables. Each such variable occupies one memory bit and carries the value 0 (false/OFF) or 1 (true/ON).

Each integer uses one memory word (2 bytes) and each float/real number uses double memory words (4 bytes). Other types of data files used include timers, counters, control, status, inputs, and outputs. Figure 4.1a shows the 256 SLC-500 data files, whereas Fig. 4.1b illustrates the 256 possible program files: two system files (file 0 and file 1) and up to 253 user program files (file 2 to file 255). Data files and program files use two dedicated separate memory areas.

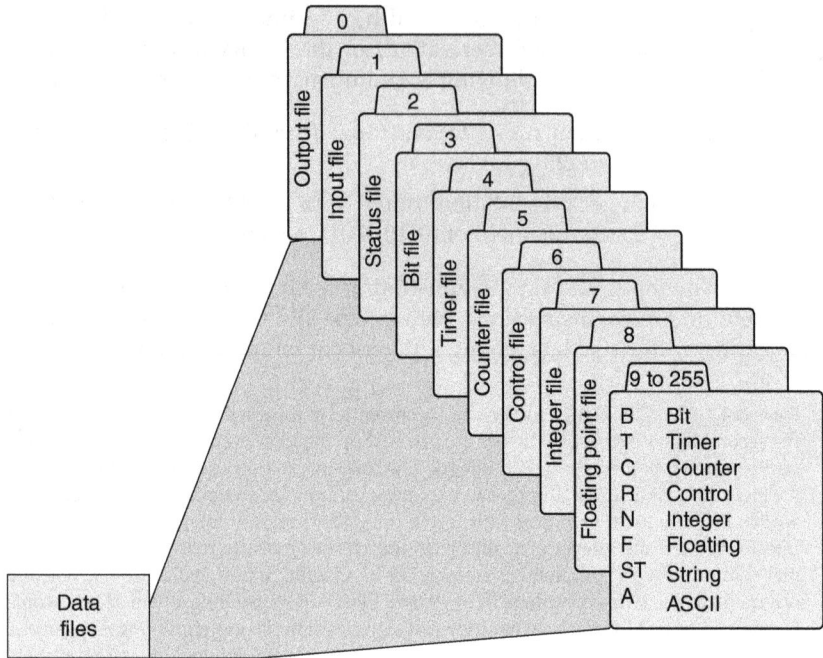

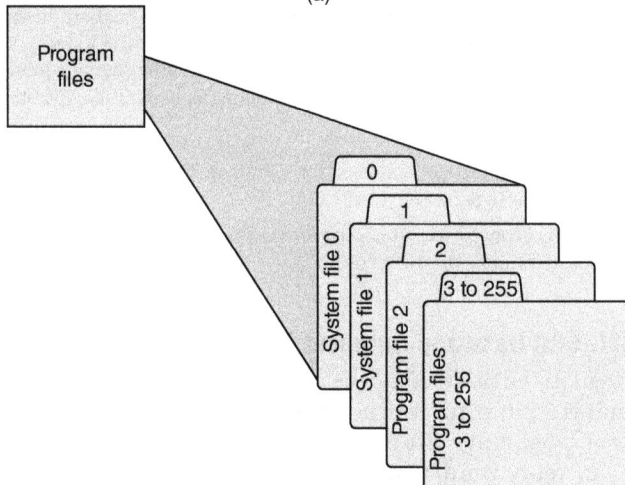

FIGURE 4.1 (a) SLC-500 data files; (b) SLC-500 program files.

The SLC-500 processor uses the little-endian microprocessor notation; the low byte of a 2-byte integer is stored first at the lower memory address and then the high byte stored at the next memory address. Table 4.1 shows available forms of data/variables representation in the AB SLC-500 PLC system. Each data type is assigned a unique mnemonic identifier. The address range defines the start address and the upper address for each type. These ranges are extremely large, but only limited subranges are used in

Data Type	Default File	Range
Output	O	4096 or more
Input	I	4096 or more
Status	S	4096 or more
Bit	B3	4096 1-bit elements
Timer	T4	256 three-word elements
Counter	C5	256 three-word elements
Control	R6	256 three-word elements
Integer	N7	256 one-word elements
Float	F8	256 two-word elements
Any of the above types	F9 to F255	As defined above per type

TABLE 4.1 SLC-500 allowed addresses and data types

most practical applications. The SLC-500 processor and associated I/O interfaces are versatile and can cover wide address ranges. Each of the 10 defined data types is discussed briefly next. The SLC 5/02 does not support floating point operations and has other limitations but still can support small process control applications at lower cost. File 9 to 255 can be defined and used by the application to allow additional addresses in any of the data categories as needed. Every file has to be defined to one type of data.

Table 4.2 summarizes the specification of the input (I) and the output (O) files in the SLC-500 system. Each slot starting with the one adjacent to the power supply (Slot 0) typically has 16 inputs or outputs, but it can have more points. The system can accommodate up to 30 I/O slots/cards. Table 4.3 lists the specification for the status file, which has 16 to 164 one-word elements, depending on the processor used. Elements or individual bits within an element are used to maintain information such as processor flags, scan time, initialization bits, and diagnostic.

Table 4.4 details the specification for the bit file. The default is file 3 (B3) with up to 4096 bits size. Elements in this file are internal memory reserved locations in addition to other defined critical discrete inputs/outputs. Additional bit files can be added using the available file numbers from 9 to 255. Table 4.5 documents the specification for the timer file (T4). There are up to 256 timers with three words allocated for each. The three words are accessible at the bit or word level. Table 4.6 shows the documentation for the counter file (C5), which is similar to the timer file memory specifications. Each timer and counter element requires the allocation of a control element from the control file (R6) discussed next.

Table 4.7 presents the same documentation for the control file (R6) with up to 256 elements. Each control element is a three consecutive memory words long. Tables 4.8 and 4.9 summarize the specification for the integer (N7) and float files (F8), respectively. Each file accommodates up to 256 elements with one and two words length each, respectively. Additional elements can be accommodated using the user-defined files 9 to 255.

The size of a PLC program in terms of the number of rungs, the number of I/O points used, and configured memory addresses defines the length of the PLC scan cycle time.

Format	Explanation		
O:e.s/b I:e.s/b	O	Output	
	I	Input	
	:	Element delimiter	
	e	Slot number (decimal).	Slot 0, adjacent to the power supply in the first chassis, applies to the processor module (CPU); succeeding slots are I/O slots, numbered from 1 to a maximum of 30
	.	Word delimiter.	Required only if a word number is necessary, as noted below.
	s	Word number	Required if the number of inputs or outputs exceeds 16 for the slot; range: 0–255 (range accommodates multi-word "specialty cards")
	/	Bit delimiter	
	b	Terminal number	Inputs: 0–15 Outputs: 0–15
Examples:			
O:3/15	Output 15, slot 3		
O:5/0	Output 0, slot 5		
O:10/11	Output 11, slot 10		
I:7/8	Input 8, slot 7		
I:2.1/3	Input 3, slot 2, word 1		
O:5	Output word 0, slot 5		
O:5.1	Output word 1, slot 5		
I:8	Input word 0, slot 8		

TABLE 4.2 Output/input file addressing format

This scan cycle time is associated with the process real-time requirements. During each scan cycle the CPU executes the entire ladder logic program and performs all hardware diagnostics. The longer the ladder programs scan time, the less frequently the process control system updates. In some cases the application dynamics are slow and scan time is not important. In other applications with fast process dynamics, scan time is critical and needs to be minimized. Also, systems with redundant PLC hardware will have greater scan time as a result of the housekeeping and duplication overhead. PLCs come in different sizes and processor speeds to accommodate different application requirements.

4.1.3 Common Math Instructions

This section covers the most commonly used mathematical instructions. This coverage includes addition, subtraction, multiplication, division, increment, decrement, and general equation calculation. Each instruction is individually covered as a block

Format	Explanation		
S:e/b	**S**	Status file	
	:	Element delimiter	
	e	Element number	Ranges from 0 to 15 in a fixed or SLC 5/01 controller, 0 to 32 in an SLC 5/02, 0 to 82 in an SLC 5/03 and 0 to 82 in an SLC 5/05, 0 to 96 in an SLC 5/04 OS400, and 0 to 163 in an SLC 5/04 OS401 processors. These are 1-word elements; 16 bits per element
	/	Bit delimiter	
	b	Bit number	Bit location within the element; ranges from 0 to 15
Examples:			
S:1/15	Element 1, bit 15; this is the "first pass" bit, which you can use to initialize instructions in your program		
S:3	Element 3; the lower byte of this element is the current scan time. The upper byte is the watchdog scan time		

TABLE **4.3** Status file addressing format

Format	Explanation		
Bf:e/b	**B**	Bit type file	
	f	File number; number 3 is the default file; a file number between 9–255 can be used if additional storage is required.	
	:	Element delimiter	
	e	Element number	Ranges from 0–255; these are 1-word elements. 16 bits per element
	/	Bit delimiter	
	b	Bit number	Bit location within the element; ranges from 0–15
Bf/b	**B**	Same as above	
	f	Same as above	
	/	Same as above	
	b	Bit number	Numerical position of the bit within the file; ranges from 0–4095
Examples:			
B3:3/14	Bit 14, element 3		
B3:252/00	Bit 0, element 252		
B3:9	Element 9		
B3/62	Bit 62		
B3/4032	Bit 4032		

TABLE **4.4** Bit file addressing format

T	Timer file	
f	File number; for SLC 500 processors the default is 4; a file number between 9 to 255 can be used for additional storage	
:	Element delimiter	
e	Element number	These are 3-word elements; the range is 0 to 255
.	Word Delimiter	Range 0 to 2
s	Word Number	
/	Bit delimiter	
b	Bit Number	Range 0 to 15
Examples		
T4:0/15 or T4:0/EN	Enable bit	
T4:0/14 or T4:0/TT	Timer timing bit	
T4:0/13 or T4:0/DN	Done bit	
T4:0.1 or T4:0.PRE	Preset value of the timer	
T4:0.2 or T4:0.ACC	Accumulated value of the timer	
T4:0.1/0 or T4:0.PRE/0	Bit 0 of the preset value	
T4:0.2/0 or T4:0.ACC/0	Bit 0 of the accumulated value	

TABLE 4.5   Timer addressing format

c	Counter	
f	File number; for SLC 500 processors the default is 5; a file number between 9 to 255 can be used for additional storage	
:	Element delimiter	
e	Element number	These are 3-word elements; the range is 0 to 255
.	Word Delimiter	
s	Word Element	0 to 2
/	Bit delimiter	
b	Bit Number	0 to 15
Examples		
C5:0/15 or C5:0/CU	Count up enable bit	
C5:0/14 or C5:0/CD	Count down enable bit	
C5:0/13 or C5:0/DN	Done bit	
C5:0/12 or C5:0/OV	Overflow bit	
C5:0/11 or C5:0/UN	Underflow bit	
C5:0/10 or C5:0/UA	Update accum. bit (use with HSC in fixed controller only)	
C5:0.1 or C5:0.PRE	Preset value of the counter	
C5:0.2 or C5:0.ACC	Accumulated value of the counter	
C5:0.1/0 or C5:0.PRE/0	Bit 0 of the preset value	
C5:0.2/0 or C5:0.ACC/0	Bit 0 of the accumulated value	

TABLE 4.6   Counter file addressing format

Format	Explanation		
Rf:e	**R**	Control file	
	f	File number; number 6 is the default file; a file number between 9 and 255 can be used if additional storage is required	
	:	Element delimiter	
	e	Element number	Ranges from 0 to 255; these are 3-word elements.
Rf:e.s/b	**Rf:e**	Explained above	
	.	Word delimiter	
	s	Indicates word	
	/	Bit delimiter	
	b	Bit	
Examples:			
R6:2	Element 2, control file 6 Address bits and words by using the format Rf:e.s/b		
R6:2/15 or R6:2/EN	Enable bit		
R6:2/14 or R6:2/EU	Unload Enable bit		
R6:2/13 or R6:2/DN	Done bit		
R6:2/12 or R6:2/EM	Stack Empty bit		
R6:2/11 or R6:2/ER	Error bit		
R6:2/10 or R6:2/UL	Unload bit		
R6:2/9 or R6:2/IN	Inhibit bit		
R6:2/8 or R6:2/FD	Found bit		

TABLE **4.7** Control file addressing format

with input enable condition, output enable signal, and input/output tags. Later in this chapter, we discuss an implemented application utilizing these mathematical instructions.

Conversion instructions are typically available with most microprocessors as part of the instruction set. They are also available in PLCs and can accommodate a wide variety of format change, which include conversions on integers, real numbers, and strings. Care should be taken when using such instructions in user's programs. The following general information applies to all math instructions:

Source is the address of the value on which the mathematical, logical, or move operation is to be performed. This can be word addresses or program constants. An instruction that has two source operands accepts constants in both operands but must have a destination register operand.

Destination is the address of the result of the operation. Signed integers are stored in two's complement format for both source and destination parameters.

Format	Explanation		
Nf:e/b	**N**	Integer file	
	f	File number; number 7 is the default file; a file number between 9 to 255 can be used if additional storage is required	
	:	Element delimiter	
	e	Element number	Ranges from 0 to 255; these are 1-word elements; 16 bits per element.
	/	Bit delimiter	
	b	Bit number	Bit location within the element; ranges from 0 to 15
Examples:			
N7:2	Element 2, integer file 7		
N7:2/8	Bit 8 in element 2, integer file 7		
N10:36	Element 36, integer file 10 (file 10 designated as an integer file by the user)		

TABLE **4.8**   Integer file addressing format

Format	Explanation		
Ff:e	**F**	Float file	
	f	File number; number 8 is the default file; a file number between 9 to 255 can be used if additional storage is required	
	:	Element delimiter	
	e	Element number	Ranges from 0 to 255; these are 2-word elements; 32 bits per element
Examples:			
F8:2	Element 2, float file 8		

TABLE **4.9**   Float file addressing format

4.1.3.1   ADD Instruction

Figure 4.2 shows the ADD instruction format. If (TAG_IN) is true, the ADD instruction is executed. The value of (N7:0) is added to the value of (N7:1). The result of the addition is stored in the destination of (N7:2). Notice that the addition repeats while TAG_IN is true, which affects the result if the destination is the same as one of the sources.

Example 4.1.3.1   This example shows the use of one-shot rising (OSR) to increment a register using the ADD instruction.

As shown in Fig. 4.3, if TAG_IN_INCR is true, the ADD instruction is executed. The values of N7:5 are incremented by 1. The OSR instruction is needed to avoid executing the ADD instruction more

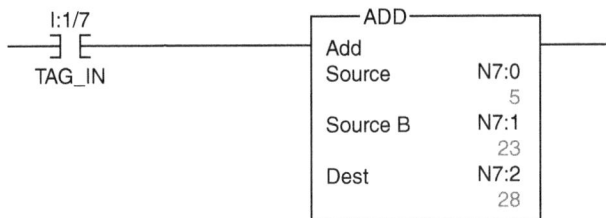

FIGURE **4.2** ADD instruction block.

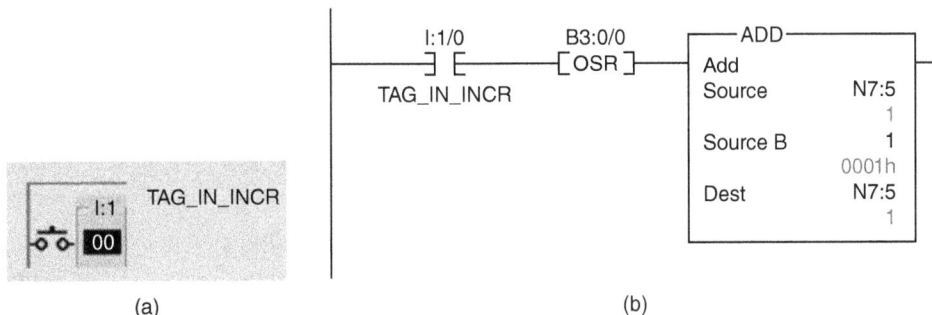

(a)   (b)

FIGURE **4.3** (a) ADD instruction panel configuration; (b) ADD instruction LogixPro 500 rung.

than once due to the bouncing action of the switch or the lasting of the pressed position for more than one scan. The OSR requires unique and unused bit address.

4.1.3.2 SUB Instruction

Figure 4.4 shows the SUB instruction format. If (TAG_IN) is true, the SUB instruction is executed. The value of (N7:1) is subtracted from the value of (N7:0). The result of the subtraction is stored in the destination of (N7:2).

Example 4.1.3.2 This example implements a register decrement using SUB instruction and an OSR. As shown in Fig. 4.5, if TAG_IN_DECR is true, the SUB instruction is executed. The value of N7:5 is decremented by 1. The OSR instruction is a must for the instruction to be executed correctly even when the input TAG_IN_DECR is configured as a push button (PB).

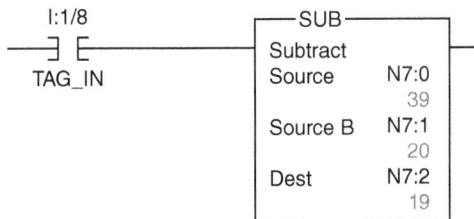

FIGURE **4.4** SUB instruction block.

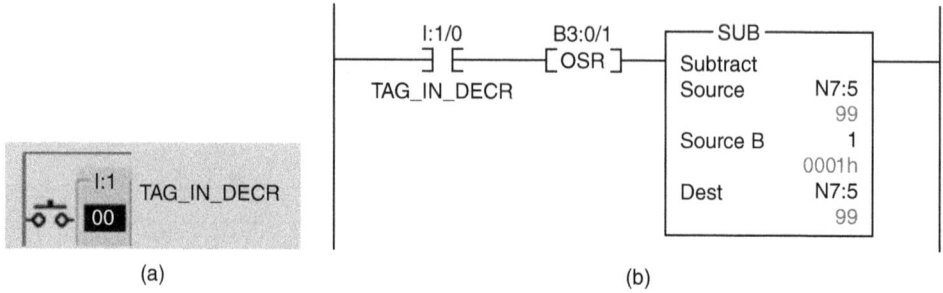

FIGURE 4.5   (a) SUB instruction panel configuration; (b) SUB instruction LogixPro 500 rung.

4.1.3.3   MUL Instruction

Figure 4.6 shows the MUL instruction format. If (TAG_IN) is true, the MUL instruction is executed. The value of (N7:0) is multiplied by the value of (N7:1). The result of the multiplication is stored in the destination of (N7:2).

4.1.3.4   DIV Instruction

Figure 4.7 shows the DIV instruction format. If (TAG_IN) is true, the DIV instruction is executed. The value of (N7:0) is divided by the value of (N7:1). The result of the division is stored in the destination of (N7:2).

4.1.3.5   Square Root Instruction

Figure 4.8 shows the SQR instruction format. If (TAG_IN) is true, the SQR instruction is executed. The positive square root of the value of (N7:10) is placed in the value of (N7:11).

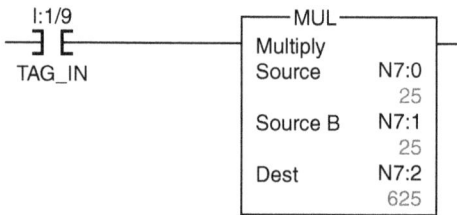

FIGURE 4.6   MUL instruction block.

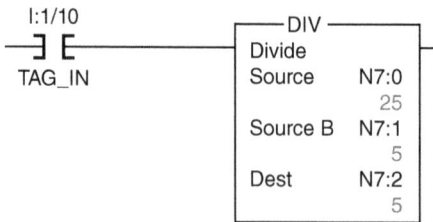

FIGURE 4.7   DIV instruction block.

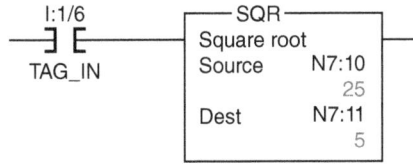

Figure 4.8   SQR instruction block.

4.1.4   Advanced Math Instruction

This section introduces additional mathematical instructions available on most PLC platforms. Few examples are discussed to enforce the reader's understanding and appreciation for real industrial automation implementation practices. Instructions covered include TOD (convert binary to BCD), FRD (convert from BCD to binary), and the SWP (swapping consecutive 2 bytes in each word of a string).

4.1.4.1   TOD Instructions

This instruction converts the binary value of a binary number that exists in the source register (N7:5) to its equivalent BCD value and places the result in the destination register O:6 (LogixPro 500 Panel BCD display).

Figure 4.9 shows the TOD instruction format. If TAG_IN is true, this output instruction converts the absolute value of a 16-bit integer source to BCD and stores it in the destination register. The instruction assumes unsigned integer source format. Figure 4.10 shows execution of the LogixPro 500 TOD instruction with the associated BCD panel display. Figure 4.11 demonstrates the data representation for the source binary value and the corresponding output BCD.

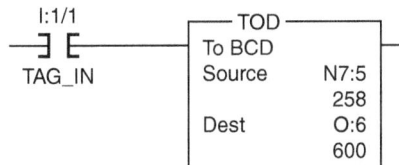

Figure 4.9   TOD instruction block.

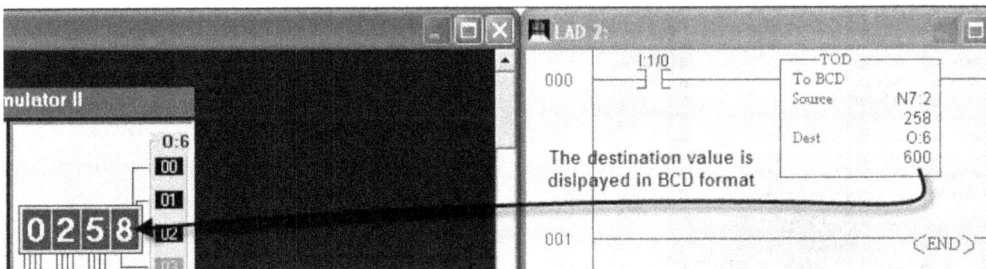

Figure 4.10   LogixPro 500 demonstration of the TOD instruction.

2	5	8	N7:2	Binary	0000	0001	0000	0010
				Decimal	258			
2	5	8	O:6	BCD	0000	0010	0101	1000
				Decimal	600			

FIGURE 4.11 Binary to BCD data format.

Example 4.1.4.1 A part is moving on a conveyor line crossing a photoelectric cell. The photoelectric cell function is to primarily count parts and display the count on a BCD panel. The conveyor line stops after 259 parts are counted. The operator can reset the system at any time by pressing the Reset PB.

Figure 4.12 shows the ladder logic implementation for the stated conveyor line application. The following is a brief explanation of the five rungs:

- Rung 0: Once START PB is pressed, the motor starts and the conveyor runs.
- Rung 1: While motor is running, the photoelectric cell (PE) triggers the part counting.
- Rung 2: The TOD instruction converts a 16-bit integer source value to the equivalent BCD and displays the result in the LogixPro 500 panel.
- Rung 3: When the total part count is equal to 259, the B3:0/0 turns true and the motor stops.
- Rung 4: Once the operator presses the manual Reset PB, the BCD panel will display a zero value and the counter will reset.

4.1.4.2 FRD Instruction

This instruction converts the value of a BCD number stored in the source register I:5 (panel BCD input) to the binary equivalent value and places it in the destination register N7:2.

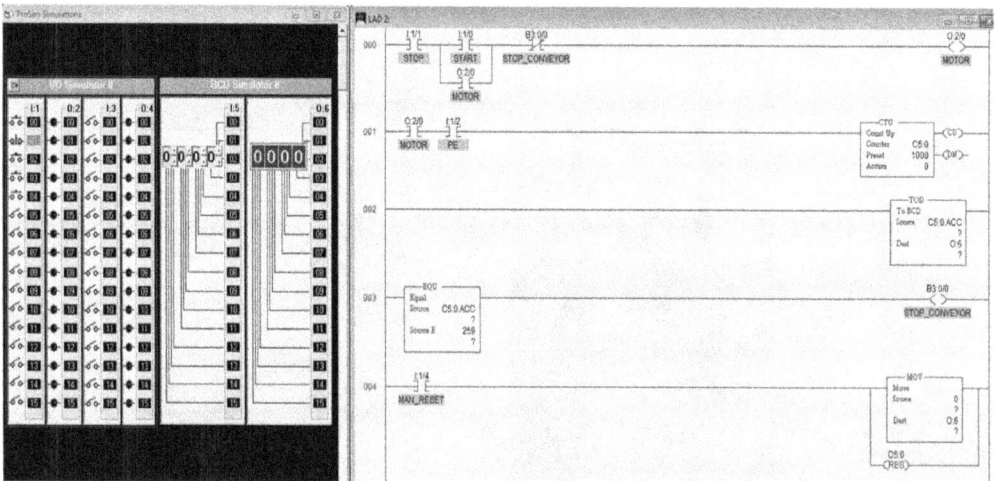

FIGURE 4.12 Example 4.1.4.1 ladder program.

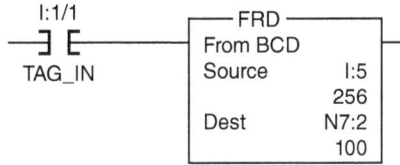

FIGURE **4.13** FRD instruction block.

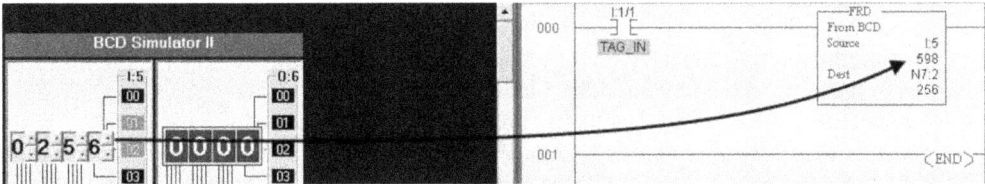

FIGURE **4.14** LogixPro 500 demonstration of the FRD instruction.

FIGURE **4.15** BCD to binary data format.

Figure 4.13 shows the FRD instruction format. While TAG_IN is true, this output instruction converts a BCD value in the source register or the input panel to a binary integer and stores it in the destination. Figure 4.14 shows the LogixPro 500 FRD instruction execution with the associated BCD panel display. Figure 4.15 demonstrates the data representation for the source BCD value and the corresponding binary output. A BCD value should be converted to binary before manipulation in the ladder program because the processor treats BCD values as binary numbers.

4.1.5 Swap Operation

It is commonly known that a 16-bit integer occupies two consecutive bytes of memory. Many people do not realize, however, that there is no standard order in which computers and instruments store these 2 bytes in memory or transmit them. As a result, computers that use one storage method cannot transmit integer data to devices that use another without first programmatically swapping each pair of bytes.

Although all 16-bit integers are stored in memory as two 8-bit bytes, these 2 bytes can be stored in two different orders. As an example, consider the decimal number 16, which in hexadecimal notation is written 0010 hex. The following 2 bytes make up this number:

00_{16}—most significant byte

10_{16}—least significant byte

The significance we have assigned to each byte is generally acceptable. However, the order in which these bytes are stored in memory varies from processor to processor. Computers based on the Intel 80 × 86 families of processors and later versions store the most significant byte in the higher address and the least significant byte in the lower address. Thus, on an Intel PC, if the integer 16 were stored at memory location 2000, the 10 hex (the least significant byte) would be stored in byte 2000 and the 00 hex (the most significant byte) in location 2001. This form of integer storage has been dubbed little-endian, because the least significant byte is stored in the low end of the memory word. Thus little-endian is known as the *Intel convention*.

Computers based on the Motorola 68000 family of processors and later versions, however, store integers in the opposite order. That is, in the preceding example, the most significant byte, 00 hex, would be stored at the lower address 2000, while the least significant byte, 10 hex, would be stored at address 2001. As you might expect, this form of storage has been named *big-endian*, because the most significant byte is in the lower address. Thus the big-endian is known as the *Motorola convention*.

If two computers exchange integer data, each must know the order in which the data will be sent and received; otherwise, the bytes might be interpreted in reverse order. In the preceding example, if an Intel machine transfers the integer value 16 in its native storage mode to a Motorola computer, which in turn interprets the 2 bytes in its own native storage mode, the Motorola computer would think it had received the integer 1000 hex, which is the decimal integer 4096. Siemens S7-1200 PLC uses the big-endian format whereas the AB SLC-500 PLC uses the little-endian format. Swap instruction can be used to reorder the 2 bytes before transmission to another processor, which uses the different-endian format.

Figure 4.16 shows the result of a swap operation on one word operand. The source value is a 2-byte integer stored in the TAG_IN_VALUE (55A2 hex). The result is shown in the destination TAG_OUT_VALUE (A255 hex). The same operation can be extended to the swapping of a two-word (4 bytes) integer. The SLC-500 uses the SWP instruction for swapping every 2 consecutive bytes (one word) in a string of arbitrary word length from 1 to 41 words.

The Swap instruction is not available in the RSLogix 500 or LogixPro 500. A simple one rung implementation of the Swap instruction is shown in the solution of homework problem 31, which is available in end of chapters HW solutions.

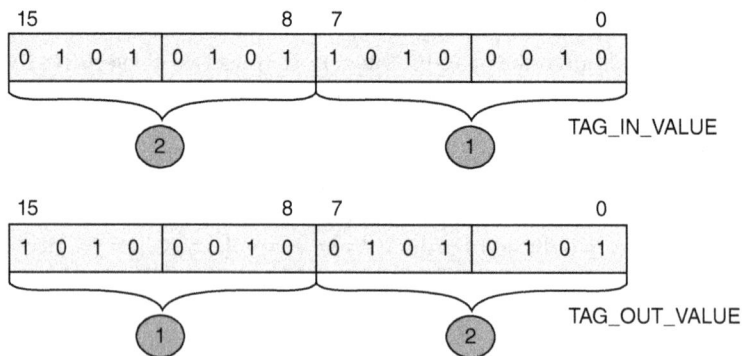

FIGURE 4.16 Swap operation result.

4.1.6   Arithmetic Status Bits

The arithmetic status bits are found in word 0, bits 0 to 3 in the controller status file as shown in Table 4.10. After an instruction is executed, the arithmetic status bits in the status file are updated (Table 4.10).

S:0/0 carry (C) sets if carry is generated; otherwise cleared. S:0/1 overflow (V) indicates that the actual result of a math instruction does not fit in the designated destination. S:0/2 zero (Z) indicates a 0 value after a math, move, or logic instruction. S:0/3 sign (S) indicates a negative (< 0) value after a math, move, or logic instruction. All arithmetic status bits can be used in the ladder program as conditions for a selected action. The four status bits are associated with the PLC processor and are available in the RSLogix software but are not available in the LogixPro 500 simulator. Next we discuss the overflow bit S:5/0 and the math register updates words: S:13 and S:14.

4.1.6.1   Overflow Trap Bit, S:5/0

Minor error bit (S:5/0) is set upon detection of a mathematical overflow or division by zero. If this bit is set upon execution of an END statement, a temporary end (TND) instruction, or an I/O refresh (REF), the recoverable major error code 0020 is declared. In applications where a math overflow or divide by zero occurs, you can avoid a CPU fault by using an output unlatch (OTU) instruction with address S:5/0 in your program. The rung must be between the overflow point and the END, TND, or REF statement.

4.1.6.2   Updates to the Math Register, S:13 and S:14

Status word S:13 contains the *least* significant word of the 32-bit value of the MUL instruction. It contains the remainder for the DIV and DDV (double division) instructions. It also contains the first four BCD digits for the Convert from BCD (FRD) and Convert to BCD (TOD) instructions. Status word S:14 contains the *most* significant word of the 32-bit value of the MUL instruction. It contains the unrounded quotient for DIV and DDV instructions. It also contains the most significant digit (digit 5) for TOD and FRD instructions.

Example 4.1.6.1   Assume that a MUL instruction execution with source A content (N7:72) equals 30,000 and source B (N7:71) content equals 2. The destination used is N7:72, then

$$N7:72 = N7:70 * N7:71 = 30,000 * 2 = 60,000$$

Notice that the result of this multiplication is >32,767 and will not fit in N7:72 (16 bits register), but the math register S14-S13 (32 bits or two words) will be able to hold on to

With This Bit		The Controller
S:0/0	Carry (C)	Sets if carry is generated; otherwise cleared.
S:0/1	Overflow (V)	Indicates that the actual result of a math instruction does not fit in the designated destination.
S:0/2	Zero (Z)	Indicates a 0 value after a math, move, or logic instruction.
S:0/3	Sign (S)	Indicates a negative (less than 0) value after a math, move, or logic instruction.

TABLE **4.10**   Arithmetic status bits

this value. So we look at the value of N7:72 = EA60 hex, which is a negative integer in a 16-bit two's complement representation of −5536 decimal.

Let us examine the value held in the math register S14-S13 (32 bits register). The value is 0000-EA60 hex, which is a positive integer in 32-bit two's complement representation of 60,000 decimal, which is the correct result.

Example 4.1.6.2 This example illustrates the DDV instruction and how it makes use of the math register. Assume that we want to divide the value held in the math register in the previous example by the DDV instruction source register N7:73 (6). The DDV instruction will take the value from the math register directly and divide it by 6. Now the result of this division (10,000) could be held in the destination register N7:74. After the execution of the DDV instruction, the unrounded quotient is placed in the most significant word of the math register S14. The remainder is placed in the least significant word of the math register S13.

Note: If the result of the DDV is > 32767 or <−32768, a minor fault will be generated.

4.2 Move and Transfer Instructions

Move instruction is used to copy a data element stored at a specified memory address to a new address location. Transfer instruction is used to move a block of data from a memory location to another in one step using declared arrays. Also, mathematical instructions can be done over arrays.

4.2.1 Move Instruction

Figure 4.17 shows the MOV instruction format. If (TAG_IN) is true, the MOV instruction is executed. The MOV instruction moves a copy of the source (N7:10) to the destination (N7:20) on each scan. The original value remains intact and unchanged in its source location.

Source is the address of the data you want to move. The source can be a constant.

Destination is the address that identifies where the data is to be moved. The destination has to be a defined register.

Example 4.2.1 This example shows a motor START/STOP using one PB. As shown in Fig. 4.18; when START/STOP PB is pressed, the ADD instruction is executed once and N7:0 will increment (least significant bit will toggle between 0 and 1), which cause the motor to change status between run and stop every time the PB is pressed.

Example 4.2.2 A flexible manufacturing conveyor system module produces three different products, but only one at any time interval. Each product is counted by a photoelectric cell PE1 mounted on the

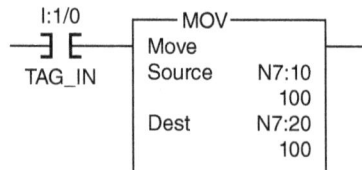

```
   I:1/0         ┌─── MOV ────┐
  ─┤ ├───────────┤ Move       ├──
   TAG_IN        │ Source  N7:10 │
                 │           100 │
                 │ Dest    N7:20 │
                 │           100 │
                 └─────────────┘
```

FIGURE 4.17 MOV instruction.

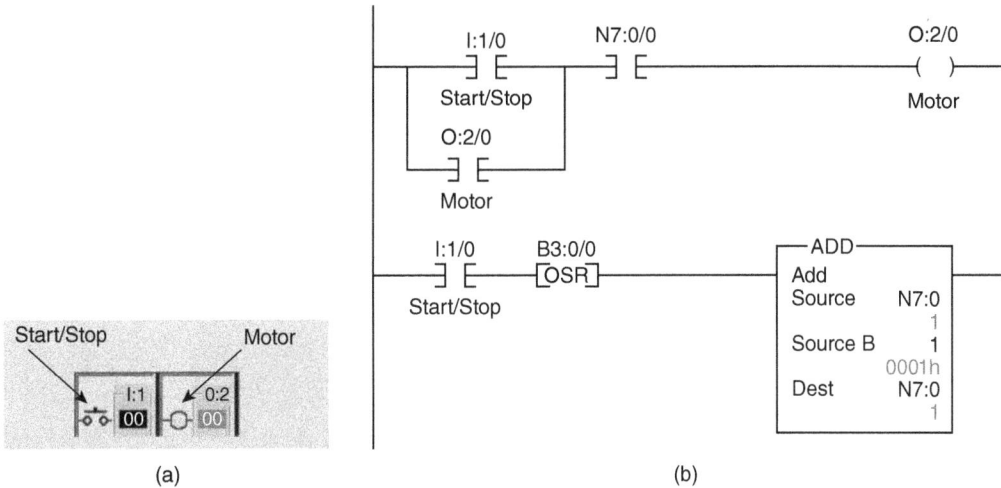

FIGURE 4.18 (*a*) Motor START/STOP panel configuration; (*b*) motor START/STOP ladder program.

conveyor line. Each batch processing is initiated through a dedicated normally open PB switches. Only one batch can be active at any time. A new batch can be started after the completion of the ongoing batch or by forcing a termination through the STOP PB. Product 1 is produced in 2000 amount, product 2 in 2000, and product 3 in 5000 total parts. A pilot light PL1 is ON after a batch product is completed. The process begins by pressing the START PB and can be terminated any time by pressing the STOP PB. A Reset PB is used to reset the counter accumulated value. Figure 4.19 shows the ladder implementation, which uses seven rungs.

4.3   Comparison Instructions

The most commonly used comparison instructions are covered in this section. Also included are two powerful instructions: the in range and the out of range.

4.3.1   Equal Instruction

As shown in Fig. 4.20, if (TAG_IN) is true, the EQU instruction is executed. The value of source A (N7:0) is compared to that of source B (N7:1). If they are equal, the output (TAG_EQU) is set.

4.3.2   Not Equal Instruction

As shown in Fig. 4.21, if (TAG_IN) is true, the NEQ instruction is executed. The value of source A (N7:0) is compared to that of source B (N7:1). If they are not equal, the output (TAG_NOT_EQU) is set.

4.3.3   Greater Than or Equal Instruction

As shown in Fig. 4.22, if (TAG_IN) is true, the GEQ instruction is executed. The value of source (N7:0) is compared to that of (N7:1); if the value of N7:0 is greater than or equal to the value of N7:1, the output (TAG_GREATER_OR_EQU) is set.

(a)

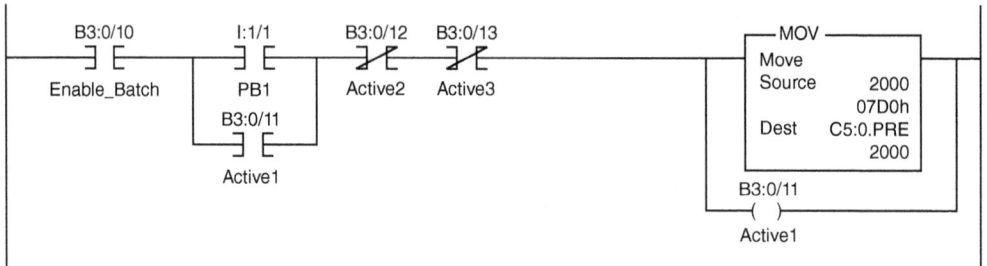

(b)

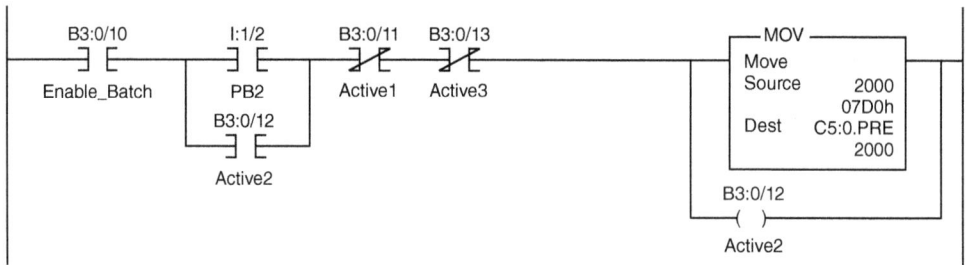

(c)

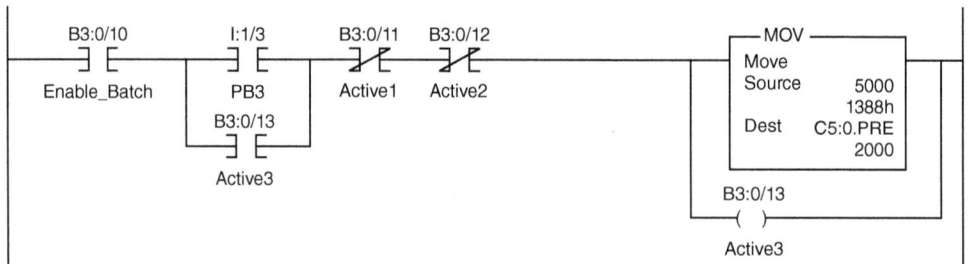

(d)

Figure 4.19   (a) Example 4.2.2 rung 1; (b) Example 4.2.2 rung 2; (c) Example 4.2.2 rung 3; (d) Example 4.2.2 rung 4; (e) Example 4.2.2 rung 5; (f) Example 4.2.2 rung 6; (g) Example 4.2.2 rung 7.

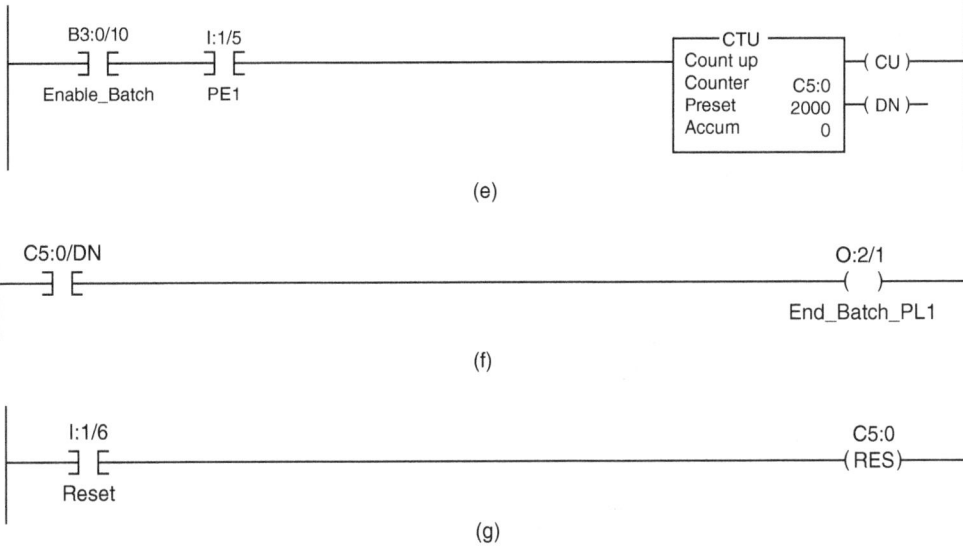

(e)

(f)

(g)

FIGURE **4.19** (*Continued*)

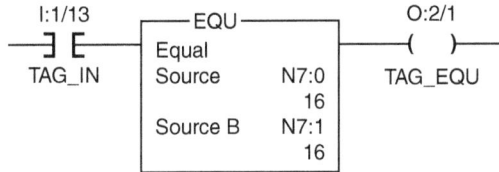

FIGURE **4.20** EQU instruction.

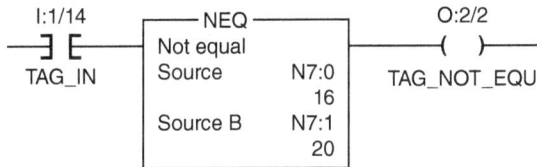

FIGURE **4.21** NEQ instruction.

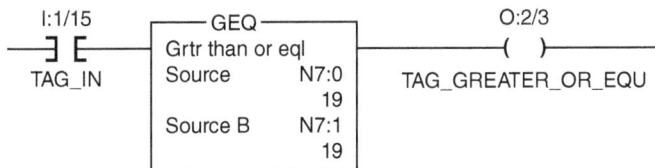

FIGURE **4.22** GEQ instruction.

4.3.4   Less Than or Equal Instruction

As shown in Fig. 4.23, if (TAG_IN) is true, the LEQ instruction is executed. The value of source A (N7:0) is compared to that of (N7:1); if the value of N7:0 is less than or equal to the value of N7:1, the output (TAG_LESS_OR_EQU) is set.

4.3.5   Greater Than Instruction

As shown in Fig. 4.24, if (TAG_IN) is true, the GRT instruction is executed. The value of (N7:0) is compared to that of (N7:1). If the value of N7:0 is greater than that of N7:1, the output (TAG_GREATER) is set.

4.3.6   Less Than Instruction

As shown in Fig. 4.25, if (TAG_IN) is true, the LES instruction is executed. The value of (N7:0) is compared to that of (N7:2). If the value (N7:0) is less than the value (N7:1), the output (TAG_LESS) is set.

Example 4.3.1   Figure 4.26 shows the formation of the MAX instruction (it does not exist in SLC-500) using the greater than and the move instructions. If TAG_IN is TRUE, the GRT and MOV instructions are executed. The value of N7:1 is compared with that of N7:2 and the maximum of the two values are stored in the value of N7:3.

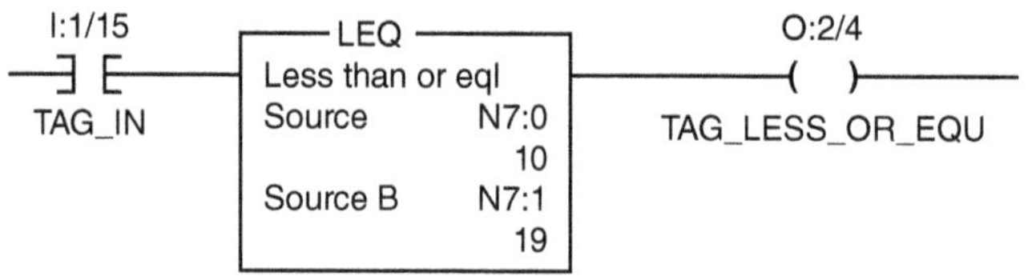

FIGURE 4.23   LEQ instruction.

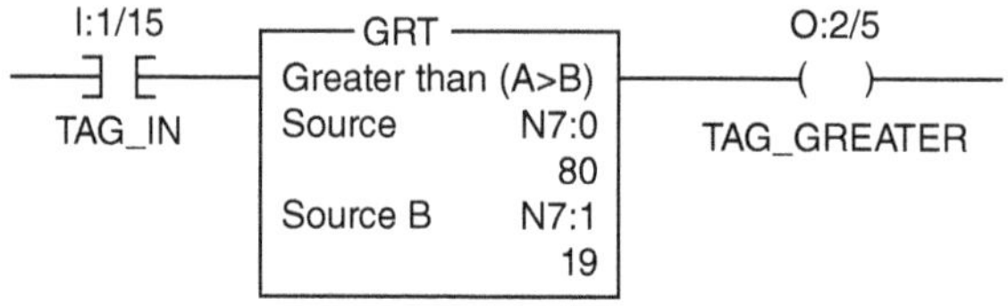

FIGURE 4.24   GRT instruction.

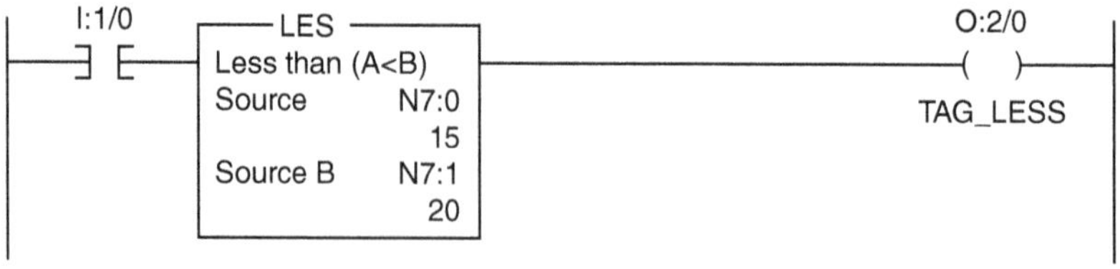

FIGURE 4.25   LES instruction.

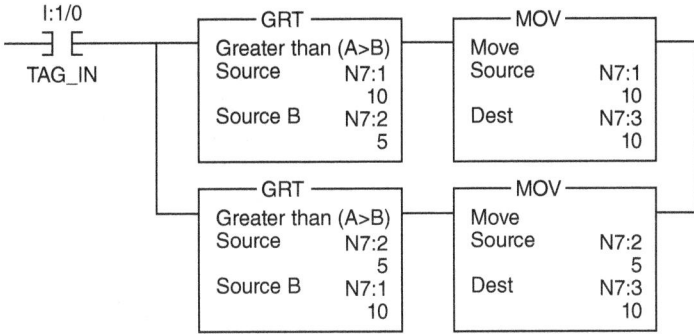

FIGURE 4.26 MAX formulated instruction.

Example 4.3.2 Figure 4.27 shows the formation of the MIN instruction (it does not exist in SLC-500) using the less than and the move instructions. If TAG_IN is TRUE, the LES and MOV instructions are executed. The value of N7:1 is compared with that of N7:2 and the minimum of the two values are stored in the value of N7:3.

Example 4.3.3 An operator set point is received from a BCD digital panel and validated between 50 and 400°F (oven temperature range). If the set point is outside the range, a pilot light flashes as an indication to the operator that the set point entered is invalid. Figure 4.28 shows the LogixPro 500 panel setup for this application. Figure 4.29 shows the ladder logic program realizing the stated requirements.

Rung 1: The FRD instruction converts the received set point entered by the operator from I:5 as a BCD (500 BCD) value and converts it to an integer and stores it in N7:10 (5*256 = 1280).

Rung 2: The greater than (GRE) instruction compares the value entered (500) to the oven high limit (400); if true, PL will flash by the rate of the status bit S:4/10. The less than (LES) instruction will also execute. The set point will be compared (less than) to the low limit (50). If true, the pilot light will flash indicating that set point is invalid.

Example 4.3.4 The previous example can be implemented more efficiently using the limit instruction instead of GRE and LES as shown in Fig. 4.30.

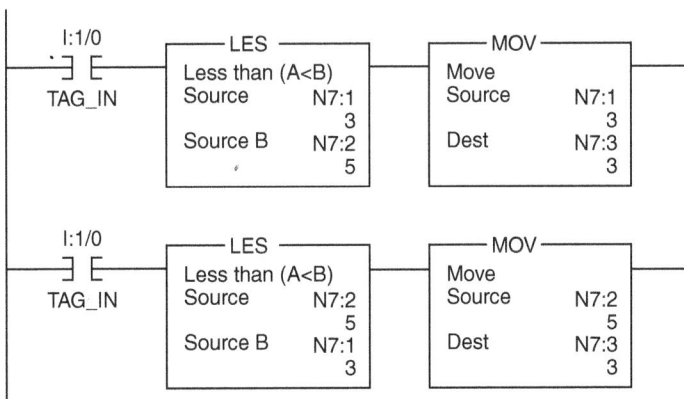

FIGURE 4.27 MIN formulated instruction.

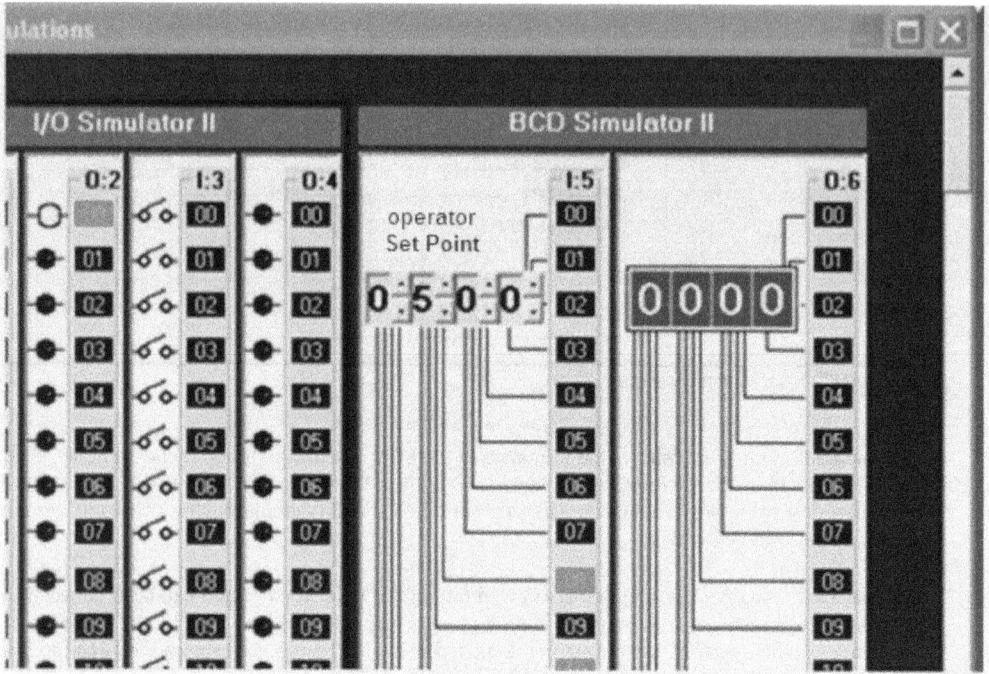

FIGURE 4.28   Oven set point validation panel setup.

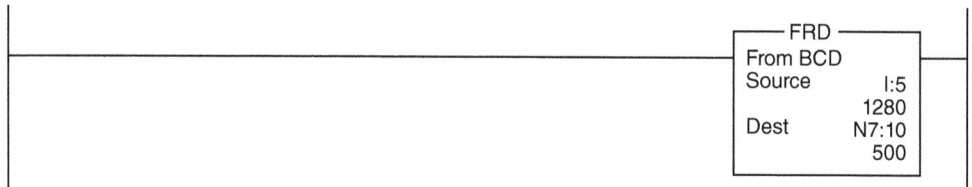

(a)

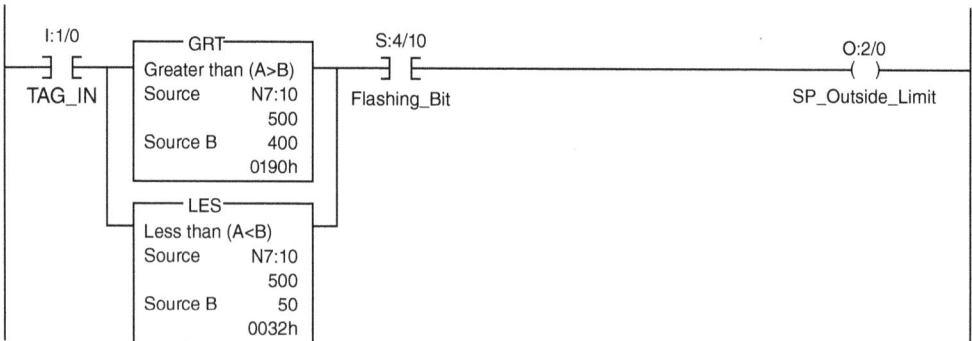

(b)

FIGURE 4.29   (a) Oven set point validation rung 1; (b) oven set point validation rung 2.

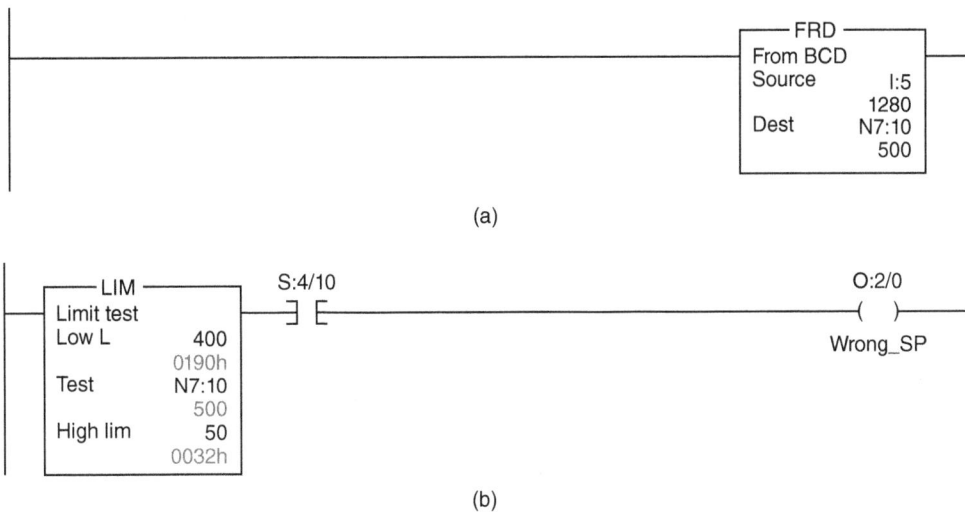

(a)

(b)

Figure 4.30 (a) Oven set point validation rung 1; (b) oven set point validation rung 2.

Rung 1: The FRD instruction converts the received set point entered by the operator from I:5 as a BCD value to a binary integer and stores it in N7:10.

Rung 2: The limit instruction will compare the set point in N7:10 (Test) to the low (Low L) and high (High Lim) limits, if set point is outside the configured range (greater than the Low or less than the High); the instruction output becomes true causing the pilot light to flash by the status bit S:4/10 rate indicating a wrong set point entry.

4.3.7 Masked Move Instruction

As shown in Fig. 4.31, while TAG_IN is true, the Masked Move (MVM) instruction is executed. Data from a source location (B3:0) is moved to a destination (B3:2), and allows portions of the destination data to be masked by a separate word (B3:1). Data at the source address passes through the mask to the destination address. The source is ANDed with the mask to produce the destination. As long as the rung remains true, the instruction moves the same data on each scan.

4.4 Program Control Instructions

Program control instructions are used to deviate from the sequential program execution, create a more efficient program, minimize scan time, and to troubleshoot a ladder program. These instructions include jump to a label (JMP, LBL), calling a designated subroutine (SBR, JSR, RET), and enabling or inhibiting a master control zone (MCR instruction) in the ladder program. The MCR instruction will be discussed in Chap. 6.

4.4.1 Jump Instruction

As in normal computer programming, it is desirable to be able to jump over certain program instructions or a section of code. The Jump instruction (JMP) is an output

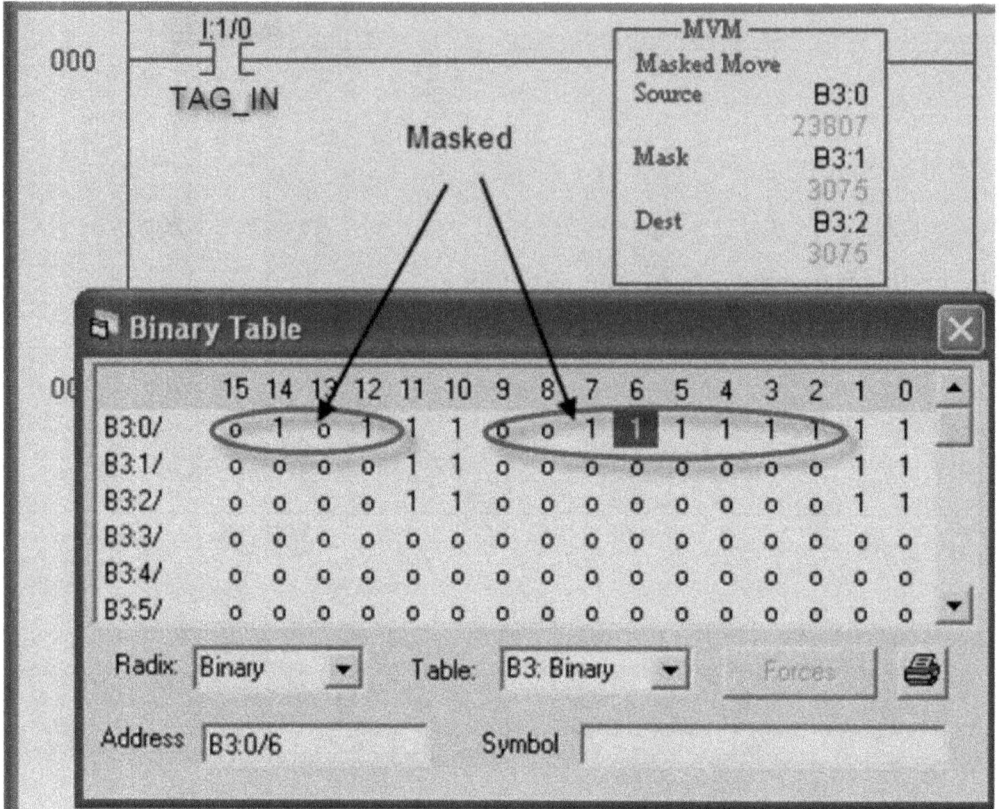

FIGURE 4.31 MVM instruction.

instruction used for this purpose. The advantages of the JMP instruction include the following:

- Sections of a program can be jumped when a production or a processor fault occurs.
- Reduce the processor scan time by jumping over instructions not pertinent to the process operation at that instant.
- The PLC can hold more than one program and scan only the program appropriate to operator requirements or instructions.

With Allen Bradley PLCs, the Jump (JMP) and the Label (LBL) instructions are used together so the scan can jump over a portion of the program. The *label* is a target for the *jump*; it is the first instruction in the rung and is always true. The *jump* and the label must have the same address, a bit address or a value from 0 to 255. You can also assign an optional same name tag for the selected address. If the JMP coil is energized, all logic between the JMP and the LBL instructions is bypassed and the processor continues scanning after the LBL instruction. By using the JMP instruction, you can branch or skip to different portions of a program and freeze all affected outputs in their last state.

Avoid jumping backward in the program too many times as this may increase the scan beyond the maximum allowable time. The processor has a watchdog timer that sets the maximum time for a total program scan, from start to end. If this time is exceeded, the processor will indicate a fault and shut down. You should never jump backward into an MCR zone. Instructions that are programmed within the MCR zone starting at the jump LBL instruction and ending at the end MCR instruction will always be evaluated as though the MCR zone is true, without consideration to the state of the start MCR instruction.

The program shown in Fig. 4.32 illustrates the JMP and LBL instructions and their application as documented in the LogixPro software. If TAG_IN in Rung 1 is true, the JMP instruction is executed and linear execution of the program is interrupted. Execution resumes in Rung 3 where label instruction exist. This will cause the OR logic in Rung 2 to be skipped while TAG_IN is true.

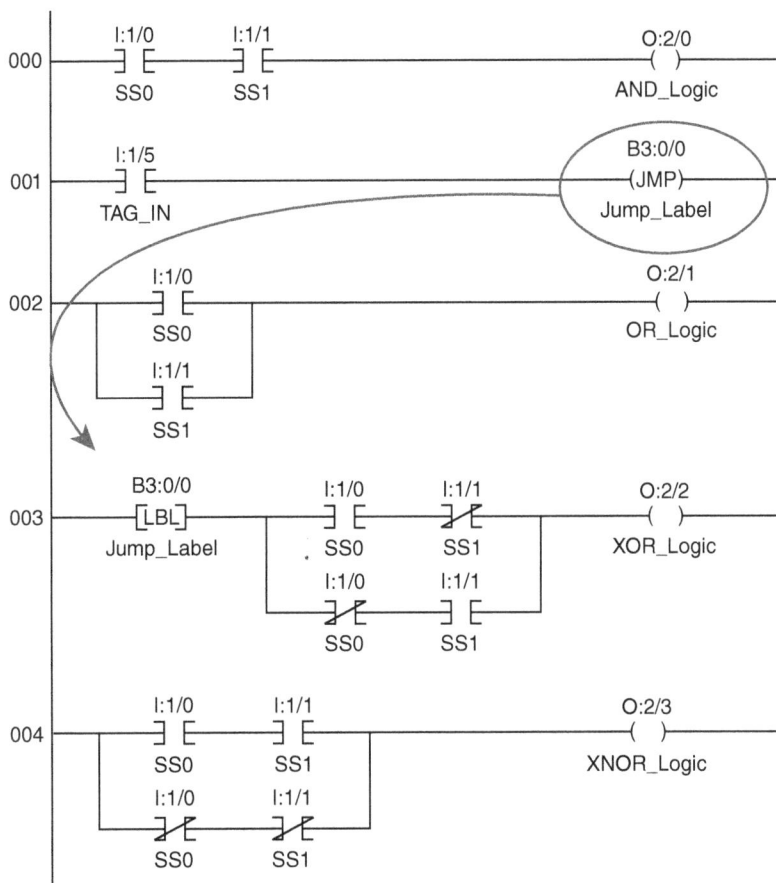

FIGURE 4.32 Program control using jump (JMP) and label (LBL) instructions.

4.4.2   Subroutine Call

A subroutine (SBR), also known in different programming languages as a procedure, function, routine, method, or subprogram, is a portion of code within a larger program that performs a specific task and is relatively independent of the remaining code. A subroutine is coded so that it can be called repeatedly from different places during a single execution of the program and then return execution to the next instruction after the call point once subroutine is concluded.

Examples of subroutine usage include the following:

- Loading and executing a specific recipe when needed
- Initializing system hardware and startup tasks
- Performing a common calculation at different points in a program
- Updating communication data and protocol parameters

Jump to Subroutine (JSR) output instruction initiates the subroutine call when rung conditions are true. The processor jumps to the targeted subroutine, in this case file number 3 as shown in Fig. 4.33. Return from Subroutine (RET) output instruction marks the end of subroutine execution or the end of the subroutine file. Subroutines can be nested, but each subroutine must be assigned unique file number for its JMP, SBR, and RET output instructions.

The program shown in Fig. 4.34 illustrates the Jump to Subroutine (JMP) and Return from Subroutine (RET) instructions and their application as documented using the

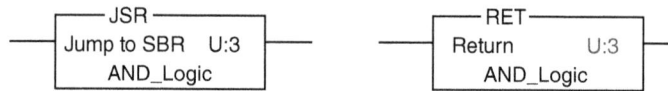

FIGURE 4.33   JSR and RET output instructions.

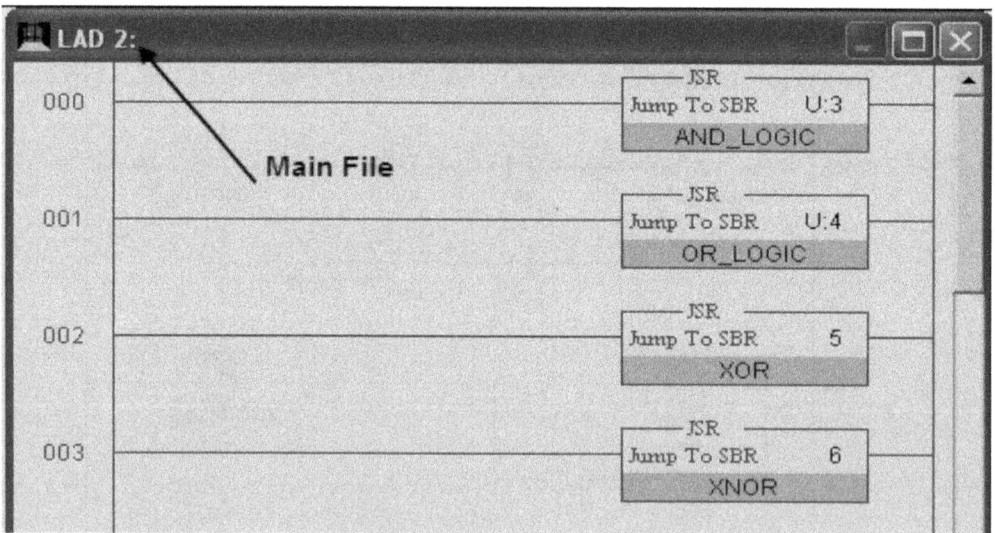

FIGURE 4.34   Structured ladder using JSR and RET output instructions.

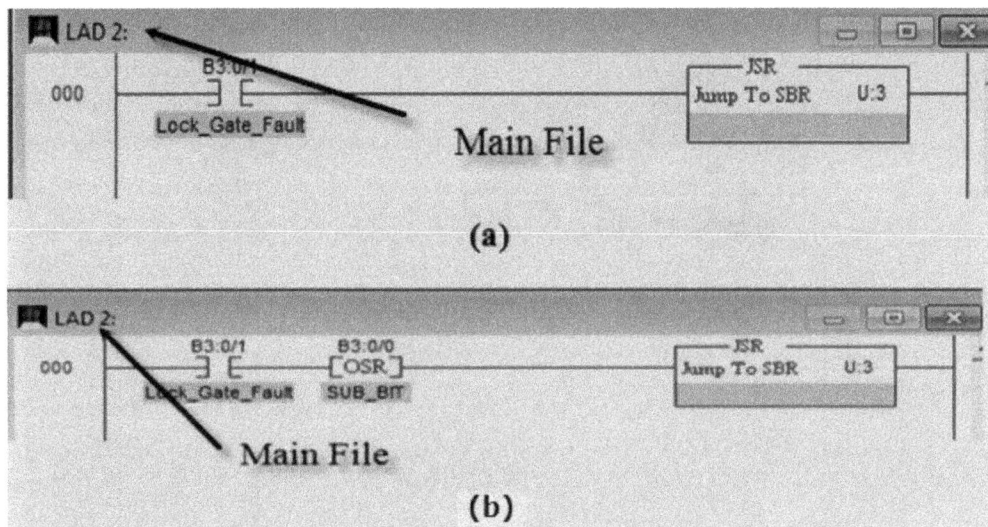

FIGURE **4.35** Conditional subroutine calls.

LogixPro software. The JSR instructions are placed in the LAD 2 file (main file) and executed once per scan. When Rung 0 executes, linear program execution is interrupted and resumes in file 3 (AND_LOGIC). This continues till RET instruction execution, which returns the program control to Rung 1 in the main file. Process will repeat for file 4, 5, 6, then 3, 4, etc.

Subroutines can conditionally execute based on true input condition/conditions for the JSR instruction. A subroutine can be called from different locations, for example when fault occurs and the subroutine is designed to alert the operator by flashing light or sounding a horn. The subroutine will be called every scan while the fault condition exists as shown in Fig. 4.35a. It can also be programmed to execute only once using the OSR instruction as shown in Fig. 4.35b.

Figure 4.36 shows an implementation of a small program using subroutine; only the relevant part of the main program is shown. Figure 4.36a shows a call to the subroutine SBR 3 (file 3) from LAD 2 file (main program). Figure 4.36b shows the subroutine code. The subroutine is called every scan while any of the three indicated production faults exists. The subroutine will turn a fault pilot light ON for 10 seconds and OFF for 3 seconds while the fault exists.

4.5 Implemented Industrial Application

This section includes a few implemented tasks that are common requirements in most process control applications. It is followed by three small industrial control module applications illustrating the instructions and concepts covered so far in this book. All implementations are done using the SLC-500 PLC hardware and RSLogix software or the LogixPro simulator. The concepts and techniques used are applicable to any other PLC environment implementation.

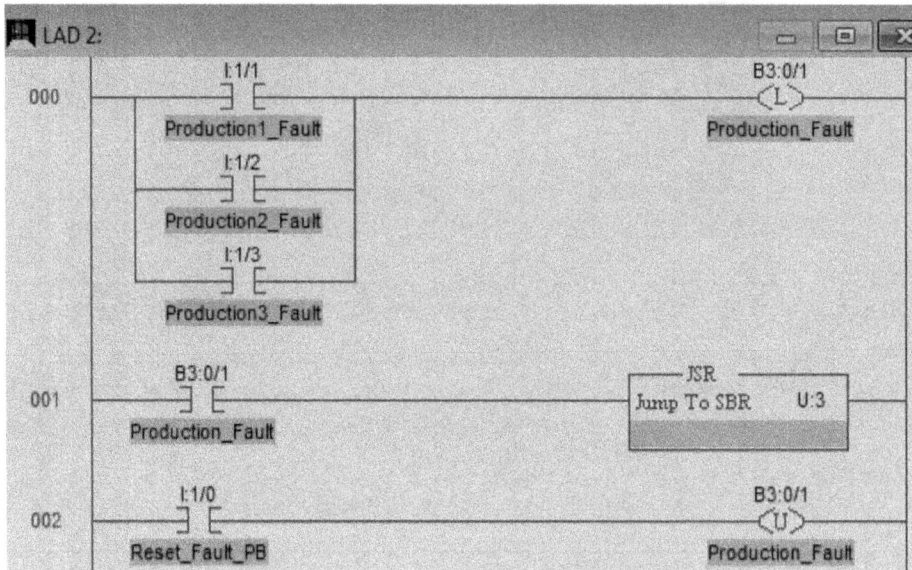
(a)

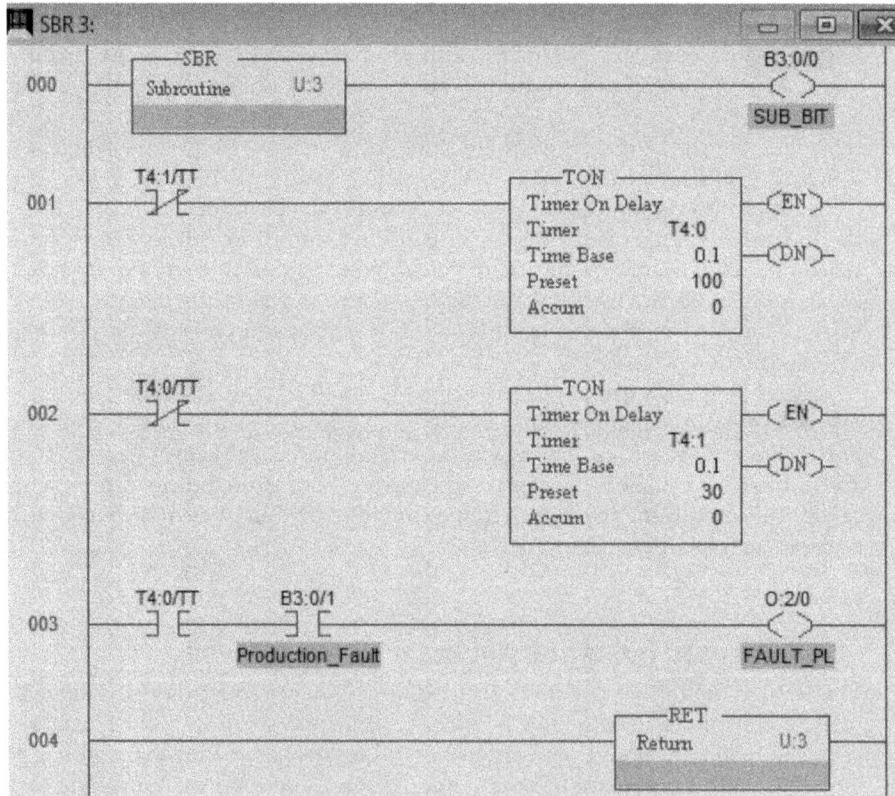
(b)

FIGURE 4.36   (a) Main program; (b) called subroutine.

4.5.1 Common Process Control Tasks

The following example calculates the average downstream water level in an irrigation canal. Two sensors, placed at two different locations downstream from a vertical gate system, provide online measurements of the water level. Sensor data validation is not part of this example and must be carried out prior to this calculation. Redundant measurements are typically used to provide sensory information for the same controlled variable, especially in cases where measurements are not as accurate and reliable at times. Sensory measurement fusion requires an understanding of both the control process and instrumentation, but it should be taken into consideration during the design and implementation phases.

Example 4.5.1.1 This application assumes two downstream water level sensors addresses, (N7:5) and (N7:20). The rung in Fig. 4.37 determines the average downstream water level using ADD and DIV instructions. The ADD instruction adds downstream level 1 to downstream level 2 and places the result in (B3:7), then the DIV instruction divides (B3:7) by 2 and places the result in (F8:2), which is the average for the two downstream water levels.

Example 4.5.1.2 The following example illustrates the use of comparison instructions in making actuators' decisions based on the process control requirements. Regulation of downstream water level is governed by both the desired water level set point and its defined dead band. Action is not required while the downstream water level stays within the dead band. This is an important condition to prevent excessive and undesired activation of control.

The rung in Fig. 4.38 determines if the downstream average level is greater than or equal to the downstream high threshold. The Greater Than or Equal instruction compares DS_AVE (B3:8) to DS_HIGH_LEVEL (B3:40). If true, power flows to the output coil and DS_HLEVEL is set.

The rung in Fig. 4.39 determines if the downstream average level is less than or equal to the downstream low threshold. The Less Than or Equal instruction compares the DS_AVE (B3:8) to DS_LOW_LEVEL (B3:41). If true, power flows to the output coil and DS_LLEVEL (O:2/1) is set.

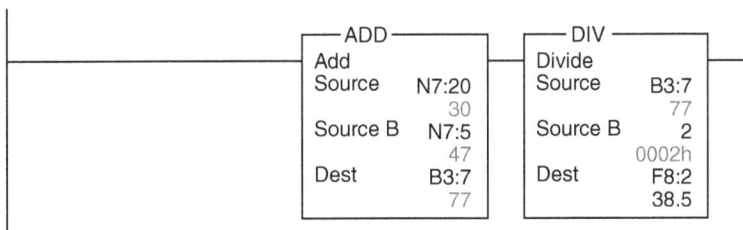

FIGURE 4.37 Downstream water level average.

FIGURE 4.38 Downstream water level greater than or equal to the high limit.

```
    ┌──────LEQ──────┐                                          O:2/1
    │ Less than or eql │                                        ─( )─
    │ Source      B3:8 │                                      DS_LLEVEL
    │               28 │
    │ Source B   B3:41 │
    │               28 │
    │ DS_AVE<=LTRShold │
    └──────────────────┘
```

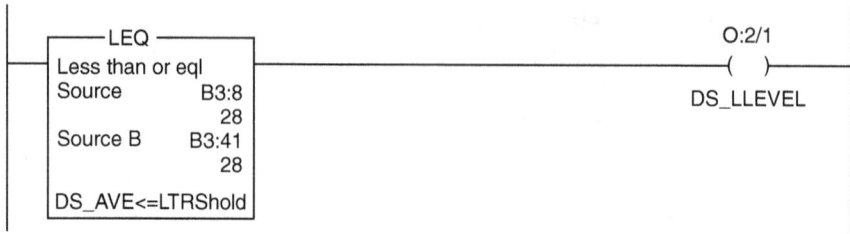

Figure 4.39   Downstream water level below or equal the high limit.

Example 4.5.1.3   A flexible manufacturing conveyor system module produces three different products, but only one at any time interval. Each product is counted by a photoelectric cell mounted on the conveyor line PE1. Each batch processing is initiated through dedicated normally open PB switches. Only one batch can be active at any time. A new batch can be started only after the completion of the ongoing batch or by forcing a termination through the STOP PB. Product 1 is produced in 2000 amount, product 2 in 2000, and product 3 in 5000 total parts. A pilot light PL1 turns ON after a batch product is completed. The process begins by pressing the START PB and can be terminated any time by pressing the STOP PB. A Reset PB is used to reset the counter-accumulated value. Figure 4.40 shows the ladder implementation, which uses six rungs.

Example 4.5.1.4   An inspection station on a conveyor system detects missing stamp on produced parts. The system is considered faulty once three missing stamps are detected in a 2-hour window (simulated by 20 seconds in our implementation). Once a fault is detected, the conveyor motor should stop running and a pilot light turns ON to indicate the faulty condition. In order to restart the motor, the operator must fix the fault before pushing the START PB. Figure 4.41 shows the five rungs used to realize the requirements of this control task.

4.5.2   Industrial Control Applications

This section covers small industrial process control application using instructions and techniques illustrated in this chapter. Only a small aspect of the application is covered, but more details will come in later chapters as more topics and techniques are covered. Implementation presented is one possible way for achieving the process control requirements. The reader is encouraged to explore other ways for basic implementation of the requirements and also for potential enhancement.

Example 4.5.2.1   Cascaded tank reactor is typical in chemical process control. In this example, three tanks are cascaded through a series of outlet solenoid valves. Material is discharged from tank 1 to tank 2 during the first 7 hours after the start of the process. At the end of the 7 hours, tank 1 valve closes and tank 2 valve opens to allow material to flow into tank 3 for 8 additional hours. In the final stage, tank 2 valve closes and tank 3 is cleared by keeping its outlet valve open for additional 5 hours. The entire reaction takes 20 hours (20 seconds simulated time) after which all valves are closed. Activating the STOP PB will terminate the process and return all valves to the default close position. The START PB initiates the entire reactor process. The listing of the four rungs used to implement this application is shown in Fig. 4.42.

Example 4.5.2.2   The rung in Fig. 4.43 compares the set point (N7:10) to a maximum limit (40) and a minimum limit (10). If the set point is greater than the maximum or less than the minimum, the output tag name (SP_OUTSIDE_LIMIT) is set.

(a)

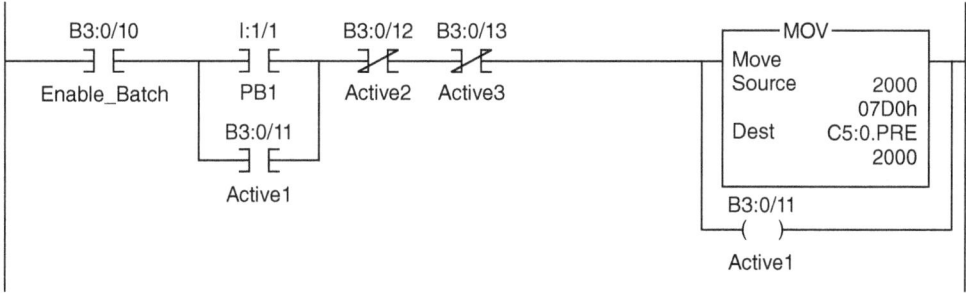

(b)

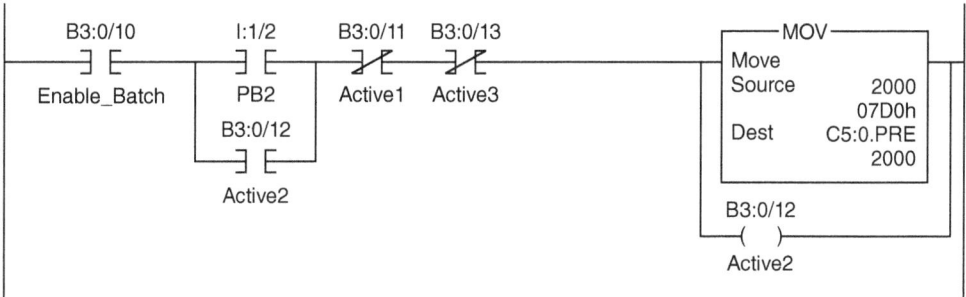

(c)

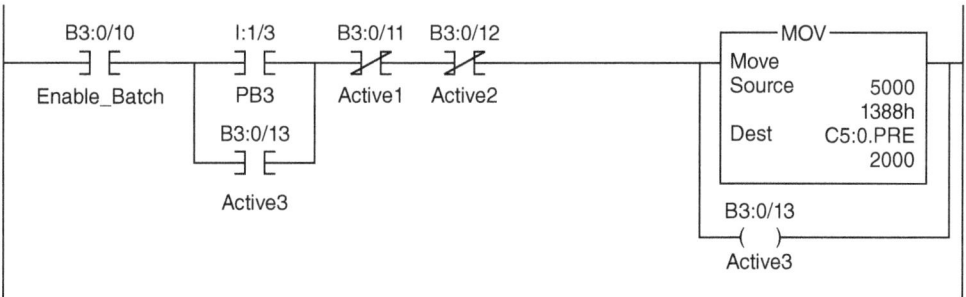

(d)

Figure 4.40   (a) Flexible manufacturing control rung 1; (b) flexible manufacturing control rung 2; (c) flexible manufacturing control rung 3; (d) flexible manufacturing control rung 4; (e) flexible manufacturing control rung 5; (f) flexible manufacturing control rung 6.

```
     B3:0/10        I:1/5                              ┌─CTU─────────────┐
   ──┘ ┌──────────┘ ┌─────────────────────────────────┤ Count up        ├──(CU)──
                                                        │ Counter   C5:0  │
     Enable_Batch   PE1                                 │ Preset    2000  ├──(DN)──
                                                        │ Accum        0  │
                                                        └─────────────────┘
```

(e)

```
     C5:0/DN                                                         O:2/1
   ──┘ ┌──────────────────────────────────────────────────────────────( )──────
                                                                 End_Batch_PL1
```

(f)

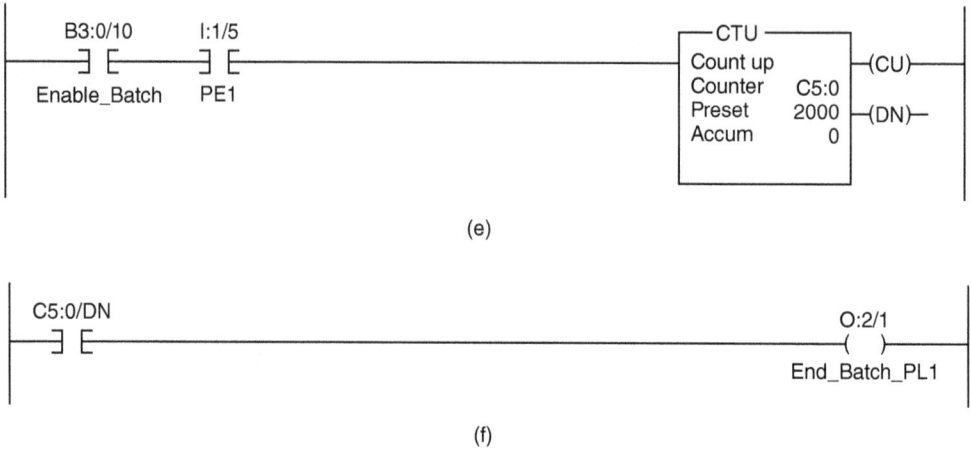

FIGURE 4.40 (Continued)

Example 4.5.2.3   The rung in Fig. 4.44 compares the set point (N7:10) to a maximum limit of 40 and a minimum limit of 10. If the set point is greater than minimum and less than the maximum, the output tag name (SP_inside_LIMIT) is set.

  As shown in Fig. 4.44, if TAG_IN is true, the GRT instruction is executed. If the source A value (N7:10) is greater than the low limit (10), the instruction output is true and the LES instruction executes. Also, if the source A value (N7:10) is less than the high limit (40), the instruction output coil (O:2/0) becomes true.

Example 4.5.2.4   This example shows more efficient way of using the LIM instruction to check whether the set point is outside the defined range. The rung in Fig. 4.45 compares the set point (N7:10) to a low limit (40) and high limit (10). If the set point is outside the range, the output tag name (SP_OUTSIDE_LIMIT) is set. Greater than (GRE) and less than (LES) instructions were previously used to accomplish the same logic. Notice that for an out-of-range indication, we reversed the low- and the high-limit assignments. If the low limit is assigned higher value than the high limit, then the LIM instruction output coil will be true if the test value is outside the defined range (less than the high *or* greater than the low).

Example 4.5.2.5   This example shows a more efficient way of using a LIM instruction to check if the set point is inside the defined range. The rung in Fig. 4.46 compares the set point to the minimum (10) and the maximum (40) limit. If the set point is inside the limit, the output tag name (SP_INSIDE_LIMIT) is set. GRE and LES instructions can be used to accomplish the same logic. If the high limit is assigned a higher value than the low limit, then the LIM instruction output coil will be true if the test value is inside the defined range (less than the high *and* greater than the low).

Example 4.5.2.6   The rungs in Fig. 4.47 show sensory input signal validation. The input signal with tag "INPUT_SIGNAL" is a 12-bit resolution A/D digital count ranging from 0 to 4095. The count is set to 0 if it comes negative and to 4095 if it is received at a value above the maximum count (4095). Scale instruction exists for SLC-500 processors: SLC 5/02, SLC 5/03, SLC 5/04, SLC 5/05, and all MicroLogix processors. The next example illustrates a typical implementation of the scaling task.

Example 4.5.2.7   This example shows the instructions used for scaling a raw analog input signals as a digital count between 0 and 4095 to engineering units, which is required by the test cell operator to monitor temperature. An oven has a range of temperature between 50 and 400°F, which is received as 4

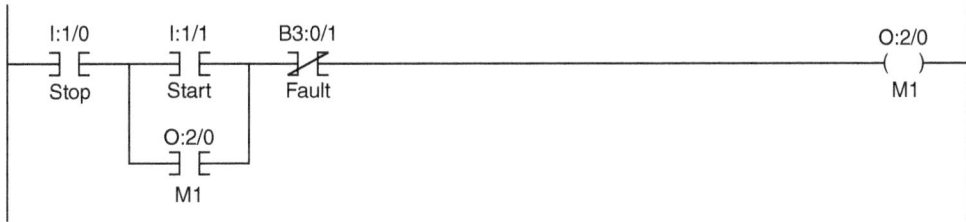

(a)

(b)

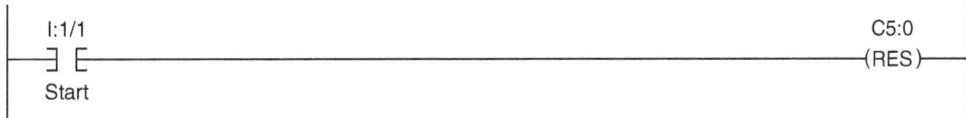

(c)

(d)

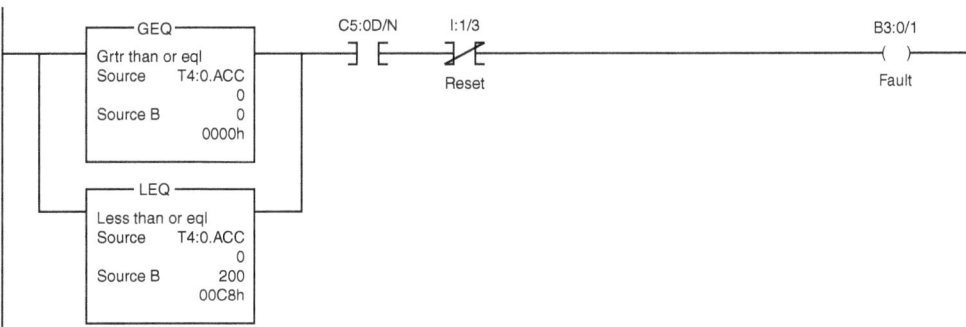

(e)

FIGURE 4.41 (a) Inspection station control rung 1; (b) inspection station control rung 2; (c) inspection station control rung 3; (d) inspection station control rung 4; (e) inspection station control rung 5.

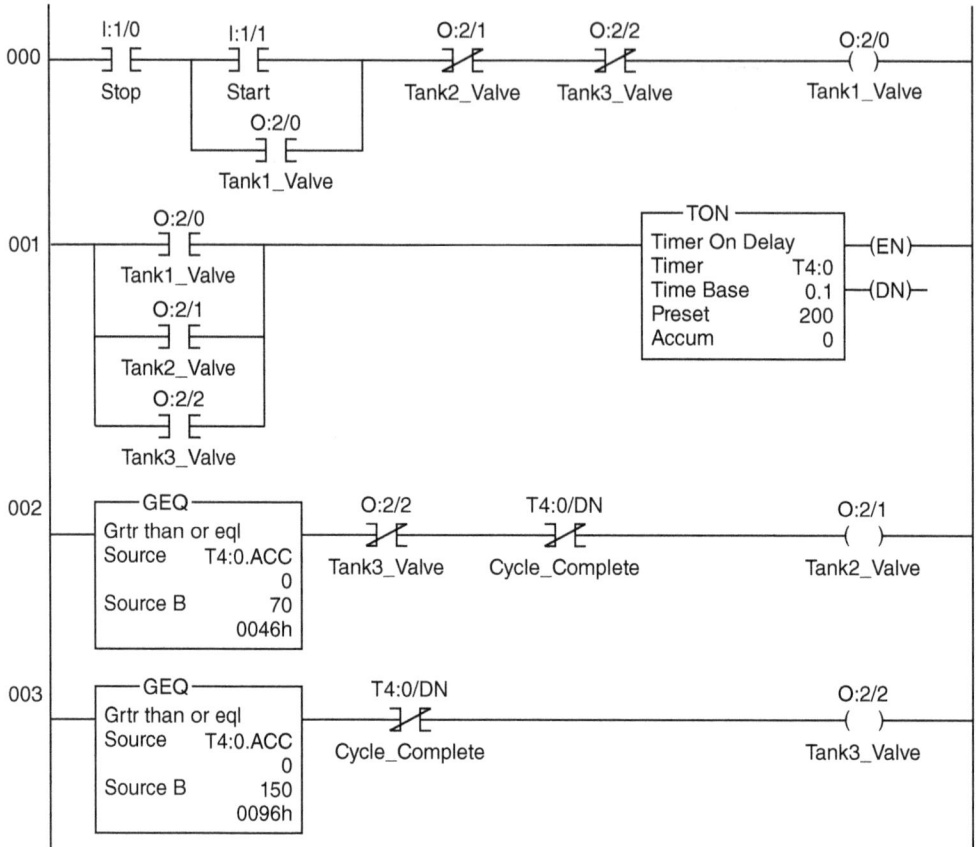

Figure 4.42 Cascaded tank control ladder program.

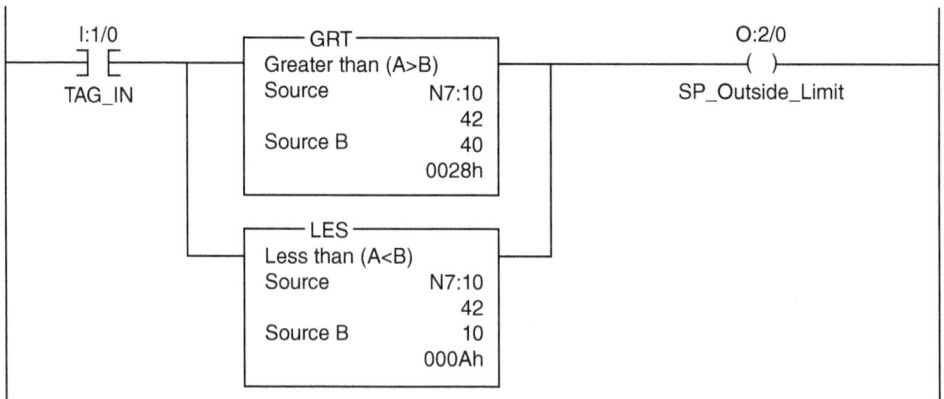

Figure 4.43 Set point validation (outside range).

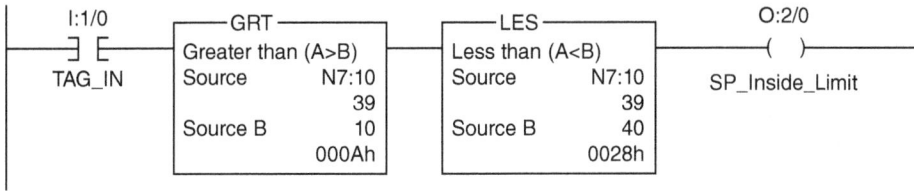

FIGURE 4.44 Set point validation (inside range).

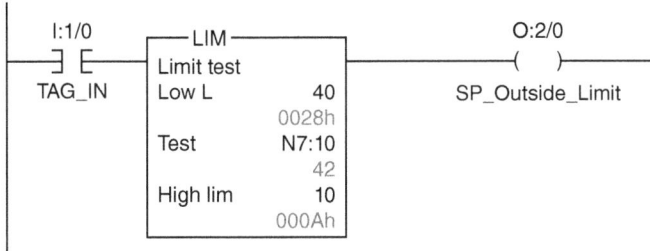

FIGURE 4.45 Outside range set point validation.

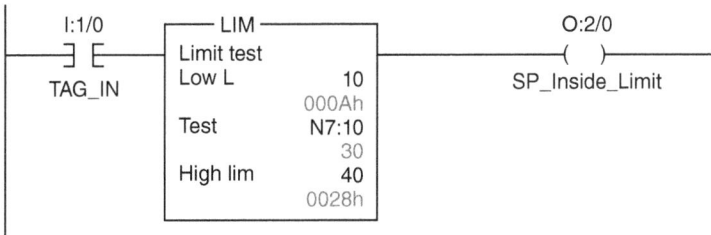

FIGURE 4.46 Downstream water level set point validation (inside the range).

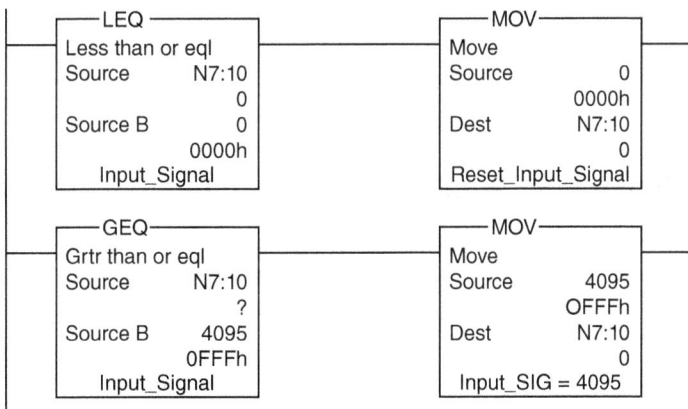

FIGURE 4.47 Inputs sensory signal validation.

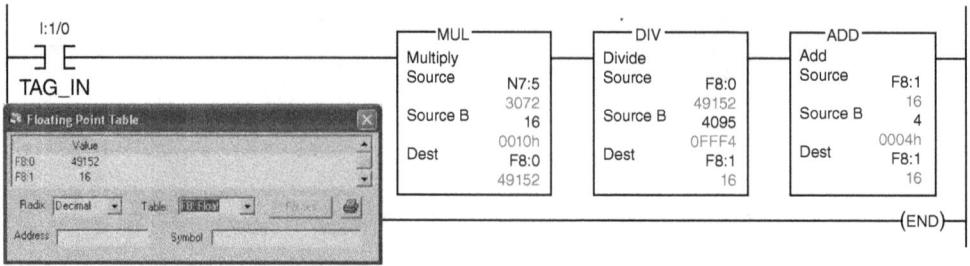

FIGURE 4.48 Analog input scaling ladder rung.

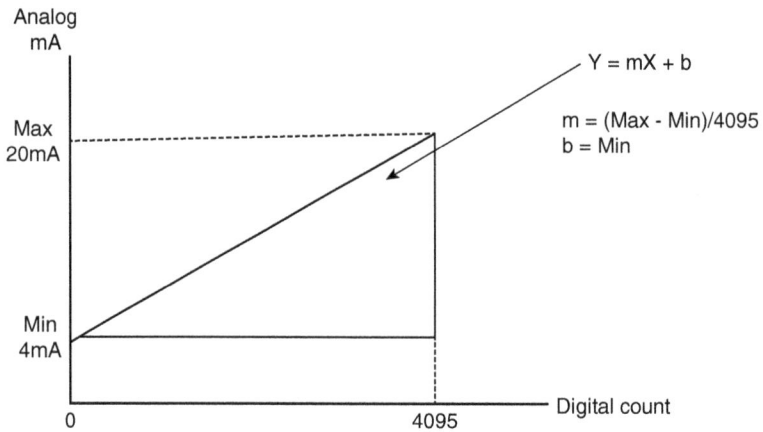

FIGURE 4.49 Analog input scaling equation.

to 20 mA signals. The oven temperature is measured using 12-bit resolution analog input module. The math instruction sequence (rung) in Fig. 4.48 is used to scale the digital counts (N7:5) to engineering units in degree fahrenheit (F8:1) to be displayed on the HMI for the operator. The scaling equation graph is shown in Fig. 4.49. Calculate or the Scale (SCL) instructions can be used to accomplish the same task in a simpler manner. Notice that LogixPro 500 does not support floating point arithmetic. Instead, it uses double-word integer representation and operations for the F8 elements. The values of F8:0 and F8:1 after execution are also shown in the window below the rung.

Chapter 4: Home Work Problems and Laboratory Projects

1. Explain the difference between file 4 and file 7 in Allen Bradley processor memory map. How many word(s) are reserved per element for each of the two files?

2. Write a ladder rung to display the content of integer register (N7:4) on output word O:2.

3. Design an up-counter to increment a word (N7:5) without using the CTU instruction.

4. How many compare instructions are available for the SLC-500? List and explain three such instructions.

5. The MAX instruction (finding the larger of two values) does not exist in LogixPro/ RSLogix 500. Write a one rung implementation of the instruction, assuming that N7:4 and N7:5 contain the two values: 5 and 10. The largest value is stored in register N7:10.

6. Assume that TAG_IN status in Fig. 4.50 is true. Words N7:4 and N7:5 contain the value for tank 1 and tank 2 levels. Under what condition(s) will MOTOR1 run?

7. Assume that TAG_IN is a normally open push button (PB1) and N7:0 has a value of 0. Answer the following questions for the rung shown in Fig. 4.51:

   a. What is the value of N7:0 if PB1 is pressed three times?
   b. What is the value of N7:0 if PB1 is replaced by a maintained switch SS1 and the switch is closed?

8. Write a ladder logic program to turn on memory bit (B3/0) if a set point value (stored in N7:2) is outside minimum and maximum limits. Assume that N7:0 and N7:1 have the minimum and maximum values.

9. In reference to the rung shown in Fig. 4.52, assume that B3:1 and B3:2 have the value of 5 and 10. Answer the following:

   a. What value will be stored in B3:4 if PB is pushed once?
   b. What value will be stored in B3:4 if PB is pushed 10 times?
   c. Remove the OSR instruction in Fig. 4.52. What value will be stored in B3:4 if PB is pushed one or more times?

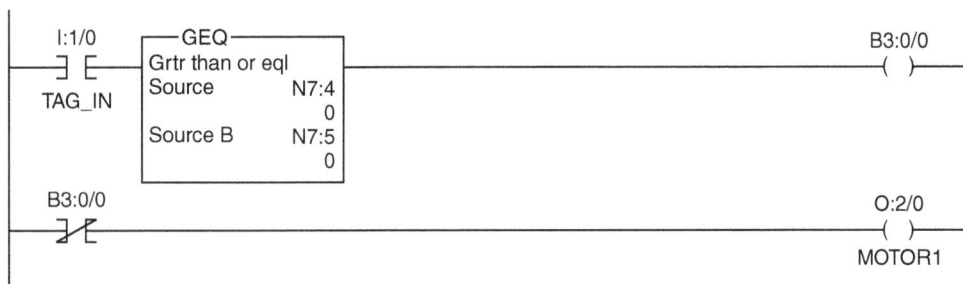

Figure 4.50   Problem 6 rungs.

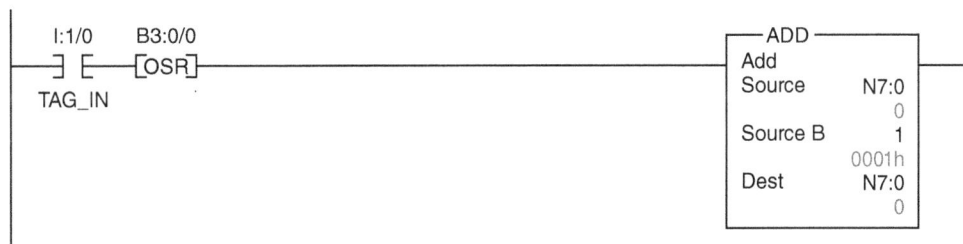

Figure 4.51   Problem 7 rung.

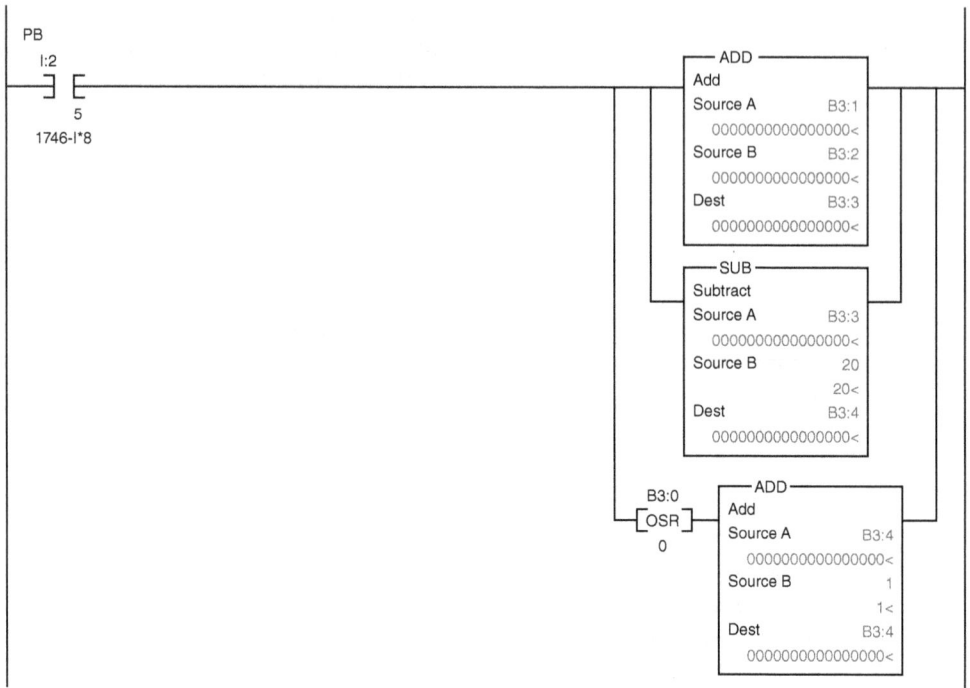

Figure 4.52 Problem 9 math instructions.

10. In reference to the math instructions shown in the ladder rungs of Fig. 4.53, assume that B3:1 and B3:2 have the values of 5 and 10, respectively. Answer the following:

    a. What value will be stored in B3:4 if PB is pushed once?
    b. What value will be stored in B3:4 if PB is pushed two times?

11. Write a ladder logic program to turn on a memory bit B3:1/2 if the counter CTU accumulated value is between 10 and 20.

12. Design PLC rungs to perform the conversion from degree celsius to degree fahrenheit. The formula for the conversion is given by the following equation:

$$F = (9 * C)/5 + 32$$

13. Study the three rungs in Fig. 4.54 and answer the following questions:

    a. What type of switch (SW1) should be used to increment the counter once on each activation?
    b. Under what condition will the solenoid valve (SV1) output be true?
    c. Under what condition will the solenoid valve (SV1) output be reset?

14. Answer the following questions for Fig. 4.55 ladder rung assuming that SS1 is a selector switch and O:4 is the address for an LED display.

    a. What is the value of O:4 if SS1 is in the closed position?
    b. Modify the rung to receive the value in N7:1 from a BCD four-digit thumb wheel switch address I:3 and display the value in O:4?

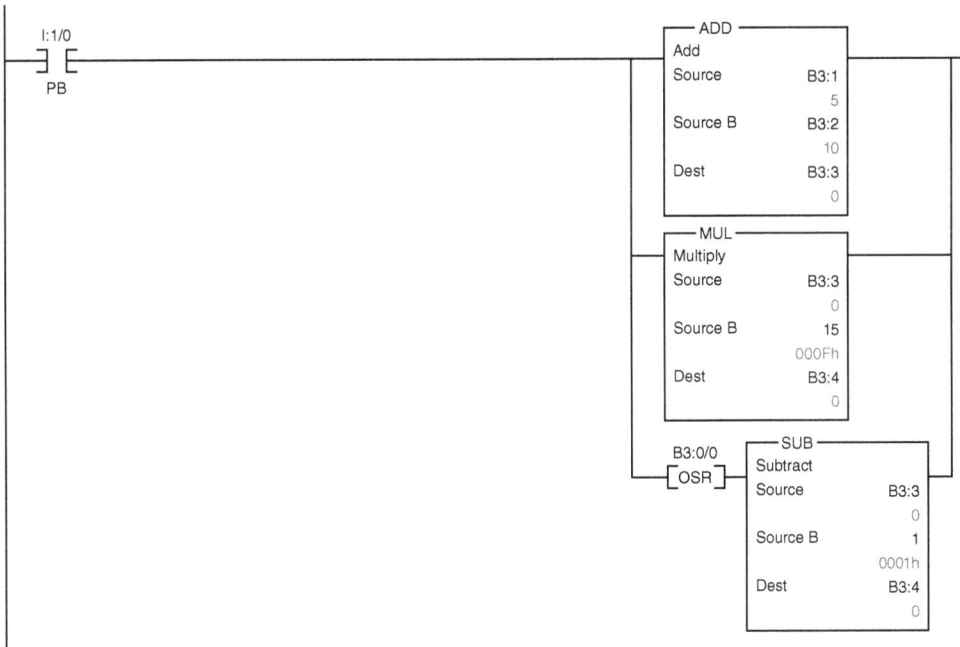

FIGURE 4.53 Problem 10 math instructions.

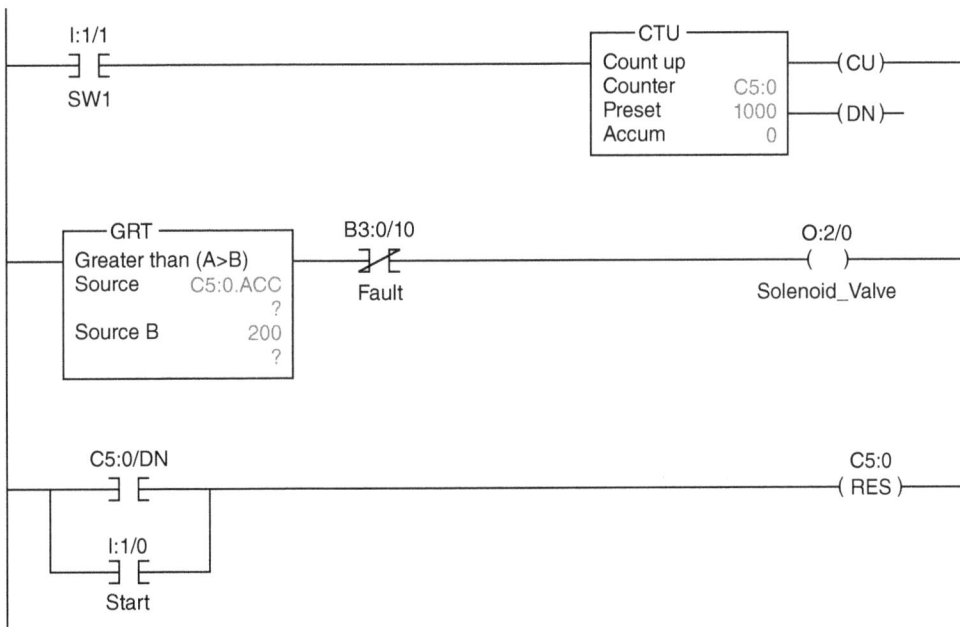

FIGURE 4.54 Problem 13 rungs.

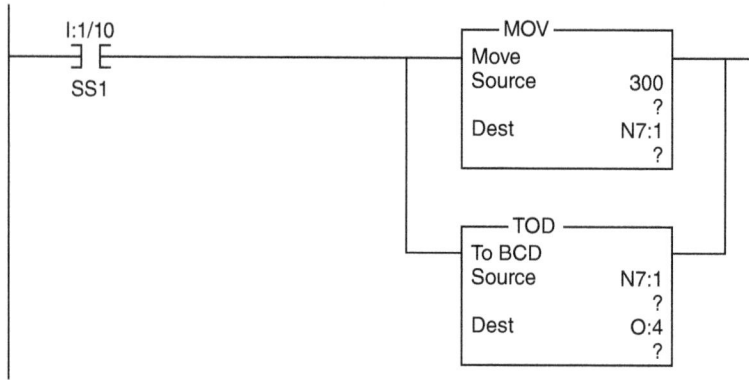

FIGURE 4.55 Problem 14 rung.

15. Assuming that N7:1 and N7:2 have the values 3 and 5, respectively, write a ladder logic rung(s) to store the minimum of the two values in N7:3.

16. Write a ladder logic program to calculate the average of three signals (Input1, Input2, and Input3), and store the result in F8:5.

17. A 0- to 10-V signal is connected to a 12-bit analog-to-digital module. Write the rung(s) required to validate the digital input counts (available in N7:4). Assuming that the count is coming in the range of 0 to 4095.

18. Write a ladder logic program using EQU and TON instructions to set motor (M1) output if the timer current value is 10 or greater and reset the motor output if the timer current value is greater than or equal to 30.

19. Using only one timer (TON) and GREATER THAN OR EQUAL instructions, program the merry-go-round example in Chap. 3 (Fig. 3.12). Four lights in a sequence are used, ON each for 5 seconds (PL1, PL2, PL3, PL4, PL1, etc.). A START PB and a STOP PB initiate and terminate the sequence.

20. Show an example of how to turn on an output coil if a TON timer accumulated value is greater than or equal to 15.

21. What is the range of values for N7:10 in the ladder rung shown in Fig. 4.56 in order to make O:2/0 MOTOR1 true?

22. Write a ladder logic program to start timer T4:0 when START_TIMER0_PB is pressed. Assume a 3000 timer preset value. When the timer accumulated value is greater than or equal to 200, a flag bit B3:1/0 is set. The RESET_TIMER0_PB activation will clear B3:1/0 and reset the timer.

23. Show an example of how to move accumulated value of TON (Timer0) to memory word N7:5 when memory word N7:2 least significant bit is true.

24. Write a ladder logic program to start counter 0 (C5:0) with preset value 1000. When a part crosses a photoelectric cell PE1 (I:1/0) on a conveyor line, the counter will increment. When counter 0 accumulated value reaches 5, start motor 1 (O:2/0), and when the counter accumulated value reaches 10, start motor 2 (O:2/1). Solenoid SV1 energizes if either of the two motors is running. A Reset PB (I:1/1) activation resets the line.

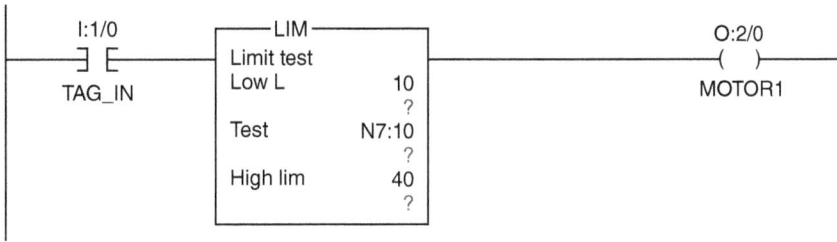

FIGURE **4.56** Problem 21 rung.

25. Show a rung to turn on memory bit B3:1/2 when the timer-accumulated value is between 40 and 60.

26. Show a rung to clear accumulated value of three timers when the START PB is pressed.

27. Write a ladder logic program to change a TON timer preset value to 19 once the timer is done counting and "Input_Value," which exists in N7:4, is less than 200.

28. Using math instruction(s), write a ladder logic program to scale an input signal from (0–4095) to (4–20 mA).

29. Write a subroutine to implement the Swap instruction for a one-word 2 bytes. Assume that the word is at address N7:1 and the answer returns in the word N7:0 address.

30. If N7:0 content is 22 and N7:1 content is 40, what is the status of Motor1 and Motor2 after executing the two rungs in Fig. 4.57?

31. Show a one rung implementation of the swap operation. Assume that the word to be converted is available in N7:1 and the swapped answer will be stored in the N7:0 word. A selector switch initiates the one time SWAP operation when turned ON.

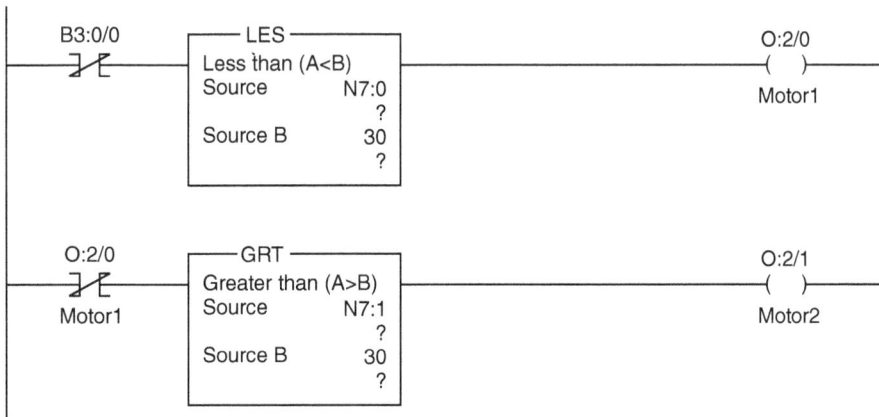

FIGURE **4.57** Problem 30 rung.

Laboratory 4.1—Tank Alarm

This laboratory covers math/comparison instructions and their use in the LogixPro/ AB SLC-500 ladder programming to produce a tank high-level alarm.

A tank (tank no_1) high-level alarm will turn on once the tank level exceeds the prespecified high threshold level (120). The alarm remains ON till the level goes below the low threshold level (100). This behavior is known as the *Schmitt Trigger action*, which is shown in Fig. 4.58*a*. The scaled analog input signal for the tank is received in memory location N7:5 (analog programming will be discussed in Chap. 8). The ladder logic program implementation is shown in Fig. 4.58*b*.

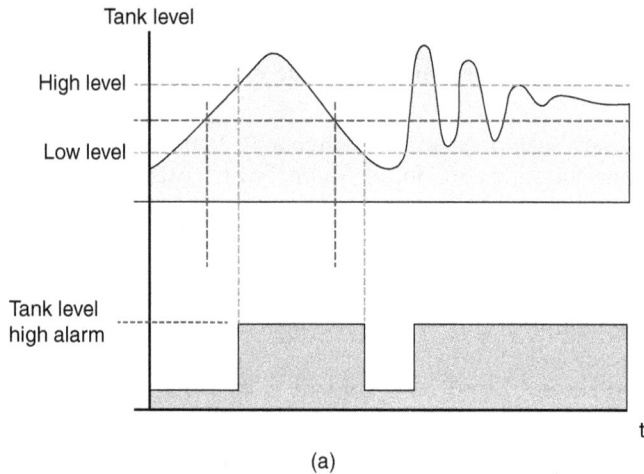

(a)

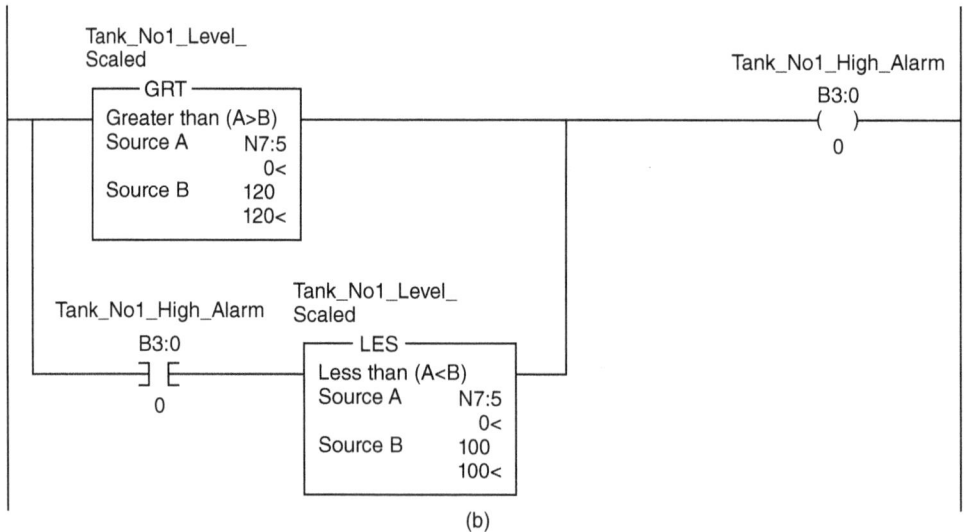

(b)

Figure 4.58   (*a*) Tank level alarm; (*b*) tank high-level alarm ladder.

Requirements:

Change the tank high-level alarm address to match the pilot light (PL1) address on your training unit or LogixPro 500 simulator.

- Simulate the alarm and monitor PL1.
- Add the logic needed to flash the PL1 ON and OFF using 3 seconds for ON and 2 seconds for the OFF duration.
- What type of logic these rungs represent?

Laboratory 4.2—Feed Flow Digester Control

In this laboratory you will learn how to use math and comparison instructions as well as counters in a real industrial application.

An up-counter (counter 0) is used to count the feed flow amount into a digester in thousands of gallons. A total of 32,000 gal needs to be discharged into the digester. A valve PV1 is calibrated to provide an accurate measure of the flow amount into the digester. Every time the PV1 valve opens, the FEED_FLOW increments based on the flow from 0 to 1000 gal. Once it reaches 1000 gal, it resets as shown in the second rung. The counter accumulated value is incremented by 1 for every 1000 gal flow as shown in rung 1 (Fig. 4.59). LS_VALVE_PV1 is the tag for the valve limit switch, which goes ON while the valve is open. High-speed counter modules are typically used to implement the amount of flow in industrial automation applications, including chemical process control. These modules provide calibrated pulse rate proportional to the actual flow in gallons per minute, which is more accurate than an analog module measurements. The pulses can be used to calculate accurate flow rate and volume. The ladder logic for the feed flow digester is shown in Fig. 4.59.

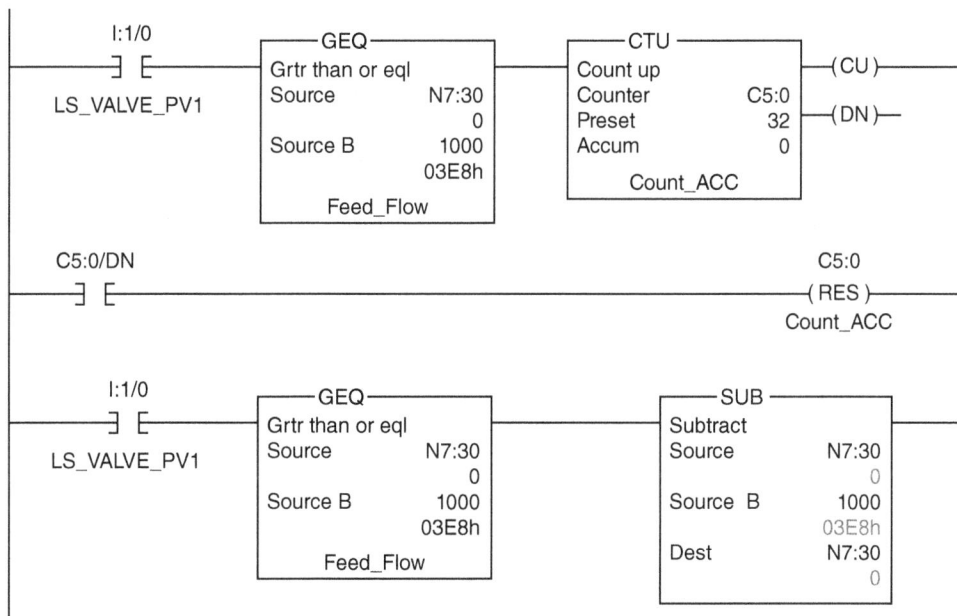

FIGURE 4.59   Feed flow ladder.

Requirements:

- Modify the counter operation to simulate the feed flow into the digester. Use a toggle switch SS1 to enable the counter. Change the 1000 gal value to 10 and the total amount from 32,000 to 32 gal.
- Toggle the SS1 and monitor the counter operation.
- When the feed flow exceeds 50 gal turn on a pilot light (PL1).

Laboratory 4.3—Merry-Go-Round Using One Timer

The objective of this laboratory is to get the students familiar with basic timing, positive edge trigger and equal instructions used to control a series of running motors simulated by four pilot lights. The same requirement was implemented in Laboratory 3.1 using four timers. This laboratory requires implementation using only one timer.

This laboratory assumes four motors represented by four pilot lights on the trainer unit. A push button START switch is used to start a sequence of pilot lights operation, which can be terminated at any time by pressing a push button STOP switch. The sequence starts by PL1, then PL2, then PL3, and then PL4. The same sequence will repeat until the process is stopped. Each selected PL will turn on for a period of 5 seconds while the other PLs are idle.

Requirements:

- Write a documented ladder logic program using only one ON-delay timer.
- Assign the I/O addresses and the bit addresses.
- Provide the laboratory checkout.

Laboratory 4.4—Pumps Alternation Using Structured Programming

The objective of this laboratory is to get students familiar with basic timing, counting, positive edge trigger, and move instructions used to control two pumps alternations based on a defined calendar. Pumps are scheduled to run according to an operator predefined calendar. This input is expected in hours of accumulated total pump run time. The two pumps must alternate till the operator press the STOP PB.

Requirements:

- Write a structured ladder logic program using a main file and three subroutines. The main file (LAD 2:) is shown in Fig. 4.60.
- Subroutine "INIT" (SBR 3:) clears timers and counters during the first scan of the program using the first scan bit from the status file (S: 1/15). Also, it initializes the calendar timer.
- Subroutine "PUMP_ALTERATION" (SBR 4:) includes the logic required to alternate between the two pumps based on the defined user calendar.
- Subroutine "PUMP_START_STOP" (SBR 5:) implements the logic required to start or stop either of the two pumps. An Emergency Shut Down (ESD) PB switch, once activated, will immediately turn off the running pump.

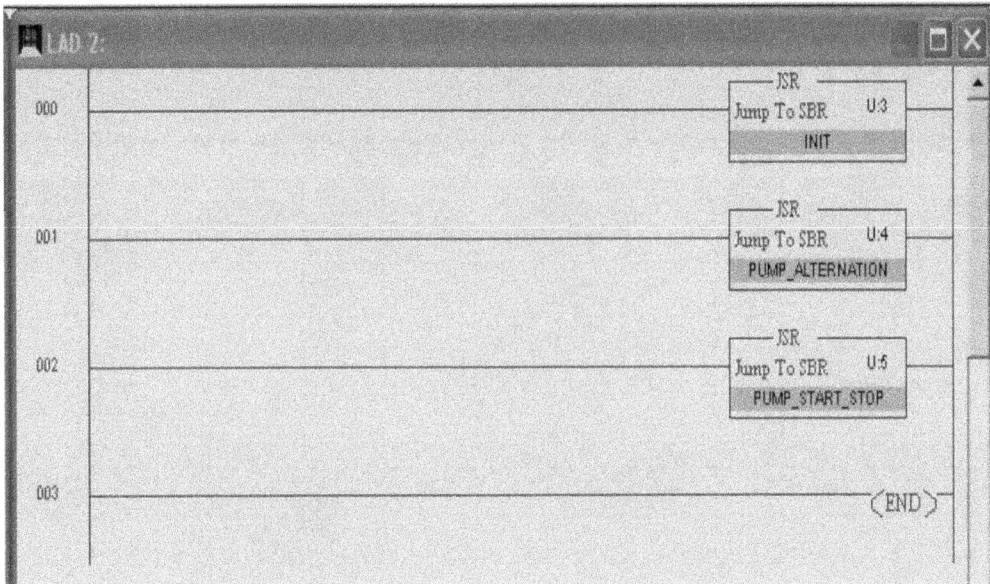

FIGURE 4.60 Pumps alteration main file.

Laboratory 4.5—Tank Fill/Refill

Refer to Example 3.5.2.3 in Chap. 3. Study the tank fill and refill process and perform the following requirements:

- Revise the ladder rungs to match the assigned addresses on your PLC hardware or simulation software.
- Document all rungs according to functionality and the assigned I/O map and variables/parameters tags.
- Try to add one or two additional requirements; provide the specifications and the associated implementation details.
- Download the program to the PLC processor and perform the needed simulation and debugging.
- Provide the final process control report documentation.

CHAPTER **5**

Human Machine Interface Design and Hardware Configuration

T his chapter introduces and details the fundamentals of Human Machine Interface (HMI), as typically used in PLC process control. It also details the steps needed to configure the PLC and associated HMI devices.

Designing safety into the machine is critical, Ropak Manufacturing.

Chapter Objectives

- Be able to implement the PLC/HMI configuration.
- Be able to perform HMI programming/communication.
- Be able to implement HMI remote control and monitoring in industrial automation.

PLCs evolved from hardwired analog control systems. They utilize a wide array of networked human machine interfaces to allow user interactions with the controlled process through visual and user-friendly tools. Figure 5.1 shows a wide array of typical PanelView HMIs used in Allen-Bradley control system applications. HMIs are primarily used for reporting/status display purposes, but they can accommodate user commands from any desired location or through the Internet anywhere in the world. The HMI infrastructure is both flexible and expandable. Its sole critical function is to allow for continuous quality improvements in process control and thus continuous enhancements in the final products produced. HMIs eliminate the need for hardwired panels and displays, which would require additional less flexible PLC I/O resources and interfaces. This chapter introduces and details the HMI fundamentals as typically used in PLC process control.

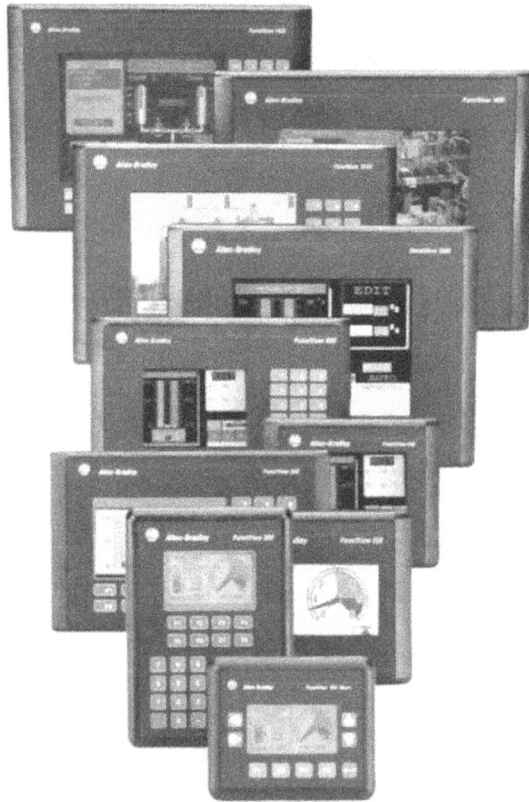

FIGURE 5.1 PanelView HMI devices. (*Courtesy of Allen-Bradley*)

Effective utilization of HMI software and associated communication tools provides the detailed and dynamic process knowledge needed to perfect product quality and process control strategy used to manage the system operation. HMI can be used for continuous quality improvements and overall system/process assessment. It can display texts, graphs, animated objects/pictures, messages, reports, alarms, trends, Boolean operations, and complex math calculation results. It can also display system messages, reports, and alarms. Data entry and process command tasks, including validation, are often implemented in HMI design. This not only eliminates the need for additional hardware and PLC interfaces but also provides additional flexibility and simplicity. HMI allows the operator and the management to view the plant's real-time behavior and improve both production quality and efficiency. Some of the world biggest chemical companies provide such access to any owned control system from any place at any time with adequate authentication control. Access is performed using a PC/laptop/smartphone or other intelligent Internet networked devices.

A brief coverage of the PanelView 600 shown in Fig. 5.2 will be given, but all concepts discussed apply to other HMI devices. Focus in this chapter is given to HMI implementation in the LogixPro 500–simulated environment using simple application examples. The same concepts can be extended to the HMI implementation on other types of display panels/devices configured on the PLC network. This includes implementation of PanelView, PCs, laptops, notebooks, or smart wireless devices.

FIGURE 5.2 Allen-Bradley PanelView 600. (*Courtesy of Allen-Bradley*)

5.1 Allen-Bradley PLC Networking Options

There are many network types available with different vendors, which are used in PLC control applications. The sole function of a network is to allow communication among PLCs, HMIs, and other external devices. Figure 5.3 shows a typical network of PLCs and HMIs in a control system. Network connectivity allows the connection of multiple HMIs and AB PLCs via Ethernet, RS-232, and RS-485 communication ports. It also provides monitoring and control of any AB PLC from one or more HMIs and read/write from one HMI to any of the AB PLCs on the local area network. The following is a brief listing of the common types of network used in AB PLC implementation:

RS232-based DF1 network: An RS232 port is present in almost all controllers manufactured by Allen-Bradley for programming purposes. It can be used for communication with a programming terminal, SCADA, or an HMI device. The driver used is AB_DF1 with speed up to 19.2 Kbps and a maximum cable length of 15 m.

RS485-based DH485 network: The DH485 is an AB industrial multidrop LAN used for factory floor applications. A single workstation or programming terminal (the master) can be used to configure and control communication among all nodes in the network. RS232 link can be used to connect nodes to the DH485 network. This type of network should be used for noncritical processes with low communication bandwidth needs. The advance interface converter (AIC) is used to connect and configure the DH485 network.

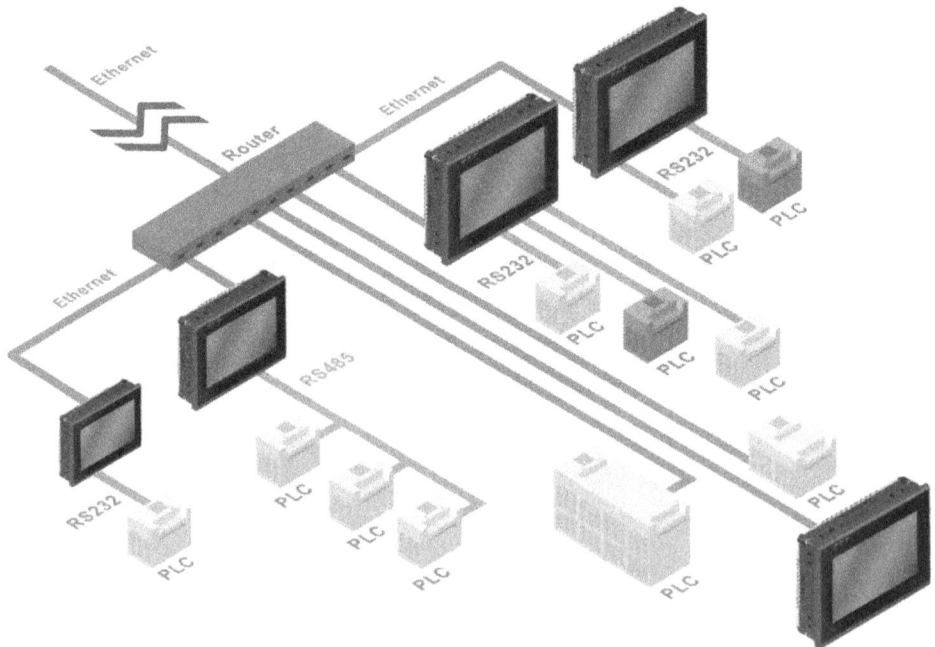

FIGURE 5.3 Ethernet PLC/HMI connectivity. (*Courtesy of http://www.maplesystems.com*)

Advanced interface converter: AIC has one D-type RS232 port, one round-type RS232 port, and one RS485 port. The RS485 port is used to link one AIC with another to form a network, while the RS232 port is used to connect external devices/node. Driver used is the 1747 PIC/AIC with up to 19.2 Kbps speed, up to 32 nodes, and a maximum length of up to 1200 m.

Ethernet network: It is the most widely used LAN topology based on Carrier Sense Multiple Access and Collision Detection (CSMA/CD). Every node in this network has its own unique class-C IP address, which consists of a Net ID and a Host ID. It supports a baud rate of 10 or 100 Mbps.

DeviceNet network: It is a simple open networking architecture that accommodates multivendor PLCs, HMIs, and smart instrumentation devices. It reduces the cost and time required to wire and install industrial automation devices and provide interchangeability of similar components from multiple vendors. It supports up to 64 nodes at a baud rate of 500 Kbps and a 487 maximum length.

ControlNet network: It is an open control network that meets the demands of real-time, high-throughput applications. ControlNet supports controller-to-controller interlocking and real-time control of I/O, drives, and valves. It also provides control networking in discrete and process applications, including high-availability applications. Up to 99 nodes at a 5 Mbps baud rate and 30 km maximum length can be supported.

5.2 PanelView Plus 600 Graphic Terminal

The PanelView Plus 600 Graphic Terminal has a 5.5-in grayscale or flat-panel color display with a display resolution of 320×240. This terminal supports operator input via keypad (10 function keys) or via keypad and touchscreen. Figure 5.4 shows the PanelView Plus 600's front, and Table 5.1 details the terminal's front features. Figure 5.5 shows the terminal's back, and Table 5.2 shows the back's features. The terminal has the following features:

- Modular design includes logic, display, and communication modules.
- Base-configured terminal available with display and logic modules.
- This terminal is optimized for Logix control architectures and it supports PLC-based and SLC-based systems.
- 18-bit color or 32-level grayscale graphics.
- RS-232 and Ethernet networks available through built-in communication ports.
- Network interface for optional communication module.
- One USB port.

5.3 Door Control Simulator/Monitor

The objectives of this section are to get students familiar with LogixPro 500 HMI simulation and the control panel used to control/monitor a garage door.

Keypad terminal

Keypad & touch screen terminal

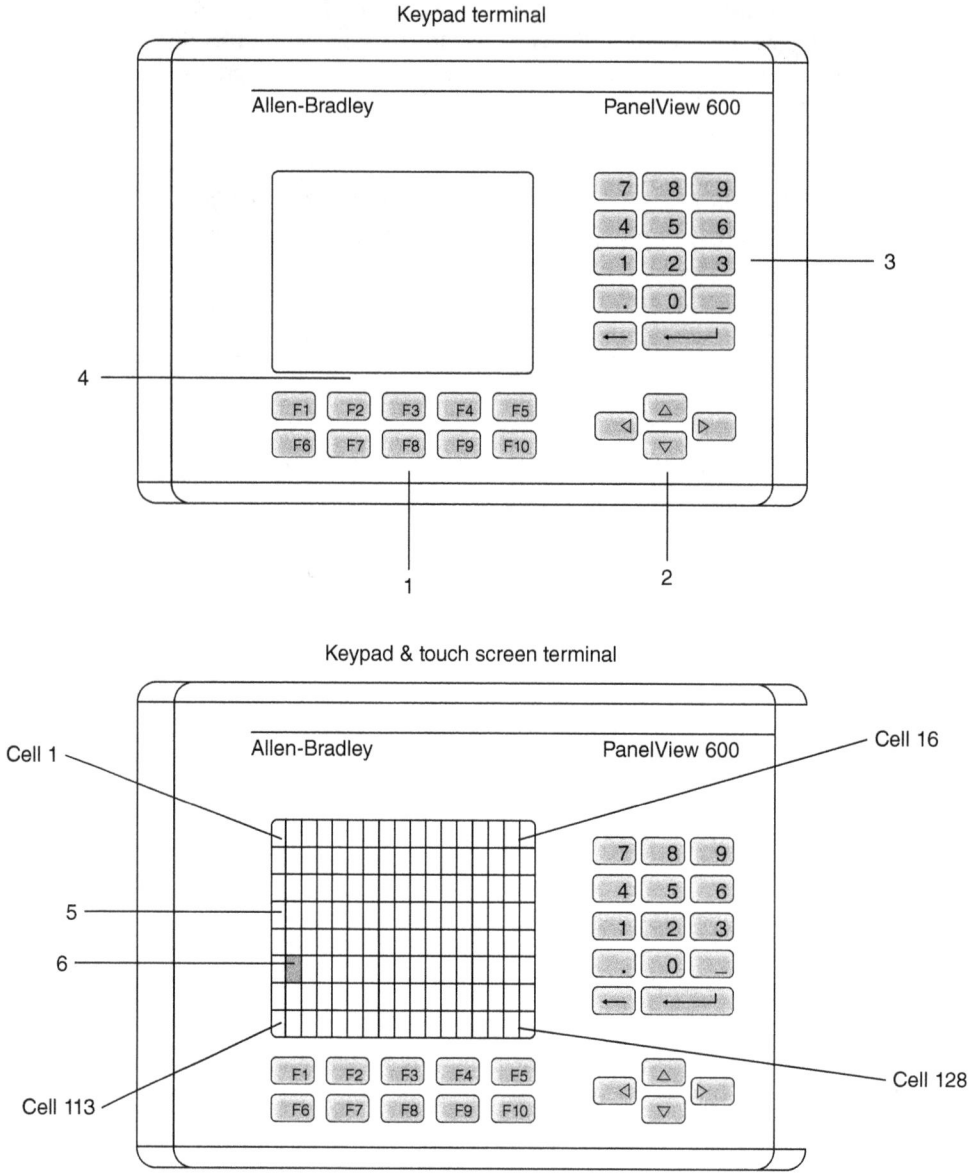

FIGURE 5.4 PanelView 600 fronts. (*Courtesy of Allen-Bradley*)

#	Feature	Description
1	Function Keys (F1...F10)	Use the function keys on keypad terminals to initiate functions on the terminal display. These keys may have custom legends. On keypad & touch screen terminals, you can initiate functions by using the function keys and/or touch screen objects.
2	Cursor Keys	Use the cursor keys to move the cursor in displayed lists, to select a numeric entry object, or to enter configuration mode.
3	Numeric Entry Keys	0 to 9 Enters numeric values. . Enters a decimal point. – Enters a negative value. ← Clears entered digits or cancels the scratchpad. ⏎ Stores an entered value.
4	Keypad Terminal Display	On keypad terminals, initiate the function of a displayed object, such as an ON or OFF push button, by pressing a function key (F1...F10).
5	Touch Screen Terminal Display	On keypad & touch screen terminals, initiate the function of a displayed object, such as an ON or OFF push button, by touching the screen object. Each interactive screen object occupies one or more of 128 cells. On keypad terminals, you can initiate functions by using the function keys and/or touch screen objects.
6	Touch Cells (Touch Screen terminal)	The 128 touch cells (16 columns × 8 rows) let you initiate functions by touching the screen. Interactive screen objects are aligned with touch cells when the application is created.

TABLE 5.1 PanelView 600's front features. (*Courtesy of Allen-Bradley*)

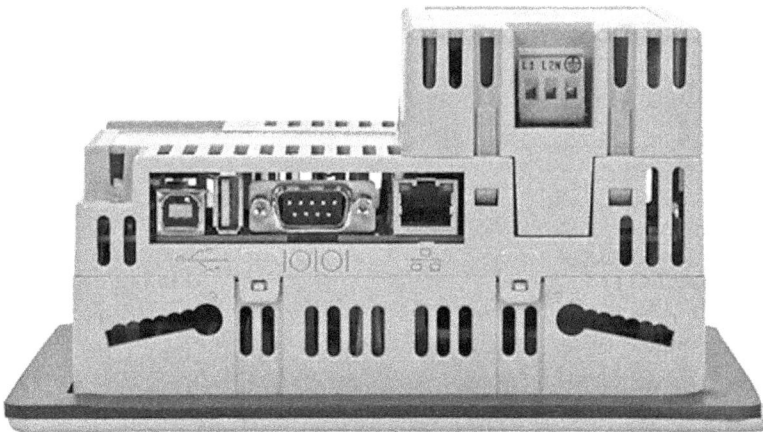

FIGURE 5.5 PanelView 600's back. (*Courtesy of Allen-Bradley*)

#	Feature	Description
1	Power Connection Terminals	Connects to external power source.
2	Nameplate Label	Provides product information.
3	Memory Card Slot	Accepts a memory card, which can store applications.
4	Sealing Gasket	Seals the front of the terminal to an enclosure or panel.
5	FAULT Status (Red) Indicator	Indicates firmware or hardware faults.
6	COMM Status (Green) Indicator	Indicates when communication is occurring.
7	DH-485 Communication Port	Connects to an SLC or MicroLogix controller, DH-485 network, or Wall-mount Power Supply (Cat. No. 1747-NP1).
8	DH-485 Programming Connector	Connects to a Personal Computer Interface Converter (Cat. No. 1747-PIC) for transferring applications. Also connects to an SLC programmer, such as the Hand-held Terminal (Cat. No. 1747-PT1).
9	RS-232 (DH-485) Communication Port	Connects to the Channel 0 port of an SLC 5/03, 5/04, or 5/05 for point-to-point DH-485 communication. Connects to a MicroLogix controller through an AIC+ Link Coupler. Also connects to the RS-232 serial port of a computer for transferring applications.
10	Remote I/O Port	Connects to a scanner or sub-scanner on a remote I/O network.
11	DH+ Communication Port	Connects to a PLC-5, SLC 5/04, or ControlLogix controller on a DH+ link.
12	DeviceNet Connector	Connects to a DeviceNet network.
13	ControlNet Connector	Connects to a ControlLogix controller (with 1756-CNB module) or PLC-5 on a ControlNet network.
14	(RS-232) DF1 Communication Port	Connects to a PLC, SLC, or MicroLogix controller with a DF1 port.
15	Ethernet Connector	Connects to a PLC-5E or SLC 5/05 controller, or a ControlLogix, MicroLogix. FlexLogix, or CompactLogix (with appropriate bridge module) on an EtherNet/IP network.
16	RS-232 Printer/File Transfer Port	Connects to a printer.

TABLE 5.2   PanelView 600 back features. (*Courtesy of Allen-Bradley*)

A motorized garage door simulation HMI is shown in Fig. 5.6. The simulator uses the following components:

- Operator station display panel and controls
- OPEN, CLOSE, and STOP push button switches
- AJAR, OPEN, SHUT panel indicators
- Fully open and fully closed limit switches
- Two door motors—one for opening and one for closing

The garage door is equipped with two motors (up/down) and two limit switches (LS1 and LS2). The door can move in the up and the down direction. Figure 5.7 shows

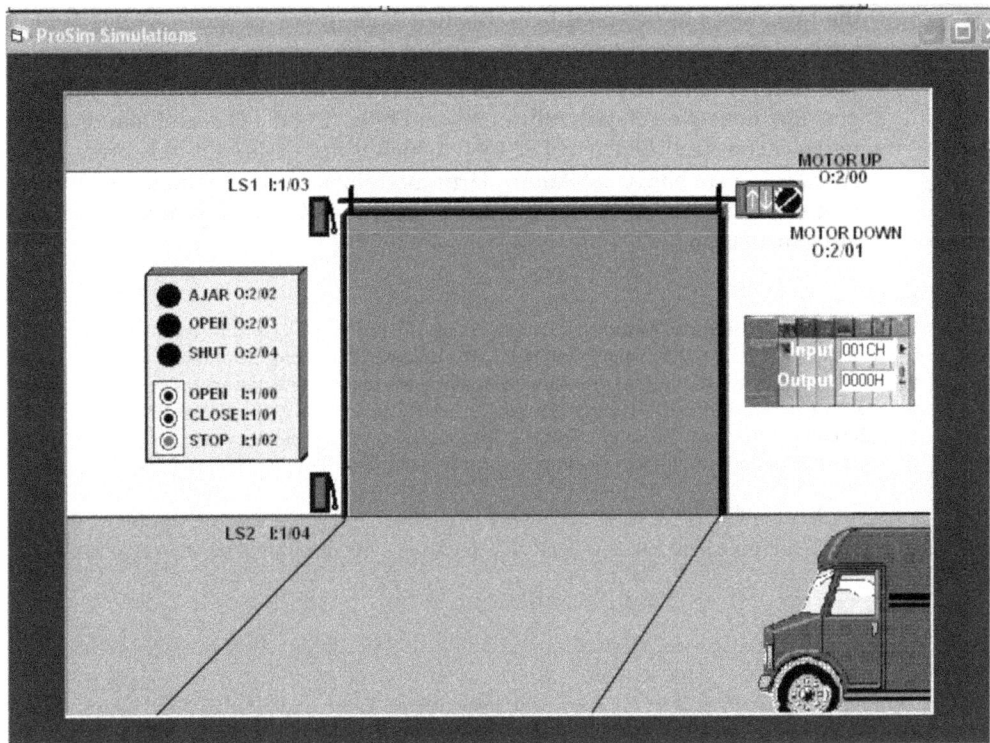

FIGURE 5.6   Garage door control/monitoring. (*Garage door simulator courtesy of the Learning Pit LogixPro 500.*)

the ladder logic implementation for the garage door control, which must accomplish the following:

- Open the door—while the operator is holding the OPEN PB switch on the operator control panel and the door is not fully open, turn on the OPEN door pilot light address O:2/3 on the control panel and MOTOR UP address O:2/0 to indicate that the door is moving up and the motor is commanded to run in the forward direction. Turn off the OPEN pilot light O:2/3 on the control panel once the door reaches the fully open position.

- Close the door—while the operator is holding the CLOSE PB switch on the operator control panel and the door is not fully closed, turn on the AJAR door pilot light address O:2/2 on the control panel and MOTOR DOWN address O:2/1 to indicate that the door is moving down and motor is commanded to run in the reverse direction. Turn off the SHUT pilot light O:2/4 on the control panel once the door reaches the fully closed position.

- Move the door in between—In each of the above cases, the status of the limit switches along with an ON AJAR light will indicate a door position in between the two limit switches. Both the OPEN and CLOSE actions are not latched. Releasing the PB switch will stop the door movement. The STOP PB is listed in the first rung but not used here. Because both open and close actions are not latched; pressing and releasing the STOP PB will not affect the door operation.

- The status of the two limit switches is translated to the associated garage door position: in between (ajar), fully open, and fully closed. One combination of the two switches (LS1 OFF and LS2 ON), which is not covered in this program, is an indication of hardware failure alarm conditions. Such situation, even though it is not expected to take place, must be included in a commercial program implementation.

Example 5.3.1 Flashing Light Indicators: Modify the program for Fig. 5.7 by adding a flasher action to the OPEN and SHUT pilot lights. We will use one of the RSLogix PLC's Free Running Timer bits in word S:4; each bit alternates at different rate with bit 0 having the highest rate. We are using bit 8 (S:4/8) for this purpose in our example. AB PLCs provide consistent rate, but with LogixPro 500 the rate varies with the computer speed. Test and select the appropriate flashing bit for your computer. Figure 5.8 shows the new ladder implementation that accommodates the following changes:

- The OPEN pilot light will flash while the door is moving up until the door is fully open.
- The SHUT pilot light will flash while the door is moving down until the door is fully closed.

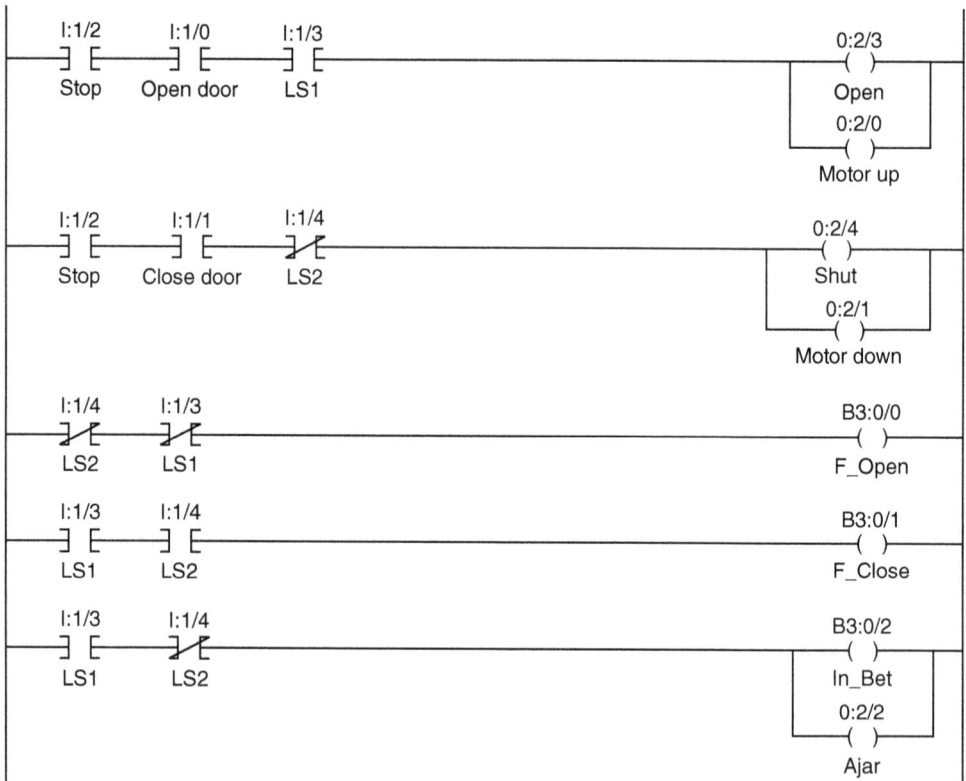

FIGURE 5.7   Garage door ladder program.

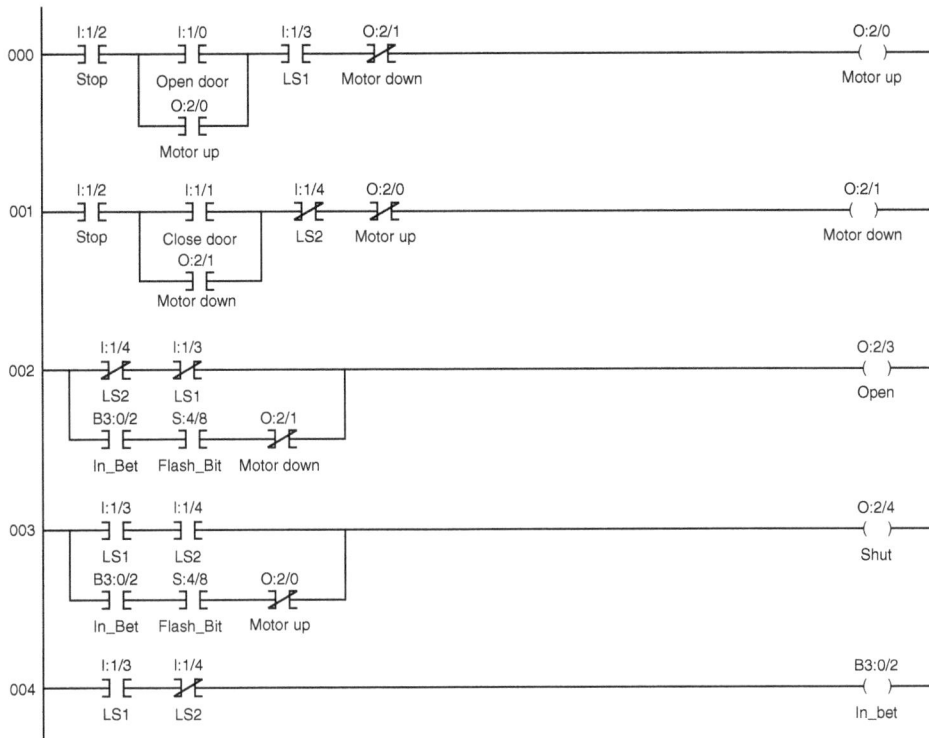

FIGURE 5.8 Flashing lights garage door ladder program.

Example 5.3.2 Painting Nozzle: A painting nozzle is controlled manually by the operator to paint a part for a certain time while the paint head moves vertically between two limit switches. The painting system includes two motors (up and down movements), a pair of limit switches (LS1 and LS2), a normally open START PB to start the motor, and normally close STOP PB. All components are connected to the PLC processor module. Using the LogixPro 500 simulation interface for the garage door control, implement a painting system control assuming that the garage door is a substitute for the painting nozzle. Table 5.3 shows the status of the two limit switches.

The OPEN or CLOSE PBs will be used to initiate the movement of the motor. First-time movement is assumed to follow the direction of the activated PB if the nozzle is not already fully up or fully down. Forward movement will continue as long as the limit switch (LS1) is true. Nozzle movements will start without activation if the nozzle is fully up or fully down.

LS1	LS2	Nozzle Location
0	0	Fully open
0	1	Alarm
1	0	In between
1	1	Fully close

TABLE 5.3 Limit Switches Status

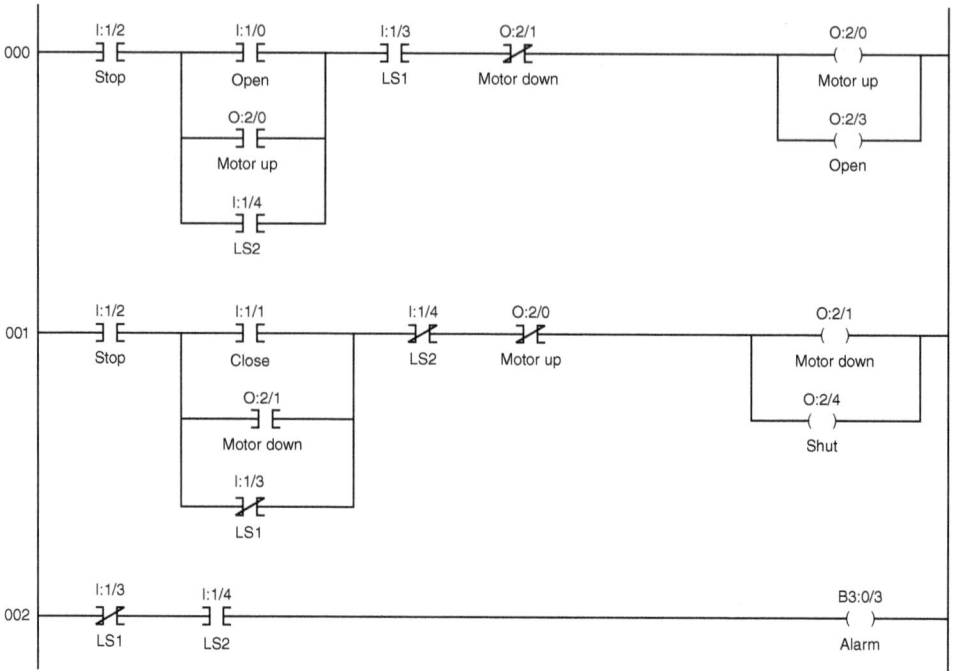

Figure 5.9   Manual painting nozzle ladder program.

Movement will be maintained in either directions and continuously run back and forth between the two limit switches till the operator pushes the STOP PB. Figure 5.9 shows the ladder program implementation of the stated painting system requirement.

The following is the detail of the three ladder rungs implementation:

- If the STOP PB is not pressed, the OPEN PB is pressed or LS2 is ON, and LS1 is ON, then the motor will run forward. The forward motion will continue while the STOP PB is not pressed until LS1 becomes OFF, which causes the motor to reverse direction.

- If the STOP PB is not pressed, CLOSE PB is pressed or LS1 is OFF, and LS2 is OFF, then the motor will run in the reverse direction. The reverse motion will continue while the STOP PB is not pressed until LS2 becomes ON, which causes the motor to reverse direction.

- An alarm (ALARM bit B3: 0/3) will be issued if the limit switches status indicates a hardware failure (LS1 OFF and LS2 ON). This condition can take place only under limit switch malfunction.

Example 5.3.3   Modify Example 5.3.2 to automatically control the painting time, assuming a simulated time of 40 seconds. The ladder program implementing the new requirements is shown in Fig. 5.10. The following is a description of the ladder rungs:

- If either the OPEN or the CLOSE PB is pressed, timer T4: 0 becomes active for 40 seconds.
- If the STOP PB is not pressed, the OPEN PB is pressed or LS2 is ON, LS1 is ON, and timer timing bit is set, then the motor will run forward. The forward motion will continue while the STOP PB is not pressed until LS1 becomes OFF, which causes the motor to reverse direction.

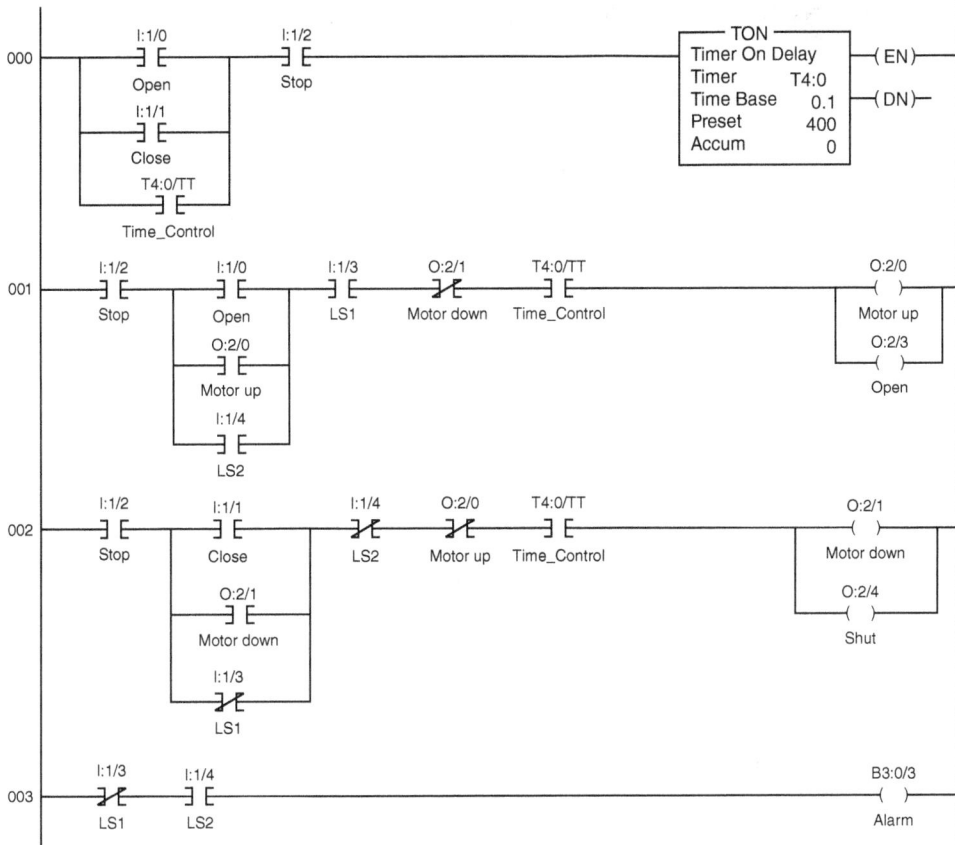

Figure 5.10 Timed painting process ladder program.

- If the STOP PB is not pressed, the CLOSE PB is pressed or LS1 is off, LS2 is off, and timer timing bit is set, then the motor will run in the reverse direction. The reverse motion will continue while the STOP PB is not pressed until LS2 becomes on, which causes the motor to reverse direction.

- Once the timer is done (timing bit is OFF), the motor stops and the vertical painting process is terminated.

5.4   Silo Conveyor Control Operation

The Silo conveyor control operation simulator shown in Fig. 5.11 will automatically position and fill the boxes that are continuously sequenced along the moving conveyor. The control ensures that the following requirements are met:

The sequence can be stopped and restarted at any time using the panel mounted STOP and START PB switches:

- The RUN light will remain energized as long as the system is operating automatically.

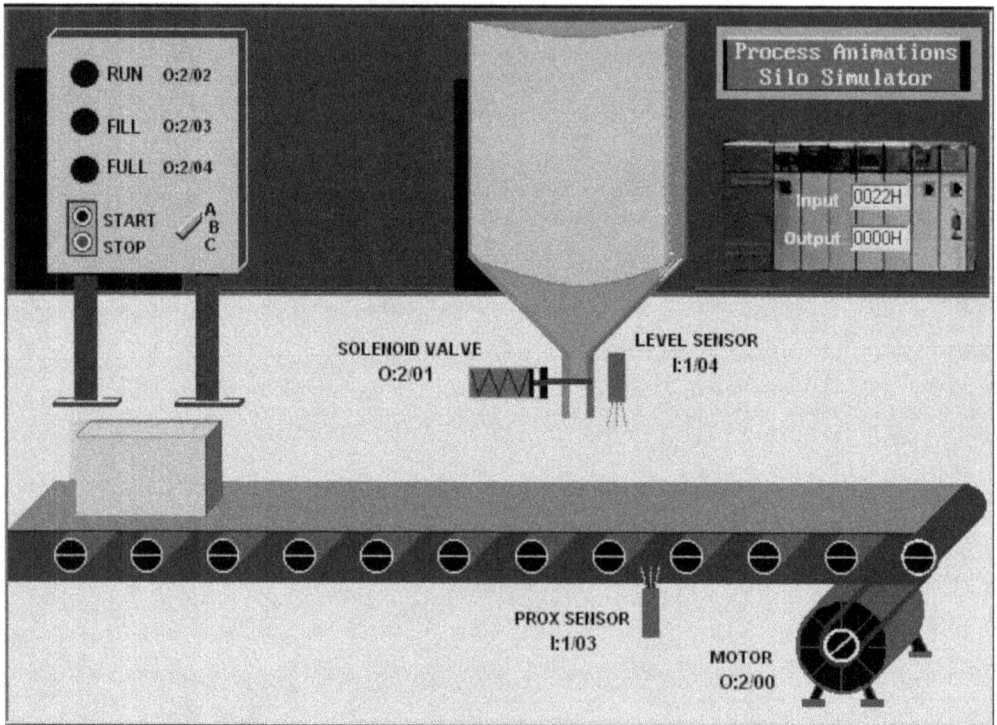

FIGURE 5.11   Silo control/monitoring. (*Silo simulator courtesy of the Learning Pit LogixPro 500.*)

- The RUN light, conveyor motor, and solenoid will de-energize whenever the system is halted via the STOP PB switch.
- The FILL light will be energized while the box is filling.
- The FULL light will energize when the box is full and will remain that way until the box has moved clear of the proximity sensor.
- Once the conveyor has begun, the sequence may be stopped and resumed at any time using the STOP PB except if the box is blocking the proximity sensor, which will cause a system alarm.

Figure 5.12 shows the ladder program used to implement the Silo continuous operation control requirements. The selector switch has to be placed on position A (I:1/5) for this mode of operation. The following is a brief discussion of the program three rungs:

1. Motor (O:2/0) and activate the RUN light (O:2/2). The motor will stop once the solenoid valve is ON during the box fill operation. Once the box FULL (O:2/4) indicator is on, the motor will start again. Activating the STOP PB (I:1/1) at any time will stop the motor.

2. Once the BOX IN PLACE indicator (I:1/3) makes the transition from false to true, the motor will stop and the solenoid valve is activated to start the box filling process and the FILL light (O:2/3) is turned ON. As the filling process

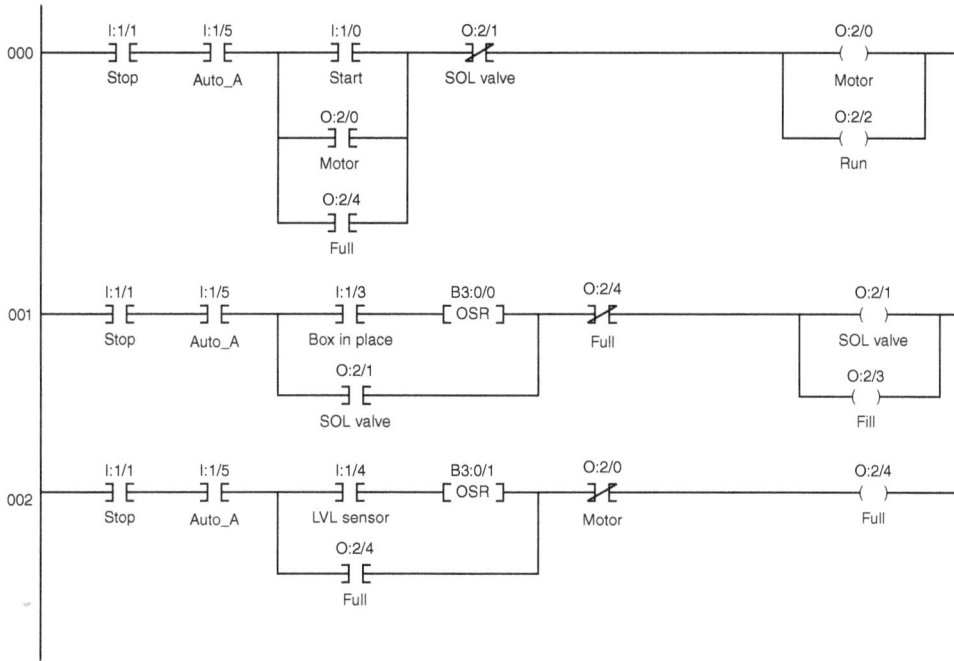

FIGURE 5.12 Silo continuous control operation ladder program.

stops once, the box FULL indicator (O:2/4) becomes true. Activating the STOP PB at any time will turn the solenoid valve OFF.

3. Once the level sensor makes the transition from false to true indicating a full box, the FULL indicator is turned on. This will cause the solenoid valve to go off (stopping the filling) and the motor to start (moving the filled box). Once the motor runs, the FULL indicator will go off.

Example 5.4.1 The selector switch must be on the B (I:1/6) position for the manual restart operation mode. Figure 5.13 shows the ladder program implementation used to alter the Silo control program for manual restart so that it incorporates the following changes:

- Stop the conveyor motor when the box is first sensed by the proximity sensor or the STOP PB is pressed or the solenoid valve is ON (open for filling).

- With the box in position and the conveyor stopped, open the solenoid valve and allow the box to fill. Filling should stop when the level sensor goes true.

- The FILL light will be energized while the box is filling.

- The FULL light will be energized when the box is full and will remain that way until the box has moved clear of the proximity sensor.

- Once the box is full, momentarily pressing the START PB switch will move the box away from the proximity switch and bring a new box into position. Forcing the operator to hold the START button down until the box clears the proximity sensor is not acceptable. The start action must be latched.

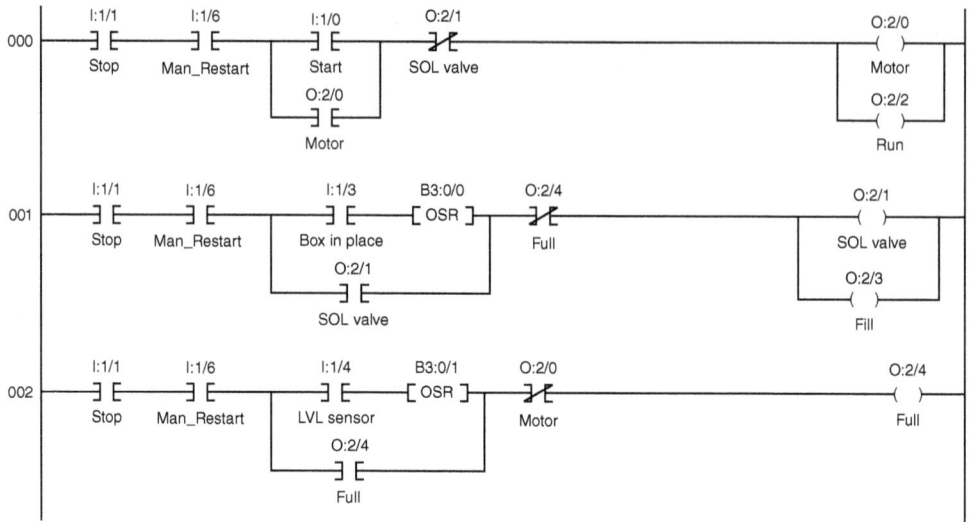

Figure 5.13 Silo manual restart control ladder program.

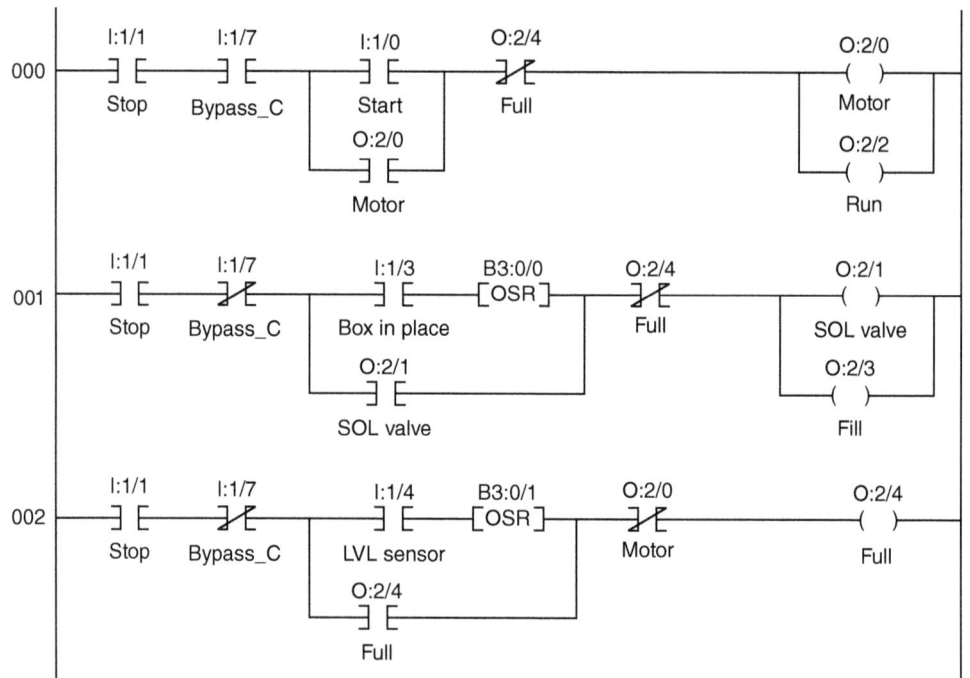

Figure 5.14 Silo bypass control ladder program.

Example 5.4.2   A bypass control mode, selector switch placed on position C (I:1/7), will continuously move boxes and bypass the filling station as shown in Fig. 5.14 ladder.

```
        I:1/1        I:1/0                    O:2/1                              O:2/0
000   ──] [──┬──────] [─────────┬────────────]/[────────────────────────────( )
        Stop  │     Start        │            SOL valve                        Motor
              │    O:2/0          │                                            O:2/2
              ├────] [────────────┤                                          ──( )
              │    Motor          │                                            Run
              │   O:2/4   I:1/5   I:1/6
              └───] [────] [─────]/[┘
                  Full   Auto_A   MAN_B

        I:1/1    I:1/7       I:1/3      B3:0/0    O:2/4                        O:2/1
001   ──] [────]/[────┬─────] [───────[OSR]─────]/[──────────────────────────( )
        Stop  Bypass_C │  Box in place           Full                        SOL valve
                       │   O:2/1                                              O:2/3
                       └──] [──┘                                            ──( )
                          SOL valve                                          Fill

        I:1/1    I:1/7       I:1/4      B3:0/1    O:2/0                        O:2/4
002   ──] [────]/[────┬─────] [───────[OSR]─────]/[──────────────────────────( )
        Stop  Bypass_C │  LVL sensor             Motor                        Full
                       │   O:2/4
                       └──] [──┘
                          Full
```

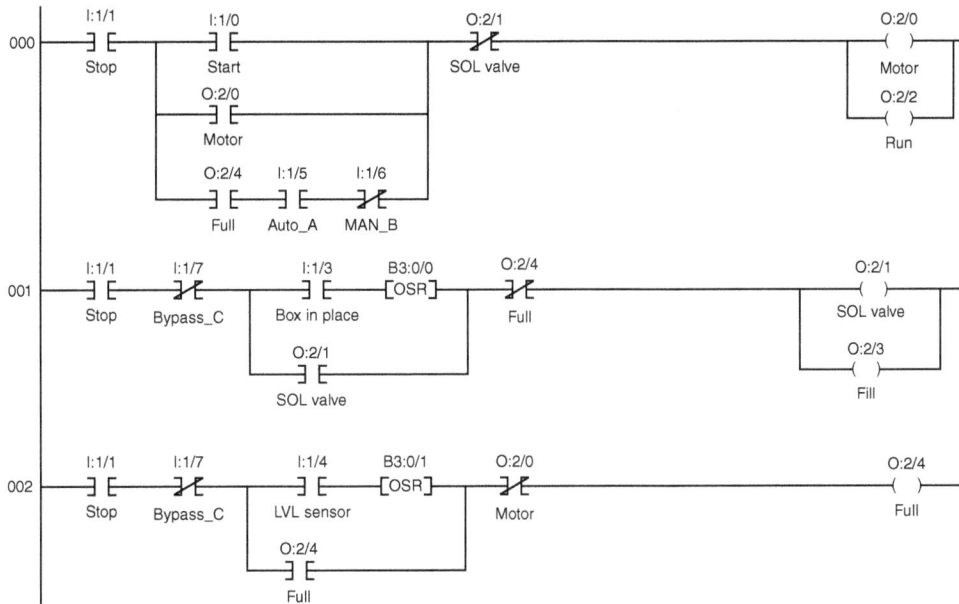

FIGURE 5.15 Silo selector switch control ladder program.

Example 5.4.3 Now it is time to give the operator the choice to run any of the three Silo operation modes discussed so far by changing the HMI panel selector switch. Position A for the continuous mode, position B for the manual restart, and position C for the bypass operation. As you can see in Fig. 5.15, only three rungs are needed to accommodate all the three modes of operation selection. Implement the program on LogixPro 500 and verify its operation using the Silo simulation panel tools.

5.5 Single Batch Mode Operation

The objective of this section is to get the reader familiar with the method used to control a batch process. Batch control often gets the correct proportion of ingredients into the batch and maintains level, flow, pressure, temperature, or other process variables at predefined set values. Such process typically requires frequently restarting, which introduces two problems. First, all selected variables for the batch control must be brought to their respective set level. Second, measurement instruments may need recalibration as recipes change in a multibatch operation. Figure 5.16 shows a chemical mixing single batch (same batch operation repeats) process.

The PLC control is expected to meet the following requirements:

- When the three-position selector switch is in position A simulating the auto position, the batch mixing process will run in a single-batch mode. The operator may start the batch mix sequence by momentarily pressing the START switch.

- Once a batch sequence has begun, the sequence may be stopped and resumed at any time using the STOP and the START PB switches.

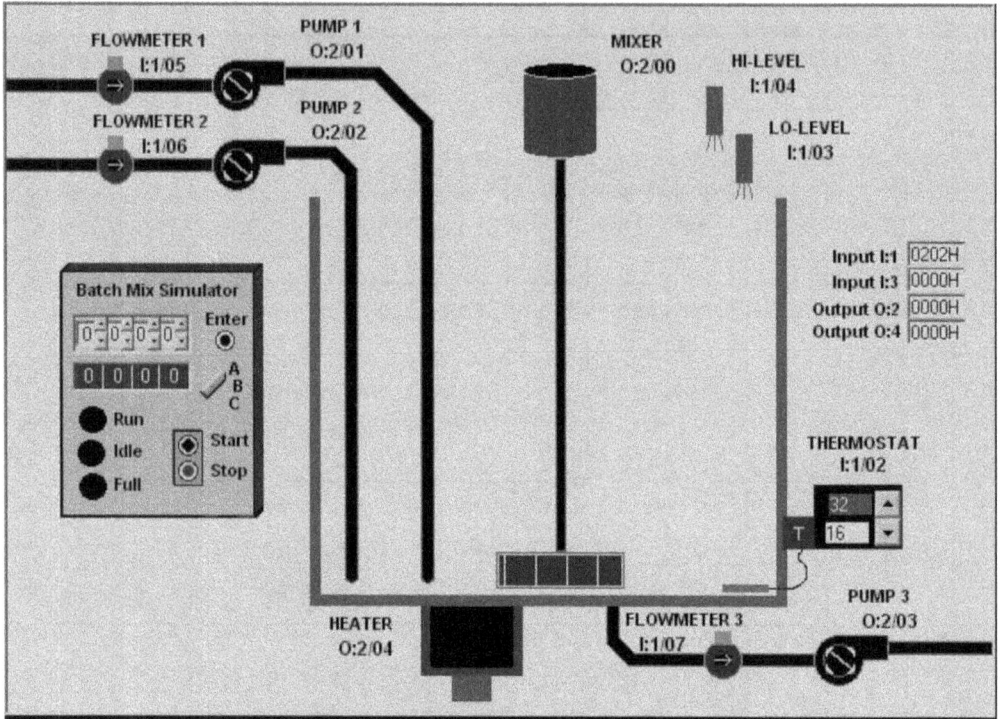

Figure 5.16 Batch process control/monitoring. (*Batch simulator courtesy of the Learning Pit LogixPro 500.*)

- The tank is to be filled with a mixture obtained from the separate fill lines utilizing fill pumps P1 and P2. When tank level reaches the high set limit, a mixture and heater will run for a period of 10 simulated seconds.

- When the mixing time is over, pump 3 will run and empty the tank. Once the tank is empty, pump 3 will stop.

Figure 5.17 shows a ladder program used to implement the single-batch process control specifications. This program only makes use of the selector switch, START PB, STOP PB, three pumps, mixer, heater, low-level sensor, and high-level sensor on the control panel. The following is the documentation for the five ladder rungs:

- Activating the START PB while the selector switch is placed on Auto (position A) will latch B3:0/10, the AUTO_START bit. Activating the STOP PB will unlatch B3:0/10.

- Activating the AUTO_START bit while the high-level sensor is OFF (the tank is not full) will start the two filling pumps (pump 1 and pump 2). The latching action is done using either of the two pumps to allow the filling while one pump is out of order or failed to start. The STOP PB can be activated to stop the filling at any time.

- Once the high-level sensor goes ON while the selector switch is on Auto, the tank full indicator goes ON. This action is latched by the reaction timer (heating

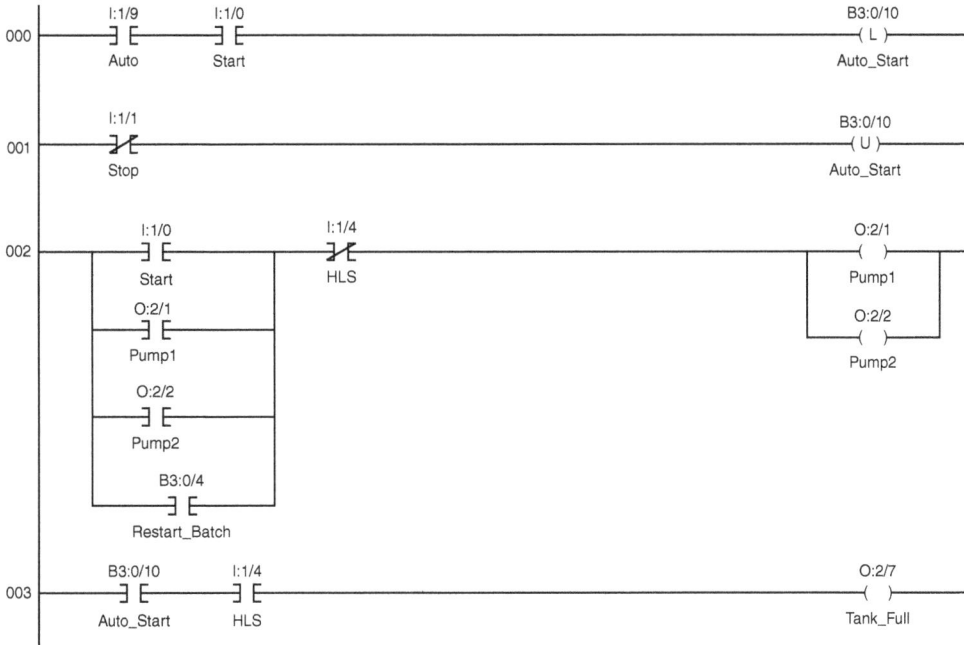

(a)

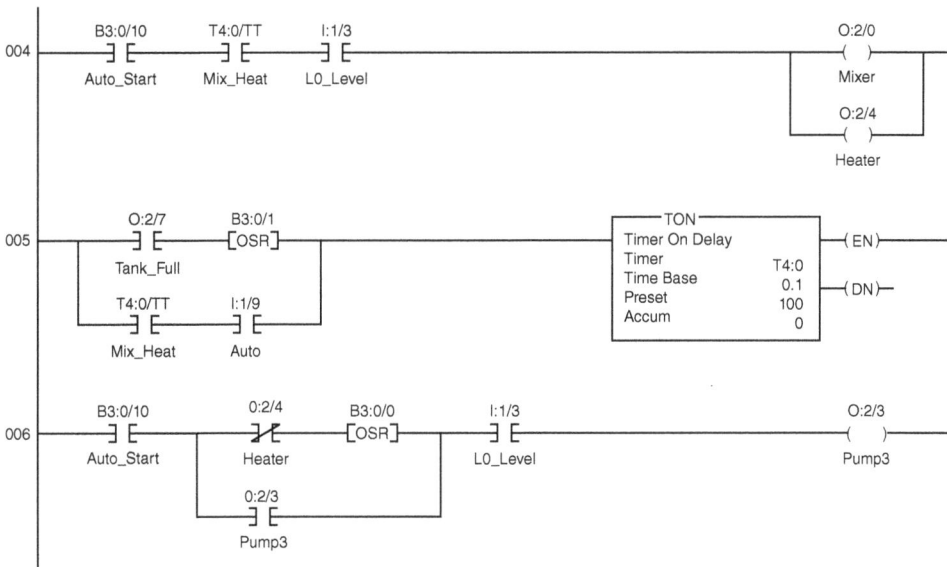

(b)

FIGURE 5.17 (a) Single batch ladder program; (b) single batch ladder program.

and mixing) to prevent fluctuation due to level transient. Activating the STOP PB will force the indicator to go OFF.

- If tank full indicator goes from low to high, a 10-second timer starts. The timing bit latches the tank full status as transient during the mixing can alter the level status.

- While the timer is timing as indicated by its timing bit, both heater and mixer are turned ON. Activating the STOP PB will turn both OFF.

- Once the filling and reaction time is over as indicated by the heater changing from ON to OFF, pump 3 starts the tank drainage process. The pumping action is latched until the low-level sensor goes ON. Activating the STOP PB at any time during the drainage process will stop pump 3.

Example 5.5.1   Modify the Fig. 5.17 program to accommodate continuous single-batch operation. The filling and emptying sequence repeats continuously once it has been started by the initial pressing of the START PB. The process can be terminated at any time by activating the STOP PB. The ladder program shown in Fig. 5.18 will implement the previously stated specifications for continuous batch operation in additional to the following requirements:

- The RUN light is ON when the mixer/heater or either pump 1 or pump 2 is running while the AUTO_START bit is latched (selector switch is ON auto position when the START PB is activated).

- The Idle light should turn ON and the process halts once the STOP PB is pressed. The Idle light will also go ON during the tank emptying operation.

- The batch process should run once the START PB is activated and the system selector switch is placed on Auto. The STOP PB activation will unlatch the AUTO_START bit and terminate the process.

- A batch restart bit (B3:0/4) is used to initiate another batch operation after the current one is finished. B3:0/4 is set once the tank low level sensor makes the transition from high to low indicating the end of the current batch. It remains set until the next mixing begins.

Example 5.5.2   Modify the continuous single-batch program to implement the following modifications shown in the ladder program of Fig. 5.19.

- Run pumps 1 and 2 once the process initiates through the START PB activation after initialization (clearing the two counters and the process variable used). The operator can disable either pump; our implementation does not verify if pumps are running.

- Assume that each flow meter pulse input represents 1 gal flow into the tank. Keep track of the amount of liquid in the tank in gallons measured from the batch start time. High-speed counters are commonly used in industry to provide pulses at a rate proportional to the flow to the PLC, which can be calibrated to adequately measure volume. Each pulse is assumed to represent 1 gal in this example.

- Allow the operator to input the desired value for the batch volume in whole gallons from the batch mix simulator panel 4 BCD input (I:3, which is set to 300 in our example).

- Fill the tank to the desired batch volume and then start the mixing/heating operation for 6 seconds followed by tank emptying. Update the amount of liquid in gallons on the batch mix simulator panel 4 BCD output (O:4) during operation: filling, reaction, and emptying.

- Notice that the tank high-level sensor is not used in this example; instead, we actually control the tank filling based on a desired liquid volume set value in gallons. High-level sensors are

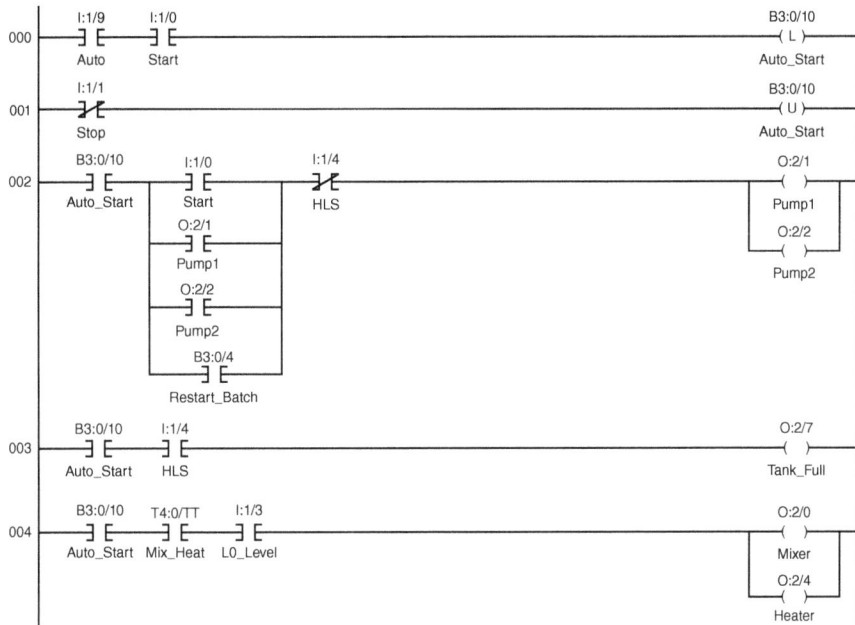

(a)

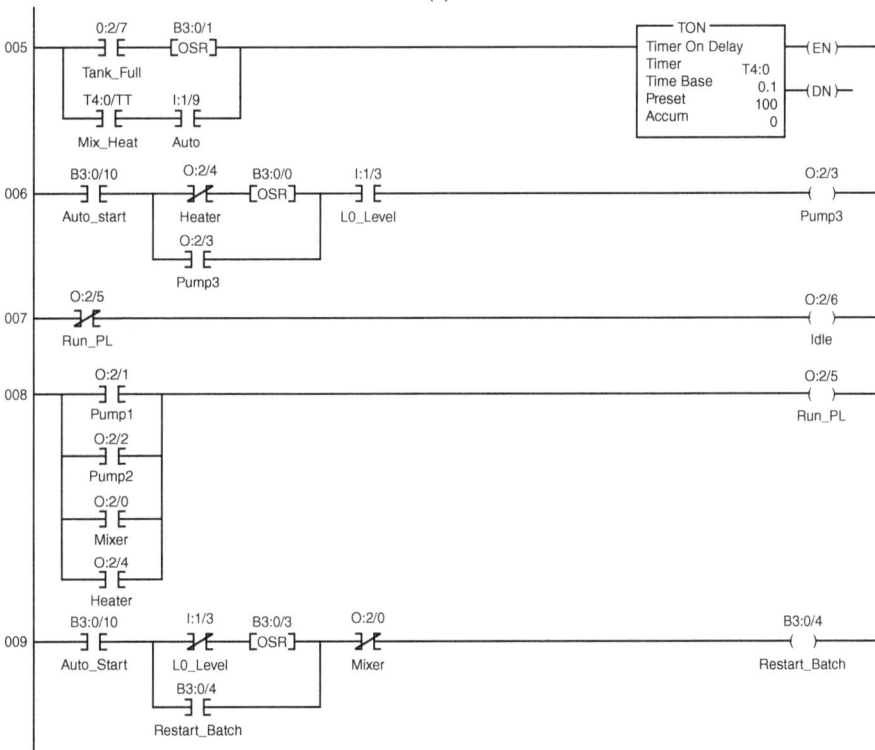

(b)

FIGURE 5.18   (*a*) Continuous single-batch ladder program; (*b*) continuous single-batch ladder program.

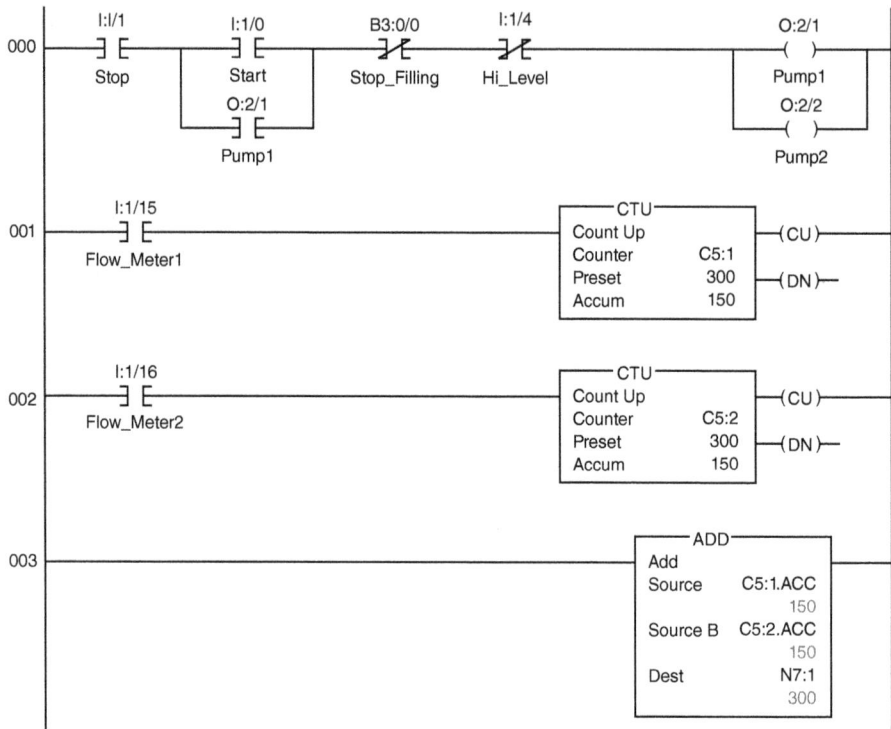

(a)

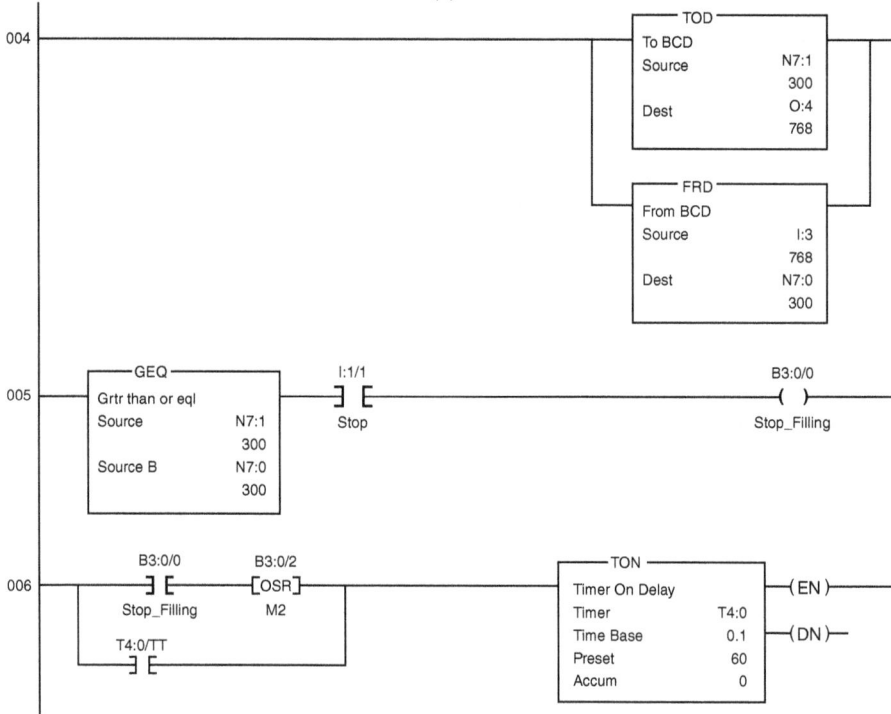

(b)

Figure 5.19 (a) Single-batch volume control ladder program; (b) single-batch volume control ladder program; (c) single-batch volume control ladder program.

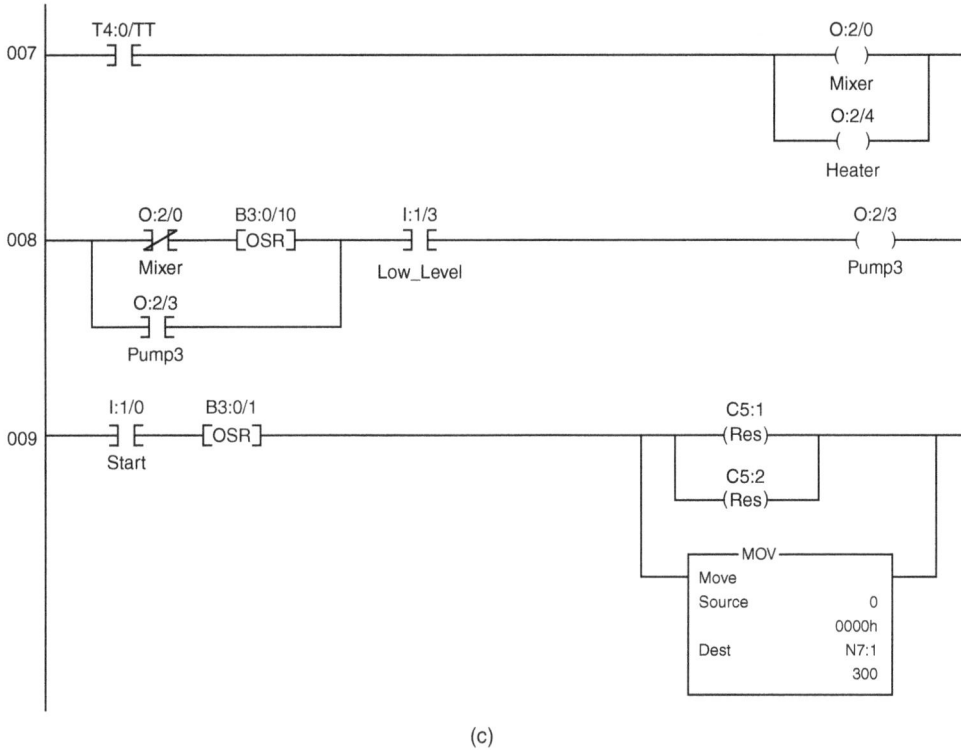

(c)

FIGURE 5.19   *(Continued)*

typically used for alarm conditions, but actual volume in a tank or reactor is normally calculated using flow measurements.

- Changing the batch volume or the mixing time can alter the process outcome, produce a failure, or cause a plant shutdown as other elements critical to a real reactor like temperature and pressure in the tank are not considered. All variables affecting the reaction outcome in a comprehensive real application must be considered, monitored, and regulated directly or through other means (see Chap. 9).

5.6   One-Way Intersection Traffic Light Control

The objective of this laboratory is to get students to learn how to use timers and HMIs in a real control application. The application presented in this section covers a one-way traffic light control. The system is simulated using the LogixPro components shown in Fig. 5.20. The simulator provides two traffic light signals at an intersection of two one-way roads. All the six lights and two cross-walks signs are connected to the PLC discrete outputs O:2/0 to O:2/7. I:1/0 and I:1/1 provide the crossing button connection to the PLC. I:1/2 connects to the S-N road car sensor, while I:1/3 connects to the E-W road car sensor. A park input can be activated to stop car movement in either of the two

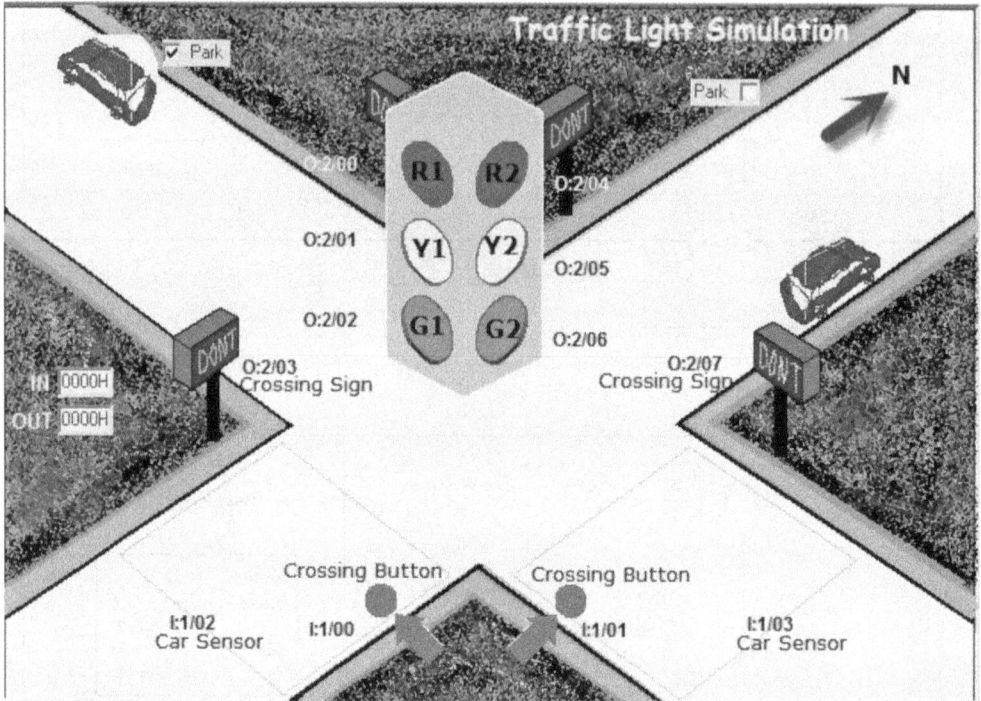

FIGURE 5.20 One-way terrific light controls/monitoring. (*Traffic light simulator courtesy of the Learning Pit LogixPro 500.*)

roads. To simplify our discussion, we will park cars on the E-W road and only control the traffic lights for the S-N road.

The sequence of traffic light operation for one road traffic is summarized in Table 5.4 and follows the following ordered timing sequence:

1. Red light R1 (O:2/00) is ON for 12 seconds.

2. Green light G1 (O:2/02) is ON for 8 seconds.

3. Yellow/amber light Y1 (O:2/01) is ON for 4 seconds.

4. The sequence repeats with the red light R1 ON again.

Figure 5.21 shows the ladder program realizing the one-way traffic signal control for the S-N road. The E-W road cars are parked. The following are details of the ladder rungs:

- A single timer is used to control the three lights T4:0 with a 24-second preset time split into the three specified intervals. The timer is programmed to continuously recycle on the done bit.

- An LEQ and GEQ comparison is used to regulate the ON time for each light and maintain the sequence: red, green, and yellow. The WALK_EW turns ON during red and the WALK-SN turns ON during green.

Red 1 (R1)	Green 1 (G1)	Yellow 1 (Y1)
12 seconds	8 seconds	4 seconds

TABLE 5.4 Traffic Light Timing Sequence

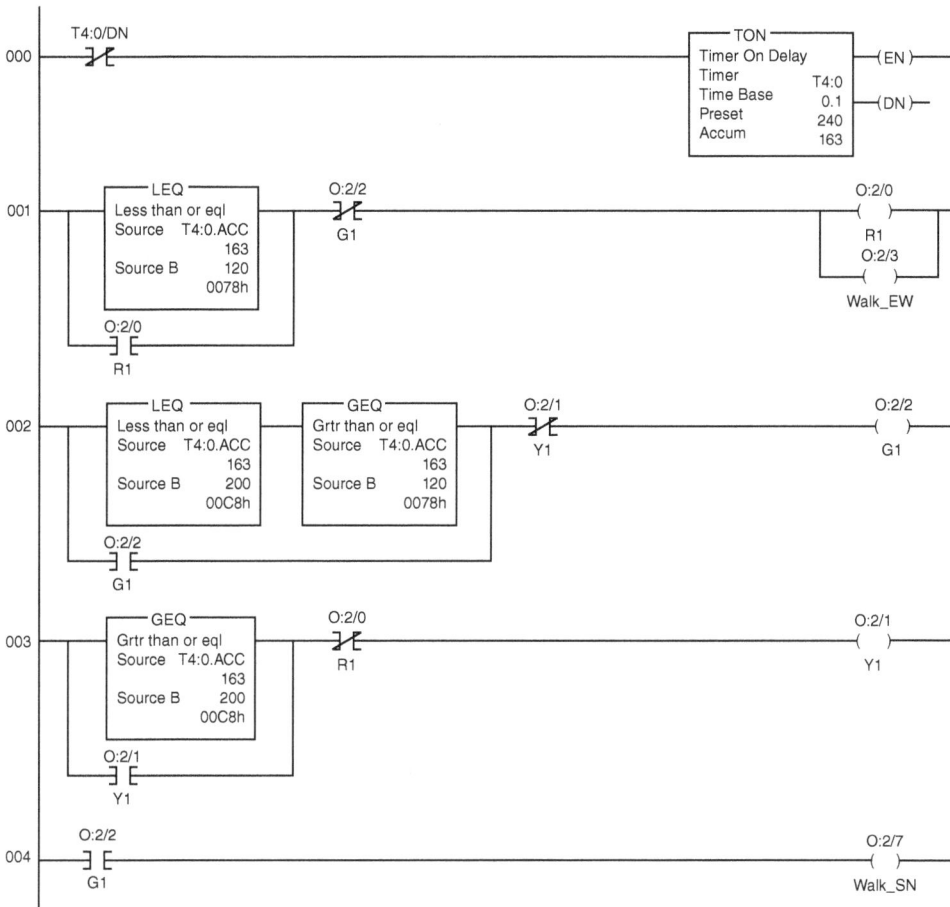

FIGURE 5.21 One-way S-N road traffic signal ladder program.

Example 5.6.1 This example shows an implementation of the one-way traffic light signal control using three timers, one for each light. It illustrates that the same application can be equally and efficiently implemented using different techniques. Figure 5.22 shows the ladder program realizing the desired control.

The ladder program for the one-way road traffic lights uses three timers: T4:0 through T4:2. One timer is assigned for each light: T4:0 is used for the red light, T4:1 is used for the green light, and T4:2 is

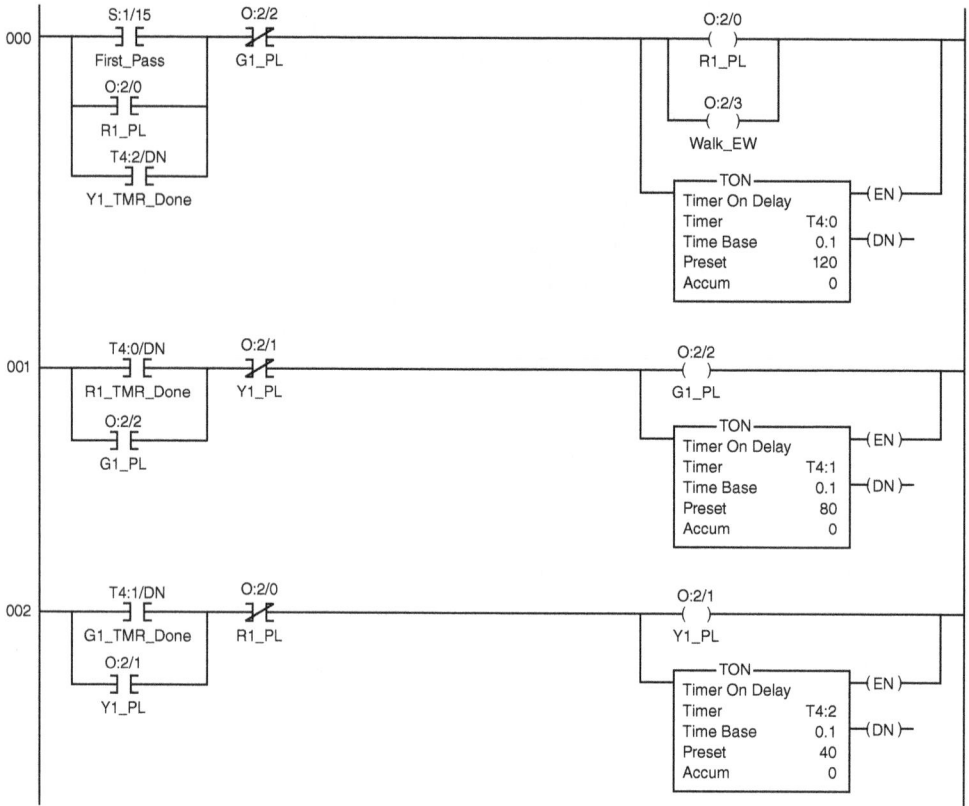

FIGURE 5.22 Two-way traffic signal ladder program.

used for the yellow light. The following is not a comprehensive coverage but an outline of the core elements used in our implementation:

- Make sure that cars are parked for both directions: S-N and E-W before starting the simulation. Also, reset simulations, rest timer and counters, and clear all data files from the simulations drop menu before running the program.

- A program enable bit S: 1/15 "FIRST_PASS" is used to start the program. This bit is toggled (OFF to ON transition) by the processor only during the first scan after power-up in order to allow program scanning and execution. The park must be cleared on the S-N direction to allow cars flow on this one-way road.

- The walk sign "WALK_EW" is turned on during a red light. The Walk push button and the car at intersection input sensor are not used in this implementation. In a real-life implementation, delays along with coordination with these sensors are needed to ensure safety at a pedestrian crossing.

- It is customary that designers get carried through the excitement of the real-time debugging process and end up with a more complex implementation than needed, which is extremely difficult to document and maintain.

Red$_1$ = O:2/00			Green$_1$ = O:2/02	Amber$_1$ = O:2/01
Green$_2$ = O:2/06	Amber$_2$ = O:2/05		Red$_2$ = O:2/04	
8 seconds	4 seconds	x	8 seconds	4 seconds

TABLE 5.5   Two-Way Traffic Light Timing Sequence

(T4:0.ACC) in Seconds	Event	End Condition	Comments
LEQ 14	R1 ON	G1 goes ON	R1 ON for the first 14 sec of cycle.
LEQ 22 and GRT 14	G1 ON	Y1 goes ON	G1 stays ON for the next 8 seconds.
GRT 22	Y1 ON	G1 goes ON	Y1 ON for the last 4 seconds of the cycle.
LEQ 12	WALK_EW	12 seconds	Goes OFF 2 seconds before R1 expires.
LEQ 8	G2 ON	Y2 goes ON	G2 stays ON during the first 8 seconds.
LEQ 12 and GRT 16	Y2 ON	R2 goes ON	Y2 stays ON during the next 4 seconds.
GRT 16	R2 ON	G2 goes ON	R2 ON for the last 14 seconds of cycle.
LEQ 24 and GRT 12	WALK_SN	12 seconds	Goes OFF 2 seconds before R2 expires.

TABLE 5.6   Two-Way Traffic Light Sequence Timing

Example 5.6.2   Expand the control implementation to the two crossing roads traffic light control using the timing sequence shown in Table 5.5. The modified ladder program will control the three lights, which represent the E-W traffic direction and coordinate their operation with the S-N 3-lights traffic signals. An "x"-second overlap time is added to the red light to prevent crashes due to cars passing the crossing at the tail end of the amber light. Stretching this time will prevent crashes at the cost of lower traffic throughput at the crossing. The specification in Table 5.5 is shown in a compact format not directly usable for programming.

This application will be implemented here using one timer and multiple comparison operations. Table 5.6 is another form for the same control timing, which lists the critical events and associated times. Two seconds R1 and R2 overlap time is used in this example implementation. The ladder logic program realizing the required specification is shown in Fig. 5.23. This implementation is simpler and easier to understand/maintain than the other solution using multiple timers.

5.7   Bottling Assembly Line Control

The LogixPro 500 bottle conveyor control simulator shown in Fig. 5.24 allows the operator to start and stop the process using the available panel-mounted switches. When the process is running, the main conveyor should be energized and bottles should

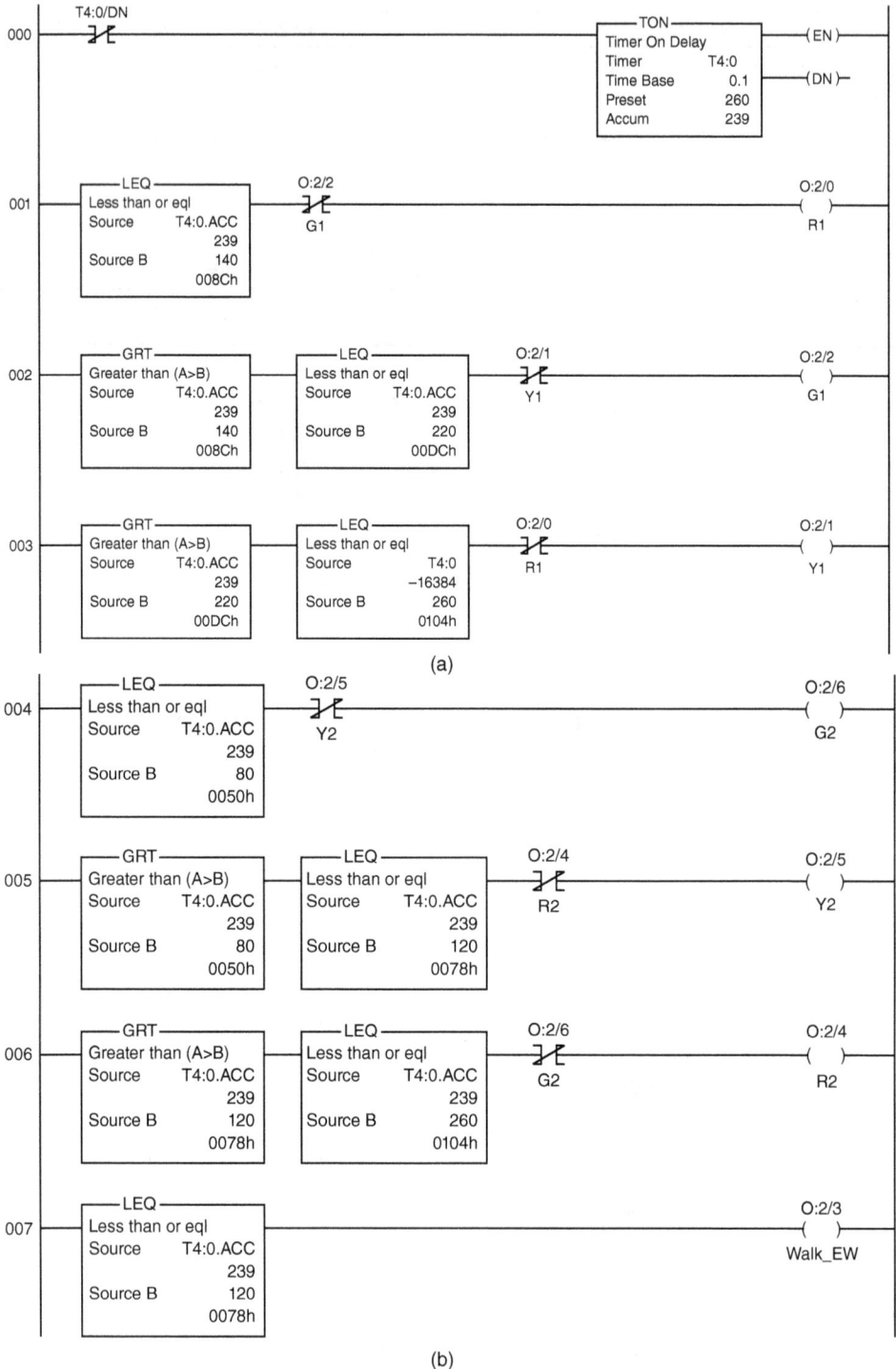

Figure 5.23 (a) Two-way traffic signal simplified ladder program; (b) two-way traffic signal simplified ladder program; (c) two-way traffic signal simplified ladder program.

```
        ┌──────GRT──────┐   ┌──────LEQ──────┐                              O:2/7
  008 ──┤ Greater than (A>B) ├─┤ Less than or eql ├──────────────────────────( )──
        │ Source    T4:0.ACC │ │ Source    T4:0.ACC │                        Walk_SN
        │              239  │ │              239  │
        │ Source B     120  │ │ Source B     240  │
        │             0078h │ │             00F0h │
        └───────────────────┘ └───────────────────┘
```

(c)

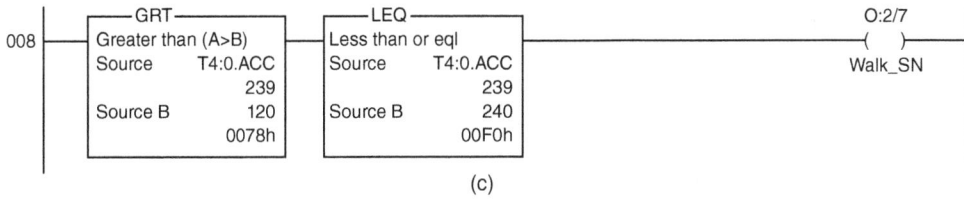

FIGURE 5.23 (Continued)

continuously enter and exit the line. The following is brief description of the key elements of the control process:

- The sequence can be stopped and restarted at any time using the panel-mounted STOP and START PB switches.

- The RUN light will remain energized as long as the system is operating and the STOP PB is not activated.

- The RUN light and conveyor motor will be de-energized whenever the system is halted via the STOP PB switch.

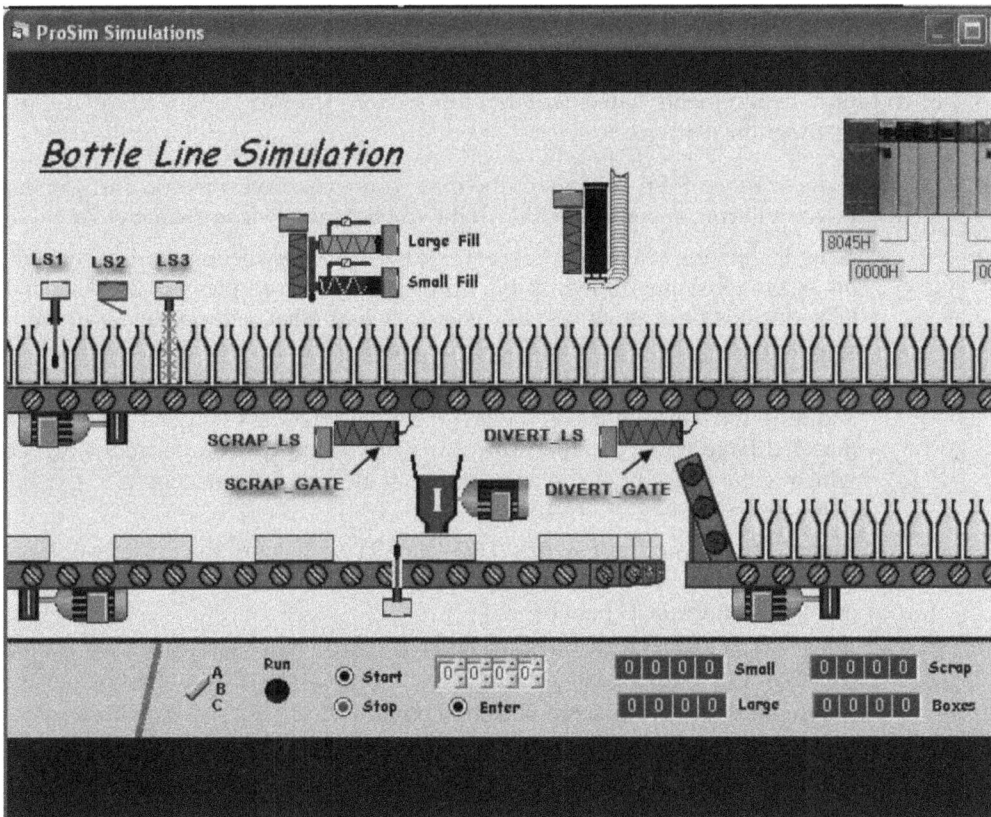

FIGURE 5.24 Bottling assembly production line. (*Traffic bottling simulator courtesy of the Learning Pit LogixPro 500.*)

- The limit switch LS8 (I:1/13) is energized every time it detects scrap bottle, which later in time (when the detected scrap bottle reaches the location of the scrap gate solenoid) activates the scrap gate solenoid (O:2/4) in order to reject the scrap bottle and divert it to the scrap conveyor.

- The limit switch LS9 (I:1/14) is energized every time it detects large bottle and later in time activates the large gate solenoid (O:2/5), which diverts the bottle and moves it to the large conveyor.

- Scrap bottles are diverted, crushed, and placed in boxes. The box is assumed to hold the scrap glass from eight consecutive bottles regardless of size. This implementation will process partially full boxes. More accurate implementation can be done knowing that a small bottle produces two-thirds of the glass of the large bottle and that the capacity of the box is 8 large bottles or 12 small bottles.

- A display of four BCD output digits is available for the total count of large, small, scrap, and boxes processed. In our implementation only processed large and scrap bottles are counted. Also, four BCD input digits and associated Enter PB are not utilized in this implementation.

- A fill station for small and large bottles is available but not used in this discussion. Also, the selector switch is not used in this implementation.

Figure 5.25 shows the ladder program used to implement the bottle conveyor operation control requirements stated earlier in this section. The following is a brief discussion of the program 10 rungs:

- Rung 0: When S PB is pressed, the main conveyor, divert motor, and grinder motor will run, and also the RUN light will light on the control panel.

- Rung 1: Utilizing LS1 (EXIT_LS) to strobe three BSL instructions to shift separate bit arrays consisting of two 16-bit words each. We used files #B3:2, #B3:4, and #B3:6 for this purpose. It will be easier to debug what is happening with your program using the data table display to monitor the selected data files.

- Rung 2: The divert bit (B3:5/4) in the array will energize the divert gate solenoid and drop the large bottles on the divert conveyor. B3:5/4 will be 1 indicating a detected large bottle 21 (16 bits or one word + 5 bits) bottle locations earlier, which is the capacity of the conveyor section from the detection point to the diversion point.

- Rung 3: The divert limit switch (LS9_DIVERT) will count the number of large bottles dropped on the large conveyors and convert these numbers to BCD displayed on the BCD panel.

- Rung 4: The scrap bit (B3: 6/8) in the scrap array will energize the scrap gate solenoid and drop the scrap bottles on the divert conveyor. B3:6/8 will be 1 eight bottles after the scrap detection point.

- Rung 5: The scrap limit switch (LS8_SCRAP) will count the number of scrap bottles dropped on the scrap conveyor and convert these numbers to BCD displayed on the BCD panel.

- Rung 6: The scrap limit switch (LS8_SCRAP) will increment the scrap bottle counter (C5:0) each time a scrap bottle is dropped in the scrap conveyor.

(a)

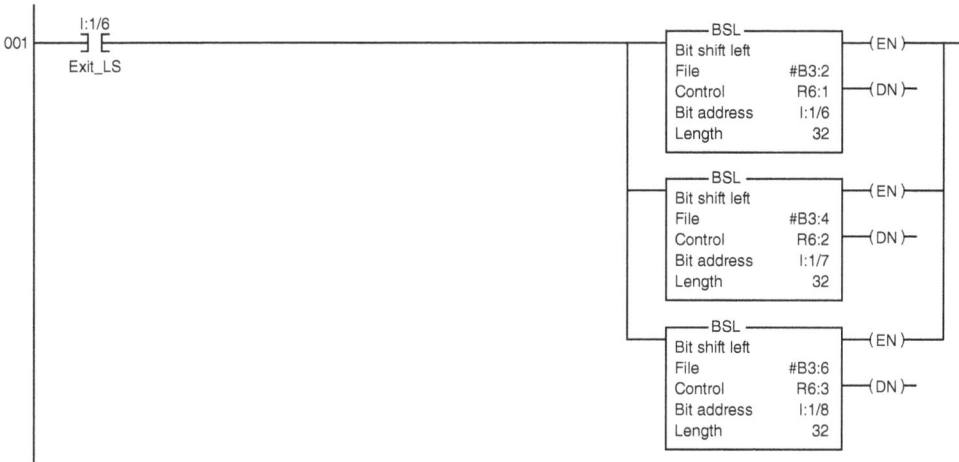

(b)

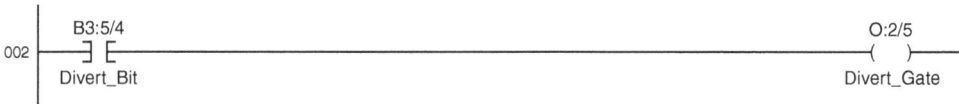

(c)

(d)

Figure 5.25 (a) Bottling control rung 0; (b) bottling control rung 1; (c) bottling control rung 2; (d) bottling control rung 3; (e) bottling control rung 4; (f) bottling control rung 5; (g) bottling control rung 6; (h) bottling control rung 7; (i) bottling control rung 8 and 9.

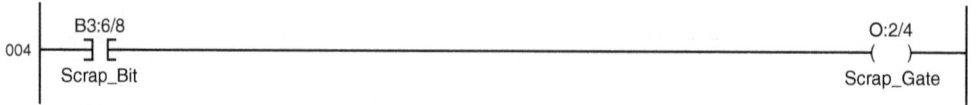

(e)

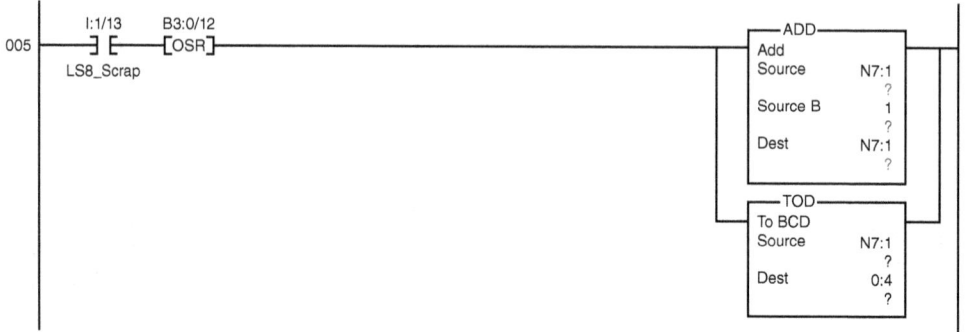

(f)

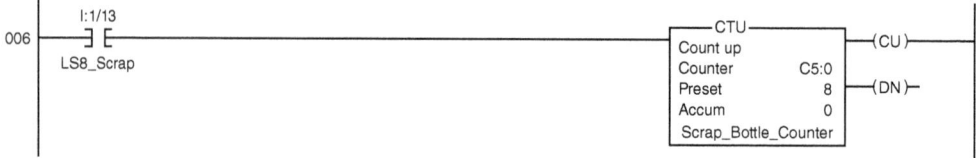

(g)

(h)

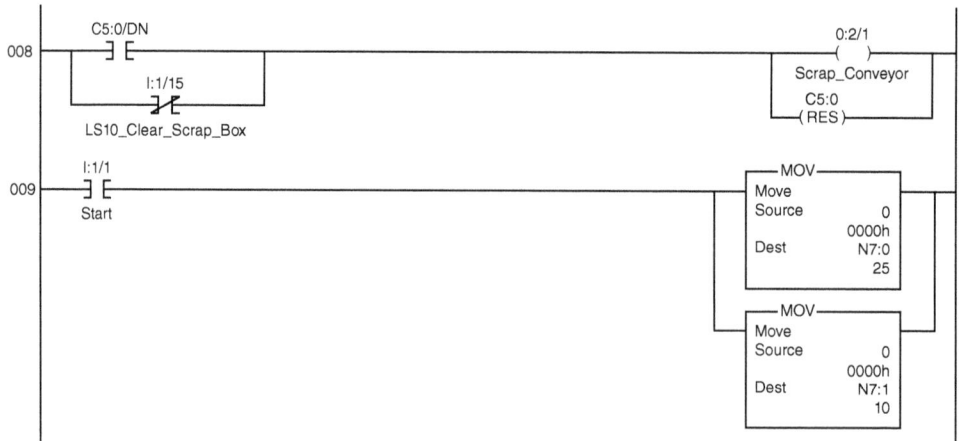

(i)

FIGURE 5.25 (Continued)

- Rung 7: The purpose of the 1-second timer (T4:0) in this rung is to delay the execution of Rung 8 till last bottle grinding and deposit in the scrap box is finished. A shorter timer preset value will also work.

- Rungs 8 and 9: Once eight consecutive bottles are scrapped or the clear limit switch (LS10) is high (indicating that scrap box is full), the scrap conveyor is turned ON. Also, the scrap bottle counter is reset.

5.8 PLC Trainer and Hardware Configuration

This section briefly introduces the AB RSLogix 500 software as the development environment for the SLC hardware and associated HMI interfaces. A small trainer utilizing a SLC 5/03 processor module, a power supply, combination analog module, a digital input module interfaced to switches, and a digital output module interfaced to pilot lights is used to accommodate our coverage in the remainder of this book. Additional modules can be added, including digital, analog, and communication to accommodate larger system needs. Expansion might require additional racks and larger power supply.

5.8.1 Book Training Unit Setup

Figure 5.26 shows the AB PLC setup used to demonstrate hardware and software concepts covered in process control and automation. The training unit is also utilized to implement, debug, and document all examples, homeworks, laboratories, and projects discussed throughout the book. The training unit consists of the following items:

1. 24-V power supply (1746-P1) with ready LED
2. SLC 5/03 PLC processor (1746-OS501) with RS486 (DHPLUS) and RS232 communication interface ports
3. Eight point input module (1746-IB8)
4. Eight point output module (1746-OB8)
5. Four channels analog input/output module (1746-NIO4V)
6. Processor status LEDs
7. Discrete inputs status LEDs
8. Discrete outputs status LEDs
9. Analog module status LEDs
10. Two analog inputs potentiometer interface
11. Eight digital inputs simulator switches interface
12. Five slots rack and din rail

The eight ON/OFF switches (I0.0 to I0.7) are used to simulate and test discrete input devices. The corresponding LEDs indicate their individual status. The eight ON/OFF outputs (O0.0 to O0.7) are used to simulate and test discrete output devices. The corresponding LEDs indicate their individual status. The two analog inputs (IW64 and IW66) are connected to two potentiometers (0 to 10 V), which simulate analog

Figure 5.26 SLC500 PLC trainer.

variable signals. The analog output is connected to a small voltmeter, 0 to 10 V. These are the tools and configurations used to establish the trainer in this book. Other configurations or commercial training units can be configured and used.

5.8.2 RSLogix500 Software

The RSLogix500 is the development portal for the AB SLC family of processors. It is used to create/load an application file, navigate through the software, configure the display view, navigate through the help system, and save a program. The RSLogix500 software display includes the following:

- Standard window toolbar
- Status window
- Ladder window
- Program/processor status toolbar
- Results window

The RSLogix project tree is shown in Fig. 5.27. It functions like the Windows Explorer software. A folder with plus (+) sign can be expanded to show its contents. A folder with minus (−) sign can be collapsed to hide its contents. Using the File toolbar, you can open, delete, copy, rename, and create new files. You need to define your target

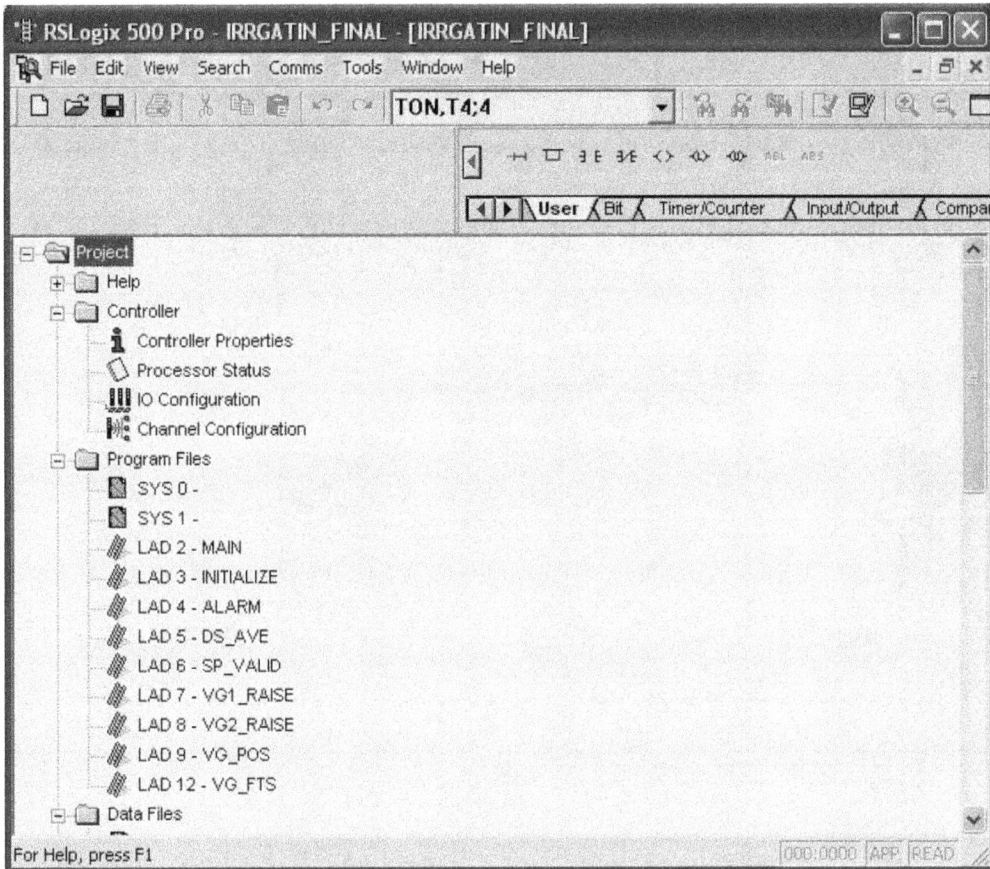

Figure 5.27 Project tree screen.

hardware before the creation of a new program. The processor unit resides in slot 0 followed by the I/O modules and additional communication or special modules as needed.

The I/O configuration screen for our training unit is shown in Fig. 5.28. As shown in the figure, the device configuration shows the hardware configured in your system. In our training unit we used a 4-Channel Analog I/O (1746-NIO4V), an 8-point Discrete Input (1746-IP8), and an 8-point Discrete Output Module (1746-OP8). The processor status link can be used to read the status, select, or change the processor used.

The Program Files folder contains the ladder logic that monitors inputs and controls outputs. The ladder window contains any open program files. Program files contain the user-defined ladder logic and all program documentation. Figure 5.29 shows the ladder window for the LAD 2, which is the default file for the main program. LAD 3, LAD 4, and higher files can be used for subroutines or other special procedures to be called from the main program. This file system encourages structural programming, which is easier to document and support. The RSLogix500 software provides detailed help features. Navigate the help menu from Overview, Contents, Index option, and the

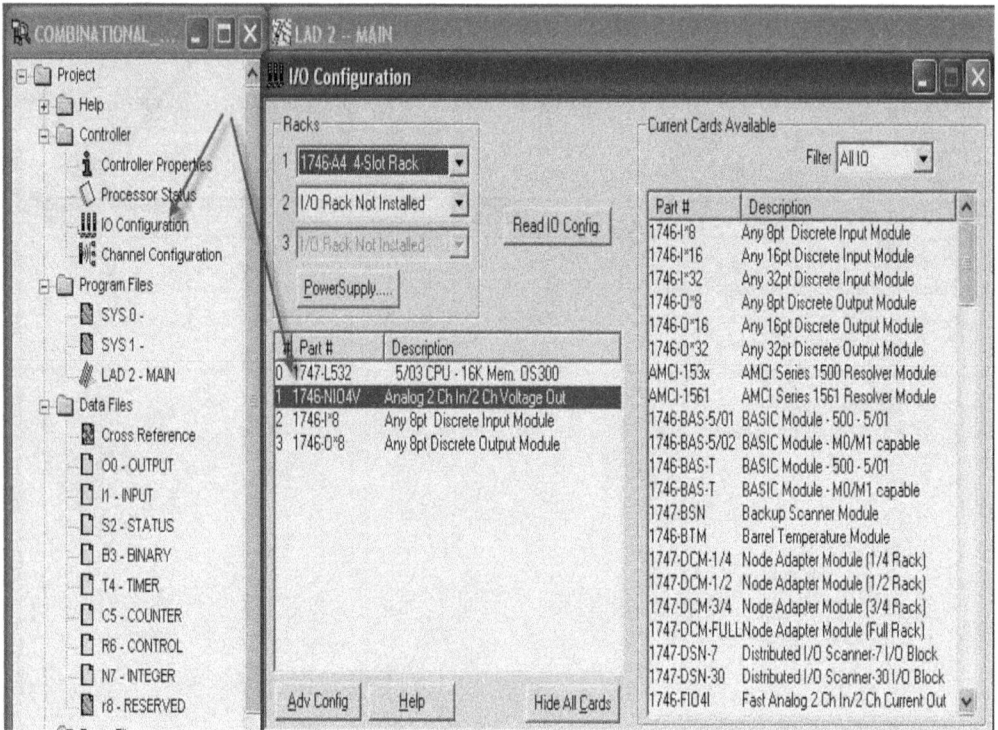

Figure 5.28 I/O configuration screen.

selection of specific needed help. Figure 5.30 shows the screen resulting from using Help sequence for the Move instructions. Online help is the best source for getting help with any instruction details. Application-related help can be found on social media and vendors websites. It is always recommended to seek help and not engage in trial and experimentation.

5.8.3 RSLinx and Online Operation

Before connecting your programming device to the PLC hardware for online operation, you must configure and set up all communication ports used. With online operation, you can configure communication drivers, download/upload program, go online, troubleshoot your program, save online program, and force ON/OFF I/O. AB RSLinx communication software must be loaded and configured on your programming device. We are using a laptop to connect to the PLC through a USP port. The port address can be detected by clicking "System" from windows control panel, then click "Hardware," then "Device Manager," then "Ports." You should identify the port address used for the connection to the PLC; in our case we are using USP COM 7 port on a laptop running Windows XP OS. You must check the address as it is assigned by the operating system and can change. Configure the port to match the PLC communication parameters as shown in Fig. 5.31. The exact port setting matching the PLC SLC5/03 Channel 0 configuration is shown in Fig. 5.32. You can select the "Auto

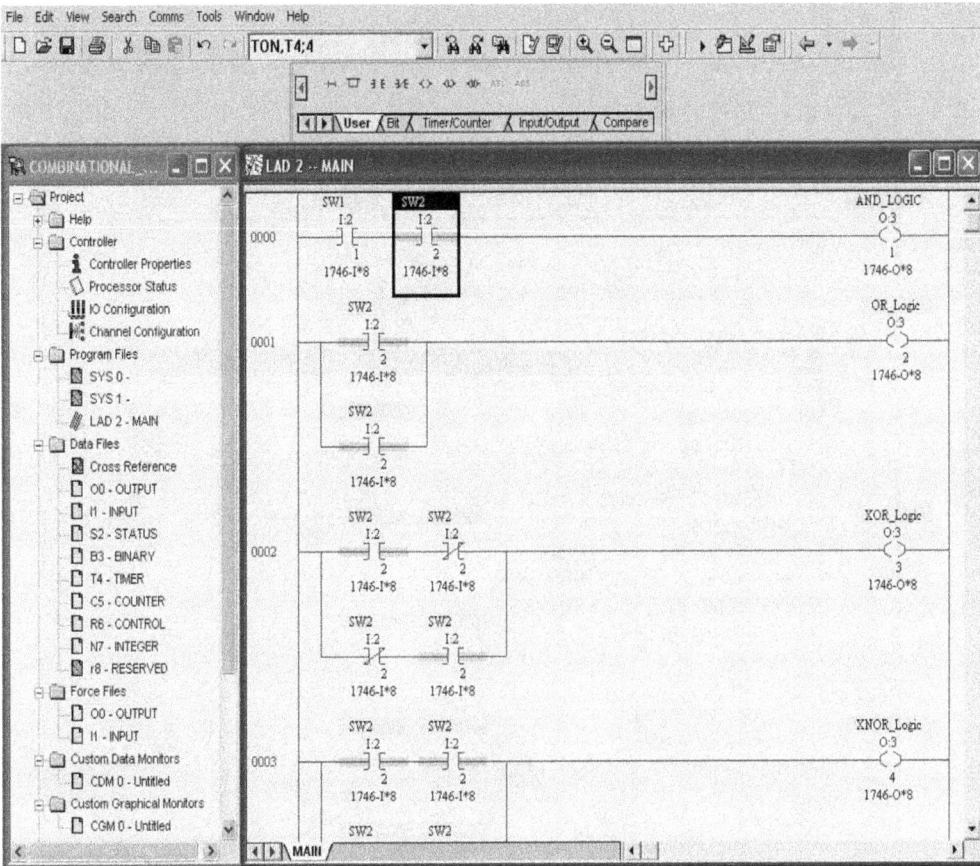

FIGURE 5.29   Ladder window screen.

Configuration" option if you do not know the PLC baud rate setting. To work with certain application in RSLogix500 system, you must establish communication with the processor before going online, downloading, editing, debugging, and saving an application program files.

Next we must configure RSLinx by selecting the correct communication driver for our connection as shown in Fig. 5.33. We are using the AB_DF1-1 DH+ driver. The driver is configured to the setting shown in Fig. 5.34. You can check active nodes on the network by clicking the RSWho icon as shown in Fig. 5.35.

Once the offline program implementation task is complete, the code needs to be compiled and transferred from the PC (programming node with active RSLogix software) to the PLC processor memory. Click the "Comms" icon, then click the "Download." If not successful, try Clearing Processor fault, checking connections, and verifying driver's configuration. If successful, click "Yes" to "Go Online." From the "Comms" drop menu click "Mode," then "Run" and conform. Figure 5.36 shows the downloaded and running program. You can also transfer the code residing in the PLC memory to the

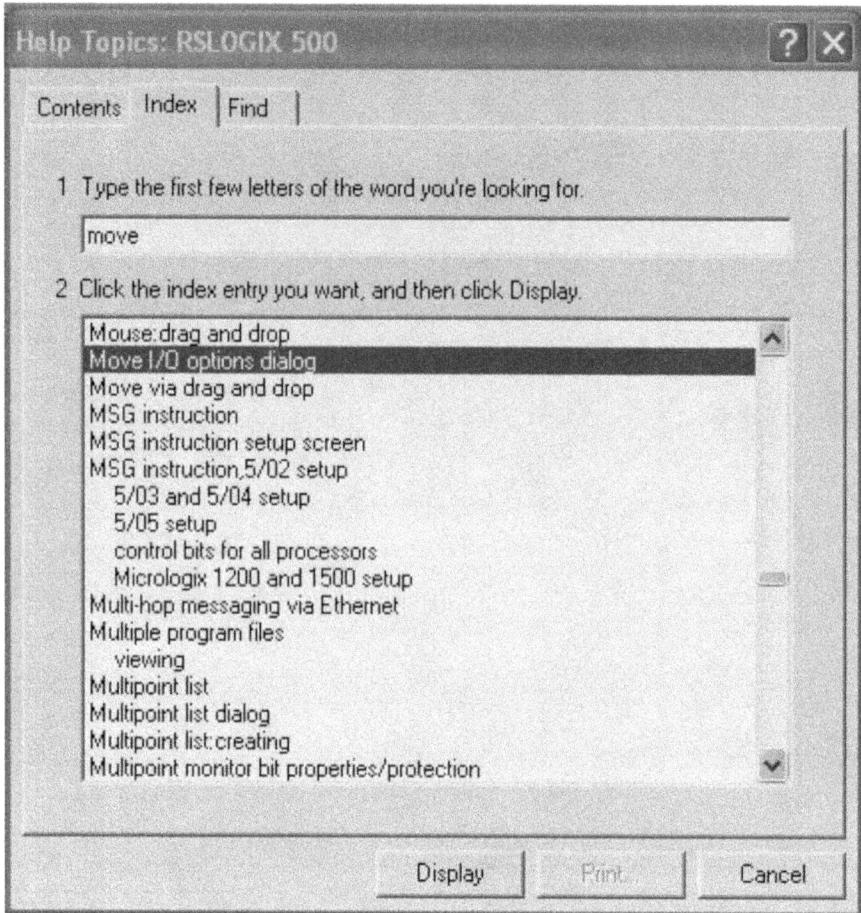

FIGURE **5.30**   Move instruction help screen.

PC by clicking the "Comms" then clicking the "Upload" command and saving the program. In the Online mode, if any modifications are made, the application is updated in the PLC memory automatically and on the PC disk in the working directory only after a save operation is performed.

The following tasks can be executed in the Online mode:

- Create/modify the ladder program.
- Modify the predefined function block parameters except the size of registers.
- Modify the number of internal words.
- Import/export a source file or variables with the PLC in Stop mode.
- Export an application with the PLC in Stop mode.
- Debug and adjust the logic.

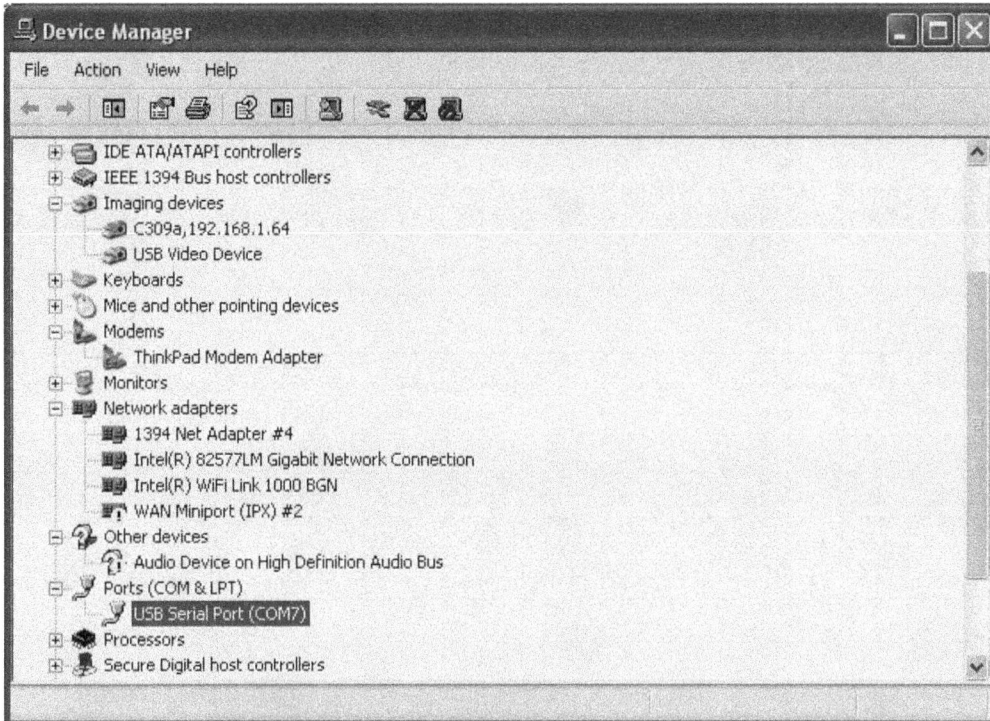

FIGURE **5.31**   Device manager ports configuration.

The following tasks cannot be executed in the Online mode:

- Add or delete a module.
- Modify the I/O channel application or specific function association.
- Add predefined function blocks.
- Modify the size of a register.
- Modify the number of internal bits and constants.
- Import an application.
- Open an application.

5.8.4   Online Application Debugging

This section shows the implementation of an application to start/stop single-speed motor. If the motor is in Auto mode and the Auto Run Bit is set by the conditional logic, or if the motor is in Maintenance mode and the Maintenance Run Bit is set by the operator, and if no fault is present, the motor will be started. If the motor is in Auto mode and the Auto Run Bit is not set, or if the motor is in Maintenance mode and the Maintenance Run Bit is not set, or if a fault is present, the motor will be stopped. Figure 5.37 shows the logic diagram for this simple application whereas Fig. 5.38 documents the ladder logic implementation in RSLogix500.

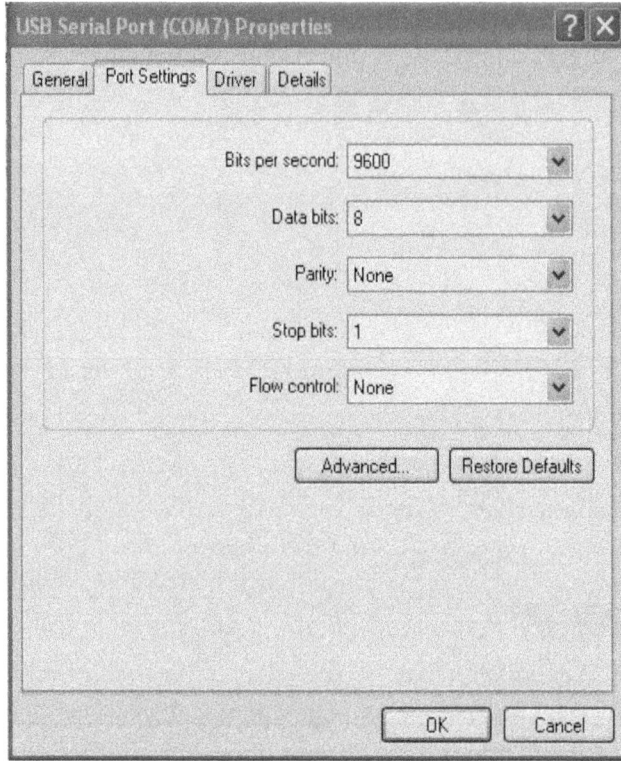

FIGURE **5.32** COM 7 port setting.

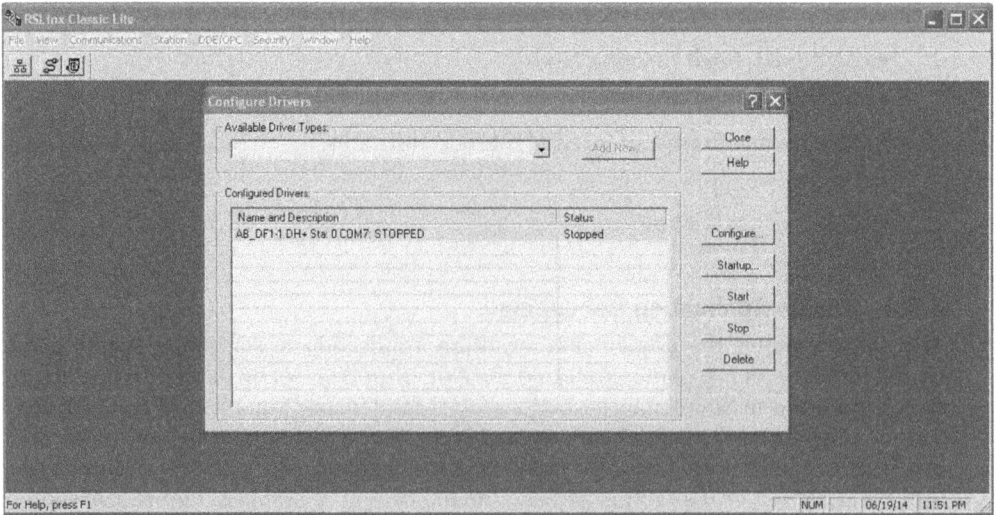

FIGURE **5.33** RSLinx driver.

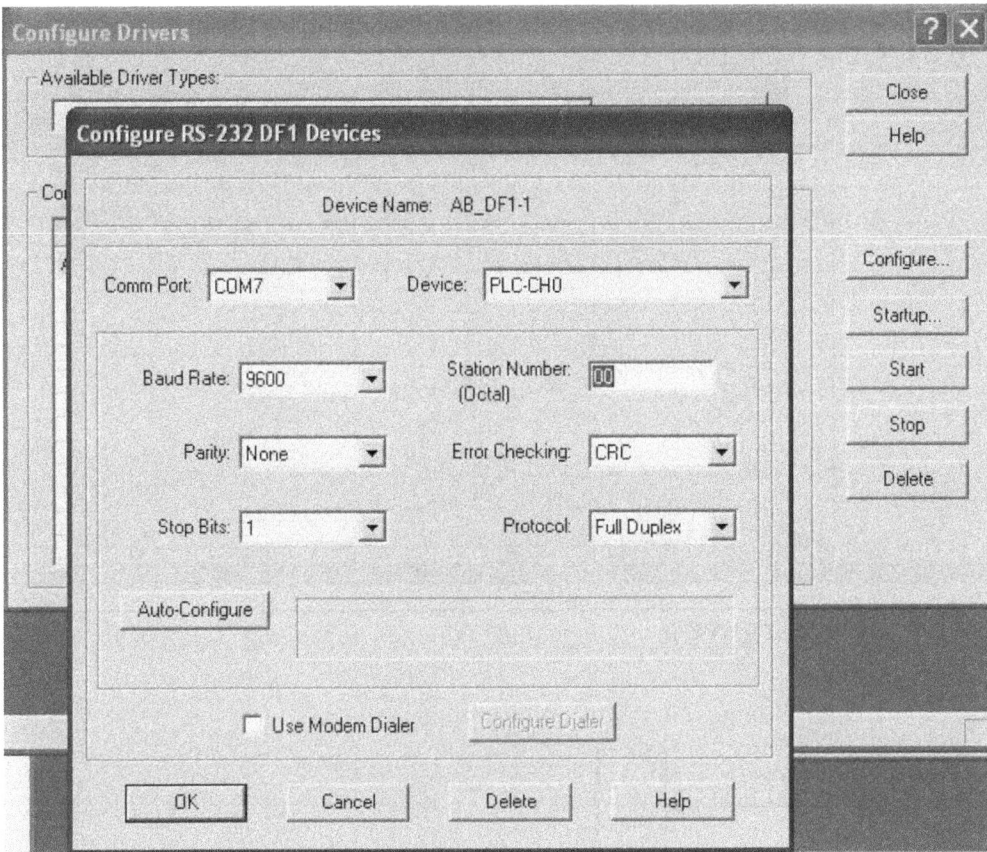

FIGURE 5.34   RSLinx driver setting.

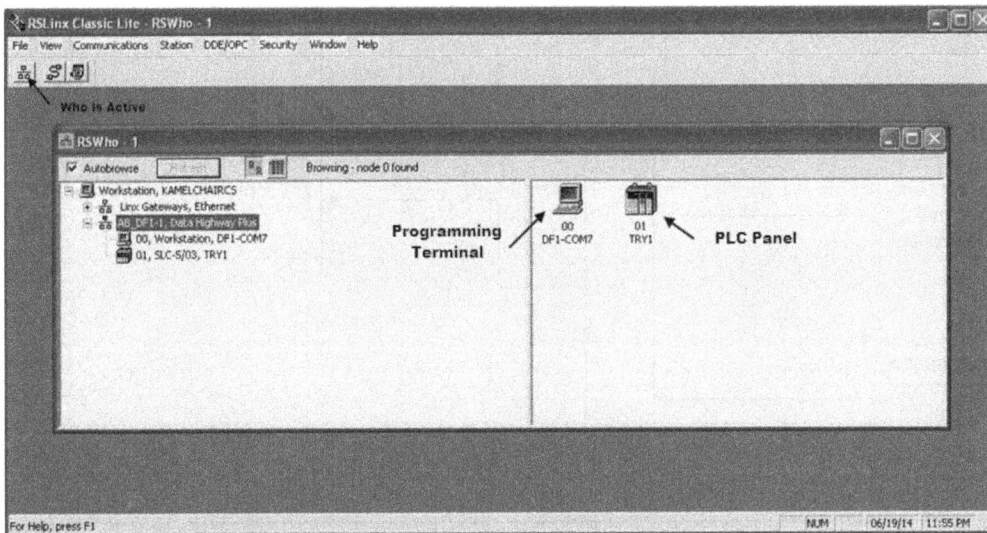

FIGURE 5.35   Active nodes listing.

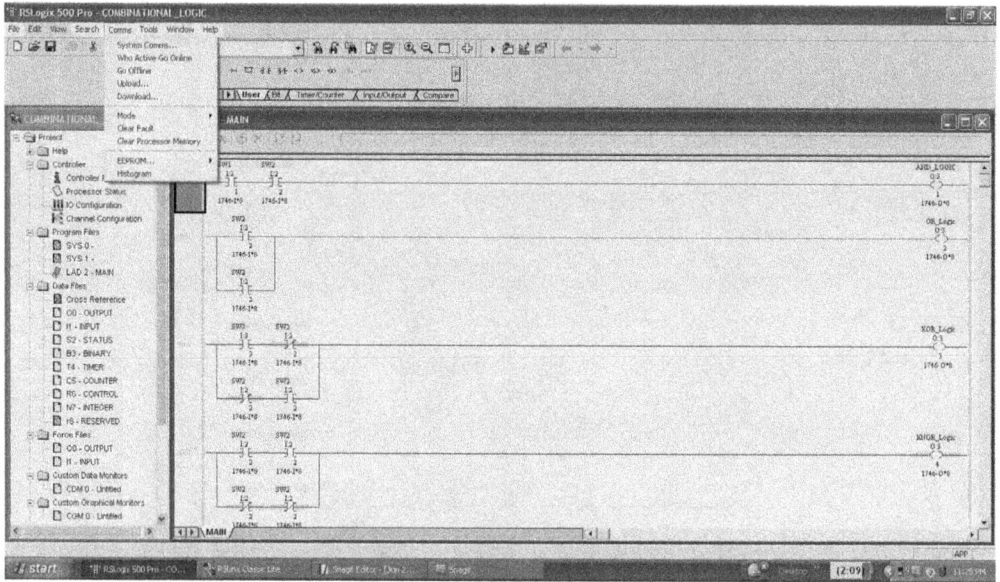

Figure 5.36   Active online program.

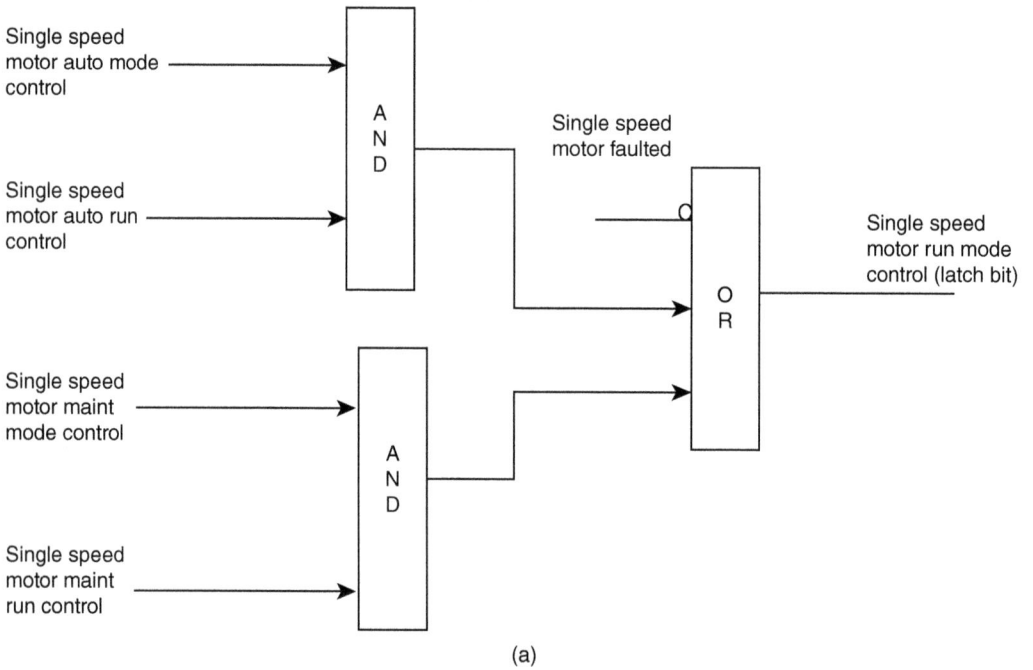

(a)

Figure 5.37   (a) Single-speed motor start logic; (b) single-speed motor stop logic.

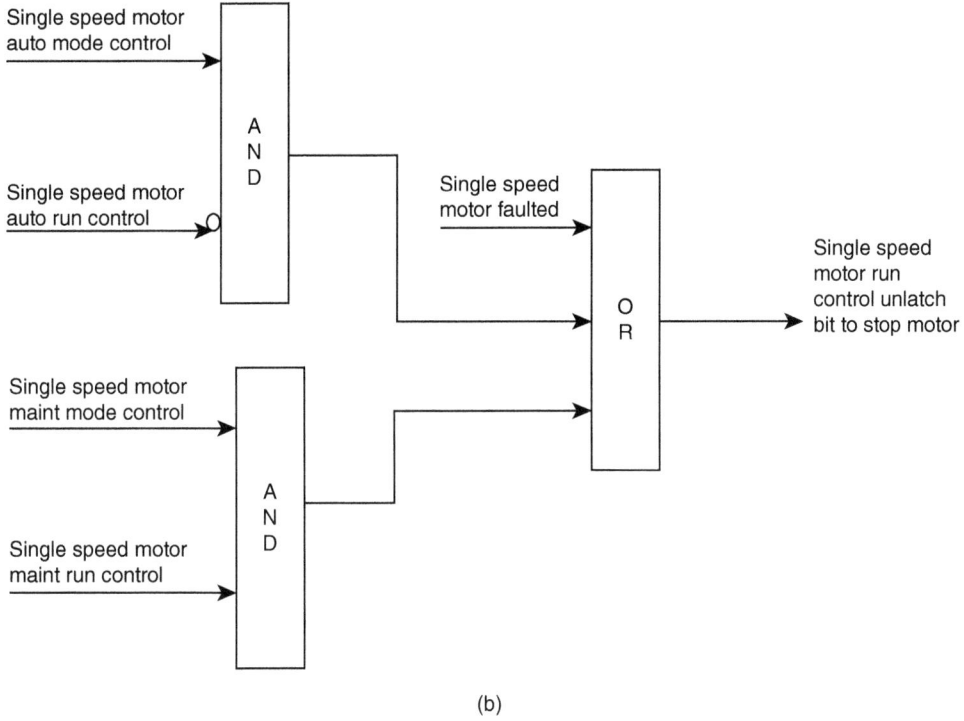

(b)

FIGURE 5.37   (Continued)

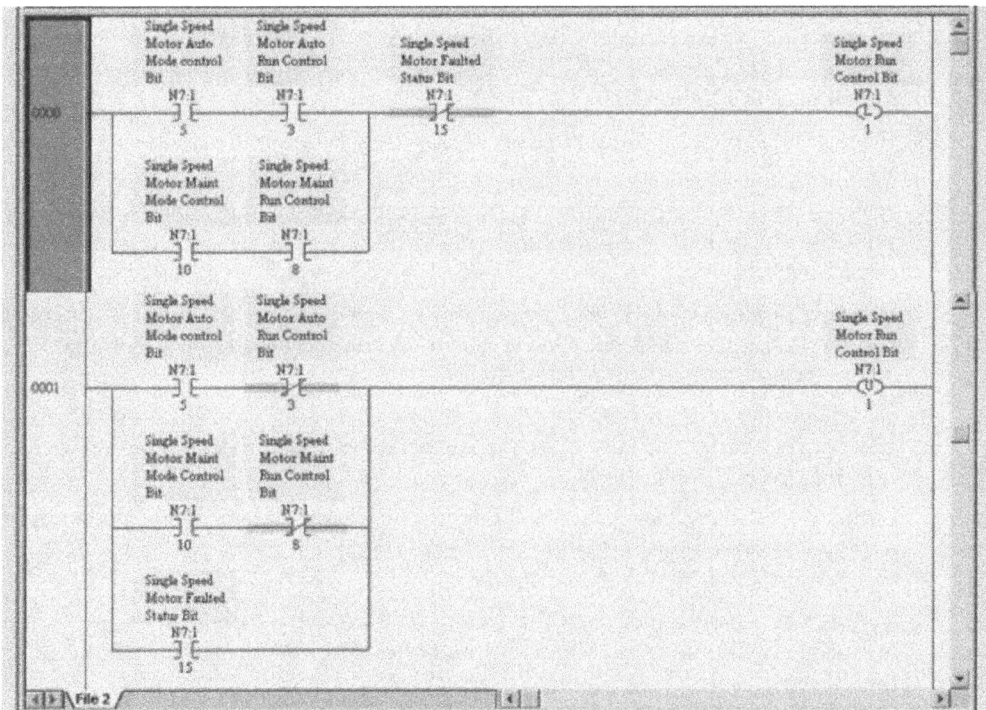

FIGURE 5.38   Single-speed motor ladder program.

Chapter 5: Home Work Problems and Laboratory Projects

1. List three benefits for the use of HMI in industrial automation and process control.

2. Explain how the PLC communicates with the HMI.

3. How many devices can be connected to the HMI?

4. What is the difference between direct communication and network communication?

5. Explain how a panel view can detect an alarm.

6. How many nodes the ControlNet network can support?

7. What is the maximum speed and length of AB_DF1?

8. What does a local area networking mean?

9. List three tasks that cannot be executed in the Online mode.

10. What is device net networking?

11. Explain the use of HMI in a small control application, similar to the ones selected in this chapter, from one of the Chap. 4 laboratories.

12. Briefly describe the steps required to configure communication parameters.

13. What is the purpose of Who Active utility in the RSLinx "RSWho"?

14. What should you check if the Who Active report shows X on the PLC panel?

15. Explain the distributed control system approach and give an example for its use.

16. Explain three uses of PanelView 600 function keys.

17. Distributed process control is becoming an attractive approach for implementing large automation projects. List and briefly discuss the technologies that support and recommend this approach.

18. When using several HMI devices and PLCs, how is communication coordination achieved?

19. HMIs are used for both status and system remote control. Other panels are typically available and provide the same functionality. How are conflict and race conditions eliminated?

20. Define a process scenario for the selector switch used in the bottling conveyor system discussed in Sec. 5.7. One of the operation modes should be assigned for Auto when the switch is in position A.

21. Data Highway Plus and Ethernet are multiple accesses with collision detect local area network protocols. How does this restrict the number of networked nodes in a real-time control application?

22. Define a process scenario for the filling and capping station use in the bottling conveyor system discussed in Sec. 5.7. Make an explicit use of all available inputs and outputs associated with this station.

23. Define a process scenario for the Enter NO PB switch "I:1/5" and associated four digits BCD input "I:3" used in the bottling conveyor system discussed in Sec. 5.7.

24. The fill station for small and large bottles discussed in Sec. 5.7 is available but not used in the textbook discussion. Devise a simple specification for implementing the filling station control in the ladder logic for the bottling system.

25. Repeat Example 5.3.1 to flash an alarm light indicator ALARM if the door is stock in between and neither of the two motors is running. The alarm flashes at a rate of 10 seconds ON and 5 seconds OFF.

26. The selector switch is not used in the bottling line implementation discussed in Sec. 5.7. Devise a simple specification for implementing the ladder logic for the selector switch by assigning a function for each of the three selections. One position might be assigned for Automatic (AUTO), the second for Manual (MAN), and the last for skipping the large/small bottle sorting.

27. The silo operation shown in Fig. 5.12 will undergo a shutdown if the operator tries to stop the conveyor system while the box is under the level sensor I:1/4. Modify the program to allow the operator to be able to stop the conveyor while the box is under the level sensor.

28. Modify the ladder shown in Fig. 5.7 for the door operation to implement continuous up and down operation once either the OPEN or the SHUT PB is activated. The door will automatically reverse the direction once it reaches either of the two limit switches. Only activating the STOP PB will terminate the door movement.

29. Referring to the door simulator, check the status of limit switch 1 (LS1) and limit switch 2 (LS2). Assuming that the status of the two limit switches are reversed. Fill the following table:

LS1	LS2	Door Location
0	0	
1	0	
0	1	
1	1	

30. Devise a simple specification for including the two car at intersection input sensors I:1/2 and I:1/3 in the ladder logic for the traffic light control. One option is to use these sensors to calculate the arrival rate of vehicles for each of the two directions: S-N and E-W way.

31. Modify the silo operation shown in Fig. 5.12 to stop the conveyor after each fill. The operator has to push the START PB to clear the box.

32. Modify the ladder program of Figure 5.9 to prevent the nozzle movement without manual activation of the Open or the Close PB's.

Laboratory 5.1—Garage Door Operation Control

The objective of this laboratory is to get students familiar with LogixPro 500 HMI simulation and the control panel used to control/monitor a garage door. A motorized garage door simulation HMI is shown in Fig. 5.39. The simulator uses the following components:

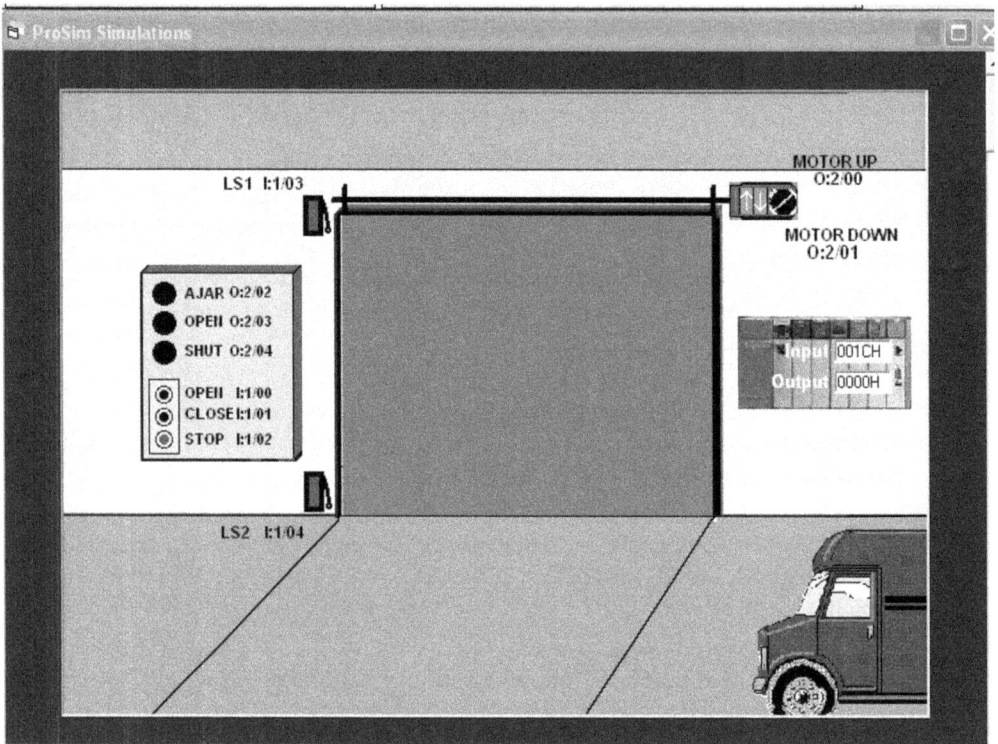

FIGURE 5.39   Garage door control/monitoring. (*Traffic bottling simulator courtesy of the Learning Pit LogixPro 500.*)

- Operator station display panel and controls
- OPEN, CLOSE, and STOP PB switches
- AJAR, OPEN, and SHUT panel indicators
- Fully open and fully closed limit switches
- Two-door motors—one for up movements and one for down movements

The garage door is equipped with two motors (MOTOR UP and MOTOR DOWN) and two limit switches (LS1 and LS2). The door can move in the up and the down directions. Figure 5.40 shows the ladder logic implementation for the garage door control.

Requirements:

- Study the ladder logic program shown in Fig. 5.40 and add the rungs documentation.
- Program the ladder logic and download to the PLC and run the program.
- Modify the ladder program to maintain the door open/close movements as the operator pushes the (OPEN/CLOSE) PBs.

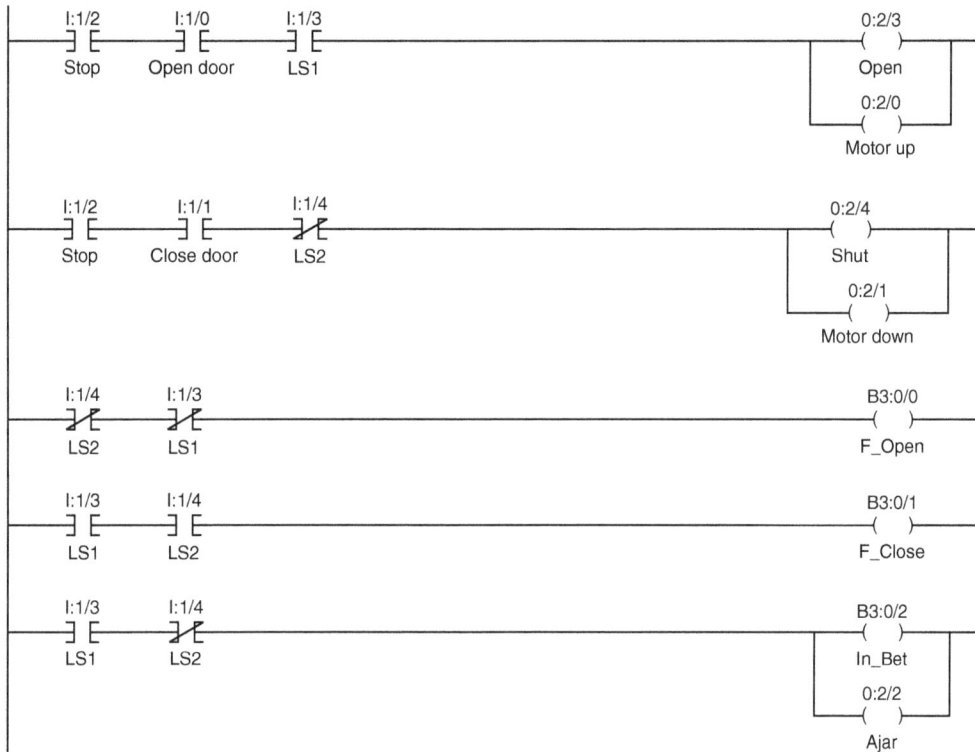

Figure 5.40 Garage door ladder program.

Laboratory 5.2—Flashing Light Indicators

In this laboratory, the program of Fig. 5.41 is a modification of laboratory 5.1 by adding a flasher action to the OPEN, SHUT, and AJAR pilot lights. We will use one of the RSLogix PLC's free running timer bits in word S:4. Word bits alternate at different rate with bit 0 having the highest rate. We are using it 8 (S:4/8) for this purpose in this laboratory. AB PLCs will provide consistent rate, but with LogixPro 500 the rate varies with the computer speed. Test and select the appropriate flashing bit for your computer. Figure 5.41 shows the new ladder implementation that accommodates the following changes:

- The OPEN pilot light will flash while the door is moving up and will stay on once the door is fully open.

- The SHUT pilot light will flash while the door is moving down and will stay on once the door is fully closed.

- The AJAR pilot light will flash when the door is not moving in either direction. The STOP PB is not used and thus not listed in this example ladder program. It is normally used to disrupt a starting and latching action.

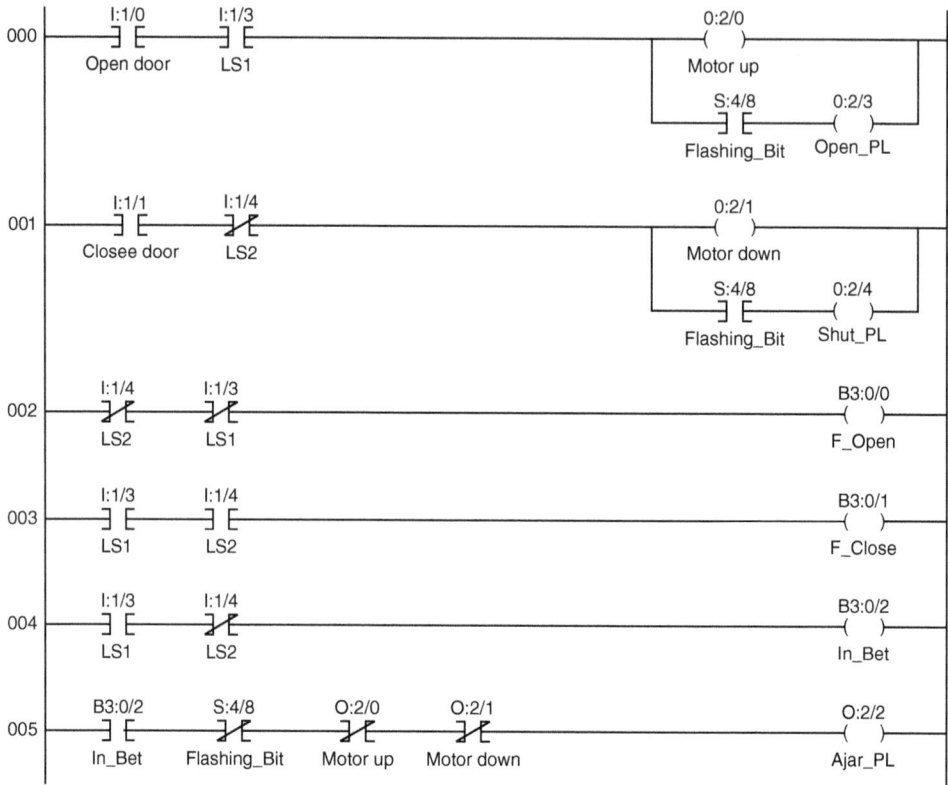

FIGURE 5.41 Flashing lights ladder program.

Requirements:

- Study the ladder logic program in Fig. 5.41 and add the rungs documentation.
- Program the ladder logic and download to the PLC and run the program.
- Modify the ladder program to maintain the door open/close pilot light indicators as the door reaches the fully open or fully close position.

Laboratory 5.3—BCD I/O Simulator

Upon completion of this laboratory, you will be familiar with FRD (convert from BCD) and TOD (convert to BCD) instructions and the comparison instruction (greater than or equal) and its use within the ladder logic to control a process.

Study and experiment with the LogixPro BCD I/O simulator shown in Fig. 5.42. Design a PLC program to perform the conversion from positive degree fahrenheit (F) to the equivalent degree centigrade (C). The degree F is limited to two BCD digits (00 to 99) to match the simulator thumb wheel switch setting. The formula for the conversion is given by the following equation:

$$C = (F - 32) * 5/9$$

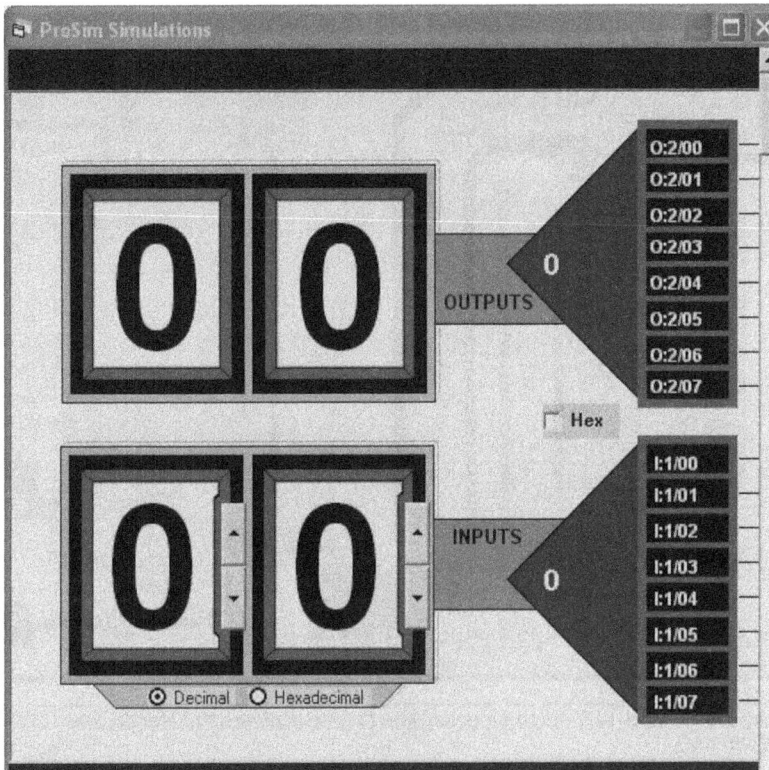

FIGURE **5.42** BCD I/O LogixPro HMI. (*Traffic bottling simulator courtesy of the Learning Pit LogixPro 500.*)

Requirements:

- Design the specification for the temperature HMI and specify the input/output range allowed by the given simulator.
- Write the ladder logic program to capture the BCD degree F entered from input I:0 and display the output O:2 in degree C using the BCD format.
- Flash output O:2 for negative degree C and maintain it for positive values.
- Modify the program design to allow for input in hexadecimal, which accommodates degree F from 0 to 255. Display the degree C in BCD format.

Laboratory 5.4—Batch Process Auto/Manual Operation Modes

The objective of this laboratory is to get students familiar with the method used to control a batch process in Auto/Manual modes. Figure 5.43 shows the LogixPro HMI interface for the batch process previously discussed in this chapter.

Requirements:

- When the three-position selector switch is in position A simulating the Auto mode, the operator can run pumps 1 and 2 by pressing the START PB, which starts the batch filling into the tank.

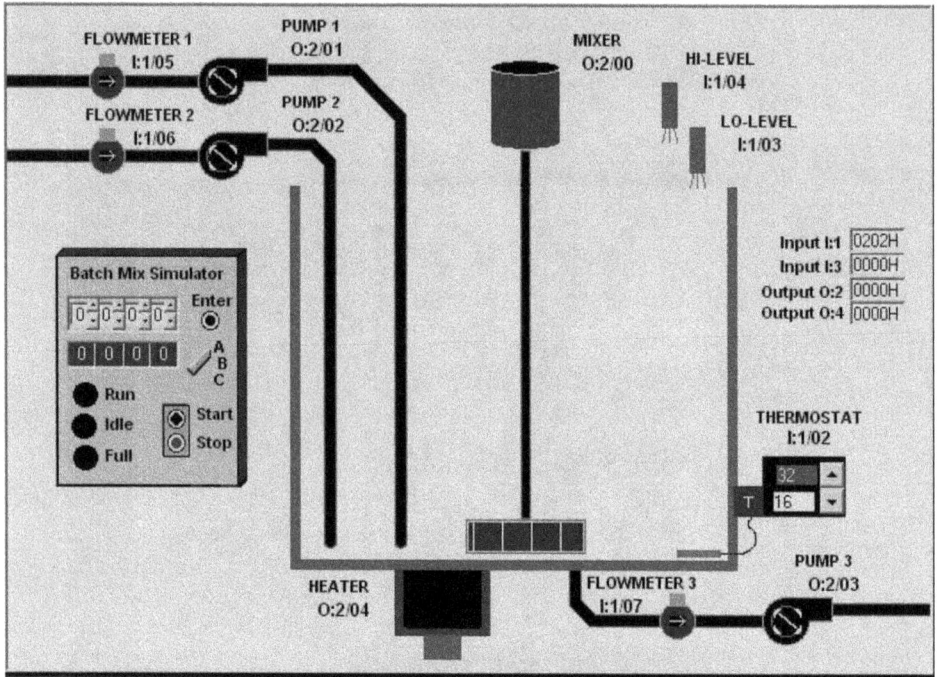

FIGURE 5.43 MAN/AUT batch process HMI. (*Traffic bottling simulator courtesy of the Learning Pit LogixPro 500.*)

- Once a batch sequence has begun, the sequence may be stopped and resumed at any time using the STOP and START PB switches.
- The tank is to be filled with a mixture obtained from the separate fill lines utilizing fill pumps P1 and P2. When tank level reaches the high set limit, a heater will run for a period of 10 simulated seconds.
- When the heater stops, a mixer will run for a period of 50 simulated seconds.
- When the mixing time is over, pump 3 will run and empty the tank. Once the tank is empty, pump 3 will stop.
- Operator can stop the process and discharge the tank by placing the selector switch on the B position (manual discharge mode). Pressing the START PB will energize pump 3 till the tank is empty or the PB is released. The operator can switch to Auto mode and continue the tank filling and the rest of the sequence.

Laboratory 5.5—Batch Process Chemical Reactor

This laboratory is a modification for the batch process shown in Fig. 5.43. The new process assumes a chemical reactor with two feed lines, each equipped with a motorized pump. One liquid is pumped through the first feed line whereas the other chemical is brought in through the second feed line. The first liquid is heated and then later mixed with the second material to produce the required chemical reaction.

Requirements:

- Liquid one is pumped through the first feed line once the START PB is activated. Pump 1 continues until the tank is half full.

- Once the tank is half full (requires experimental timer preset value definition), the heater is run for simulated 20 seconds, after which the heater will be stopped.

- Next, the second material is brought to the tank using pump 2, which will stop once the tank is full.

- Once the tank becomes full, the mixer is run for simulated 30 seconds to produce the desired chemical reaction.

- Next, pump 3 runs and empties the tank through feeder 3 outlet line. Once the tank becomes empty, the whole process stops.

- Activating the STOP PB at any time will stop the entire process, which will require manual handling to continue the sequence or clear the tank before starting another batch.

Laboratory 5.6—Structured Program for One-Way Traffic Lights Control

The objective of this laboratory is to get students to learn how to use subroutines in a real control application. This laboratory is a remake of the ladder program listed in Example 5.6.1, which controlled the traffic in the S-N one way. I:1/2 connects to the S-N road car at intersection sensor whereas I:1/3 connects to the E-W road car sensor. A park input can be activated to stop car movement in either of the two roads. To simplify our discussion, we will park cars on the E-W road and only control the traffic lights for the S-N road. The sequence of traffic light operation for one road traffic is summarized in Table 5.4 and obey the following timing:

1. Red light (O:2/00) is ON for 12 seconds.
2. Green light (O:2/02) is ON for 8 seconds.
3. Yellow/amber light (O:2/01) is ON for 4 seconds.
4. The sequence now repeats with the red light ON again.

Requirements:

- Use three subroutines to implement the required sequence: one for initialization, second for the timing control, and the third for counting cars crossing the S-N intersection per hour.

- Initialization subroutine will execute once during the first pass utilizing the S:1/15 status bit (first pass). The routine will clear all used timers (four timers) and counters (one counter).

- Traffic lights control subroutine will provide the functions previously shown in Example 5.6.1.

- The counting subroutine will account for and report the number of cars crossing the intersection every hour.

- Debug your documented program using the LogixPro traffic lights simulator.

Laboratory 5.7—Batch Process Using ON-OFF Level Control

The objective of this laboratory is to acquaint readers with hands-on knowledge of the ON-OFF process control in commercial and industrial application. Refer to chapter 7 Section 7.5.1 for more detailed coverage of the ON/OFF process control.

Two limit sensors provide real time measurements of a tank level; the low level and the high level. The tank level is desired to maintain a Set Point by sending an ON / OFF signal to Pump1/Pump3. The dead band allowed for the ON / OFF level control is 10 centimeter (+ or − 5 centimeter from the Set Point). The batch mix simulator's "RUN" light is turned ON while Pump1 is running.

Requirements:

- Start the LogixPro batch simulator.
- Enter the ladder logic program shown in Figure 5.44.
- Add the documentations for each rung.
- Download the program and go on line.
- Enter different set points from the Batch Mix Simulator and verify that Pump1/ Pump3 is turning ON/OFF and rectifying the error.
- Repeat part 5 with smaller dead band and verify Pump 1/Pump3 operation.
- Document your observations/comments.

Laboratory 5.8—Conveyor Inspection Station Control

This lab was discussed in Chap. 3 without the user interface/HMI configurations and implementation. The objective of this laboratory is to get more familiar with counter instructions and associated HMI's design. If possible, an actual HMI can be configured and used. Otherwise the LogixPro I/O simulator can be used.

The up-/down-counter can be used to keep track of the good parts count in a conveyor system. The setup used in the laboratory assumes a simple conveyor system configuration with two photocells, sorting station, and an elevated belt for rejected parts

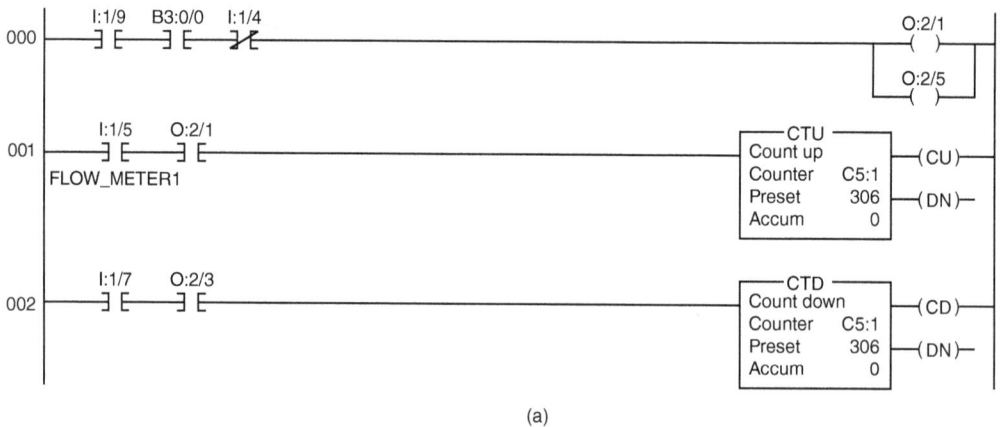

(a)

Figure 5.44   (a) 1 Batch process ladder program; (b) 2 Batch process ladder program; (c) 3 Batch process ladder program

003
```
                                    ┌──────TOD──────┐
                                    │ To BCD        │
                                    │ Source  C5:1.ACC │
                                    │              ?  │
                                    │ Dest        O:4 │
                                    │              ?  │
                                    └───────────────┘
                                    ┌──────FRD──────┐
                                    │ From BCD      │
                                    │ Source      I:3 │
                                    │              ?  │
                                    │ Dest       N7:0 │
                                    │              ?  │
                                    └───────────────┘
```

004
```
                                    ┌──────ADD──────┐
                                    │ Add           │
                                    │ Source     N7:0 │
                                    │              ?  │
                                    │ Source B      1 │
                                    │              ?  │
                                    │ Dest       N7:1 │
                                    │              ?  │
                                    └───────────────┘
```

(b)

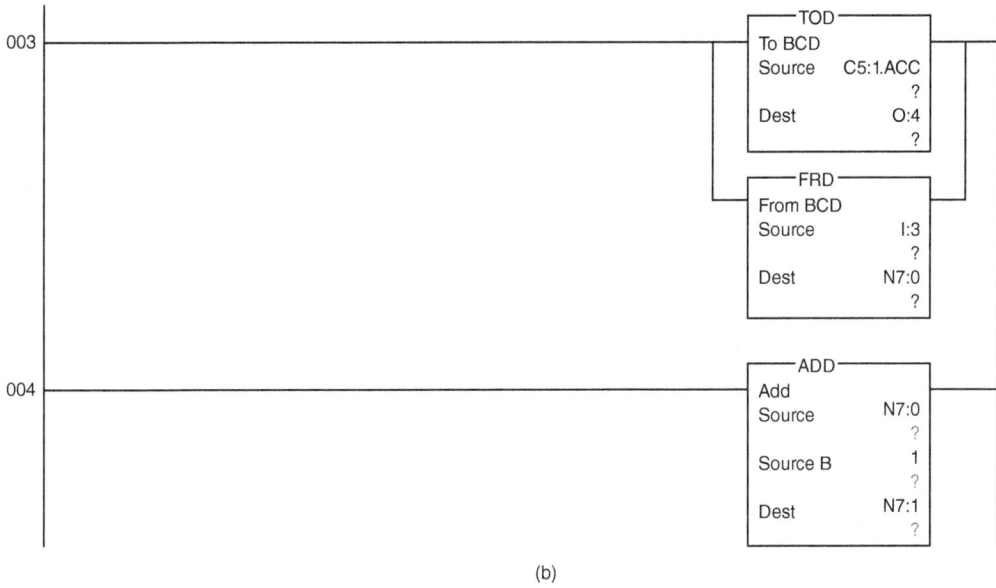

FIGURE 5.44   (*Continued*)

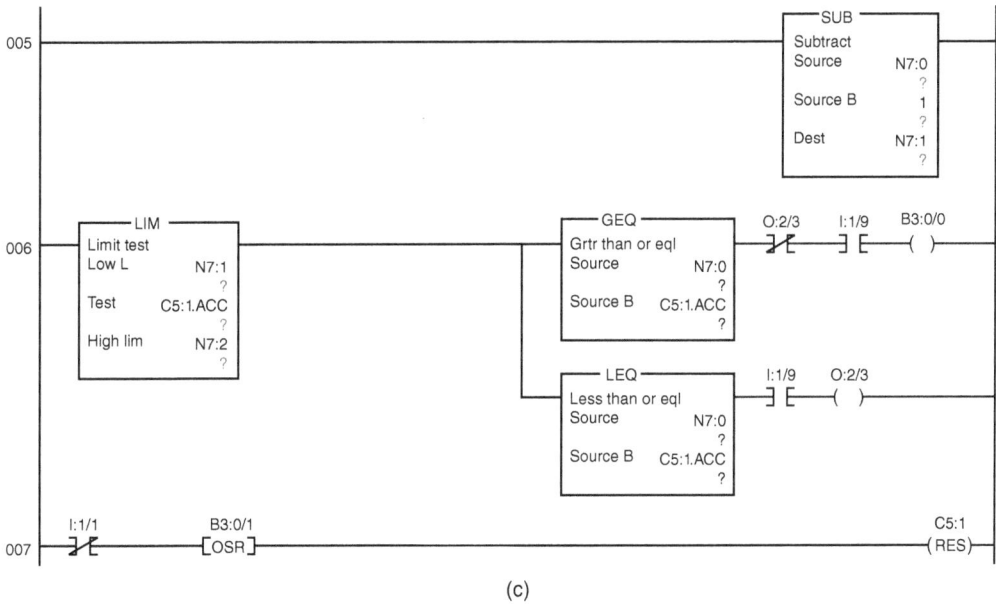

005
```
                                              ┌──────SUB──────┐
                                              │ Subtract      │
                                              │ Source     N7:0 │
                                              │              ?  │
                                              │ Source B      1 │
                                              │              ?  │
                                              │ Dest       N7:1 │
                                              │              ?  │
                                              └───────────────┘
```

006
```
  ┌──────LIM──────┐              ┌──────GEQ──────┐  O:2/3   I:1/9   B3:0/0
  │ Limit test    │              │ Grtr than or eql │  ─]/[─   ─] [─   ─( )─
  │ Low L     N7:1 │              │ Source     N7:0 │
  │           ?    │              │              ?  │
  │ Test   C5:1.ACC │              │ Source B C5:1.ACC │
  │           ?    │              │              ?  │
  │ High lim  N7:2 │              └───────────────┘
  │           ?    │              ┌──────LEQ──────┐  I:1/9   O:2/3
  └───────────────┘              │ Less than or eql │  ─] [─   ─( )─
                                  │ Source     N7:0 │
                                  │              ?  │
                                  │ Source B C5:1.ACC │
                                  │              ?  │
                                  └───────────────┘
```

007
```
  I:1/1      B3:0/1                                              C5:1
  ─] [─     ─[OSR]─                                             ─(RES)─
```

(c)

FIGURE 5.44   (*Continued*)

navigation. The photoelectric cell PE1 keeps track of the incoming parts count whereas PE2 counts the number of bad parts exiting the inspection station through the rejection belt. The counting is implemented with the up-counter enabled through PE1 and the down-counter enabled by PE2. The counter accumulated value, tag name

"GOOD_PARTS" contains the number of good parts. The process terminates once 1000 good parts are produced.

Requirements:
Implement the above specifications and do the following:

- Configure the HMI or the I/O simulator status/control page with three PBs (PE1, PE2, and RES).
- Configure the count status for the GOOD_PARTS on the HMI.
- Simulate six incoming parts by pushing PE1 six times.
- Simulate two rejected parts by pushing PE2 two times.
- Monitor the number of good parts and report the results.

Laboratory 5.9—Structured Programming for the Bottling Assembly line

The objective of this laboratory is to get students to learn how to perform structured programming and use subroutines in a real control application. This laboratory is a remake of the ladder program listed in Figure 5.25 and detailed in Section 5.7, which controlled the bottling assembly line simulator shown in Figure 5.24. Five subroutines will be used to implement the required process control as shown in Figure 5.45 ladder program.

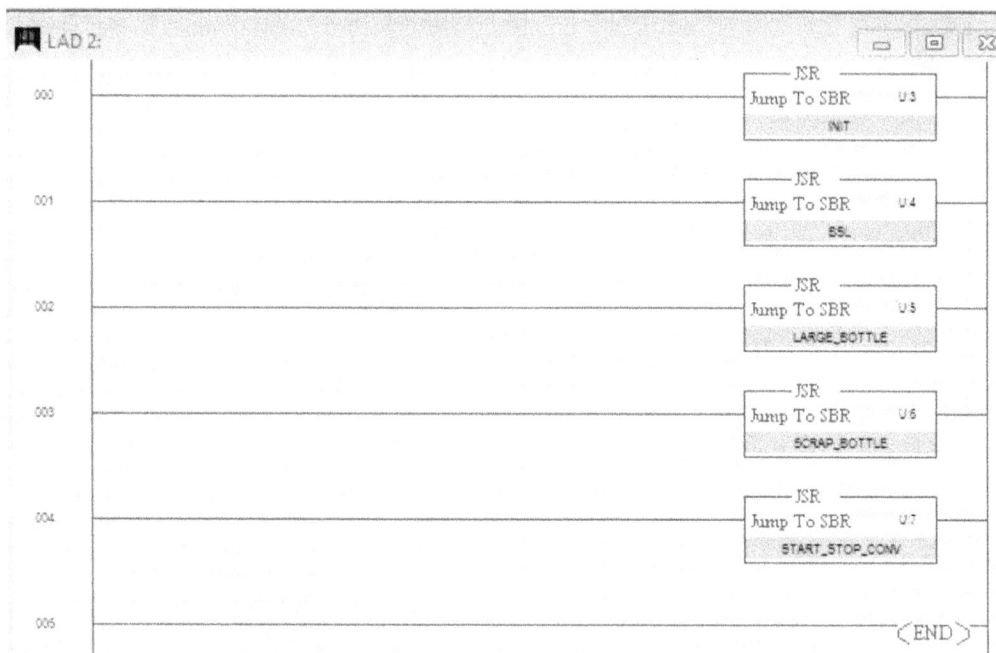

FIGURE 5.45 Ladder file 2 for laboratory 5.9

Requirements:

- Use five subroutines to implement the required sequence; one for Initialization (INIT), the second for the Bottle Tracking (BSL), the third for the large bottles sorting / control (LARGE_BOTTLE), the fourth for the scrap bottles sorting / control, and the fifth for the conveyors control (START_STOP_CONV).

- Initialization subroutine will execute once during the first pass utilizing the S: 1/15 status bit (First Pass). The routine will clear all used counters and associated displays.

- Subroutines will implement the exact specifications stated earlier in Section 5.7 and replace the ladder rungs listed in that section. Implement an accurate display of Large, Small, scrap, and boxes in BCD using the simulator output O:4 and the associated multiplexor bits.

- Debug your documented program using the LogixPro simulator.

CHAPTER 6

Process Control Design and Troubleshooting

T his chapter covers PLC process control design and troubleshooting. The system presented here is formed of three layers: process control overview, process control implementation, and process control checkout and startup.

Automated car manufacturing plant.

Chapter Objectives

- Document process description, components, and control.
- Implement process logic diagrams/function blocks.
- Be able to perform process control checkout/startup.
- Understand safety standard.

This chapter covers the fundamentals of process control/automation design techniques and industry accepted standards. As with other established and successful computer systems, and as in computer operating systems and computer networks protocols, a layered hierarchal system is used. The simplified system presented here is formed of three layers: process control overview (layer 1), process control implementation (layer 2), and process control checkout and startup (layer 3). The general guideline for the three layers is discussed in the first three sections of this chapter. Design reviews and approval are required for each layer before proceeding to the next level and committing requested resources. Special attention in this chapter is given to standard checkout techniques (layer 3) as the design and implementation issues (layers 1 and 2) were adequately covered in previous chapter's examples and are reinforced in the comprehensive case study included in Chap. 8. A brief overview of the RSLogix 500 software, training unit, hardware configuration, and offline/online operation was given in Chap. 5.

6.1   Process Control Overview, Layer 1

Layer 1 provides a process control/automation overview for a project, including a comprehensive definition of the level of automation and what will be controlled. It is used as the basis of preliminary cost, resources needs, and overall project schedule estimates. Specific issues to be covered in this layer include the following:

- Control system requirements
- Start and stop procedures
- Disturbances/alarms handling strategies
- Process constraints
- Regulatory and safety requirements

The documentation in this layer 1 will serve as the basis for later development of the detailed process control strategy, layer 2. Layer 1 describes the intended scope and purpose of the project. It includes relevant information on the current process if the project is an enhancement, replacement, or revision to an existing system. Specialized or abbreviated terms should be avoided unless fully defined since this document is intended for use by professionals who are stakeholder of the overall system operation and well-being but not necessarily versed in process control/automation. Reference sources for the information supplied and used in this document should be made whenever possible. Necessary details of layer 1 include process descriptions, level of control/automation, and control system components; all are discussed in the next three subsections.

6.1.1 Process Descriptions

This subsection conveys in a very general way the functionality and details of this particular automation system or process control. It should clearly identify all components/ processes of the system and list the important attributes of each process. Significant concerns must be documented, such as quality issues, safety concerns, energy consumption, and possible disturbances. Relevant information that may enhance the general understanding of the process must be included or referenced. The information in this section is typically the result of collaborations between the process control lead and other process/operation experts. It can also include references to already existing documentations.

Describe the specific process control automation objectives for the project. The following are typical objectives:

- Minimize manual interactions—a standard objective of automation is to eliminate or minimize manual operations. This includes the automation of container loading or unloading, the use of computerized data acquisition, automating actuation control, adding safety functions, and other quality enhancement functions.

- Protection of equipment—this includes operational restrictions for key equipment, safeguards during startup and shutdown, and other standard operation and maintenance procedures.

- Quality control—this ensures correct raw materials handling, control recipes implementation, instrumentation adequacy for measurements, meeting product specifications, and adequate monitoring for continuous quality-control improvements.

- Safety and the environment—this includes handling of hazard and toxic materials, operations at high pressure or temperature, leakage and spill avoidance, risk minimization, and safety procedures.

- Energy use and recycling—this includes reduction in energy use, improvement in reuse/recycling yield, elimination or reduction in waste, and wastewater/ material treatment and recycling.

- Asset utilization—this objective increases capacity, reduces process cycle time, adds flexible production capabilities, improves process performance, and reduces system down time.

- New technology—this objective implements new technologies safely and effectively to enhance process performance and product quality.

6.1.2 Level of Control/Automation

This subsection explicitly defines the recommended level of automation for the control system. It includes process start, run, stop, and potential upsets/disturbances. Recommendations must take into account the complexity, integration level, safety requirements, hazards, and quality needs of the required process flexibility/restrictions during the whole cycle of operation. Recommended level of control must add value to the existing system and its associated cost/benefit must be justified.

This section defines the startup philosophy, which can be either human driven, totally automated, or a mixture of both. The following are examples of issues to be considered:

- Process control utilities and associated functionality
- Subsystems coordination needs and sequencing
- Alarm and interlocking mechanisms and associated procedures
- Basic constraints in process control and overriding procedures
- Emergency shutdown and power failure restarting and recovery functions

This document can include additional details of both the functionality and primary objectives for the "Run" and "Shutdown" states of the process control system. It might also include some specific constraints or limitations that need to be applied for successful and safe system operation. The level of control/automation should also identify any governmental requirements or professional standards that apply to the control and operation of the system. Standards include applicable internal organization practices and guidelines as well as specific contractual agreements. Social implications of the automation project and its impact on all stakeholders must be documented.

A wide variety of national and international standards apply to the design and implementation of automation and process control projects. This includes the US Food and Drug Administration (FDA) regulations, The Toxic Substances Control Act (TSCA), International Standard for developing an automated interface between enterprise and control systems (ISA-95), Good Automated Manufacturing Practice (GAMP-4), the Code of Federal Regulations (CFR), or the American National Standards Institute (ANSI) other specialized regulations.

6.1.3   Control System Components

This section includes a high-level scope description of the process control project and supporting systems. The type of control system and vendor should be identified with its specific components, if decided; otherwise general requirements of the system must be given. Include any special requirements, such as interfaces to vendor or customer systems, operator panels, remote I/O, field stations, control room displays, and communication with other control systems. Define the basic system architecture, identifying and explaining any deviations from standard models.

The following is a sample of typical control system components:

- Complete listing of control system inputs and associated sensors.
- Complete listing of control system outputs and associated final control elements.
- Basic system control structure and alarming parts.
- Safety control system requirements.
- Human Machine Interface (HMI) display and control interface.
- Process information system and data acquisition.
- Procedural control recipes, coordination, and sequencing.
- Communication and networking interfaces.

This section also identifies known software requirements with the version selected for the control system, for the operator interface, HMIs, process modeling or advanced control, data acquisition, and simulation activities. The protection of the process from inadvertent operator actions needs to be addressed. Specific considerations need to be employed in applying protection to the control system, including manual intervention procedures.

6.2   Process Control Implementation, Layer 2

This document includes all information associated with the implementation of the process control system, including software design, hardware configuration, and communication protocols collaboration/communication avenues to all project stakeholders. This collaboration is essential and often produces changes and enhancements in the implementation. Stakeholders often include domain experts (electric, mechanical, chemical, civil, environmental, and industrial engineers/technicians), operation and maintenance personnel, and the owner.

Assumptions made during the implementation of this section (layer 2) that require clarification, confirmation, or additional information in the later stages of the project in layer 3 must be fully documented. Outstanding issues that require follow-up or consultation must be clearly highlighted. These issues must be resolved and removed from the layer 2 section prior to start of the final checkout and startup, layer 3. The remainder of this section details the steps carried during the process control implementation; which include I/O detailed listings, data acquisition tasks, closed loop control, project logic diagrams/flow charts, specification of ladder program function blocks, and overall system documentation.

6.2.1   I/O List

All inputs and outputs associated with the process control system must be listed in the I/O worksheet/action table along with the associated instrumentation. Required accuracy and resolution of relevant I/O are also documented in this table. Process control designer identifies any special issues associated with the instrumentation in the I/O list; including safe fail position for outputs, calibration requirements for inputs, and redundancy consideration. The I/O action table captures most details necessary for the coding of all inputs and outputs, both digital and analog, as briefly described as follows:

- **Digital outputs:** Identify the position of each digital output in all process phases and against all possible interlock actions. Fail position and disconnect action must be documented. Show the template used to create the code and program the output. The purpose and action for each digital output needs to be understood and communicated.

- **Digital inputs:** Identify the position of each digital input in all process phases and against all possible interlock actions. Fail position and disconnect action must be documented. Show the template used to create the code and program the input. The purpose and action for each digital input needs to be understood and communicated.

- **Analog outputs:** Analog outputs are listed in the I/O action table, which captures most details necessary for the coding. Identify the control action of each analog output in all process phases and against all possible interlock actions. The analog output action table is also used to identify fail position, disconnect action, template used to create the code, and other additional logic required to program the output.

- **Analog inputs:** Analog inputs are listed in the I/O action table, which captures most details necessary for the coding. Identify the control action of each analog input in all process phases and against all possible interlock actions. The analog input action table is also used to identify fail position, disconnect action, template used to create the code, and other additional logic required to program the input.

Special types of input/output are common in some process control systems. This includes high-speed pulse counter for measuring flow rate and digital stepper motors output count. All special I/Os must be clearly configured and documented. Also, associated interfaces and specification must be listed with any deviation from the default standards highlighted. Scaling must take into consideration instruments mis calibration and potential signal deviation from the prespecified range.

6.2.2   Data Acquisition and Closed Loop Control Tasks

Data acquisition is the process of sampling analog input signals that measure real-world physical conditions and converting the resulting samples into digital numeric values that can be manipulated by the PLC or most commonly by a computer system more situated for modeling, simulation, and database tasks. Most analog inputs are part of the data acquisition task with smaller numbers used in closed control loop control. Every closed loop has at least two analog variables assigned: the controlled values and controlling values. The components of data acquisition and closed loop control systems tasks include the following:

- Sensors that constitute a simple data acquisition task must be documented in both the functionality and the detailed programming implementation. This includes details of the method used, as in redundant sensory data validation or complementary sensory data fusion techniques.

- Data acquisition tasks associated with HMIs must be detailed with the associated communication and graphics tasks. Linkage between the ladder program and the HMIs is very critical for system operation, future enhancements, and overall system maintenance.

- Some data acquisition applications/tasks are often developed using various general purpose programming languages or software tools, such as Visual Basic, Java, C++, and Lab VIEW. These tasks need to be documented as part of the overall control system design.

- Closed loop control tasks should clearly identify the controlled variable, the controlling variable, digital count format, engineering units, type of control, and the algorithms used to achieve control. ON/OFF and PID control are the most common process control techniques in use today with an increasing trend toward fuzzy logic control in recent years.

Detailed coverage of process control techniques and instrumentation is given in Chap. 7. A case study in Chap. 8 will demonstrate the design and implementation of basic data acquisition and process control tasks for a real-time industrial project.

6.2.3 Project Logic Diagrams and Ladder Function Blocks

Historically flowcharts were often used to document logic flow prior to actual programming implementation. With the advancement of PLCs and distributed digital process control, pseudo codes and logic diagrams are becoming the tool of choice. Logic diagrams are used in this book and were introduced in Chap. 2. A structured approach must be enforced when a given task is divided into a number of smaller interconnected subtasks. Logic diagrams are effective visual documentation that directly map into the ladder program and can be used with any PLC/control system platform. Pseudo codes tools are platform dependent but can automatically generate the ladder or executable control code.

Although ladder programming may be the most widespread language for the implementation of PLC control, function block diagrams/logic diagrams are probably the second most used language. This graphical language resembles a wiring diagram with the blocks wired together into a sequence that is easy to follow. It uses the same notations/instructions as ladder diagram, but visually is more understandable to a viewer who is not versed in relay logic. Logic diagrams are not ideal for a very large programming task, which includes the use of special functions and I/Os. Implementing a process control program using ladder diagrams requires more preparation upfront to ensure full understanding of the program structure and the control flow before any code is committed.

Allen-Bradley ladder program structure is based on subroutine blocks, as was described in Chap. 5. Typically a process control application uses a varying number of subroutine blocks, including initialization, data acquisition, HMI, alarms, communication, diagnostic, or other major processing blocks. Chapter 9 covers these concepts in a comprehensive industrial process control case study. A small example is used earlier in this chapter to illustrate the concept.

6.2.4 Control System Preliminary Documentation

The purpose of the system preliminary document is to communicate the process control implementation details to all stakeholders before the final checkout and startup. The document essentially serves as the base for critical design reviews and potential revisions/enhancements. It can also be used by a wide range of audiences, including facilities, partners, customers, process leaders, or anyone involved in any of the processes outlined in the document. Always be aware of potential liabilities by documenting and seeking validation or clarification on all safety-/hazard-related issues.

The following are some of the basic tasks involved in creating your process control preliminary documentation:

- Process scope and goal statement
- Inputs and outputs map
- Hardware configuration
- Communication protocol and configuration
- Process logic diagrams and ladder program function blocks
- Safety and operational hazard issues
- Process control management system procedures
- Exception management process
- Ladder and HMI listing

Each of these tasks will be a section within your document, which you can be created in Microsoft Office Word or other word publishing program. It is recommended using a program that automatically generates a table of contents based on the use of heading styles. Also, it's important to assign a clear numbering system within your document. For instance, the process scope and goal statement might be 1, while items included with that statement are 1.1, 1.2, 1.3, and so on. The inputs and outputs map might be 2. A numbering system accomplishes two goals, in that it makes information easy for readers to find and allows easy tracking of changes to the document by different individuals. Changes will be documented by date, person, and the revisions made. This is particularly critical for effective collaboration among a team of participants and stakeholder in the design and implementation process. Only authorized users will be allowed to finalize/incorporate changes.

6.2.5 Program Documentation Using Cross Reference

A cross reference report lists all logical addresses/symbols in a project with the location of every occurrence of each address documented. The report includes the address, symbol, instruction mnemonic, file # (and name), and rung #. You can sort the cross reference report by address or by symbol.

The cross reference function searches the entire project and finds every occurrence of the selected element address only. The cross reference function can be performed both offline and online. You can perform cross reference searches on:

- An entire project—sorted by address or symbol
- A specific instruction in the ladder
- An address in the custom data monitor, the multipoint monitor, or the data file dialog

To search an entire project sorted by address, right click the Cross Reference from the project tree to open the Sorted by Address screen as shown in Fig. 6.1.

A program titled "COMBINATIONAL_LOGIC" is used to illustrate the program documentation using cross references. The program implements four logic operations (AND, OR, EXOR, and EXNOR). AND logic is used in Fig. 6.2 to illustrate the use of cross reference information. From portal view, double-click Show Cross References. From the project view, you can monitor the AND_LOGIC address, date created, and last modified. To search a specific instruction in the ladder, right click the address, then choose Cross Reference from the drop menu.

6.3 Process Control Checkout and Startup, Layer 3

The final step in the design is the checkout and startup of the process control system, layer 3. Preliminary testing of hardware, software, communication, and user interfaces is a major part of the system implementation, layer 2. This layer is concerned with final system troubleshooting and commissioning. Technology transfer to the owner/customer is a key issue in completed automation projects. It must be planned and budgeted in the early stages of the project, layer 1. All checkout activities are performed in the field using the system physical resources in real time. Stakeholders, including operation and maintenance personnel, are all critical part of this step, which is led by the

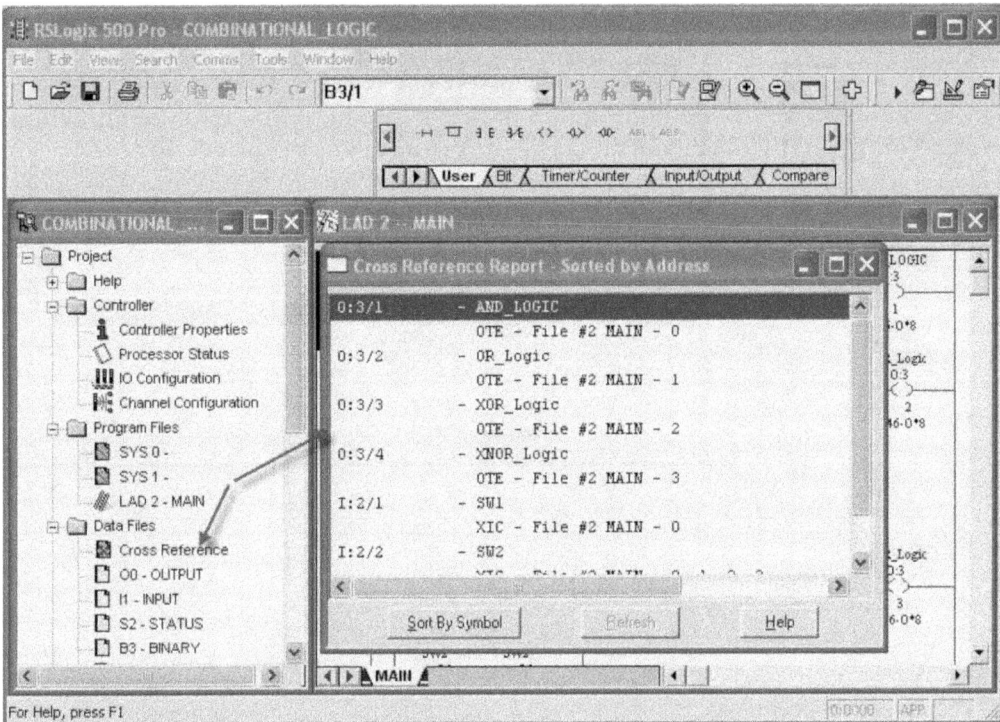

FIGURE 6.1 Documentation using cross reference.

process control/automation lead that designed and implemented the system. The process includes standard procedures and steps, which are detailed in this section. Also, standard safety procedures will follow our checkout discussions.

The AB SLC500 PLC provides three different methods for the testing of the user program: testing using program status and system diagnostics, testing using the watch table, and testing using the force table. Each of the three techniques is briefly discussed as follows:

- **Testing with the program status and system diagnostics:** The program status allows users to monitor the running of the program. You can display the values of operands and the results of logic operations (RLO), which allows for the detection and fixing of logical errors in the program. The PLC system diagnostic tools allow for the use of a wide variety of information during checkout.

- **Testing with the watch table:** With the watch table, current values of individual tags in the user program or on a CPU can be monitored and modified. Values can be assigned to individual tags for testing and program runs in a variety of different situations. Fixed values can be assigned to the I/O tags of a CPU in the STOP mode, which is typically used during the system static checkout of wiring.

- **Testing with the force table:** With the force table, current values of individual tags in the user program or on a CPU can be monitored and forced. Forcing an individual tag will overwrite that tag with the specified value. This allows program testing and run under various situations and scenarios.

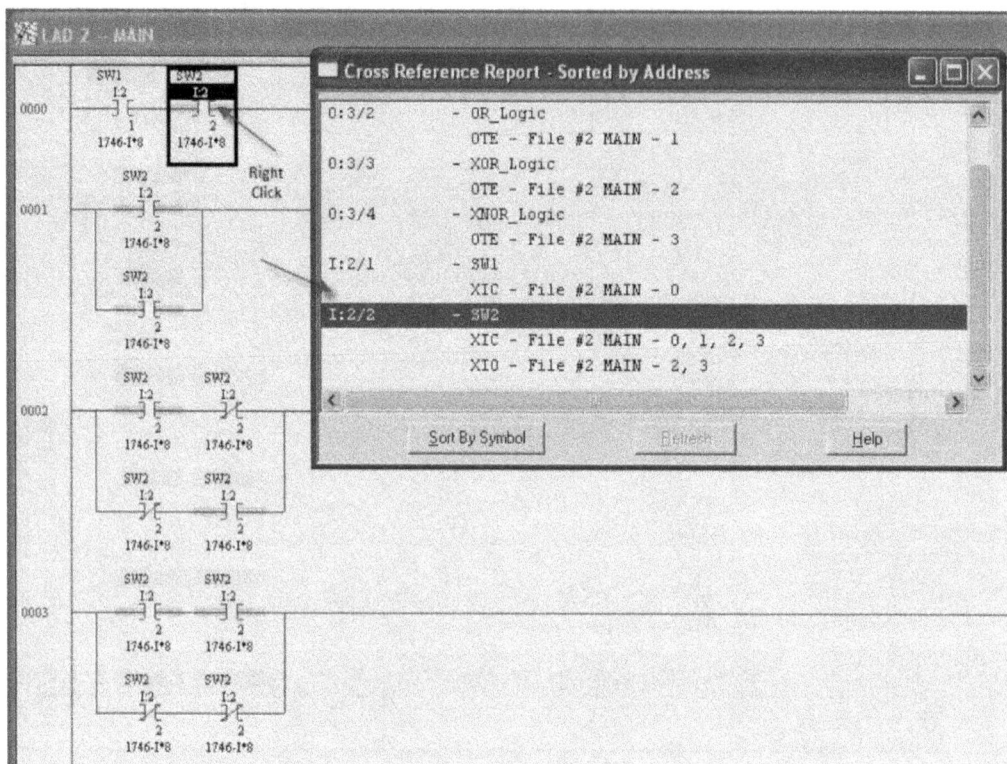

FIGURE **6.2**   Checkout using cross-reference monitoring.

6.3.1   Debugging Using Find and Replace Function

The Find option allows user to locate instructions, addresses, and symbols (if they have been defined) in the ladder program. The Find and Find All functions can be performed both offline and online. Access these options by selecting "Search" then "Find" or "Replace." Type the mnemonic (XIO, CTU, etc.) followed by the desired address (N7:6, etc.). *Debugging using* "Find" and "Replace" functions can be initiated by selecting "Search," then "Find," and then "Replace" from the main menu. Debugging using "Find" and "Replace All" functions can also be initiated by selecting "Search," then "Find All," and then "Replace" from the main menu as shown in Fig. 6.3.

6.3.2   Debugging Using Program Data File

In order to create a data file (bit file number 3); choose "Search" from the tool menu, then "Go to," and enter the file number as shown in Fig. 6.4*a*. Repeat the above steps to open multiple files as shown in Fig. 6.4*b*. All values of interest included in open data files can be monitored online in real time during program execution. This is similar to monitoring ladder online for selected inputs and outputs during program scanning as we did in previous chapters.

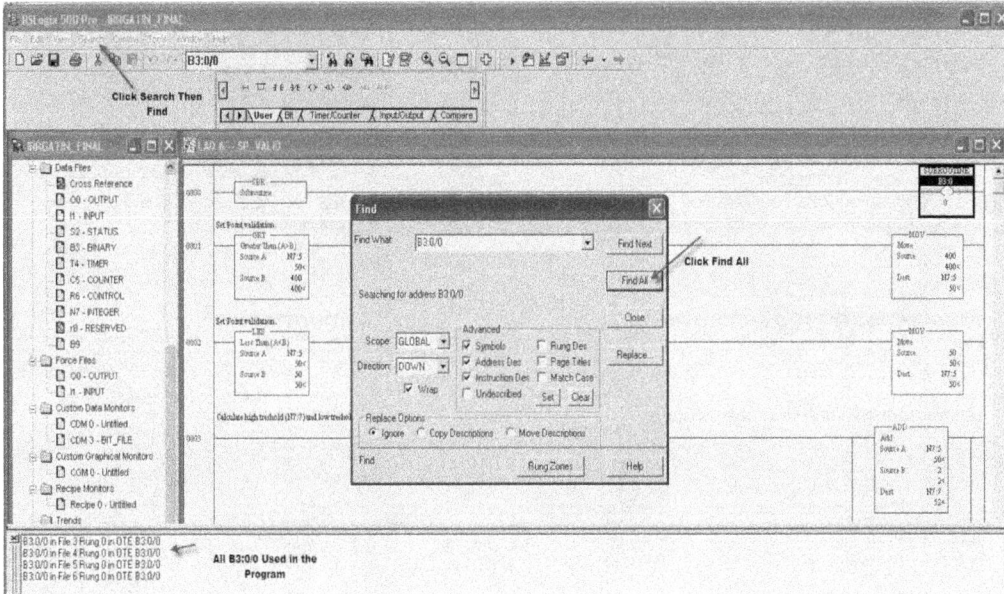

FIGURE 6.3   Checkout using find and replace functions.

6.3.3   Debugging Using Advance Diagnostic

Advanced diagnostics is a method of locating a section of ladder logic quickly by searching the page titles in your project. If you have not documented page titles, the advanced diagnostic search will not be helpful to you. To initiate this task, select "Search" then "Advanced Diagnostics" as shown in Fig. 6.5. Select the program file that contains the ladder logic of interest. Click the "Expand" to list the page titles in the program file you selected that describes the logic you are looking for.

6.3.4   Checkout Using Forcing Functions

Great care and precautions must be taken during the use of PLC forcing functions. Because the forcing function allows the user to intervene permanently in the process by assigning an arbitrary status/value to a control variable, observance of the following notices is essential:

- Prevent potential personal injury and assets damage. An incorrect action while executing the "Force" function can harm persons or pose a health hazard. It can also produce damage to machinery or the entire plant.

- Before starting the "Force" function, you should ensure that no one else is currently executing this function on the same CPU.

- Forcing is not affected by previous rung logic.

- Forcing can only be stopped by clicking the Stop Forcing icon or using the "Online > Force > Stop Forcing" command. Closing the active force table does *not* stop the forcing.

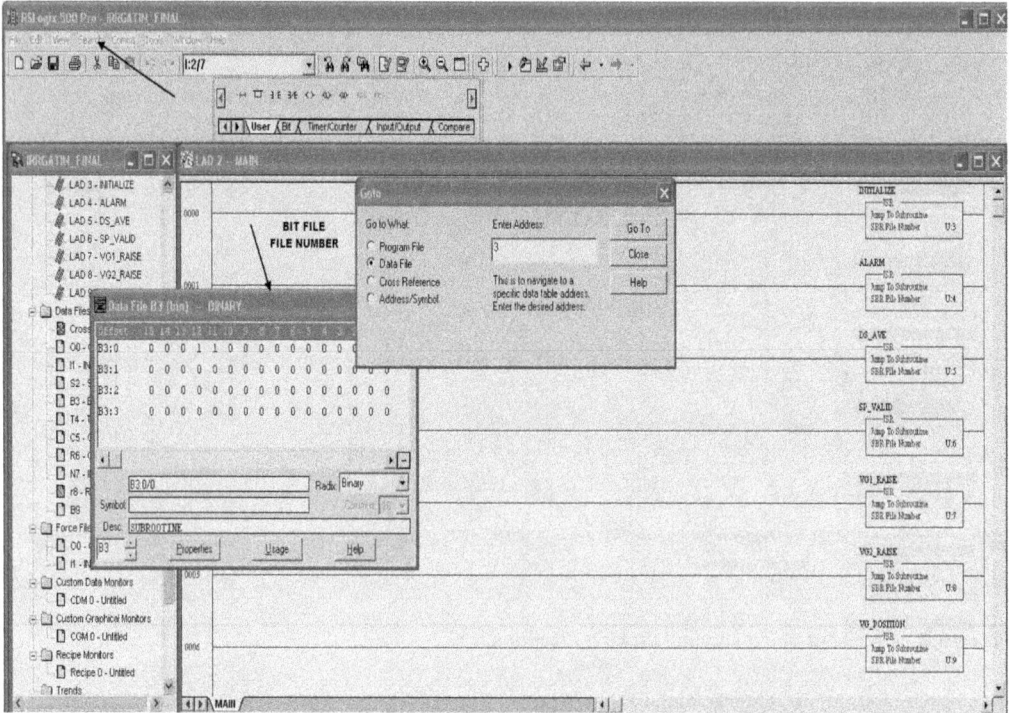

(a)

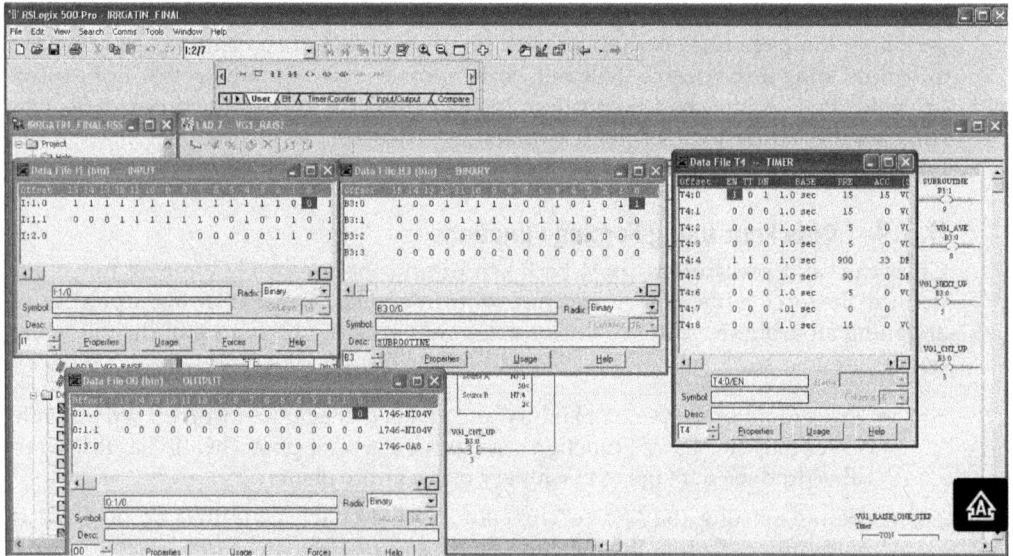

(b)

Figure 6.4   (a) Checkout using data file; (b) checkouts using multiple data files.

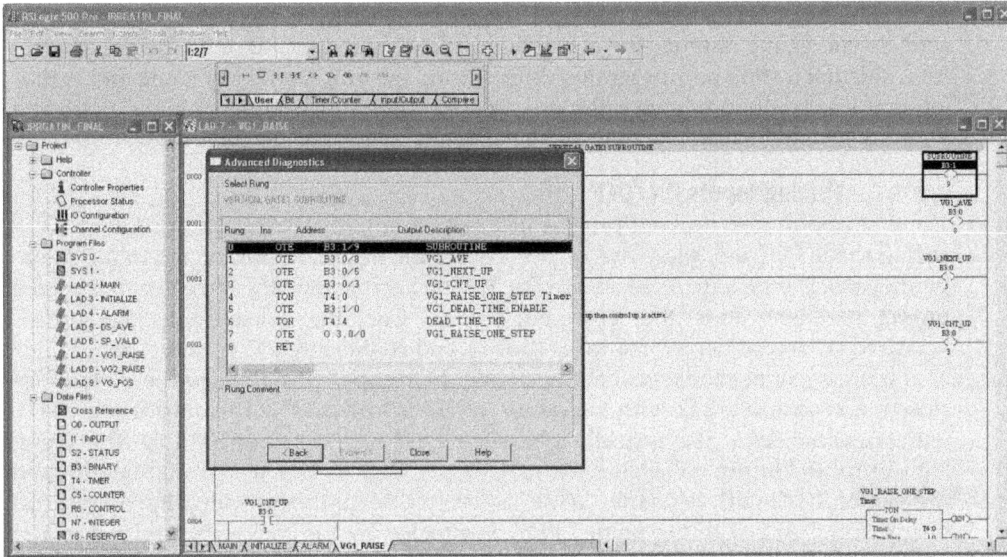

FIGURE 6.5   Checkout using advanced diagnostic.

- Forcing cannot be undone through the program logic execution.
- Modifying tags are different from forcing tags. The modifications can be changed by the program logic or through the updated I/O memory image, while forcing is permanent. Users should check Allen-Bradley technical manuals before using the force function.

Force tables can be used to monitor and force the current values of selected I/O in the program. When you force an input or an output, you overwrite the status of the selected I/O. This allows the logic designer to test the program while the system is running.

The RSLogix 500 software performs diagnostic tasks every ladder program scan and updates the hardware status. It also provides a log for all important and relevant events. This information is available, and users need to be aware of the best way to make use of such information. Program status and cross reference information can be accessed and used during program, hardware, communication, and graphical user interface debugging. This section introduces these tools, but more details are available in the technical manuals or online using the Help menus.

System checkout is divided into two parts: static and dynamic checkout. Static checkout is used to verify that the input/output wiring of the system is implemented correctly to verify that a force ON and force OFF functions properly. This allows user to turn input and output instructions ON or OFF from the programming terminal and verify that I/O devices are functioning correctly. When an output address is forced ON or OFF, only the output is affected. Any examine instructions with the same address as the forced output will remain unchanged. All forces ON or OFF can be removed from the programming terminal.

The SLC-500 family and other AB PLC platforms have a test mode, which allows the user to test the program without affecting the output status. In test mode, the processor scans all inputs, solves logic, and updates output instructions as if the program were actually energizing outputs, but actual output devices are not affected. This is very useful when first testing new programs. After testing the program in test mode and verifying that it is running properly, the processor should be placed on running mode and program should be tested with the actual outputs energized/de-energized.

6.3.4.1 Forcing Inputs ON/OFF

The first step in the checkout process includes testing of all input and output devices wiring to the PLC modules. This section covers the input/output forcing in the Allen-Bradley SLC500 system. As shown in Fig. 6.6, you can initiate the force action from the project tree menu under the "Force File Folder." Click the "Force," enter the address you need to force, enter a force value (1 or 0), and enable force.

Figure 6.7 shows the physical switch SW1 in the open position and the examine-if-closed instruction I: 2/2 with tag name "SW1" is forced ON. The output coil O:3/0 with tag name "PL1" is physically ON when 0 to 1 is detected on SW1 by RLO (rung logic output). Output coil O:3/1 with tag name "PL2" is OFF as the examine-if-open instruction I:2/2 with tag name "SW1" is forced ON. In the last rung, the output coil instruction O:3/2 with tag name "PL3" is affected by the force ON, and a 0 to 1 is detected by RLO causing output coil O:3/3 with tag name "PL3" to turn ON, which physically turns the PL3 ON. Siemens PLC handles the forcing differently; the forced input does not affect the rest of the ladder logic. In this case, PL3 will remain OFF.

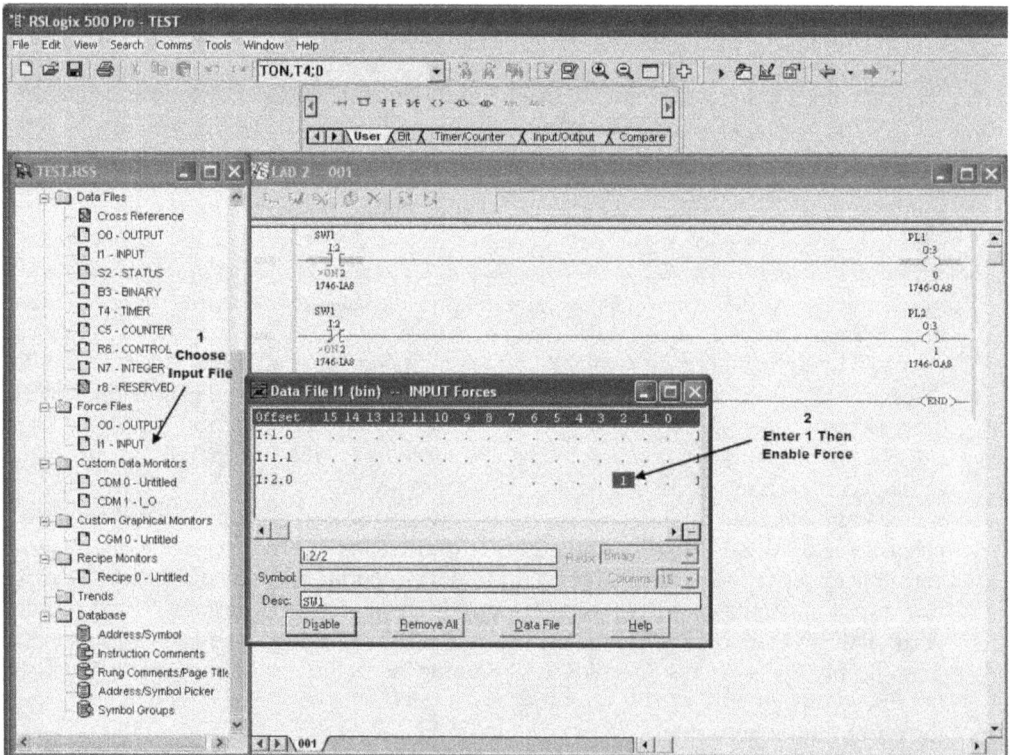

FIGURE 6.6 Forcing a PLC input in RSLogix 500.

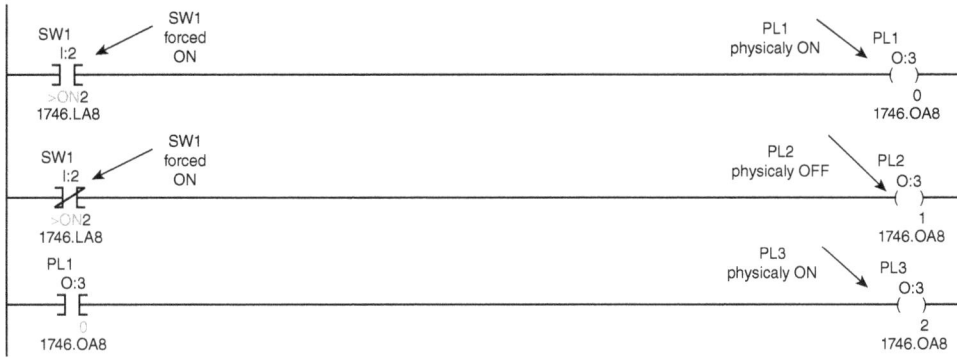

FIGURE 6.7 Forcing PLC inputs.

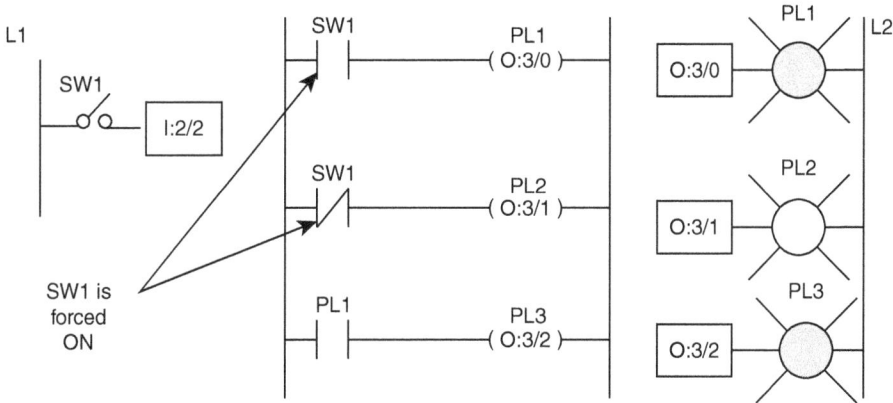

FIGURE 6.8 Forcing inputs in relation to physical inputs and outputs.

Figure 6.8 shows the status of the pilot lights in relation to the input SS1 and the ladder program as described in Fig. 6.7.

The static checkout process is entirely performed from the PLC programming terminal with assistance from a technician or an operator in the field where actual sensors and actuators are located. Communication between the person in the field and the PLC operator is essential to the checkout process using the input/output forcing functions. It is important to execute the forcing in sequential order and remove the forcing of the completed item before moving on to the next I/O in the checkout list. There is no value in checking the implemented control program logic before completing a successful static checkout.

6.3.4.2 Forcing Outputs ON/OFF

Figure 6.9 illustrates the output forcing g operation. It shows the physical switch SW2 open and output coil O:3/1 with tag name "PL2," which is forced ON. Physical pilot light (PL2) in the first rung illuminates as indicated. The normally open contact O:3/1 connected to PL4 in the third rung is not affected by the force ON applied to the output

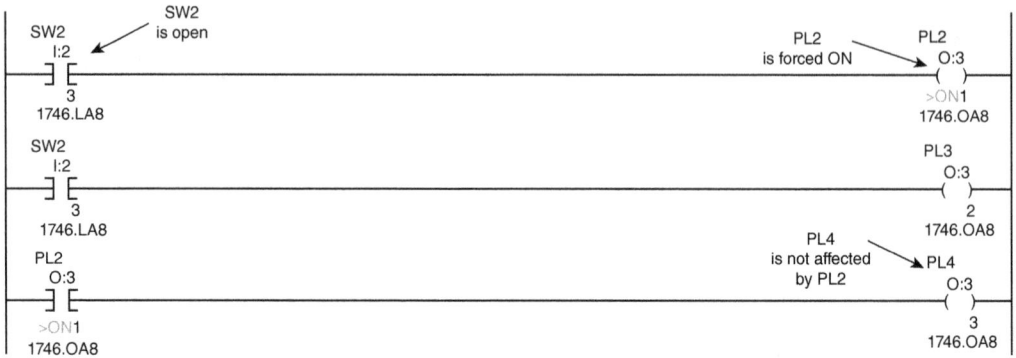

FIGURE **6.9** Forcing a PLC outputs.

O:3/1 (PL 2) in the first rung and thus PL4 is turned OFF. The figure also shows PL3 in the second rung OFF since SW2 is open.

Figure 6.10 demonstrates the output forcing example in relation to the actual physical inputs and outputs used. The figure shows the status of the pilot lights (PL2, PL3, and PL4) in relation to the input switch SW2 and the ladder program. The forcing function is intentionally designed to only affect the selected output but none of the remaining logic in the program will be affected. This function is primarily used during static checkout of all outputs wiring.

Notice the status of the normally open contact SW2 with address I:2/3, which is open. Output coil O:3/1 created internal contact is physically connected to PL4, and this output/coil is being forced ON/true. PL2 turns ON as a result of the forcing of coil O:3/1 to true status. The examine-if-closed contact associated with PL2 stays OFF, only the forced output O:3/1 is ON, thus PL4 stays OFF. PL3 is OFF as SW2 is also OFF, which does not allow power to propagate to the coil associated with PL3 with address O:3/2. Of course, switching SW2 ON will cause all lights to go on regardless of the forcing action.

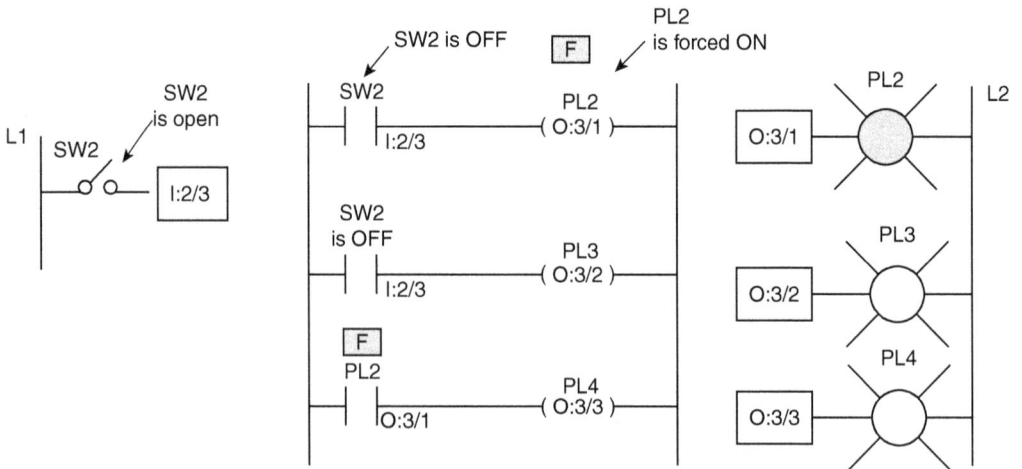

FIGURE **6.10** Forcing output in relation to physical inputs and outputs.

6.3.5 Checkout Using Custom Data Monitors

The custom data monitor (CDM) is a feature that allows a user to create a list of addresses from various data table files within your project and monitor or change the values for the addresses in the list. To create a custom data monitor, right-click "Custom Data Monitors" from the project tree, then click "New," and enter the file number, name, and title. You can also drag and drop the addresses you want to monitor on the table (the number is 4 and the title is Timer) as shown in Fig. 6.11.

Using output coils in more than one location in a program can be a source of trouble during checkout. In this section, we will demonstrate the checkout using Custom Data Monitor in conjunction with this problem. Output coil addresses cannot be repeated due to the way the processor scans the program (left, up, down, right) as shown in the ladder of Fig. 6.12 and in the CDM table of Fig. 6.13; PL1 will be OFF when SW1 is ON, SW2 is ON, and SW3 is OFF.

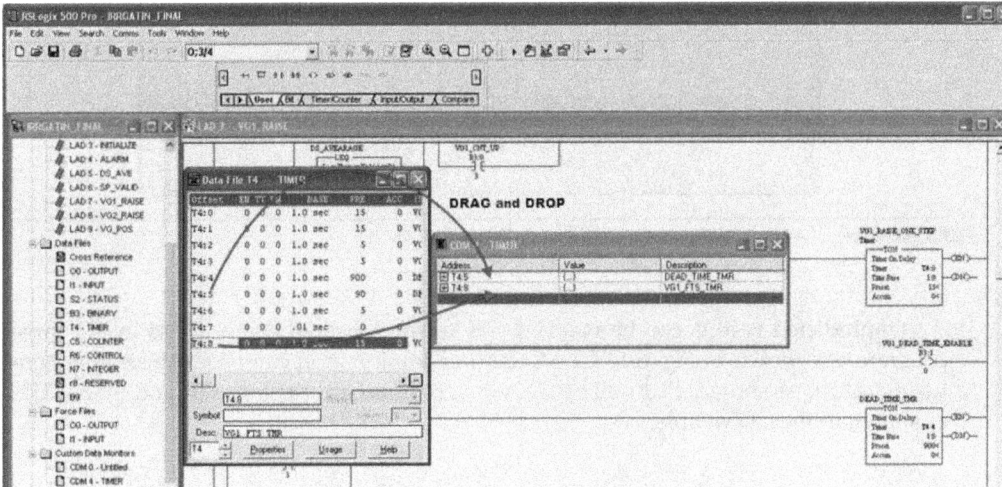

FIGURE 6.11 Creating a CDM.

FIGURE 6.12 Repeated coil ladder.

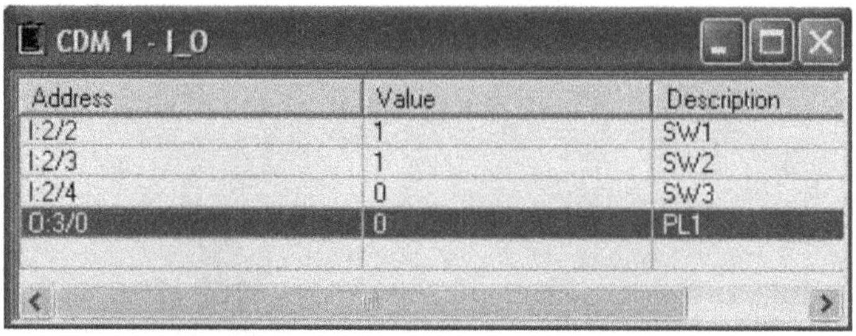

FIGURE **6.13**   Custom Data Monitor (CDM configuration 1).

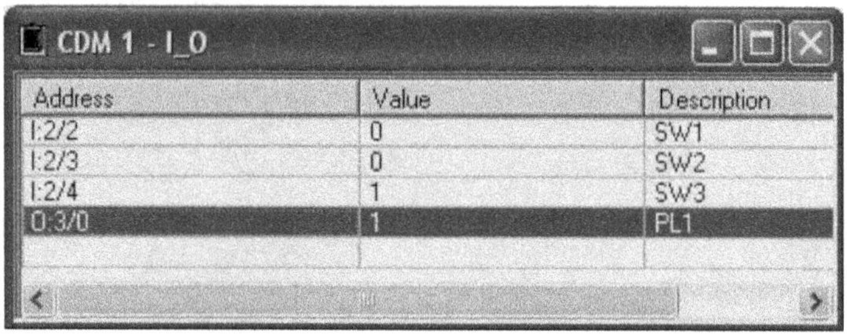

FIGURE **6.14**   Custom data monitor (CDM configuration 2).

Unpredicted results can be reached due to the repeated coil address in the same program. As shown in Fig. 6.14 CDM, another combination can produce another confusing result. As shown, PL1 will be ON when SW1 is OFF, SW2 is OFF, and SW3 is ON as shown in the CDM table.

6.4   System Checkout and Troubleshooting

System checkout and troubleshooting is a critical tasks that needs to be carried out before the final deployment and commissioning of any process control/automation system. This task must be performed according to well-established rules and documented procedures. These rules/procedures are designed carefully to prevent any potential problems that might cause personnel injuries, damage to resources, or compromise any of the predefined process/end product qualities. Simulation and emulation techniques are used throughout the implementation phases of the project prior to final checkout. Simulation tools allow designers to examine system performance situations under varying scenarios before or aside of the actual implementation. This step also can be used to efficiently verify design concepts prior to the commitment of resources. Emulation phases are accommodated in PLC software development tools, which allow the user to emulate I/O hardware, HMI interfaces, control logic, and communication facilities. The initial phase of emulation checks covers only the control logic where the second phase includes actual and simulated hardware/interfaces. The final emulation phase utilizes actual system hardware, interfaces, and communication resources.

6.4.1 Static Checkout

Static checkout is a process used to verify that input/output wiring is implemented correctly. It also verifies the continuity of current to and from the PLC/instrumentation. This process must use the following steps:

- Provide proper grounding and electrical noise protection.

- Implement safeguards that are independent of the PLC system to protect against possible personal injury or equipment damage.

- Maintain safe status of the PLC low-voltage circuits, external connections to communication ports, analog circuits, all 24 V nominal power supply, and I/O circuits. All must be powered from approved sources that meet all safety requirements.

- Provide over-current protection, such as a fuses or circuit breakers, to limit fault currents from damaging supply wiring.

- Inductive loads should be equipped with suppression circuits to limit voltage rise due to inductive reactions.

- Identify any equipment that might require hardwired logic for safety.

- Ensure fail safe conditions for devices to prevent unsafe manner failure, which can produce unexpected startup or damage to equipment.

- Provide the appropriate status information to the PLC including detailed alarms and fault information to ensure safe guards as implemented in the control logic software.

- Provide the appropriate system status information to the assigned HMIs to enforce implemented operators safeguards interactions.

Static checkout of all instrumentation must be conducted before any program logic debugging. It is a time-consuming operation and requires collaboration between the instrumentation technician and the PLC operator. PLC software tools greatly reduce the effort needed to accomplish this task. The following are typical steps used to checkout I/O connections:

- Turn each discrete input ON/OFF and observe the associated LED indicators on the input module switch panel.

- Check the assigned discrete input addresses in the PLC and confirm that input change between 0 and 1 as the input signal change from OFF to ON.

- Outputs should be tested while system is in Manual mode. Use push button (PB) switches as an example for motor START/STOP. The motor should be checked using NO/NC (normally open/normally close) momentary switches and by verifying that the motor respond correctly.

- Outputs should be tested by using the PLC forcing function. As the bit is forced from the PLC from 0 to 1, output should turn from OFF to ON. Safety precautions should be taken before forcing I/O.

- Check all analog input signals and make the needed adjustments in the sensor offset and span values. This analog signal calibration requires coordination and is time consuming but essential to achieving accurate data acquisition and overall process control.

6.4.2 Wired Master Control Relay Safety Standards

Master control relays (MCRs) are used as a standard hardware safety technique to protect the system under emergency conditions. An MCR is operated by START and STOP PB switches. Figure 6.15 shows a typical industrial-type MCR whereas Fig. 6.16 documents the power distribution connection to the I/O modules. Pressing the START PB energizes the MCR and provides power to the I/O modules. Pressing the ESD (not shown in the figure) or E-stops PB, as shown in the figure, will disconnect the power to the I/O modules, but the processor will continue to operate allowing operator access to the PLC programs and software tools with complete isolation from all I/Os. MCR is a hardwired relay that is used to provide a controlled safe stop. The master control relay de-energizes the supply power to the outputs and the machine power to the controlled devices.

The following are few examples of safety standards routine practices used in implementing process control systems. As stated, these are universal standards not optional actions or choices and must be followed:

- A step-down transformer provides isolation from high main power to the 120 V ac going to the PLC power supply (L1, L2).

- Normally closed contact Emergency Stop PB is wired in series with the MCR coil to de-energize the coil in case of an emergency and disconnect the power to all I/O modules (CPU still receive power and its LEDs provide status information.

- NO/NC momentary push button contacts are used for motor START/STOP operations in order to protect against accidental restart of motors after power outage is restored.

- Two START push buttons are simultaneously used to start a machine to protect the operator hands during start-up.

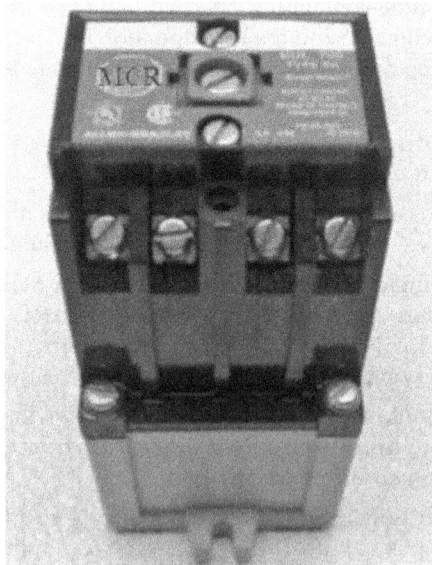

FIGURE 6.15   Typical industrial type master control relay.

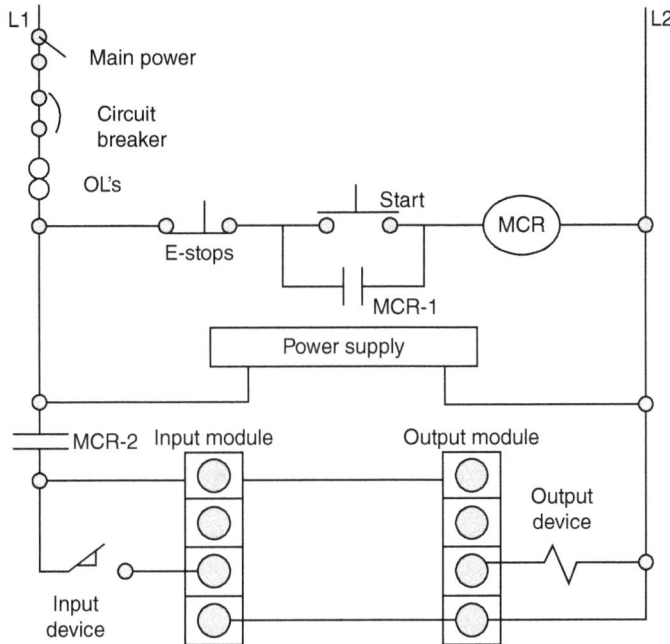

Figure 6.16   Master control relay circuit.

- Interlocking switching mechanism is used for motors running in forward/reverse directions. This system requires motor stopping before the reversal of running direction and must have the highest possible reliability.

- The use of Emergency Shutdown (ESD) switch with all power connection to PLC modules conducted through its contacts. Activation of the ESD stops all outputs in case of any emergency situation.

- Fault detection hardware is deployed in critical areas, as in a hazardous chemical possibly overflowing a reactor tank, to provide backup in case of sensors or control logic failure.

- PLCs have hardware and online impeded diagnostic tools including a combination of both software and hardware. LEDs are used to indicate the status of the PLC CPU and I/O modules.

- Diagnostic PLC software is an essential part of every program scan cycle. The CPU provides the following diagnostic status indicators:

  1. STOP/RUN
     - Solid orange indicates the STOP mode.
     - Solid green indicates the RUN mode.
     - Flashing (alternating green and orange) indicates that the CPU is starting up.

2. ERROR
- Flashing red indicates an error; such as an internal error in the CPU, an error with the memory card, or a configuration error (mismatched modules).
- Solid red indicates a defective hardware.

6.4.3 Master Control Reset Instruction Control Zones

The master control reset (MCR) output instruction, sometimes known as *zone control*, is used to set up areas or *zones* in the ladder program where all nonretentive outputs can be disabled at the same time for a specific duration or under set conditions. MCR is part of the "Program Control" instruction category as shown in Fig. 6.17. Typically, these instructions are used to minimize scan time, deviate from the sequential program execution, create a more efficient program, and troubleshoot a ladder program.

The instruction is used in coil pair, one MCR coil to define the start of the ladder area to be affected and another MCR coil to define the end of the zone. An input condition is programmed on the rung of the first MCR to control the rung logic continuity. When the MCR coil goes OFF, all nonretentive outputs within the controlled zone are disabled. When the MCR goes on, all elements are scanned according to their normal rung conditions, disregarding the zone control instruction.

The MCR instruction is not a substitute for the hardwired master control relay that provides emergency stop capability. Precaution must be taken when MCR instruction is used for zones containing timers, counters, or latched outputs. Timers and counters will not run during the zone release duration (MCR is OFF)—except for OFF-delay timers (TOF), which will run if the zone is released when the timer-enabling input is ON. All latched outputs (retentive) will maintain their last status regardless of the MCR control status, which is different from the hardwired MCR, where all outputs get disabled regardless of type once the MCR action is activated. Also, the zone extends to the entire ladder program in regards to the hardwired MCR. Use of the MCR instruction can be confusing; thus it is recommended to avoid including timers, counters, or latched output elements in an MCR-controlled zone.

Figure 6.18 illustrates the operation of the MCR instruction. It shows two motors (M1 and M2) controlled from two separate switches (left and right), an ON-delay timer (T4:0), and an OFF-delay timer (T4:1). Motor 1 uses latch and unlatch instructions whereas motor 2 uses XIC, XIO, and OTL instructions. Input switch I:1/0 "CONTROL_ZONE" when ON, it sets up the zone and the program runs under normal conditions. When the switch is OFF, all nonretentive outputs are OFF and retentive output latches their last state. TOF when placed in the zone will start

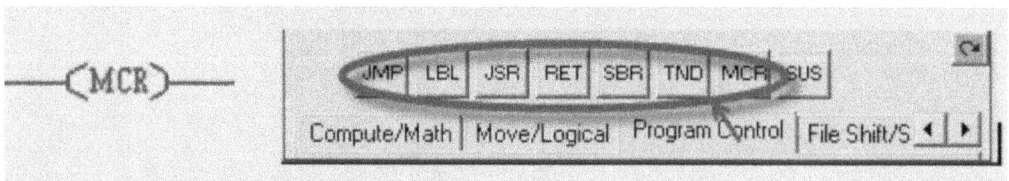

Figure 6.17 MCR instruction.

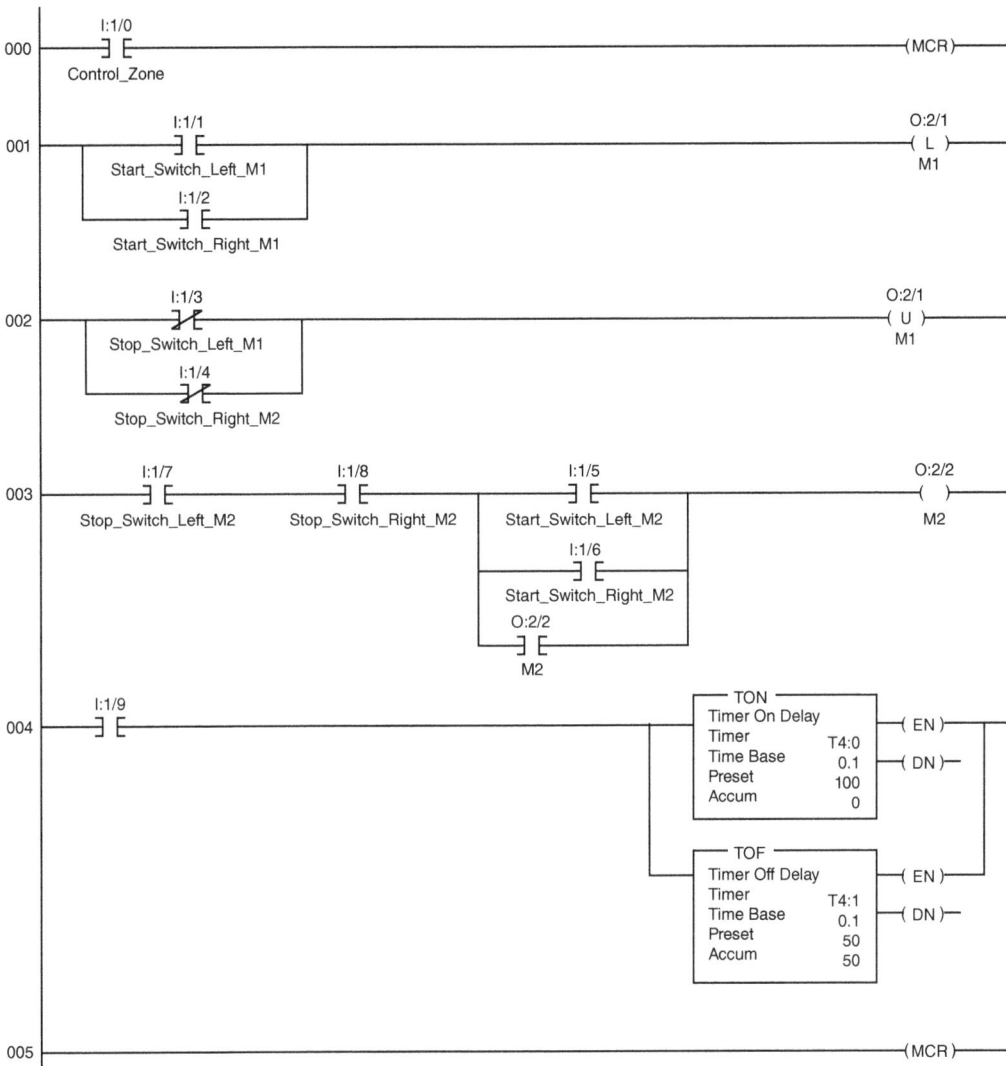

FIGURE 6.18 MCR motors control zone.

timing if its input instruction is active. Do not use conditional logic before an ending MCR instruction rung. The ending MCR instruction must be the only element on that rung.

You should experiment with the shown ladder program on your LogixPro simulator or SLC-500 trainer and monitor the behavior of the zone rungs under different simulated input conditions. Notice the behavior of the TON and the TOF timers under all combinations of the MCR control input (I:1/0) and the timers input control (I:1/9). Also, observe the difference in behavior between M1 and M2 under the different input switches scenarios.

6.5 Safeguard Implementation Examples

This section presents a few examples showing safety implementations in the Allen-Bradley SLC500 PLC systems. All principles included in these examples are practical situations independent of the PLC platform used for the process control implementation. A few minor syntax modifications might be needed for other vendors PLC environment.

Example 6.5.1 A given machine tool process requires an operator to start the conveyor system by pushing two START PB switches at the same time to restrict the location/position of the operator hands and thus guarantees maximum human protection during the startup process. The ladder rung in Fig. 6.19 assumes two normally open START1 and START2 PBs, one normally close STOP PB, and one output coil for MOTOR1 starter.

Once the START1 and START2 PBs are pressed, I:2/1 and I:2/2 become true, which in turn makes O:3/1 true and MOTOR1 runs. The next scan O:3/1 will latch the START1 and START2 PBs and maintains the rung true status, which keeps MOTOR1 running. Once the STOP PB is pressed, MOTOR1 losses power and goes OFF.

Example 6.5.2 Figure 6.20 illustrates a line diagram of magnetic reversing starter, which uses an interlock to control the motor in forward and reverse direction. Pressing the Forward PB completes the forward coil circuit from L1 to L2, causing coil F to be energized. Energizing coil F in turn energizes two auxiliary contacts: F-1 and F-2. The normally open contact F-1 provides a latch around the Forward PB maintaining coil F energized. The normally closed contact F-2 will prevent the motor from running in the reverse direction if the Reverse PB is pressed while the motor is running in the forward direction. The following is a summary of the operation of this system:

- When FWD PB is pressed, I:2/4 (FWD) becomes true, I:2/3 (STOP) is still true, the examine-if-open O:3/3 (R) is true, and output coil O:3/2 (F) is set. The next scan O:3/2 will latch the FWD PB and maintains rung 1 true status. This will keep the motor running in forward direction.

- To stop the motor, operator presses the STOP PB. This causes I:2/3 (STOP) to go false momentarily, which drops the latch around I:2/4 (FWD) causing the output O:3/2 (F) to de-energize.

- When REV PB is pressed, I:2/4 (REV) becomes true, I:2/3 (STOP) is still true, the examine-if-open O:3/2 (F) is true, and output coil O:3/3 (R) is set. The next scan O:3/3 will latch the REV PB and maintains rung 2 true status. This will keep the motor running in reverse direction.

- To run motor in the reverse direction while it is running in forward, the operator must press the STOP PB before pressing the REV PB. Also, to run motor in the forward direction while it is running in reverse, the operator must press the STOP PB before pressing the FWD PB.

- The F and R coils are automatically interlocked with the examine-if-open O:3/2 (F) and the examine-if-open O:3/3 (R).

Example 6.5.3 The two rungs shown in Fig. 6.21 were part of wet well pumps control discussed in Chap. 5. Students can refer to this chapter for further description. It shows the use of emergency

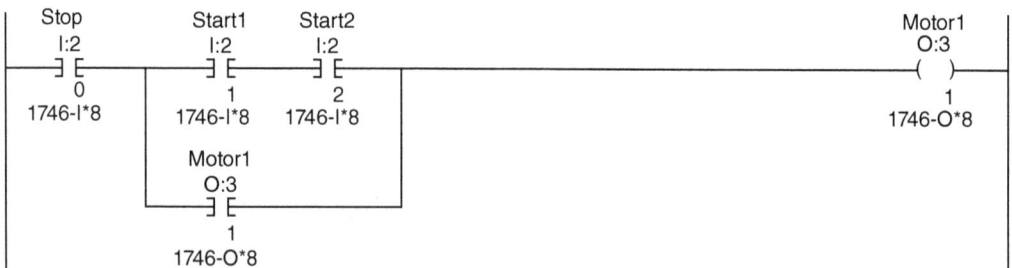

Figure 6.19 Two simultaneous START PBs of a motor.

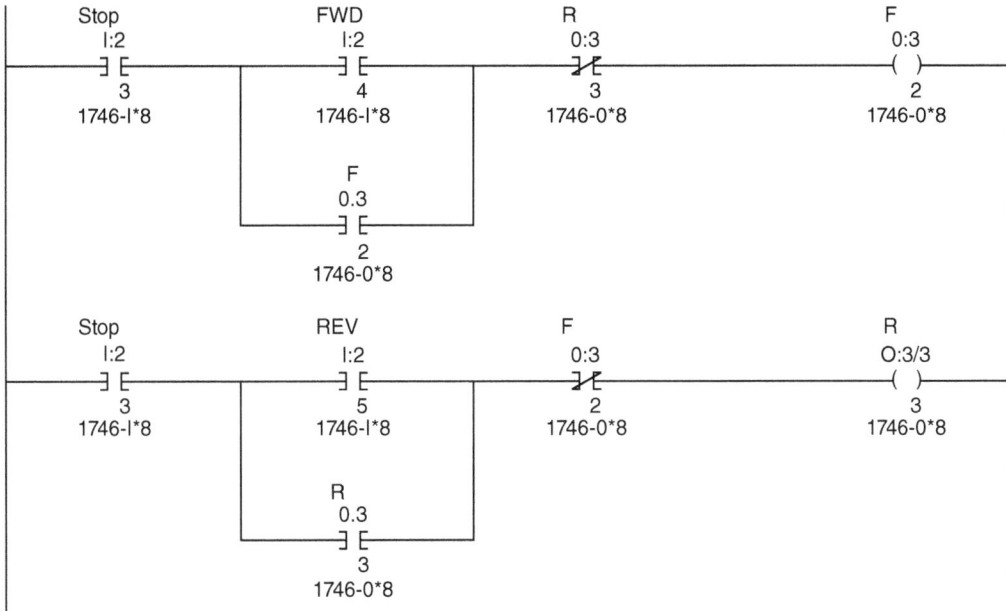

FIGURE 6.20 Interlocked forward and reverse motor actions.

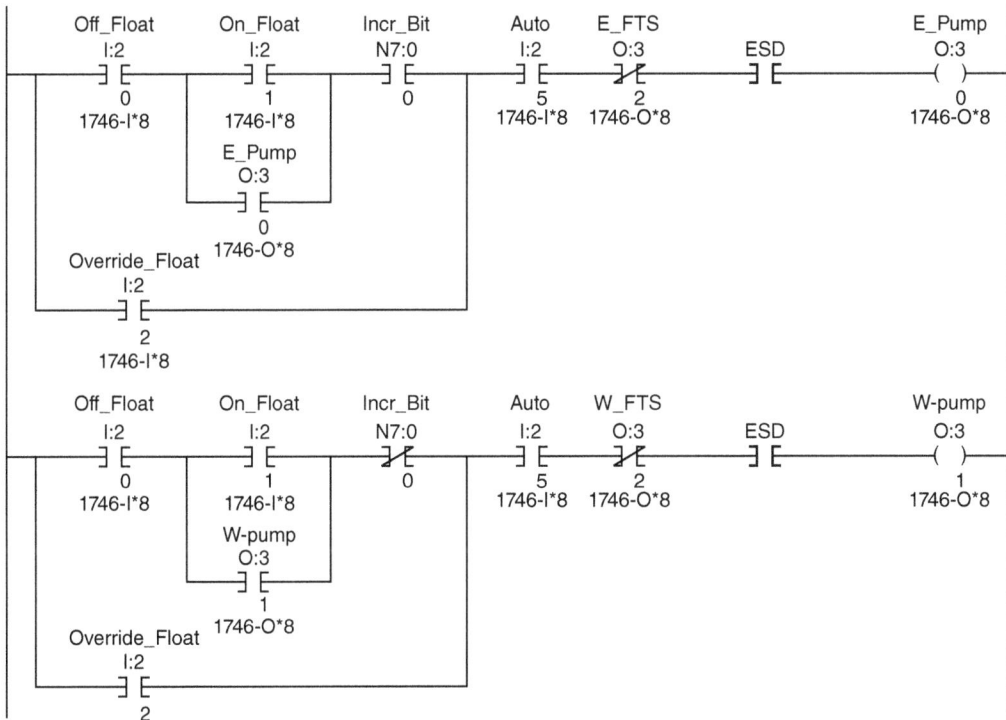

FIGURE 6.21 ESD use for output signal isolation.

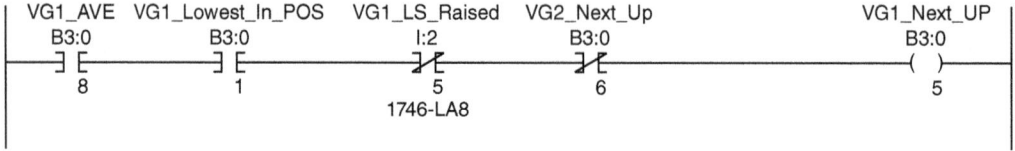

Figure 6.22 Limit switch use for output signal isolation.

shutdown (ESD) for the protection against unusual conditions. Note that ESD is wired in series with the output module motor signal. Once ESD is activated, it interrupts the power to the motor regardless of the action of the PLC scanned logic. Two rungs are designed: one for starting the east well pump "E_PUMP" and the other is used for the west well pump "W_PUMP." The two pumps require similar conditions for each to be selected to run. Common conditions for both require the system to be on Auto mode, ESD is not activated, and that the pump motor does not fail to start. Notice that ESD has to stay OFF for either pump to run even with all other conditions satisfied. Also, notice that the ESD is wired high, meaning it is a closed contact while it is not active. Once ESD is activated, its contacts will open. ESD is highlighted in the rung display indicating power continuation or a closed contact.

Example 6.5.4 Limit switches are used as safeguard for the protection of assets and personnel in addition to redundant measures in the PLC logic. Figure 6.22 shows one use of the limit switch, which is placed in the upper position of the vertical gate 1 used to regulate the downstream water level in an irrigation canal. This limit switch "VG1_FULLY_RAISE_LS" ensures that the power to the motor will disconnect (VG1_NEXT_UP is the input to the VG1_RAISE) when vertical gate 1 reaches the upper position. This will ensure protection for both the motor and the structure from overload. Similar logic is required for the vertical gate 1 next to go down using another limit switch. Large industrial motors have internal protection against overloads and other undesired conditions but the PLC hardware/software must prevent the motor from reaching such conditions.

Chapter 6: Home Work Problems and Laboratory Projects

1. List three issues covered in the process control layer 1 and explain what "process constraints" means.

2. Explain briefly what are the main tasks included in process control layer 2.

3. What is the most standard analog I/O signal? Give few examples for typical analog input and output process control variables.

4. Is a PLC placed in the run mode during static check out?

5. List at least four of the basic tasks involved in creating the process control preliminary documentation.

6. Explain briefly what are the main tasks included in process control layer 3.

7. What are the basic three modes used in the testing the SLC500 user program?

8. A program malfunctioned and force output method was used to check out and debug its operation. The problem was fixed, but later during another run, the program malfunctioned again. What do you think might have caused this failure?

9. Is it possible to account for all possible future process scenarios during system checkout? How do HMIs contribute to the system overall continuous enhancement?

10. Why do some machines require two START PBs to start manually, whereas only one is necessary to start the machine?

11. How is an ESD switch wired and included in the PLC system functionality?

12. What is the advantage of 24-V dc I/O modules over 120-V ac modules?

13. Why does the Electrical National Code require NO (normally open) PBs to start the motor and NC (normally close) PBs to stop the motor?

14. What are the components of a data acquisition system? Explain briefly the function of each component.

15. A vertical gate is traveling between two limits: fully raised and fully lowered positions. One motor is used, which runs in the forward direction for raising the gate and in reverse direction for lowering the gate. Write a ladder logic program to protect the motor from moving in either direction if either of the two limits (forward/reverse) is reached. Add sensors as needed.

16. Explain the function of an HMI. How does it enhance the overall automation system operation and maintainability?

17. As shown in Fig. 6.23, the physical switch SS1 is open and in the ladder logic is forced ON. What is the status of PL1, PL2, and PL3?

18. Study the ladder program shown in Fig. 6.24 and answer the following:
    a. Explain the function of each rung as programmed and documented.
    b. Check each rung and verify that it implements the desired specification as shown in the documentation.
    c. Troubleshoot the program and reprogram the rung(s) to work as it should, based on the correct rung documentation.
    Hint: rung 3 has a critical error!

19. In Fig. 6.25, the physical switch SS2 is open and output coil Q0.1 with tag name "PL2" is forced on. What is the status of the physical outputs PL2, PL3, and PL4?

20. What is the advantage of alarm acknowledge page on the HMI. If the alarm sounds, how would you proceed to solve the problem?

21. What kind of problems arise if the analog input signal is coming to the PLC in a nonstandard format? Explain how you can solve this problem using scaling.

22. What is the most important tool(s) in a control system to identify a fault? List three such tools.

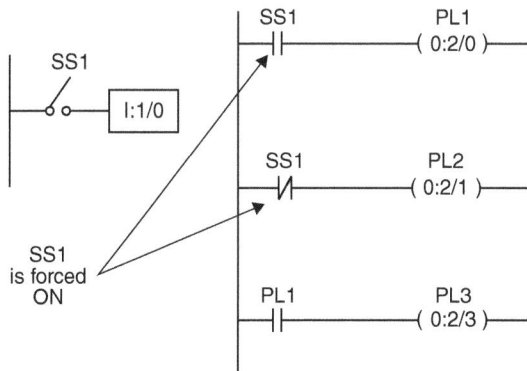

FIGURE 6.23 Problem 17 ladder.

Rung 1:

```
                                                              VG1_Fully_Raised_
   LS1_Fully_Raised      LS1_Fully_Lowered                         Green_PL
          I:2                   I:2                                  O:3
───────] [──────────────────]/[────────────────────────────────────( )───────
           0                    1                                     0
        1746-I*8             1746-I*8                              1746-O*8
```

Rung 2:

```
                                                              VG1_Fully_Lowered_
   LS1_Fully_Raised      LS1_Fully_Lowered                        Orange_PL
          I:2                   I:2                                  O:3
───────]/[──────────────────] [────────────────────────────────────( )───────
           0                    1                                     0
        1746-I*8             1746-I*8                              1746-O*8
```

Rung 3:

```
     LS1_Fully_Raised    LS1_Fully_Lowered     Flash_Bit
            I:2                 I:2                S:4        ┌──────TON───────┐
    ┌─────] [────────────────] [─────────────┬──] [─────────┤Timer On Delay  ├──(EN)─
    │        0                  1             │   8          │Timer      T4:0 │
    │     1746-I*8           1746-I*8         │              │Time Base   1.0 ├──(DN)─
    │  LS1_Fully_Raised  LS1_Fully_Lowered    │              │Preset      15< │
    │        I:2                 I:2           │              │Accum        0< │
    └─────]/[────────────────]/[──────────────┘              └────────────────┘
             0                  1
          1746-I*8           1746-I*8
```

Rung 4:

```
                                                              VG1_Alarm_Red_PL
        T4:0                                                        O:3
───────] [──────────────────────────────────────────────────────────( )───────
         DN                                                          2
                                                                  1746-O*8
```

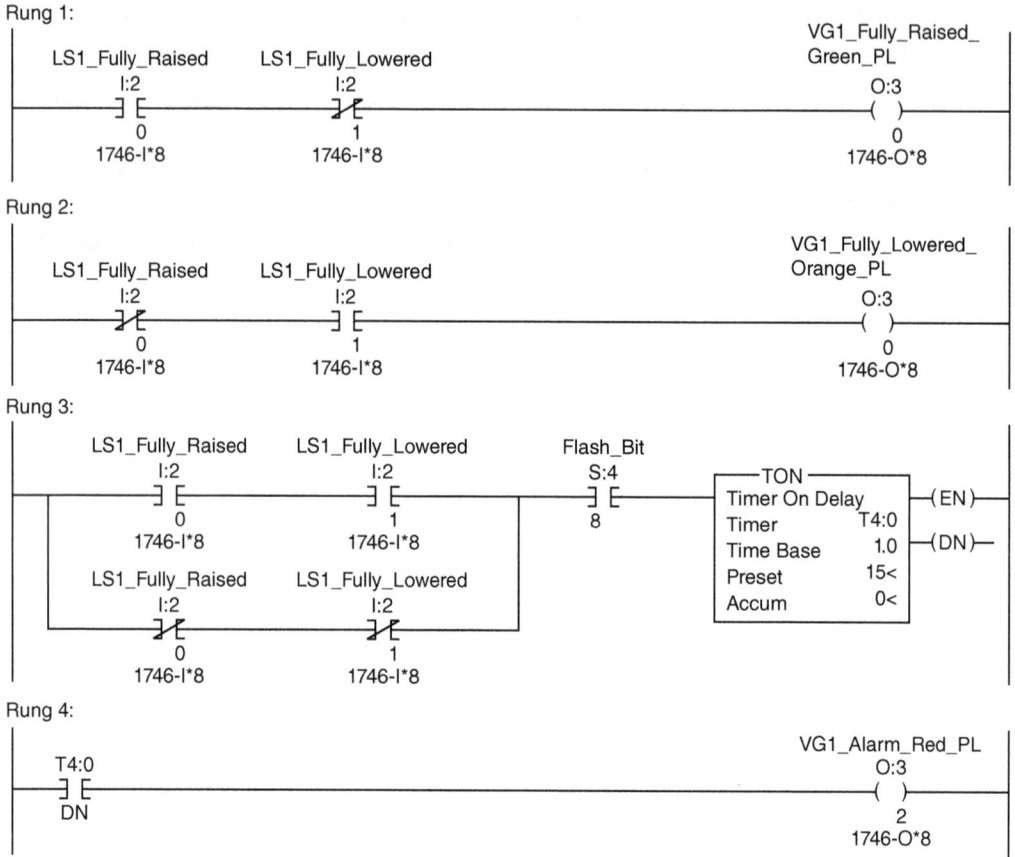

FIGURE 6.24 Problem 18 ladder program.

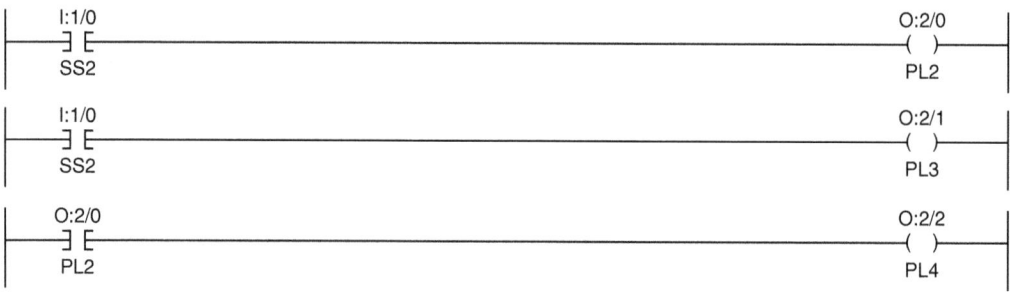

```
     I:1/0                                                        O:2/0
────] [──────────────────────────────────────────────────────────( )───────
     SS2                                                           PL2

     I:1/0                                                        O:2/1
────] [──────────────────────────────────────────────────────────( )───────
     SS2                                                           PL3

     O:2/0                                                        O:2/2
────] [──────────────────────────────────────────────────────────( )───────
     PL2                                                           PL4
```

FIGURE 6.25 Problem 19 ladder.

Laboratory 6.1—Conveyor System Control

The main objective of this laboratory is to learn about the methods of troubleshooting used in industrial application. A part is moving on a conveyor line crossing a photoelectric cell. The photoelectric cell's primary function is to provide pulses for counting parts. The conveyor line stops after 100 parts are counted.

- The START PB starts the conveyor motor after a 5-second delay. The motor must only start if the Auto/Manual switch is placed in the Auto position. The photoelectric cell should only count when the motor is running.
- When the count reaches 100, stop the conveyor after a 3-second delay.
- Switch ON pilot light indicating the end of the sequence.
- STOP switch, when activated, takes the system to initial condition.
- Operator can restart the same sequence by activating the START switch.

Use SS1 for the photoelectric cell input signal, SS2 for the motor running input indication signal, PL1 for the motor starting output signal, and PL2 for the end of sequence pilot light output signal.

Requirements:

- Assign the system inputs.
- Assign the system outputs.
- Enter the rungs shown in Fig. 6.26.
- Add comments to each rung.
- Download the program and go online.
- Simulate the program operation using the training units I/O or the LogixPro simulator and verify the program is running according to the process descriptions.

Testing Inputs and Outputs

The following are typical steps used to checkout I/O connections:

- Turn each discrete input ON/OFF and observe the associated LED indicators on the input module switch panel.
- Check the assigned discrete input addresses in the PLC and confirm that input change between 0 and 1 as the input signal change from OFF to ON.
- Using push button switches as an input for motor start/stop. The motor should be checked using NO/NC (normally open/normally close) momentary switches and by verifying that the motor responds correctly.
- Outputs should also be tested by using the PLC forcing function. As the bit is forced from the PLC from 0 to 1, output should turn ON from the OFF state. Safety precautions should be taken before forcing I/O.

Laboratory 6.2—Irrigation Canal Sensors, Validation and Calibration

An irrigation canal is equipped with three redundant sensors measuring the level inside the canal. The three sensors are expected to provide values with deviation less than 5% of the average level currently recorded. A sensor reading outside the previous average range is rejected, not used in calculating the next average, and must be reported (alarm condition) as malfunctioning or needs recalibration. The program used for this application is just a small part of a larger process control implementation.

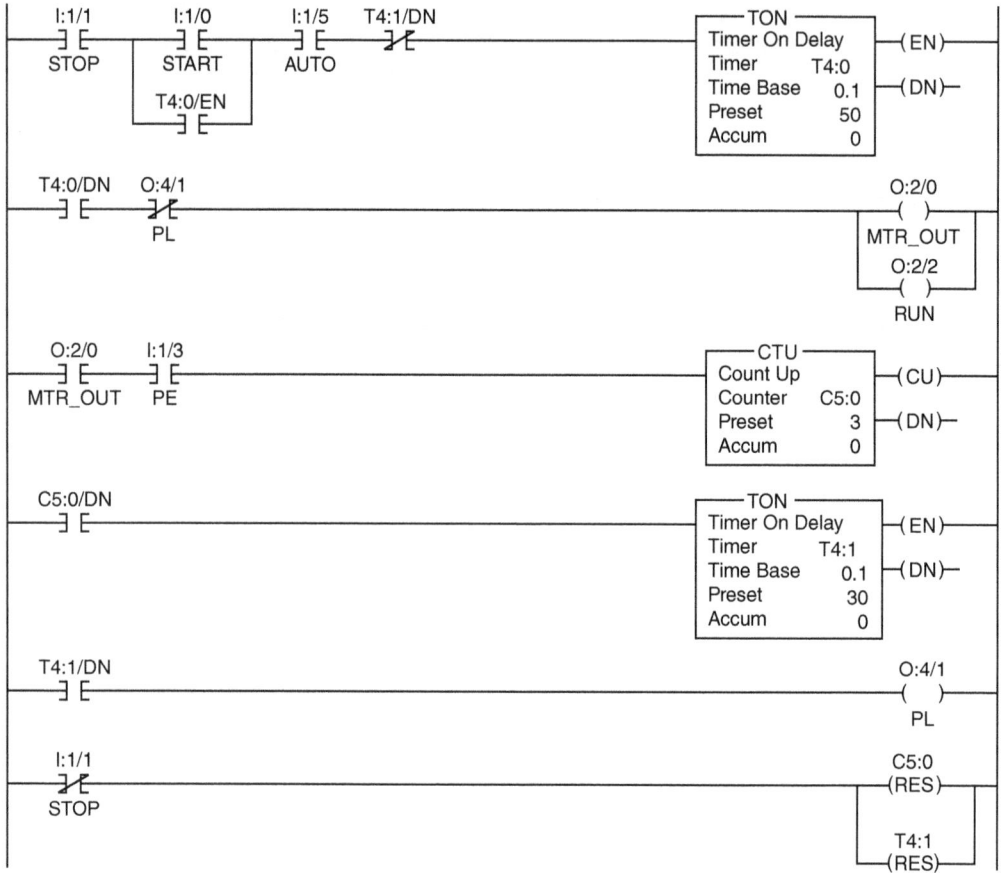

FIGURE **6.26** Laboratory 6.1 ladder program.

Each of the level sensors has an associated dead band value, which is defined as 5% of the current average level reading. Under normal operation, each sensor's reading should be within the previous average reading by the dead band margin. If a sensor is malfunctioning or outs of calibration, its reading will deviate and thus will not be used in defining the average value. If the sensor is left without calibration, eventually the average reading will become excessively incorrect. A program module/subroutine must be designed and included in the final control program to diagnose and report a malfunctioned or out of calibration sensor.

Requirements:

- Configure your analog input module for three input voltage signals from 0 to 10 V. Three potentiometers can be used to simulate the three signals.
- Design and implement the module for calculating the average level, validating each sensor, and reporting alarm conditions. Three subroutines can be used, one for each task called from the main module (main subroutine).

- Values below the low-limit count (average level minus the dead band) cause a software low-level alarm. Values above the high-limit count (average level plus the dead band) cause a software high-level alarm.

- Two hardware float limit switches can cause redundant hardware alarms. One limit switch triggers at the lowest permissible value, while the other becomes true at the highest allowable level. Simulate these two alarms and document their effect on a pump start/stop operation.

- Simulate one or more level sensor failure or out of calibration, using different inputs and dead band values. Monitor the corresponding level alarm/fail display on the I/O panel or your training unit HMI.

CHAPTER 7

Instrumentation and Process Control

This chapter covers PLC process control and associated instrumentation fundamentals. Process control strategies, control modes, and common types of control are covered. Instrumentation and system performance are examined.

Nuclear Ship Savannah Control Room

Chapter Objectives

- Understand instrumentation basics: digital and analog.
- Identify process control elements and associated use.
- Understand signal conversion and quantification errors.
- Understand process control types and common techniques.

A PLC is designed to regulate real-time processes having both discrete and analog variables. So far we have covered the discrete/digital programming side; we will do the same for analog programming in this chapter. Analog signals are brought to the PLC from sensors through an analog-to-digital (ATD) converter interface, while analog output signals are produced by the PLC through a digital-to-analog (DTA) converter interface. All analog signals are connected to the PLC through standard I/O modules, which incorporate the ATD and DTA conversions, signal conditioning/scaling, and electronic isolation. Sensors measure and provide small current or voltage signal to the analog input module, while actuators receive the output analog signal from the PLC output module. Sensors, actuators, and analog I/O modules are available in different standard formats and can meet the requirement of any process.

7.1 Instrumentation Basics

An important part of a control system is the incorporation of sensors. Sensors translate the physical real-time world into the standardized world of PLCs. This section explains the common types of sensors used in process control and their basic operation fundamentals. It will briefly cover common analog and digital sensors.

7.1.1 Sensor Basics

Sensors translate physical process attributes into values that the I/O modules can accommodate. The translation produces some sensor's standard output value that interfaces to the PLC. In general, most sensors fall into one of two categories: analog sensors and digital sensors. An analog sensor, such as a thermostat, might be wired into a circuit and calibrated to produce an output that ranges from 0 to 10 V. The analog signal can assume any value within the available range, 0 to 10 V in this case, as defined by the sensor resolution. For the PLC to deal with analog signals, they must be converted to a digital format in a simple and transparent way. The transformation from analog to digital and digital to analog must also follow predefined universal standards.

Digital sensors generate discrete signals, which typically have a stair step shape with every signal having a predefined relation with the values preceding and following it. A push button switch is one of the simplest forms of sensors with two discrete signal values: on or off. Other discrete sensors might provide a binary value in a given range. A stepper motor position encoder, for example, may provide the motor current position by sending a 10-bit value with a range from 0 to 1023. In this case, the discrete signal has 1024 possibilities. Much of our discussion assumes a digital/discrete signal to be binary.

7.1.2   Analog Sensors

Analog sensor signals must be converted into a digital format. Sensor output circuits are designed to be connected to an ATD converter port. Most standard microcontrollers and PLCs, such as Allen Bradley SLC 500 system, have built in ATD ports in the I/O modules interface.

An often overlooked, but extremely popular sensor, is the potentiometer. A potentiometer is a resistive sensor. Almost all resistive sensors are wired in a similar fashion as part of a voltage divider. Figure 7.1 shows an example of a potentiometer, which is connected to the Vcc (assumed to be +10 V in this section) and the GND (assumed to be 0 V). The potentiometer must be carefully selected to ensure adequate current limitation. Notice in this circuit the current limiting resistor R3 is connected to limit the output current when the sweep on the potentiometer is turned all the way to the top position. There are two types of potentiometers on the market: audio and linear. A linear potentiometer changes its value at a linear rate. An audio potentiometer changes its value on a logarithmic scale.

Using the resistors values shown, the voltage ranges the potentiometer will provide can be calculated. With the sweep all the way to the top position, the value for R2 at the top sweep is 10 kΩ. The voltage drop across R2 = Vcc * [R2/(R2 + R3)] = 10.0 * [10k/(10K + 330)] = 9.68 volts. Assuming an ATD converter resolution of 0.01 V, the highest digital value will be 9.68/0.01 = 968. The lowest value should be zero, since with the sweep all the way to the bottom, the ATD port will be connected to GND. Thus, the limiting resistor has reduced the useful range of the potentiometer to 9.68 instead of 10 V. The range can be increased by selecting a larger R2. For example, using a 100-kΩ potentiometer means a top sweep voltage across R2 of 10.0 * [100k/(100k + 330)] = 9.967 V. Thus the highest digital count value is 9.96/0.01 = 996.

7.1.3   Digital Sensors

There are many different types of digital input sensors. Many of them are wired in the same form, which uses a pull-up resistor to force the line voltage high and limit the amount of current that can flow to the ATD converter circuit. One of the most basic of all sensors is a simple switch. Switches are used to detect limits of motion, proximity to an object, for user input, and a whole host of other things.

Switches come in two types: normally open (NO) and normally close (NC). Many micro switch designs actually have one common terminal and both an NO terminal and an NC terminal. The PLC wiring diagram for a switch is simple as was discussed in Chapter 2. NO switches are recommended to limit the amount of power

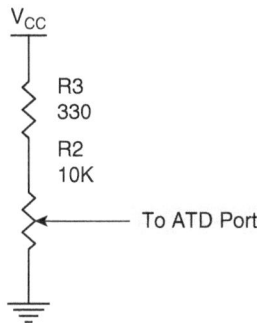

FIGURE 7.1   Potentiometer sensor wiring diagram.

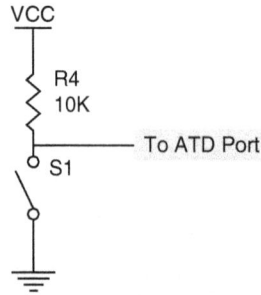

VCC

R4
10K

To ATD Port

S1

FIGURE 7.2 NO switch sensor connection.

consumed (Fig. 7.2). With a 10-kΩ pull-up resistor, the amount of current is small, but many switches can add up to some noticeable power.

7.2 Process Control Elements

A simple process control loop consists of three elements: the measurement, the controller, and the final control element. Measurement is one of the most important elements in any control processing plant. Decisions made by the controller are based on the real-time measurements information received. Regardless of system type, all controller decisions are similarly based on measurements, control strategy, and the desired process response/performance.

Final control elements can refer to actuators like control valves, heaters, variable speed drives, solenoids, and dampers. In most chemical process plants, a final control element is often a control valve. In manufacturing assembly lines, final control elements will mostly include variable speed drives, solenoids, and dampers. Final control elements receive command signals from the controller/PLC in real time to bring about the desired changes in the controlled process. As with sensors/measurement elements, final control devices interface with the PLC output modules in a similar way. The PLC digital signal outputs are transformed to the actuator required digital or analog signal format, which might require a DTA conversion or coupling isolation.

7.2.1 Basic Measurement System

A basic instrument/measurement system consists of three elements:

1. **Transducer/sensor:** The transducer is the part of the measurement system that initially converts the controlled variable into another form suitable for the next stage. In most cases, conversion will be from the actual variable into some form of electrical signal, although there is often an intermediate form, such as pneumatic.

2. **Signal conditioning:** In computer process control, signal conditioning is used to adjust the measurement signal to interface properly with the ATD conversion system.

3. **Transmitter:** The transmitter has the function of propagating measurement information from the site of measurement to the control room where the control function is to occur. Usually pneumatic or electronic signals are used.

A simplified block diagram of a basic measurement system is shown in Fig. 7.3.

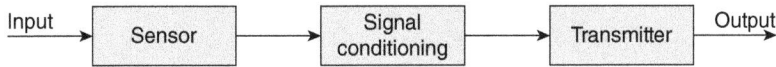

Figure 7.3 Instrument block diagram.

Most modern analog instruments utilize the following standard signal ranges:

- Electric current of 4 to 20 mA.
- Electric voltage of 0 to 10 V.
- Pneumatic pressure of 0.2 to 1.0 bar, the bar is a unit of pressure equal to 100 kPa and roughly equal to the atmospheric pressure on earth at sea level.
- Digital with a binary digital encoder built in to give a binary digital output. Having a standard instrument range or using digital signals greatly contributed to the advancement of digital process control and the evolution of modern PLCs.

The following are few of the primary advantages of such instruments:

- All instruments can be easily calibrated.
- Produced signal is independent of the physical measurement. For example, the minimum signal (temperature, speed, force, pressure, acidity, and many other measurements) is represented by 4 mA or 0.2 bar and the maximum signal is represented by 20 mA or 1.0 bar.
- Same PLC hardware interface modules are used for all measurements.
- Users can select instruments from large number of competing vendors; all must comply with universal standards.

7.2.2 Process Control Variables

Process control variables commonly, either measured by sensors or regulated through actuators (final control elements), include temperature, pressure, speed, flow rate, force, movement, velocity, acceleration, stress, strain, level, depth, mass, weight, density, size, volume, and acidity. Sensors may operate simple ON/OFF switches to indicate certain events, to detect objects (proximity switch), empty or full (level switch), hot or cold (thermostat), high or low pressure (pressure switch), and other overload conditions.

German physicist Thomas Johann Seebeck discovered the conversion of temperature differences directly into electricity in 1821. The block diagram of a temperature sensor is shown in Fig. 7.4. This phenomenon takes place when two wires with dissimilar electrical properties (A and B wires) are joined at both ends and one junction is made hot (T_1) and the other cold (T_2). A small electric voltage is produced proportional to the difference in the temperature between the two junctions. This voltage can be calibrated to indicate the measured temperature value as shown. A typical industrial temperature probe with a flexible extension and standard plug is shown in Fig. 7.5.

The final or correcting control element is the part of the control system that acts to physically change the process behavior. In most processes, the final control element is a valve used to restrict or cutoff fluid flow, pump motors, louvers used to regulate air flow, solenoids, or other devices. Final control elements are typically used to increase

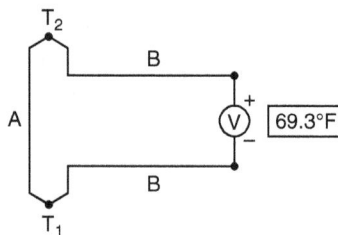

FIGURE 7.4 Temperature sensor.

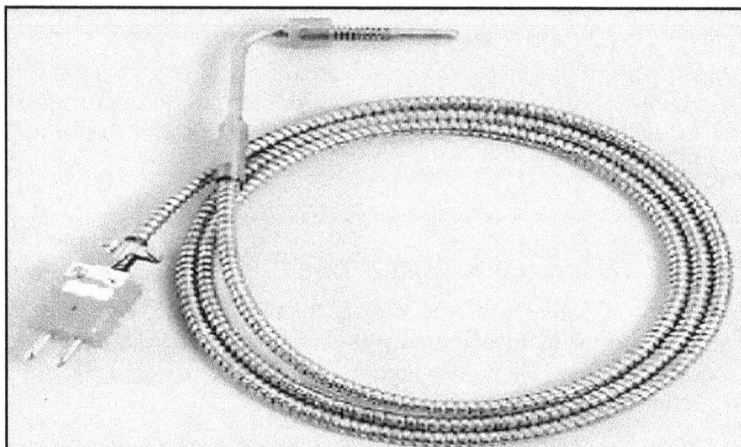

FIGURE 7.5 Industrial temperature probe.

or decrease fluid flow. For example, a final control element may regulate the flow of fuel to a burner to control temperature, the flow of a catalyst into a reactor to control a chemical reaction, or the flow of air into a boiler to control boiler combustion. In any control loop, the speed with which a final control element reacts to correct a variable that is out of set point (SP) is very important. Many of the technological improvements in final control elements are related to improving their response time.

7.2.3 Signal Conditioning

The signal conditioning element changes the characteristic of the sensor/transducer measured signal. One example is the square root extractor. For example, differential pressure flow meters/sensors produce an output that is directly proportional to the square of the flow. A signal conditioning/processor is used to extract the square root so that the resulting signal delivered to the transmitter element is directly proportional to the actual flow rate. Other signal conditioning elements include integrators, differentiators, and a wide variety of signal filters. Filters are used with electric signals to remove unwanted parts, noise, or interferences. For example, a signal may contain unwanted frequencies or undesired dc component due to external sources or due to the inherent characteristic of the sensor technology. Some signal conditioning is performed in order to prepare for successful transmission.

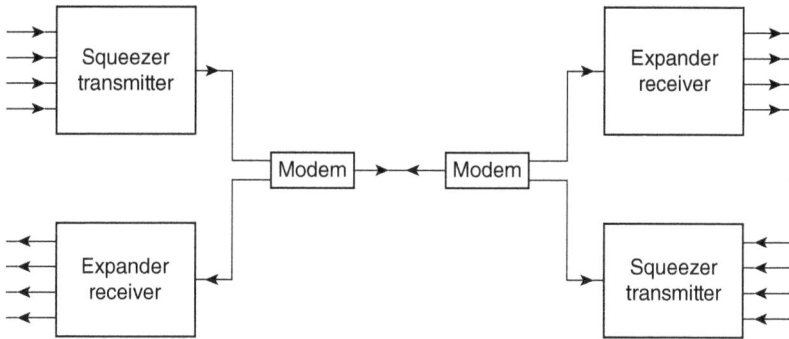

FIGURE 7.6   Communication frequency division multiplexor.

7.2.4   Signal Transmitters

Industrial sensory information needs to be sent from one place to another using either a wireless or wired transmission media. Transmitters are used to send the measured and conditioned signals to the controller using single frequency of transmission or a single pair of wires, such as the twisted pair used in wired telephone lines. Modems are simple devices that can receive and transmit information through direct connection or the telephone network. They require a marker signal to let each other know when to receive or transmit. They use handshaking to regulate and enforce the predefined communication protocol. A time-division multiplexing (TDM) for digital signals or frequency-division multiplexor (FDM) for analog signals can be used to accommodate four sensory channels communications as shown in Fig. 7.6.

7.3   Signal Conversion

The real word is a mix of analog and digital variables combining to describe a process or a system behavior in real time. The PLC side uses digital format in both variables measurements and control. This section covers simplified ATD and DTA converters. It will also briefly discuss issues associated with these converters, including resolution, sampling rate, and quantification errors.

7.3.1   Analog-to-Digital Conversion

Analog-to-digital (ATD) conversion can be achieved in different ways. One type utilizes a synchronous counter as shown in Fig. 7.7. The output of the counter, which is driven by a clock of a fixed frequency, is a digital pattern. This count is converted back into an analog signal by the DTA converter and compared with the input analog signal. Once the two signals match, the counter is disabled by the end of conversion signal and the digital count is latched. The counter stops and the digital value of the counter output, which is equivalent to the analog signal, is acquired by the PLC/computer input interface. The higher the number of counter bits, the higher the resolution/accuracy of the ATD converter and the longer the average conversion time.

The PLC always deals with discrete/digital values. An important part of using an analog signal is being able to convert it to a discrete signal, such as a 10-bit digital value. This allows the PLC to do things like compute values, perform comparisons, and execute logic operations. Fortunately, most modern PLCs/controllers have a resource

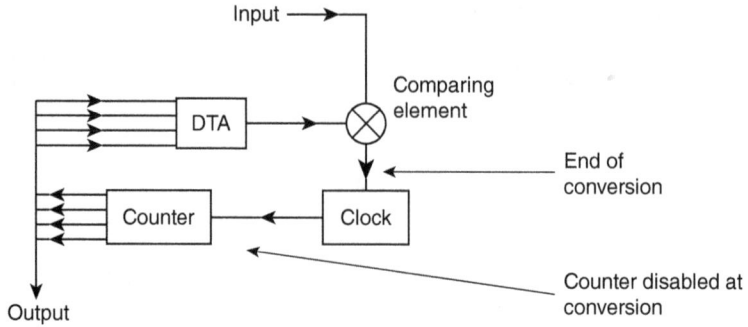

FIGURE 7.7   Analog-to-digital converter.

called *analog-to-digital* (ATD) *converter*. The function of the ATD converter is to convert an analog signal into a digital value. It does this with a mapping function that assigns discrete values to the entire range of voltages or currents. It is typical for the range of an A/D converter to be 0 to +10 V or 4 to 20 mA.

The ATD converter will divide the range of values by the number of discrete combinations. For example, Table 7.1 shows eight samples of an analog signal that have been converted into digital values. The range of the analog signal is 0 to +10.24 V. It is a 10-bit ATD converter, which has 1024 discrete values. Therefore, the ATD converter divides 10.24 V by 1024 to yield approximately 0.01 V per step. The table shows how voltages map to specific conversion values. The values shown only include the first six samples, but the table would continue up to the conversion value of 1023.

There are many types of ATD converters on the market. An important feature of ATD converters is its resolution, which is proportionate to the number of bits used by the ATD converter to quantify and store the analog signal samples. An 8-bit converter is widely used on microcontrollers whereas 12- and 16-bit ATD converters are common in PLCs. A 16-bit ATD converter will use 65356 discrete values. The resolution required for an application depends on the accuracy your sensor requires and the transient nature of the process. The higher the resolution, the greater is the accuracy of the signal representation in the digital format and the lower the quantization error. In this example, the worst case quantization error is 0.01 V whereas the average value is 0.005 V.

From (V)	To (V)	Conversion (decimal)	Conversion (binary)
0.00	0.01	0	0000000000
0.01	0.02	1	0000000001
0.02	0.03	2	0000000010
0.03	0.04	3	0000000011
0.04	0.05	4	0000000100
0.05	0.06	5	0000000101
0.06	0.07	6	0000000110
0.07	0.08	7	0000000111

TABLE 7.1   10-Bit ATD conversion pattern

Example 7.3.1.1 A 10-bit ATD converter has a 10-V reference (V_r) and a digital output count of 0010100111.

  a. What is the ATD resolution R in volts per bit?

  b. What is the digital output count N in hex for an analog input of 6 V?

  c. What is the average quantization error?

Solution:

  a. $R = V_r/2^n = 10/1024 = 0.0098 \text{ V/bit}$

  b. $N = 2^n * V_{in}/V_R = 1024 * 6/10 = 614 = 266 \text{ H}$

  c. $R/2 = 0.0098/2 = 0.0049 \text{ V/bit}$

7.3.2 Digital-to-Analog Conversion

Wikipedia, states that a digital-to-analog (DTA) converter is a device that converts "a binary digital code to an analog signal (current or voltage)." An ATD performs the reverse operation. Signals are easily stored and transmitted in digital form, but a DTA is needed for the signal to be recognized by human senses or other nondigital systems. A common use of DTA converters is generation of audio signals from digital information in music players. Digital video signals are converted to analog in televisions and cell phones to display colors and shades. DTA conversion can degrade a signal, so conversion details are normally chosen so that the errors are negligible.

Due to cost and the need for matched components, DACs are almost exclusively manufactured on integrated circuits (ICs). There are many DAC architectures that have different advantages and disadvantages. The suitability of a particular DAC for an application is determined by a variety of measurements, including speed and resolution. As documented in the Nyquist and Shannon sampling theorems, a DTA can accurately reconstruct the original signal from the sampled data provided that its bandwidth meets certain requirements. Digital sampling introduces quantization error that manifests as low-level noise added to the reconstructed signal. Figure 7.8 shows a simplified functional diagram of an 8-bit DTA converter.

PLCs use a wide variety of analog input and output modules accommodating different current, voltage, and high-speed pulse signals. These modules come in different sizes (the number of I/O points per module) and also ATD/DTA resolutions. A typical analog I/O module uses a 12-bit resolution with signed or unsigned integer's representation. The internal operation of the PLC analog modules is independent of the type of physical sensors/actuators interfaced. This simplifies the analog module configuration and its associated application programming/deployment.

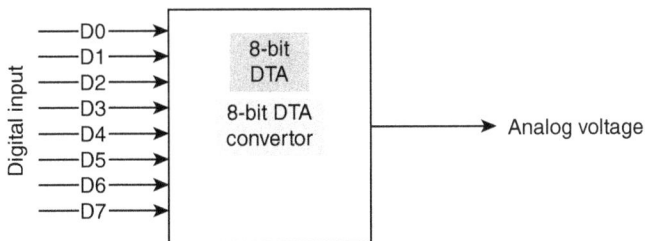

FIGURE 7.8 Simplified DTA conversions.

Example 7.3.2.1   A 12-bit DAC with 10 V reference voltage is used to convert digital counts to analog output voltage. Calculate the following:

    a. What is the analog output voltage for a digital input = 0A3h H?

    b. What is the input digital count N for an analog output voltage of 8 V?

Solution:

    a. $V_{out} = (N/2^n) * V_R = (163/4096) * 10 = 0.398$ V

    b. $N = (2^n * V_{in})/V_r = (4096 * 8)/10 = 3276$

7.3.3   Quantification Errors and Resolution

The resolution of the ATD or DTA converter indicates the number of discrete values it can produce over the range of analog values. The values are usually stored electronically in a fixed length binary form, so the resolution is usually expressed in bits. The number of discrete values or levels available is a power of two. For example, an ATD with a resolution of 8 bits can encode an analog input to one in 256 different levels, since $2^8 = 256$. The values can represent the ranges from 0 to 255 (unsigned integer) or from -128 to 127 (signed integer), depending on the application. Figure 7.9 illustrates the DTA conversion for a 3-bit resolution and a normalized 1-V range. The binary count ranges from 000 to 111, which represent 2^3 or eight different levels equivalent to the analog range from 0 to 1 V. The LSB or the DTA converter resolution is 0.125 V, which is equal to the worst case quantization error. The average quantization error in this case is 0.0625 V.

Resolution can also be defined electrically and expressed in volts or current. The minimum change in voltage required to guarantee a change in the output code level is called the least significant bit (LSB) voltage. The resolution R of the ATD is equal to the LSB voltage. The voltage resolution of an ATD is equal to its overall voltage measurement range divided by the number of discrete voltage intervals as shown here:

$$R = F_{full\ scale\ range}/N$$

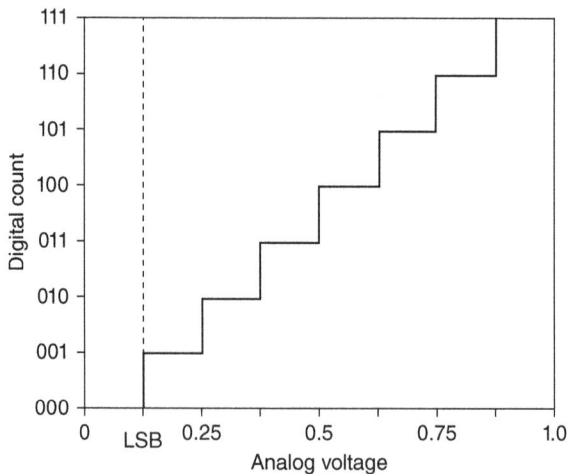

FIGURE 7.9   Three bits DTA conversion.

where N is the number of voltage intervals and $F_{\text{full scale range}}$ is the difference between upper and lower extremes, respectively, of the voltages that can be coded.

The number of voltage intervals is given by

$$N = 2^M$$

where M is the ATD's resolution in bits.

Quantization error, also known as *quantization noise*, is the difference between the original analog signal and the digitized binary count. The magnitude of the average quantization error at the sampling instant is equal to half of one LSB voltage. Quantization error is due to the finite resolution of the digital representation of the signal, and is an unavoidable imperfection in all types of ATD converters.

Example 7.3.3.1 An 8-bit ADC with a 10-V reference converts a temperature of 0°C into 00000000 digital outputs. If the temperature transducer outputs 20 mV/degree, answer the following:

  a. What is the maximum temperature that the converter can measure?

  b. What is the resolution of the ADC in mV/bit?

  c. What is the worst case quantization error in degrees Celsius?

Solution:

  a. Max. temperature = 10,000/20 = 500°C

  b. Resolution = 10,000/256 = 39.06 mV/bit

  c. Worst case quantization error = 500/256 = 1.952°C/bit

7.4 Process Control System

In a process control system, the controller is the element linking the measurement and the final control element. Traditionally, closed-loop proportional integral derivative (PID) controllers are used. These controllers are designed to execute PID control functions. Other types of control are also common, including ON/OFF and fuzzy logic. Advancements in computer hardware and software have led to a wide deployment of PLCs and distributed digital control systems as a replacement of the hardwired relay analog controllers. This section focuses on controller types typically implemented using PLCs.

Analog controllers use mechanical, electrical, pneumatic, or other type devices that cause changes in the process through the final control element. The controller's moving mechanical parts are subject to wear and tear over time and that affects the response/performance of the process. Also, analog controllers regulate the process by continuously providing signal to the final control element. Digital controllers do not have mechanical moving parts. Instead, they use processors to calculate the output based on the measured values. Since they do not have moving parts, they are not susceptible to deterioration with time. Digital controllers do not regulate continuously but they execute at very high rate, usually several times every second. Pneumatic controllers use instrument air to pass measurement and controller signals instead of electronic signals. These controllers have the disadvantage of longer dead time and lag due to the compressibility of the instrument air.

7.4.1   Control Process

In the industrial world, the word *process* refers to an interacting set of operations that lead to the manufacture or the development of some product. In the chemical industry, process means the operations necessary to take an assemblage of raw materials and cause them to react in some prescribed fashion to produce a desired end product, such as gasoline. In the food industry, process means to take raw materials and operate on them in such a manner that an edible product results. In each use, and in all other cases in the process industries, the end product must have certain specified properties, which depend on the conditions of the reactions and operations that produce them. The word control is used to describe the steps necessary to assure that the conditions produce the correct properties in the product.

A process as shown in Fig. 7.10 can be described by an equation. The process has m variables: v_1 to v_m. Suppose we let a product be defined by a set of properties, P_1, P_2, ..., P_n. Each of these properties must have a certain value for the product to be correct. Examples of properties are things like color, density, chemical composition, and size.

$$P_i = f(v_1, v_2, ..., v_m, t)$$

where P_i is the ith property and t is time.

7.4.2   Controlled Variables

To produce a product with the specified properties, some or all the variables must be maintained at specific values. Figure 7.11 shows an unregulated tank with open flow coming in and out. Fluid level in the tank is expected vary as long as a difference in flow

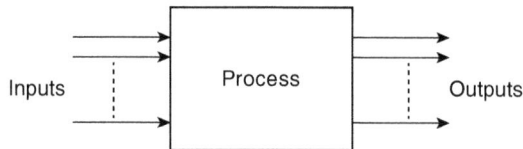

FIGURE **7.10**   Process block diagram.

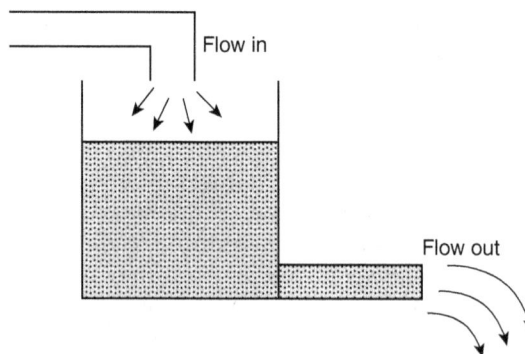

FIGURE **7.11**   Unregulated tank process.

in and out rates exist. A controlled/controlling variable can be used to exert control over this simple tank process. Choices of such variable can include the flow rate in or flow rate out of the tank. The controlled variable must be accessible and easy to change. It must also have adequate influence on the regulation of the selected control variable, the tank level in this case.

Some of the variables in a process may exhibit the property of self-regulation, whereby they will naturally maintain a certain value under normal conditions and very small disturbances. Control of variables is necessary to maintain the properties of the product within specification.

7.4.3   Control Strategy and Types

The value of a variable v_i actually depends on other variables in the process, and also on time. Typically one or few variables dominate and define the dependency relationship. This relation can be expressed as shown in the following equation:

$$v_j = g\ (v_1, v_2, \ldots, v_c, \ldots v_m, t)$$

Two types of control are common: mainly the single-variable control and the multivariable control. These two types are briefly defined as follows.

7.4.3.1   Single-Variable Control

We will demonstrate single-variable control using the tank process with minor modification. Two valves are added: one at the inlet flow to the tank and one at the output. The level in the tank varies as function of the flow rate through the input valve and the flow rate through the output valve. The level in the tank is the control variable, which can be measured and regulated through either inlet or output valve control and adjustment. Figure 7.12 illustrates the modified tank process, which can accommodate the implementation of single variable control. Only one of the two valves available in this mode can be selected as a controlling variable to regulate the tank level, the control variable. The next mode of control, multivariable control, is more complex and can utilize more

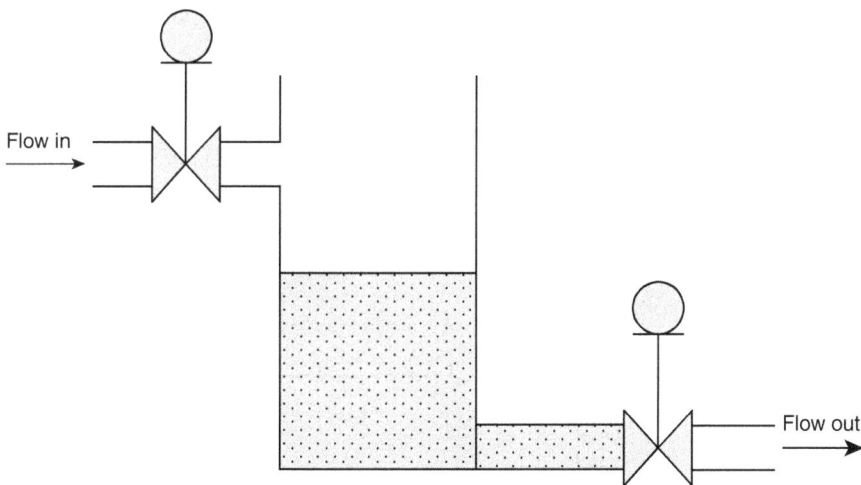

FIGURE 7.12   Regulated tank process.

than one controlling variable to regulate the control variable, the tank level in our process. For example, regulation of the two valves can be used to control the tank level at a desired set of values at different times.

7.4.3.2 Multivariable Control

Figure 7.13 shows a schematic diagram for an oven used to bake crackers in a system under multivariable control. The control variables, which may be used, include the feed rate, conveyor speed, oven temperature, cracker color, and cracker size. Other variables, such as the temperature outside the oven, are difficult to measure, control, and use in the control system strategy. Multivariable control is more complex due to the strong and nonlinear interactions between variables.

7.4.4 Process Control Loop

Process control loops are the core of all control systems and process automation tasks. A schematic for a typical control loop is shown in Fig. 7.14. The loop consists of three blocks: process, measuring elements, and the final control element. What comes from the process is what we referred to as the control variable, the tank level in our previous example, which can be easily measured and quantified in time. What goes as input to the process is the controlling variable, the inlet valve opening or flow rate in the tank process. The valve opening can be changed manually or automatically using available final control element. Load disturbances refer to external process influences affecting the behavior of the process. The *set point* (SP) refers to the desired control variable value as defined by the system end user. Actuation of the final control element is based on the SP, control variable measurements, characteristic of the process, type of final control element, and the implemented control strategy. The man shown in the loop represents

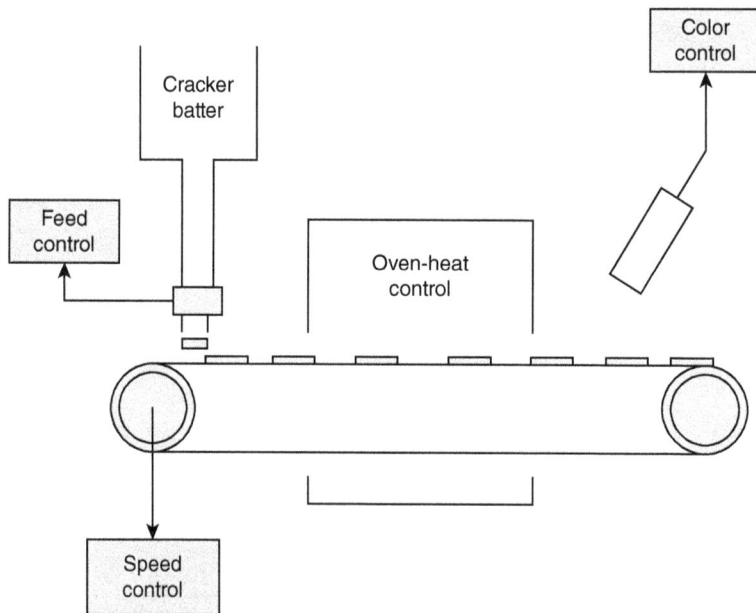

Figure 7.13 Multivariable oven control.

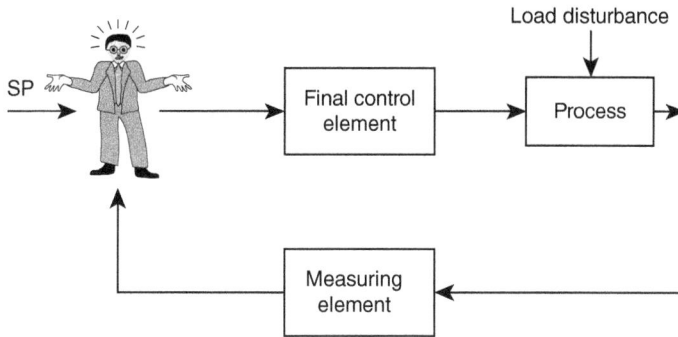

FIGURE 7.14 Process control loop.

two primary loop functions: mainly error detection and the control tasks. Most of the human involvement in the control loop can be replaced by automated means, including the deployment of PLCs, PCs, and other types of automated control.

In manual control, the operator is expected to perform the task of error detection and control. Observations and actions taken by operators can lack consistency and reliability. In automated systems, the operator is removed and replaced by electronic controllers as in the wide use of PLCs and specialized digital computers. The operator still plays an important role in an automated system, including input of control variable SPs and required human interactions with the control system. Automated systems, as the one shown in Fig. 7.15, allow for greater flexibility, higher degrees of process control, quality, and consistency. The following are the steps used in implementing single-variable closed-loop control:

1. Select one variable to be a controlled variable (the controlled variable in this case is labeled V_c).

2. Make a measurement of the controlled variable V_c to determine its present value. The desired value is called the *set point* (SP) of the controlled variable.

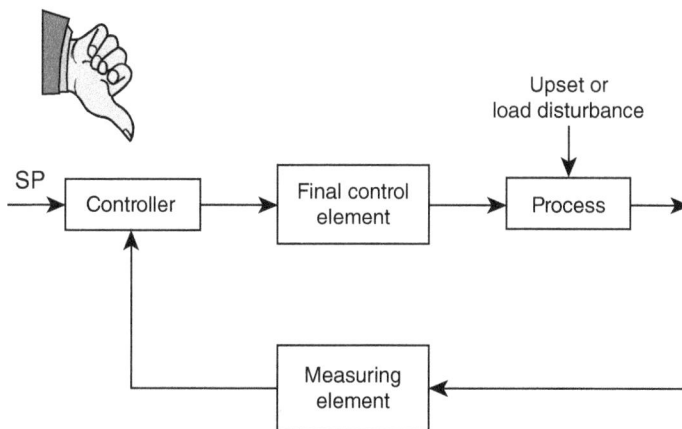

FIGURE 7.15 Automated process control loop.

3. Compare the measured value of the controlled variable with the desired value (SP).

4. Determine a change in the controlling variable that will correct any deviation, or error in the controlled variable.

5. Feedback this changed value of the controlling variable to the process through the final control element to create the desired correction in the controlled variable.

6. Go to step 2 and repeat.

7.4.5   Control System Error Quantification

Perfect regulation of a process variable by any control system is not possible. Errors can be measured in the following three ways:

1. **Variable value:** Set point = 230°C, measured value = 220°C, range = 200 to 250°C, and error = 10°C.

2. **Percent of set point:** The error is expressed as a fraction or percent of the controlled variable SP. Error = (10/230) * 100 = 4.4%.

3. **Percent of range:** The error is expressed as a fraction or percent of the controlled variable range. Error = [10/(250 − 200)] * 100 = 20%.

Two types of errors are of great importance in system performance as observed in all selected control variables: mainly the steady-state residual errors and the transient dynamic of such errors. Both residual (the steady-state error) and dynamic errors (errors during the transient behavior) are used to evaluate a control system implementation and design. In most control applications, the steady-state error is the primary goal. It is desired that the error would be small or show a rapid decay in magnitude with time. The residual error is expected to be reached after a small transient time from a load change or when a system offset takes place. The quantification of the value of the error is always subject to a small tolerance.

The transient response for a process includes the time interval prior to and leading to a steady-state conditions from the time of a load change or an offset (a change in the SP). Primary interest during the transient time includes error values, frequency of oscillation, and duration of oscillation for the controlled variable under consideration. Oscillations in controlled variables are expected but can lead to oscillatory instability, which is often caused by an ill-designed controller or control strategy. Hardware failure can cause monotonic instability, which is materialized in the continuous increase or decrease in the value of the controlled variable resulting in an overall system failure. For example, if the transmitter, for the tank level signal, fails, it can cause the level to increase and eventually overflow the tank. The tank can run dry if the last correct measurement before the transmitter failure required a decrease in tank level. This type of failure can be detected and prevented by using backup hardware, such as limit switches and redundant sensors.

7.4.6   Control System Transient and Performance Evaluation

The quality of a control system performance is based on many factors, including transient response, steady-state errors, stability, scalability, user interface, continuous quality improvements, and ease of maintenance. Controller response to errors depends on

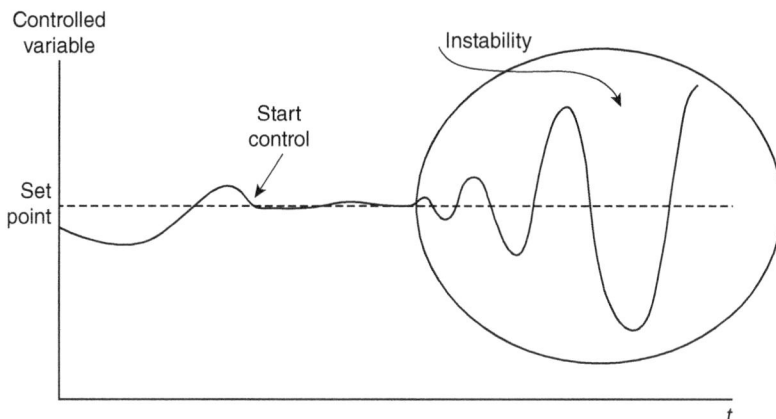

FIGURE 7.16   Oscillatory instability.

the control system strategy. The objective of a control system is to minimize, not to eliminate, the error without affecting the overall system stability and performance.

As stated in Sec. 7.4.5, perfect regulation of a process variable by any control system is not possible, which means we have to live with some errors, hopefully small. Control strategy and adequate tuning of existing control loops are very critical to the performance of the whole system. This task is one of the most challenging, which can be easier to perform if the error behavior is well understood.

Figure 7.16 shows an ill-designed control system behavior or one completely out of tuning. As shown previously, the controlled variable has gone wild with increasing amplitude oscillations, ultimately leading to instability and system shutdown. This type of behavior is called *oscillatory instability* and can be eliminated during the implementation of the controller and the overall strategy.

As stated in Sec. 7.5.1, a tolerance around the SP where no controller action needed is essential to protect the final control element (actuator) from failure due to excessive controls actions. This is true in all kinds of process regulations: manual, PID, or on/off control. Figure 7.17 shows the transient behavior of an overdamped process, where T_D (transient/delay time) is the total delay/transient time. There are no overshoots or oscillation in this case, but the controller action is slow. Figure 7.18 shows a faster controller for an underdamped process, resulting in controlled variable oscillations and shorter T_D. Configuring the controller to be very aggressive in reacting to real-time controlled variable errors can lead to instability and the situation shown previously in Fig. 7.16. The most-desired controlled variable response is known as *quarter decay*, which means that the ratio of consecutive over shoots is approximately 4, which represents a 0.25 decay ratio. Guidelines and techniques for tuning PID controllers for such response exist in the literature and in most PLCs technical manuals.

7.5   Closed-Loop Process Control Types

Closed-loop process control can be implemented using a wide variety of techniques and strategies. Three commonly used techniques are covered briefly in this section: on/off, proportional, PID, and supervisory control. Each type

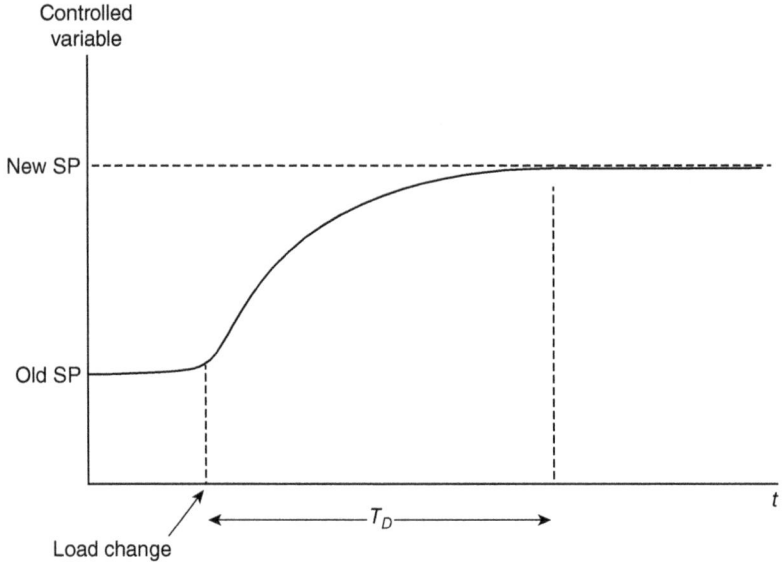

FIGURE 7.17   Overdamped system controller action.

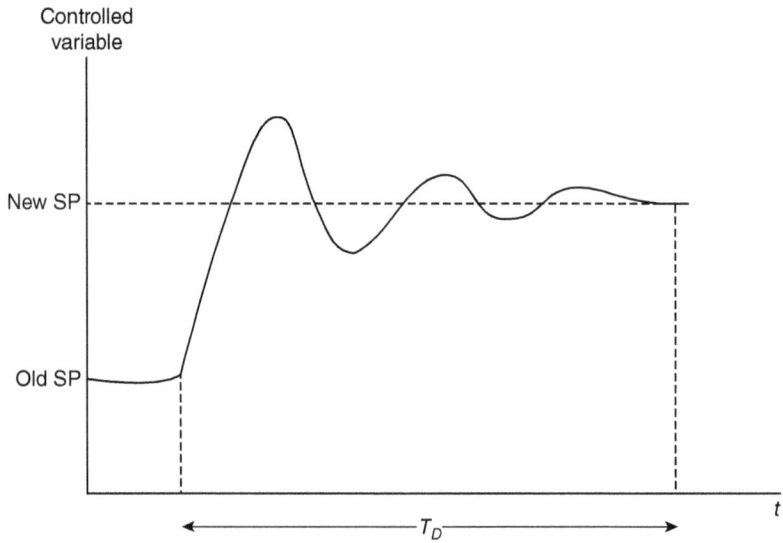

FIGURE 7.18   Underdamped system controller action.

of process control has unique characteristics and can be the best strategy for the correct application. Depending on the process and the associated control requirements, the most appropriate and simple technique possible is selected.

7.5.1 On/Off Control Mode

An on/off control is the simplest modes of closed-loop control, but it is a good match for many applications. The final control element is either ON or OFF depending on the controlled variable measured value. A dead band (a tolerance around the SP where no control action is needed) is used to prevent the final control element from being switched ON/OFF excessively. The controller output is ON if the error value is greater than the (SP + ε), and is OFF if the error value drops below the (SP − ε), where ε is one-half the dead band. No control action is needed while the controlled variable measurement lies within the implemented dead band. Figure 7.19 illustrates the general behavior of an on/off control action.

> **Example 7.5.1.1** Assume a temperature cooling control process with a SP of 80°C and a dead band of 6°C. The system cools at −2° per minute once the controller output is ON. The system heats at +4°C per minute when the output is OFF. The on/off controller response is shown in Fig. 7.20. Control action is ON while the temperature is higher than the upper threshold of the dead band (83°C) and turns OFF once the temperature dips below the low threshold of the dead band (77°C).
>
> The figure shows two functions of time: the measured temperature variation and the controller output. The first function is continuous in time whereas the second is discrete; controller either ON or OFF. Notice that the control action only takes place while measured temperature is outside the specified dead band. More precise control can be achieved, if needed, using other more sophisticated techniques.

7.5.2 Proportional Control Mode

Proportional control mode assumes a correction strategy based on the calculated error, the difference between the measured controlled variable and the SP. The controller

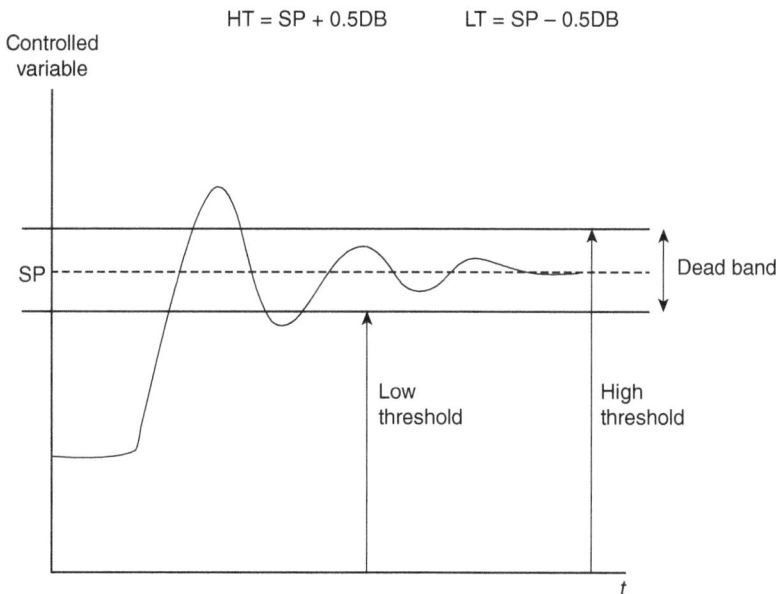

HT = SP + 0.5DB   LT = SP − 0.5DB

FIGURE 7.19 On/off controller action.

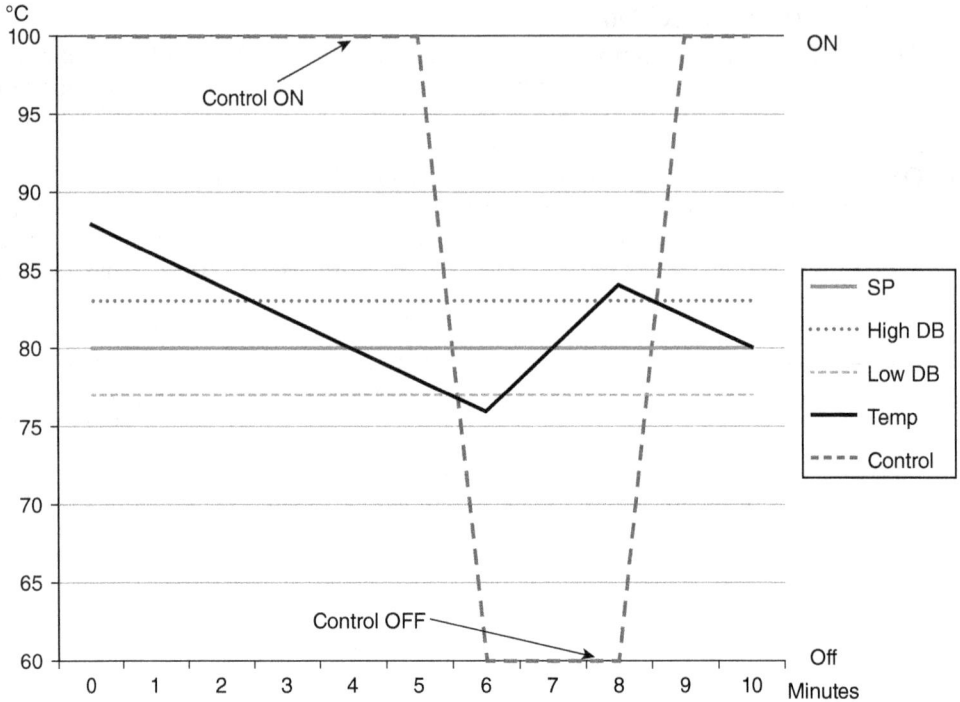

Figure 7.20 On/off temperature control.

output (controlling variable) is proportional to the amount of error in addition to the fixed amount needed to support the process during the time the control variable stays within the defined dead band. This amount is known as the controller output with zero error. The following is the mathematical formulation for the proportional control mode:

$$C_p = K_p * E_p + C_o$$

where

C_p = controller output in percent

K_p = proportional gain in percent of output/percent of error

E_p = error in percent of range

C_o = controller output with zero error

Example 7.5.2.1 A control system is to control pressure in a range from 120 lb/in² (psi) to 240 psi with a 180-psi SP. If the proportional gain is 2.5 %/% and the zero error output is 65%. The error as a percentage of range is given by the following formulation:

$$E_p = (P - 180)/(240 - 120) * 100$$
$$= 0.833 * (P - 180)$$

where P is the measured pressure.

The controller output is given by the following formulation:

$$C_p = 2.5 * E_p + 65$$
$$= 2.5 * 0.833 * (P - 180) + 65$$
$$= 2.0825 * (P - 180) + 65$$

The controller output varies from 0 to 100%, which corresponds to an error range defined as in the following formulation:

$$E_p \text{ (at } C_p = 0) = -26\% \quad \text{and} \quad E_p \text{ (at } C_p = 100\%) = 14\%$$

The range of errors covering the entire available controller output is known as the controller proportional band, which is calculated in this example as follows:

$$\text{Proportional band} = 14\% - (-26\%) = 40\%$$

Notice that the higher the proportional band the lower the controller proportional gain as the following relation is true at all times:

$$\text{Proportional band} * \text{proportional gain} = 1$$

This type of control is more complex relative to on/off control as it requires experience in implementing and tuning both the proportional gain and the controller zero error output.

7.5.3 Composite Control Mode

Closed-loop process controllers can be designed to respond to the history of an error during a prespecified time period (integral mode), the forecast of the error behavior in the near future (derivative mode), and the current instantaneous value of errors (proportional mode). These controllers are commonly labeled in the following three types:

1. Proportional-integral (PI) mode

2. Proportional-derivative (PD) mode

3. Proportional-integral-derivative (PID) mode

Figure 7.21 shows a simplified schematic for the PID control while the Auto/Manual mode switch is placed on Auto, which is the universal format for the composite controller. All composite controllers must include the proportional mode.

7.5.4 PLC/Distributed Computer Supervisory Control

The initial use of computers in process control was in support of the traditional analog system process control. This type of application of computers still exists since many industries use analog control systems and will no doubt continue to do so. Generally, large- or medium-scale computers provide the support activities, including data acquisition, human interface, simulation/modeling, communication, and digital control. A simplified block diagram for a supervisory control system is shown in Fig. 7.22.

Distributed computer control/supervisory control have evolved as the choice for automation and process control implementation in the past 20 years. This revolution greatly benefitted and made use of huge advancements in technologies, including universal standards, digital hardware, real-time operating systems, communication and networking, Human Machine Interfaces (HMIs), remote sensing, sensory fusion, redundancy and safety tools, and the widespread of open system architectures. Some of the

Proportional + Integral + Derivative controller

100

0

M

Output

Set point

+

−

[Calculation]

Auto/manual
mode

Process
variable

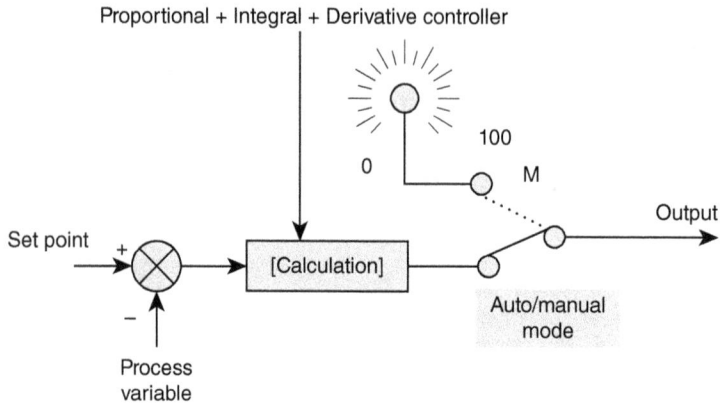

FIGURE 7.21   PID composite controllers.

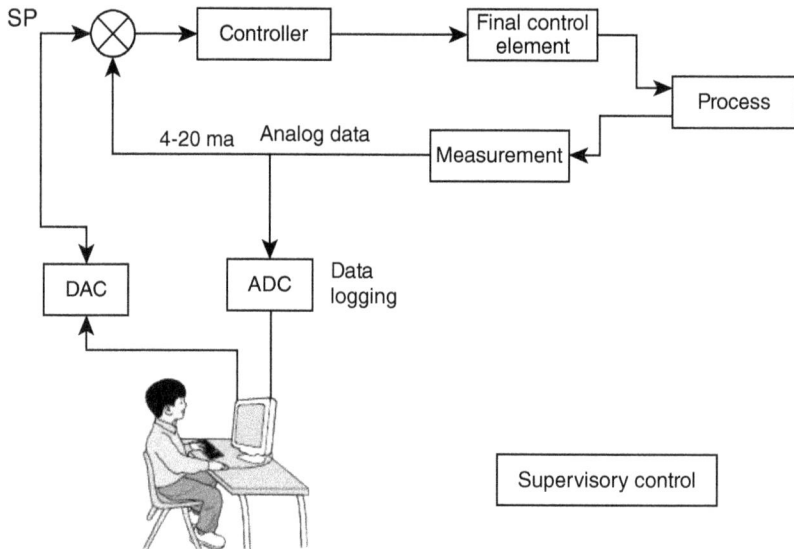

SP

Controller

Final control
element

Process

4-20 ma   Analog data

Measurement

DAC

ADC

Data
logging

Supervisory control

FIGURE 7.22   Supervisory control.

world's largest international chemical and petroleum corporations own and operate the largest global computer networks. Every process controller, data acquisition system, HMI, actuators, sensors, and communication devices have to be accessible from any location in the world with proper authorization.

Supervisory distributed control allows for a more efficient modular design and far easier overall system cost in all phases of system development, implementation, deployment, enhancements, expansion/scalability, and maintainability. It also allows for greater operators, designers, and users interactions leading to the realization of an overall system continuous quality improvement. Most large systems are composed of several highly interactive and connected sub systems. The SP for a given single-variable closed-loop control might be a function of other variables belonging to different sub systems. Distributed control systems demand larger and more complex development with higher cost, but they are more effective and less expensive in the long term.

Chapter 7: Home Work Problems and Laboratory Projects

1. Define the following terms:
   a. Signal conditioner
   b. Transmitter
   c. Multiplexors
   d. Modems
   e. Quantization error

2. What is the difference between the following?
   a. Digital sensors and analog sensors
   b. Sensors and actuators

3. Which statement is true about instruments?
   a. All instruments can be easily calibrated.
   b. All instruments produce signal, which is independent of the physical measure.
   c. Users can select instruments from a large number of competing vendors; all must comply with universal standards.
   d. All of the above.

4. What are the elements of basic measurement system?

5. What does term "seebeck effect" refer to?

6. Draw the basic elements of process control loop, and describe the function of each element.

7. List two standard analog signals.

8. For an 8-bit analog input module (A/D), what is the range of values it can represent (signed/unsigned)?

9. What does multi variable's control mean?

10. What methods are used in the process-control loop to provide optimal control to a process control system?

11. Define the following terms:
    a. Data logging
    b. Digital process control
    c. Supervisory control
    d. ATD converter
    e. DTA converter

12. What will be the output of a reverse acting controller when the process changes from 50% to 75%, the proportional band is set at 50%, and set point is set at 50%?

13. What PID control mode generates a controller output proportional to the rate of change in the process error? Explain.

14. Explain what the word *process load* means and give an example.

15. Which statement is true in proportional control?

    a. Proportional control usually produces a zero error when stability is reached after a change in the load.

    b. Proportional control usually produces an offset when stability is reached after a change in the load.

16. Define the following terms:

    a. Process transient time

    b. Process load time

    c. Process regulation time

    d. Process lag time

17. A temperature sensor is used to measure an oven temperature between 50 and 300°F. The output of the sensor is converted to a signal in the range of 0 to 5 V. The signal is connected to a 12-bit-resolution ATD converter of a PLC system. Answer the followings:

    a. What is the sensor resolution in °F/volt?

    b. What is the ATD resolution in °F/bit?

    c. What ATD digital count corresponds to a 100°F?

    d. What is the resolution of the ATD in Volt/bit?

    e. Calculate the average quantization error of the ATD converter.

18. A sensor provides real-time temperature measurements of an oven in the range of 0 to 10 V, which represents engineering unit values of 50 to 400°F. The oven temperature is desired to maintain a 200°F SP by sending an on/off signal to the oven heater. The dead band allowed for the on/off oven control is 2°F. Calculate the high threshold and low threshold around the dead band.

19. In a home heating system, what is the effect of increasing or decreasing the dead band?

20. What is the advantage of SP automatic generation in supervisory control system?

21. What is the proportional band of a temperature controller having a 0.75 proportional gain and a SP of 300°F?

22. A 12-bit ATD with 10 V reference has an input signal of 2.69 V. What is the digital count for this input signal? What is the equivalent analog input signal to a 3A5 hex digital count?

23. A sensor provides real-time temperature measurements of an oven in the range of 4 to 20 mA, which represents engineering unit values of 40 to 350°C. An analog input module is interfaced to the sensor output. The oven temperature is desired to maintain a 250°C SP using an on/off control and an oven heater. The dead band allowed for the on/off oven control is 4°C. Answer the following:

    a. What is the sensor resolution in °C/mA?

    b. What is the ATD resolution in °C/bit?

    c. What is the digital count of the ATD for a 159°C measurement?

    d. Calculate the high and the low thresholds for the given SP. Calculate the maximum quantization error of the ATD converter.

24. A sensor provides real-time measurements of a tank level in the range of 4 to 20 mA, which represents engineering unit values of 20 to 500 m. An analog input module with a 12-bit resolution is used to acquire this signal. What range of level a $(256)_{10}$ count represent?

Laboratory 7.1—On/Off Temperature Control

The objective of this laboratory is to get hands-on knowledge of the on/off process control in commercial and industrial application. Refer to Sec. 7.5.1 for more detailed coverage of on/off process control.

A sensor provides real-time temperature measurements of an oven in the range of 0 to 10 V, which represents engineering unit values of 50 to 400°F. The oven temperature is desired to maintain a 300°F SP by sending an on/off signal to the oven heater. The dead band allowed for the on/off oven control is 2°F (+ or −1°F from the SP).

Requirements:

- Configure the PLC analog module attached to the CPU for a 0- to 10-V input signal range.

- Use a 10-V potentiometer to supply the analog input signal (0 to 10 V). Apply the signal to the analog input module connected to the main CPU of the training unit.

- Change the potentiometer setting in the range from 0 to 10 V to represent an oven temperature in engineering units in the range of 50 to 400°F.

- Configure a new HMI function keys page. This page will allow the operator to access two other pages: status and control pages. Define this page as the HMI start page.

- Configure a Status page in the HMI to display the oven temperature in °F as you change the potentiometer from the minimum to the maximum voltage (0 to 10 V) representing the engineering units (50 to 400°F). Also, define two text objects: heater on and heater off.

- Configure a Control page in the HMI to allow the operator to enter the oven SP in the range of 50 to 400°F.

Laboratory Requirements

- Assign the system inputs.
- Assign the system outputs.
- Program the required rungs.
- Download the program and go online.
- Simulate the program using the training units or the LogixPro simulator. Configure the data window to display the raw input digital count (0 to 32767) and temperature in engineering units (50 to 400°F) as the potentiometer setting is adjusted from 0 to 10 V.
- Simulate the program and verify that the control is running according to the system specification.

Laboratory Modifications

Document all the rungs for the program shown in Fig. 7.23. The program consists of a main file and three subroutine files. Add needed rungs in a fourth subroutine to validate that the operator entered SP is within limits (50 to 400°F). This subroutine will be called from the main program like the case with all other subroutines but must only execute once if the operator enters new SP. A normally open PB switch can be assigned

(a)

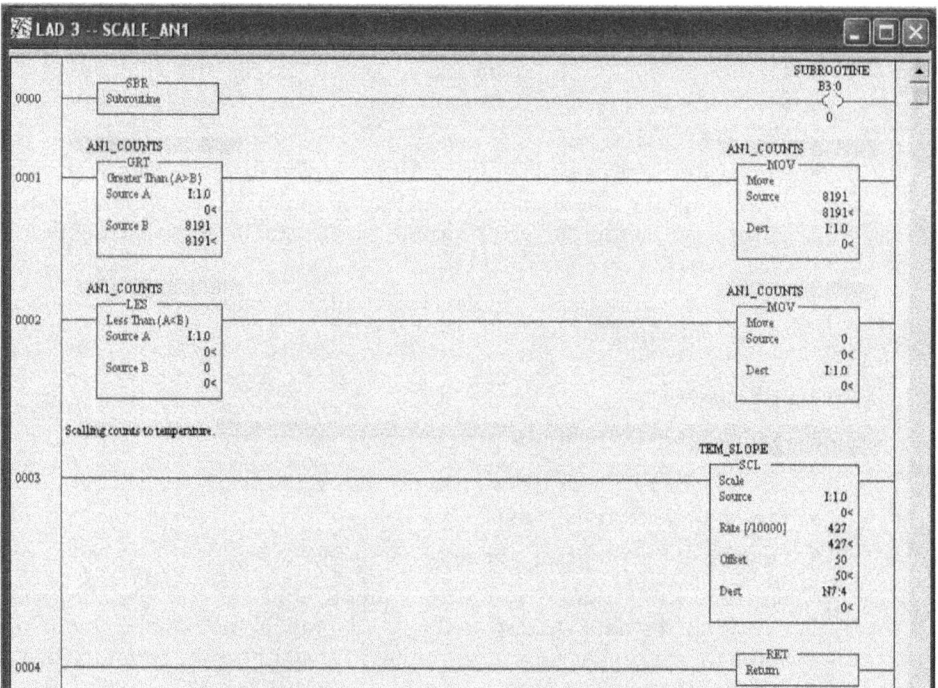

(b)

Figure 7.23   (a) Main program; (b) subroutine file 3; (c) subroutine file 4; (d) subroutine file 5.

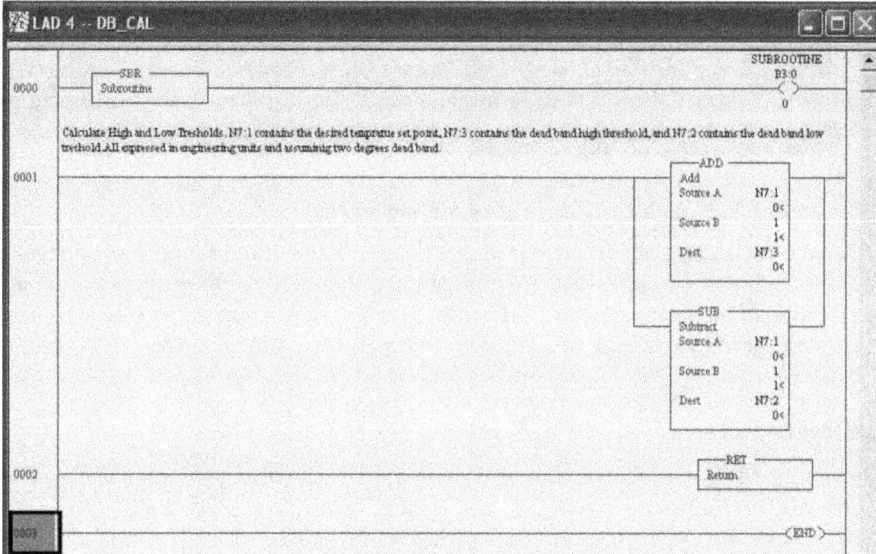

(c)

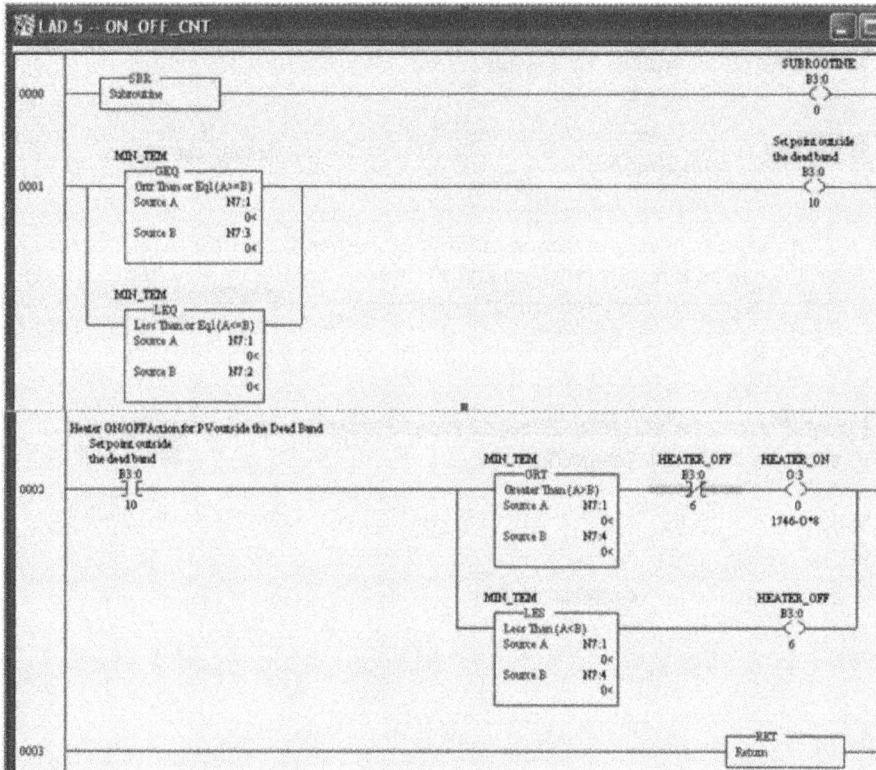

(d)

FIGURE 7.23 *(Continued)*

as an indication of a new SP value or more realistically the SP can be entered from a configured HMI. If SP is outside the limit, send a message to the HMI "Invalid Set Point! Re-Enter" or flag a designated light on the I/O panel if using the LogixPro simulator. Implement the HMI requirements and make the needed modifications to the RSLogix 500 ladder program. Download the program to your processor and perform the checkout.

Laboratory 7.2—Tank Level and Flow Rate Monitoring

A tank maintains a chemical liquid at a desired level. Liquid enters and leaves the tank at a variable flow rate values. An analog input signal represents the flow rate going into the tank in the range of 0 to 10 ft³/min. The second analog input signal represents the flow rate out of the tank in the range of 0 to 10 ft³/min. The tank has a constant area of 10 ft².

Requirements:

- Configure your analog input module for two input voltage signals in the range of 0 to 5 V.
- Write a PLC program to read the two input signals. Calculate and display the rate of change of the tank height in "x" in/min.
- The rate of change of the tank height should be displayed in integer register, utilizing one of the following light indications:
  - Pilot light #1 on: The tank level is decreasing at the rate of x in/min indicated in register F8: 10.
  - Pilot light #2 on: The tank level is increasing at the rate of x in/min indicated in register F8: 11.
  - Pilot light #3 on: The tank level is stable.
- Test your program for an inflow rate greater than outflow rate, an inflow rate less than an out flow rate, and an inflow rate equal an outflow rate. Provide the check out and document the final ladder listing.

CHAPTER 8

Analog Programming and Advanced Control

T his chapter examines the process of interfacing analog input and output variables to the Allen Bradley SLC 500 processor. It covers the fundamentals of analog I/O programming and its use in process control. Advanced industrial process
control implementation is discussed briefly.

Port of Houston.

Chapter Objectives

- Perform analog module configuration and diagnostic.
- Perform and debug Allen Bradley SLC 500 analog I/O programming.
- Perform AB SLC-500 PID configuration/programming.
- Understand PID control structure and performance.

A PLC is the evolution of hardwired analog control systems. Analog is a keyword in the world of process control and automation. Most physical entities we would like to control in the real world are analog in nature. This include variables like temperature, pressure, speed, acidity, position, level, flow rate, viscosity, displacement, weight, frequency, and many others. All can be measured and quantified in time, and some exhibit self-regulation ability. Self-regulation is the ability of the control variable to make adjustments and arrive at a stable state under small disturbances, which is a desired characteristic in the selected control variables. This chapter examines the process of interfacing analog input and output variables to the AB SLC-500 processor. It covers the fundamentals of analog I/O programming and its use in process control. Advanced process control techniques will be discussed with small industrial implementation.

8.1   Analog Input/Output Configuration and Programming

Standard modules are available off the shelf and can be interfaced to the AB SLC-500 PLC. The CPU used in this chapter is the SLC 5/03 with an analog composite input/output analog module with two inputs and two outputs. Available analog inputs and the analog output are configured and used in our coverage of analog programming and the implementation of closed-loop process control. The analog I/O modules used in this chapter are described followed by the steps to configure the module, scale the I/O, and program it.

8.1.1   Analog Input/Output Modules

Analog modules are available in a variety of specifications to accommodate all types of applications and sensors/actuators interfaces. The reader has to refer to the specific technical manual for the analog modules used. In this section, we provide details on the NIO4V composite analog module selected for our training unit and used to demonstrate analog programming in this chapter. This module provides two voltage analog input channels and two voltage analog output channels. Figure 8.1 shows the wiring interface to an analog signal source and a load. Only the unused input channel is jumper to the module analog common point.

The analog module does not provide loop power for analog inputs. A power supply that matches the transmitter specifications must be provided. Figure 8.2 shows the module's input interface wiring for a two-, three-, and four-wire sensor transmitter.

8.1.2   Configuring Analog Input/Output Modules

Signal modules, including analog input and output, can be added to the PLC processor. Figure 8.3 shows the device configuration screen. To start the configuration process,

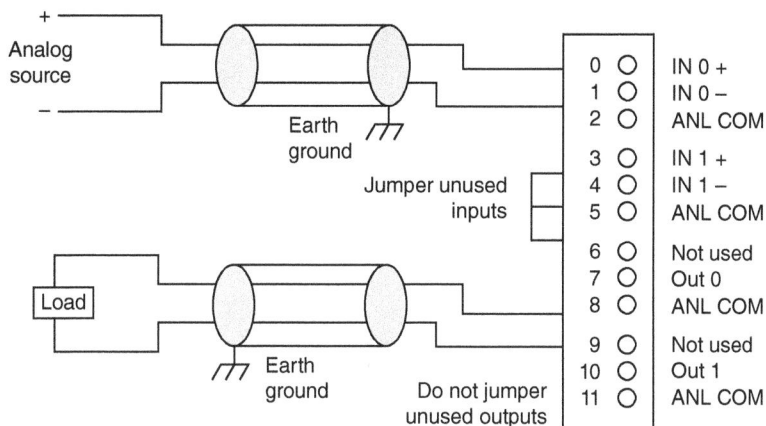

FIGURE 8.1 NIO4V analog module interface. (*Courtesy of the Allen-Bradley SLC 500.*)

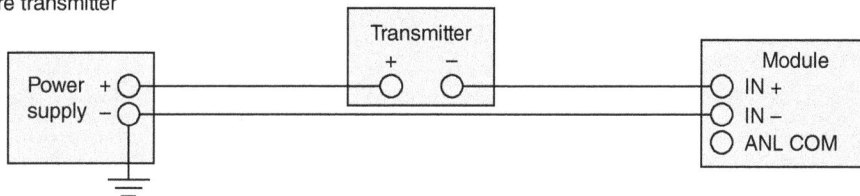

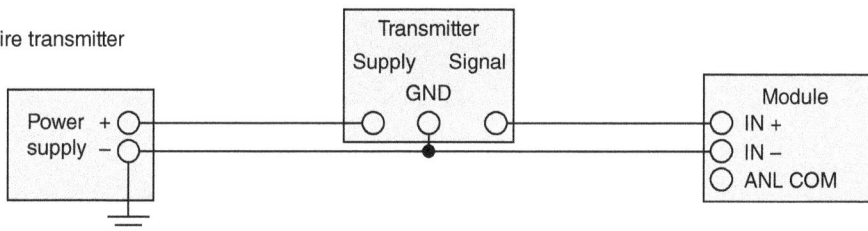

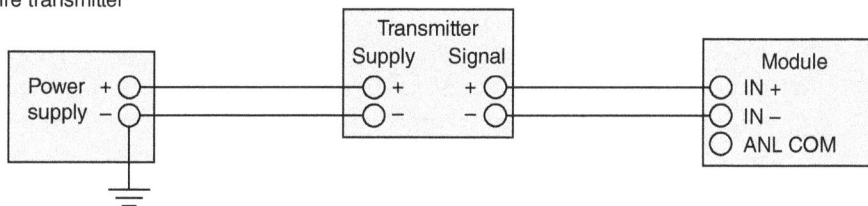

FIGURE 8.2 Input sensor transmitter interface. (*Courtesy of the Allen Bradley SLC 500.*)

open the "I/O configuration" from the project tree. Add your module and then configure each of the channel's properties.

8.1.3 Scale Instruction

SCL instruction shown in Fig. 8.4 is used to scale signal count produced by your analog
module and placed in designated PLC memory. For example, SCL is used to convert a

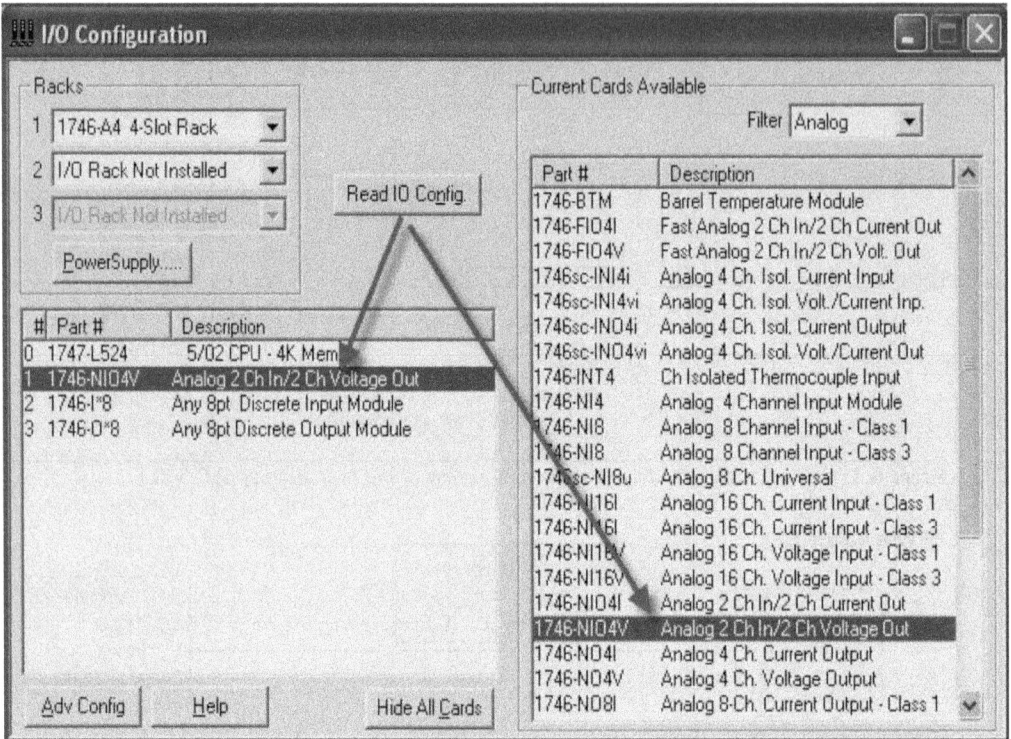

FIGURE **8.3** Device configuration.

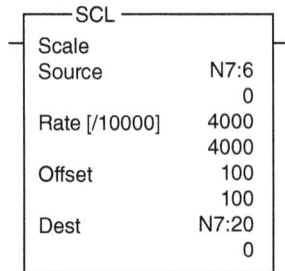

FIGURE **8.4** Scale instruction.

4- to 20-mA input signal to a desired PID process variable format. The instruction is also used to scale an analog input to a desired range suitable for controlling an analog output device. When rung conditions are true, SCL instruction multiplies the "source" by a specified "rate." The rounded result is added to an "offset" value and placed in the destination "dest." Other scaling instructions for processors above the SLC 5/02, which support floating point arithmetic, are discussed and used later in this chapter.

The following equation is used in calculating the linear scaling relationship:

Scaled value (dest) = (source * rate) + offset

Rate = scaled value range/input signal range

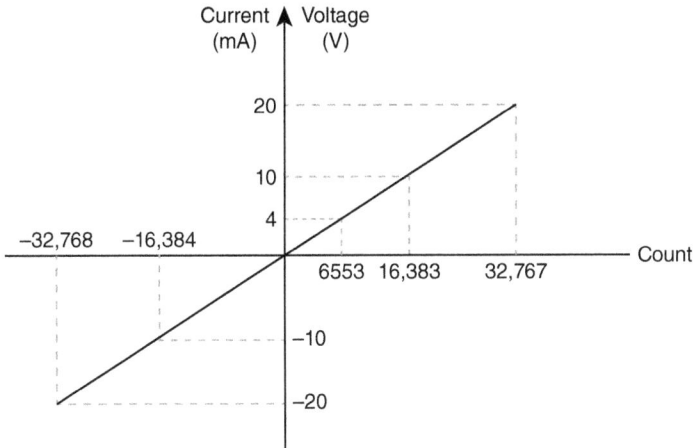

FIGURE 8.5 Scaling counts to voltage/currents.

$$\text{Range} = \text{maximum value} - \text{minimum value}$$
$$\text{Offset} = \text{scaled minimum} - (\text{input signal minimum} * \text{rate})$$

The SCL parameters are specified as follows:

- Count or source values must fit in a 16-bit integer representation (−32,768 to +32,767).
- Rate (or slope) is a positive or negative value you enter divided by 10,000.
- Rate and offset can be a program constant or a value stored in a word address.
- Destination is the address where the result of the operation is stored.

If either the result of the source times or the rate divided by 10,000 or "dest" value is less than −32,768 or greater than 32,767, the SCL instruction overflows causing error 0020 (minor error bit) and places −32,768 or 32,767 in the destination. This occurs regardless of the current offset. If this error occurs, you must reset the error associated bit S:5/0 within the ladder program before the end of the current scan or else a major error will be declared.

Figure 8.5 shows a generic analog module linear relation between different voltage or current signals and the associated digital counts. For example, the 4 mA is mapped to a "6553" count and the 20 mA to the 32,767 count using a linear scaling function. Figure 8.6 shows the initialization of the watch table tags for monitoring analog variables for the 1746-NIO4V module. The display format can be adjusted by selecting the desired Radix. The monitor icon at the top of the screen, shown shaded, must be activated to start the real-time monitoring. All monitored or forced values must have a prior definition in the PLC tags with the correct addresses and the associated matching format. The software will validate user's inputs and prompt for corrections, if needed.

8.1.4 NIO4V Composite Module Bit Addressing and Data Conversion

This section provides more details on the NIO4V composite analog module, which is the one used in this book. The operation of other analog modules is similar but may

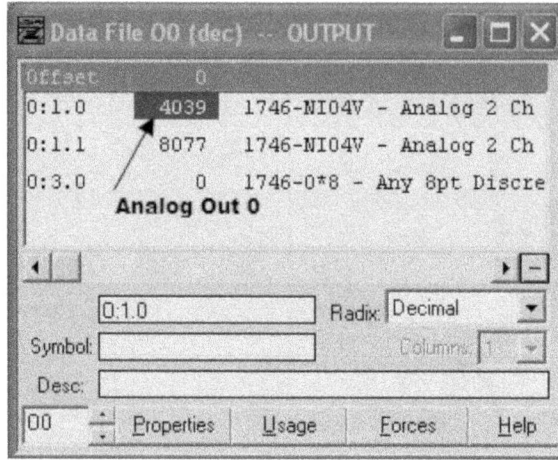

FIGURE 8.6 Monitoring analog signal count.

have slightly different specifications. Figure 8.7 bit maps show bit-level addressing for the analog inputs and outputs. The input channel converter resolution is 16 bits, which is equivalent to one word. The output channel converter resolution is 14 bits and is loaded from the most significant 14 bits of the associated output word. The "XX" and "YY", the two least significant bits (O:e.0/0 and O:e.0/1) of the output word have no effect on the actual output value. "e" is the element address designation. The "XX and YY" value of the two least significant bits is irrelevant.

The analog input and output data is updated by the processor once during each scan of the user program. The analog input and output data can be monitored in several different radices using your programming software. Viewing the radix as decimal allows the analog input and output data to be viewed as decimal representations of integer words. When monitoring in binary radix, data is viewed in two's complement representation for negative values. Analog inputs convert current and voltage signals into 16-bit two's complement binary values. Table 8.1 identifies the current and voltage input ranges for the input channels, the number of significant bits for the applications using input ranges less than full scale, and their resolution. Channel resolution is the amount of voltage or current equivalent to a count value of +1, the least significant bit value.

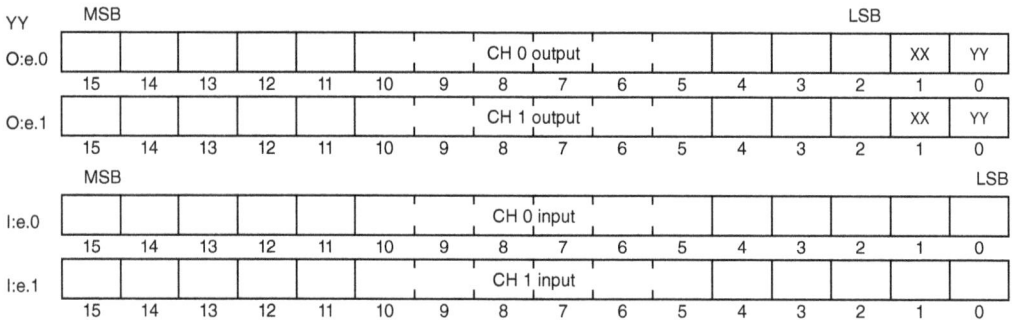

FIGURE 8.7 Inputs and outputs bit addressing. (*Courtesy of the Allen-Bradley RSLogix 500.*)

Voltage/Current Range	Decimal Representation	Number of Significant Bits	Resolution per LSB
−10 to +10 V dc − 1 LSB	−32,768 to +32,767	16 bits	305.176 µV
0 to 10 V dc − 1 LSB	0 to 32,767	15 bits	
0 to 5 V dc	0 to 16,384	14 bits	
1 to 5 V dc	3,277 to 16,384	13.67 bits	
−20 to +20 mA	−16,384 to +16,384	15 bits	1.22070 µA
0 to 20 mA	0 to 16,384	14 bits	
4 to 20 mA	3,277 to 16,384	13.67 bits	

* Courtesy of the Allen-Bradley RSLogix 500.

TABLE 8.1   Input Channel Analog Resolution

The relation between the input analog signal and the associated digital count is linear. The following is such relation for an input channel configured from 0 to 10 V:

$$\text{Input signal voltage} = (10/32{,}768) * \text{input count}$$

where input count is decimal value of the input image word corresponding to the analog input.

Example 8.1.4.1   If a count value of +16,021 is in the input image, the calculated input voltage is (10/32,768) * 16,021, which equals 4.89 V. It should be noted that the calculated value may be slightly different from the actual value because of the accuracy/resolution limitations of the analog module conversion process.

The analog output channel converts a 16-bit two's complement binary value into an analog output signal. Because the analog output channels have a 14-bit digital-to-analog converter, the 14 most significant bits of this 16-bit integer are the bits that the output channel converts. The NIO4V supports two voltage outputs, ranging from −10 to +10 V dc. Table 8.2 identifies the voltage output ranges for the output channels, the number of significant bits for the applications using output ranges less than full scale, and their resolution.

The following equation is used to determine the decimal value for the voltage output:

$$\text{Output voltage count} = \text{desired output voltage} * (32{,}768/10)$$

Voltage Range	Decimal Representation for Output Word	Number of Significant Bits	Resolution per LSB
−10 to +10 V dc − 1 LSB	−32,768 to +32.764	14 bits	1.22070 µV
0 to +10 V dc −1 LSB	0 to +32,764	13 bits	
0 to 5 V dc	0 to +16,384	12 bits	
1 to 5 V dc	+3,277 to +16,384	11.67 bits	

* Courtesy of the Allen-Bradley LogixPro 500

TABLE 8.2   Output Channel Analog Resolution

Example 8.1.4.2 If an output value of 2 V is desired, the digital count value to be stored in the corresponding word in the output image table is 2 * (32,768/10), which is equal to 6554. Notice that the actual resolution for the analog output is 1.22 mV per LSB, where the least significant bit is bit number 2 of the output word. This output resolution is one-fourth the one for the analog input channels. Refer to Tables 8.1 and 8.2 to verify the answer and have more understanding of the resolution difference.

8.1.5 Analog Scaling and Mapping

Analog signals are typically either a voltage or current signals at the device level: transducer/sensor or actuator/final control element. The physical entity being measured or manipulated is typically expressed in engineering units. Inside the PLC, all values are expressed in digital counts using integer two's complement representation. Often, users like to express the same values in different format suitable for a particular interface or use: HMI, BCD interface, and data communication. Mapping and scaling from one form to another is a common task in analog programming. Same value might be stored as volts, count, engineering units, BCD, ASCII, percent of range, or percent of set point. This section briefly covers the main issues involved in analog scaling, mapping, and programming. The following assumptions are being made for an analog module located in slot 3 of a modular PLC system:

- A temperature transducer with a 0- to 10-V dc output is wired to the second input channel on the analog module.

- The transducer voltage signal is proportional to a range of 100 to 500°C (212 to 932°F).

- The process temperature must stay between 275 and 300°C (527 and 572°F). If the temperature deviates from this range, a flag is set and this out-of-range value is not processed. The data is presented in °C for monitoring and display purposes.

The scaling operation is displayed in Fig. 8.8's plot. It shows the linear relationship between the input value (counts/volts) and the resulting scaled values in °C. The process desired operating range is highlighted.

The following equation expresses the linear relationship between the input value and the resulting scaled value:

$$\textbf{Scaled value} = (\textbf{input value} * \text{slope}) + \text{offset}$$

where slope $= (500 - 100)/(32{,}767 - 0) = 400/32{,}767$

$$\text{Offset} = \text{scaled min.} - (\text{input min.} * \text{slope}) = 100 - [0 \times (400/32{,}767)] = 100$$

$$\textbf{Scaled value} = (\textbf{Input value} * (400/32{,}767)) + 100$$

The following equation is used to calculate the low- and high-limit input values, which determine the status of the out-of-range flag:

$$\textbf{Input value} = (\textbf{scaled value} - \text{offset})/\text{slope}$$

$$\text{Low limit} = (275 - 100)/(400/32{,}767) = 14{,}344$$

$$\text{High limit} = (300 - 100)/(400/32{,}767) = 16{,}393$$

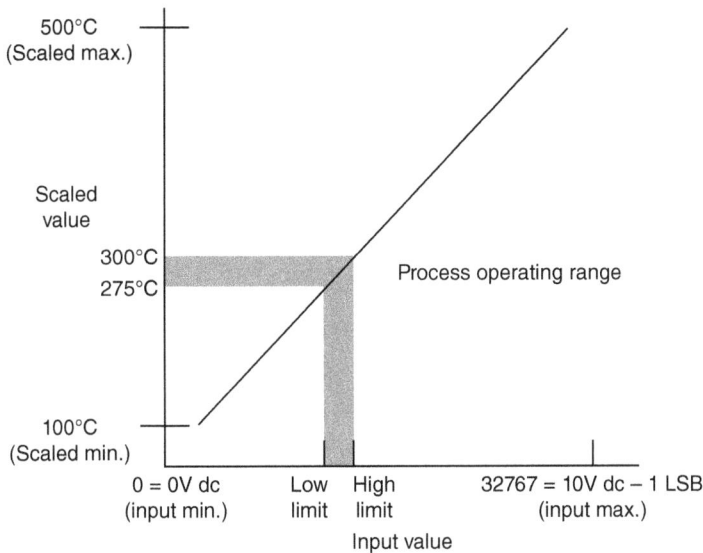

FIGURE **8.8**   Analog inputs linear relation. (*Courtesy of the Rockwell RSLogix 500.*)

Example 8.1.5.1   This example uses the linear relationship of Fig. 8.8 and the calculated out-of-range flag values to implement the following small process control task:

- Turn ON a below-range or above-range alarm if the process temperature is outside the desired range of 275 and 300°C.
- It is not necessary to scale input counts that are outside the desired range but associated alarms and flags must be set.
- *Scale* the process temperature only if the input count is within the temperature desired operating range.

Figure 8.9 ladder rungs show how to implement the required specifications. It uses standard math instructions available in any SLC-500 processor. The ladder program prevents a processor fault by unlatching the mathematical overflow bit S2:5/0 before the end of the scan. The input analog count is assumed in I:1.1 and the scaled output value in engineering units (°C) is stored in the integer word N7:0.

Example 8.1.5.2   This example implements the same requirements using the scaling instruction (SCL) available in the 5/02 and higher processors. The "rate" parameter is calculated by multiplying the "slope" of the linear relation by 10,000. This implementation is the preferred for the SLC 5/02 processor since it does not support floating point arithmetic.

$$\textbf{Rate} = (400/32{,}767) * 10{,}000 = 122$$

The offset value for the scale instruction is 100 in this example. If the result of the source times for which the rate, divided by 10,000, is greater than 32,767, the SCL instruction overflows, causing error 0020 (minor error bit), and places 32,767 in the destination. This occurs regardless of the current offset. Figure 8.10 shows the ladder listing for this implementation.

Example 8.1.5.3   This example implements the same requirements using the SCP (scale with parameters) instruction available in the SLC 5/03 (OS302 or later), SLC 5/04 (OS401 or later), and SLC 5/05 only. You could certainly insert the SCP instruction using the toolbar. Also, you can use the

(a)

(b)

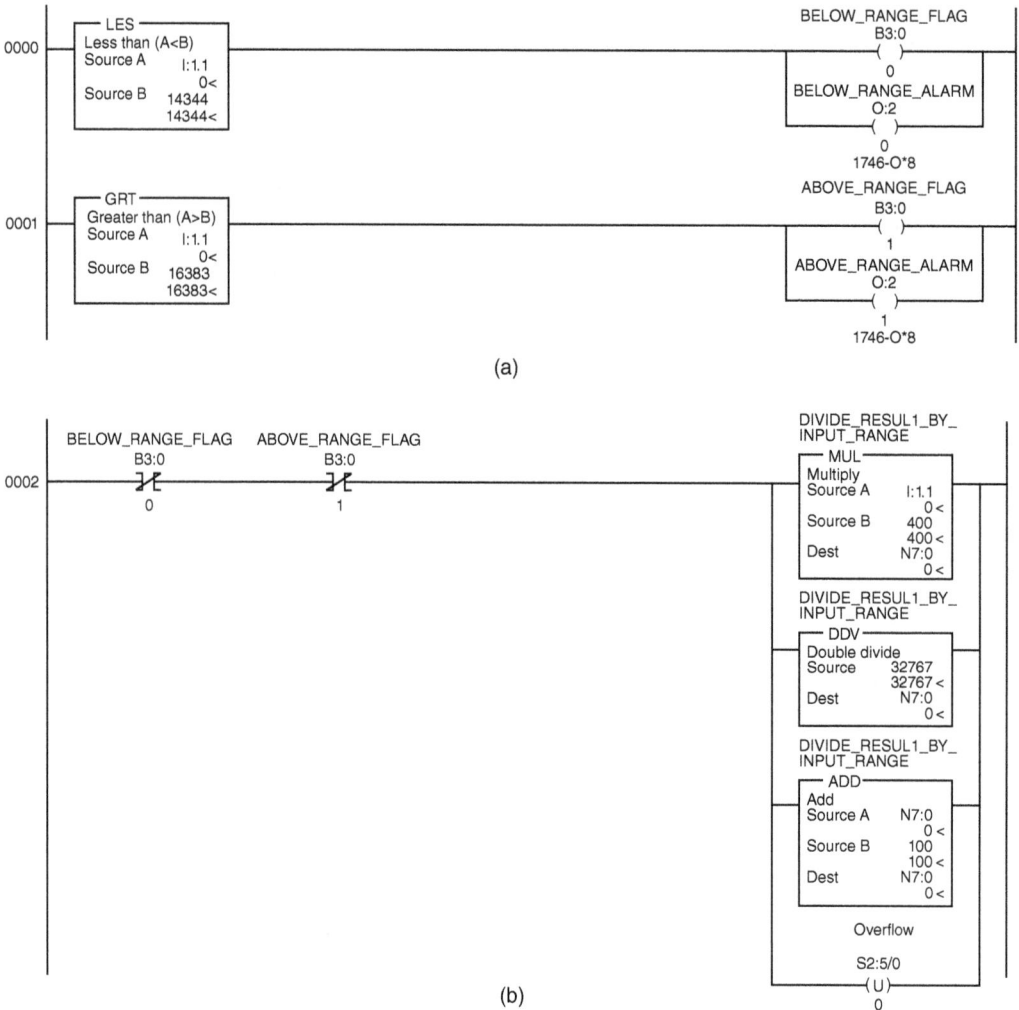

FIGURE 8.9   (a) Example 8.1.5.1 ladder program; (b) Example 8.1.5.1 ladder program.

ASCII editor to insert an instruction by double-clicking on the rung address in RSLogix 500, then type SCP in the box and press Enter. One of the common uses of the SCP instruction is to take the count value from an analog input and scale it to an engineering value. Figure 8.11 shows the ladder listing for this implementation.

Example 8.1.5.4   This example demonstrates addressing of analog I/O, scaling, and range checking of analog input and output values. A four-point analog composite module is placed in slot 1 of an SLC-500 system. A 0- to 200-psi pressure sensor provides input as a 4- to 20-mA (3277 to 16,384 from Table 8.1) signal to input channel 0. The input value is checked to ensure that it is, within the specified range. It is then scaled and output as a 0 to 2.5 V (0 to 8192 from Table 8.1) signal to a panel pressure meter display connected to output channel 0. If an out-of-range condition is detected, a flag bit will be set. The scaling equation is plotted in Fig. 8.12, which displays the linear relationship between the input value and the resulting scaled value.

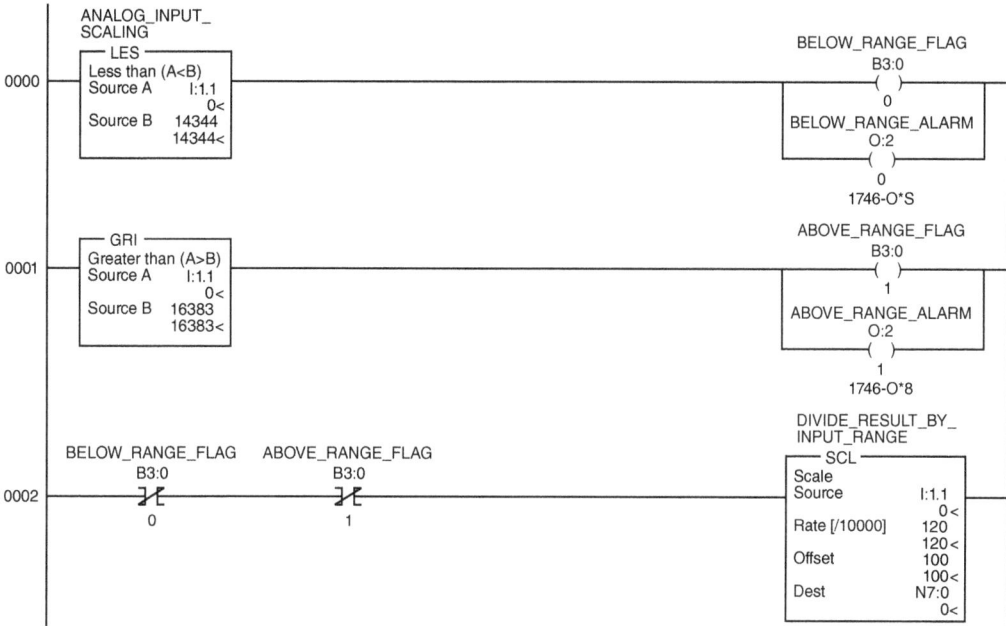

FIGURE 8.10   Example 8.1.5.2 ladder program.

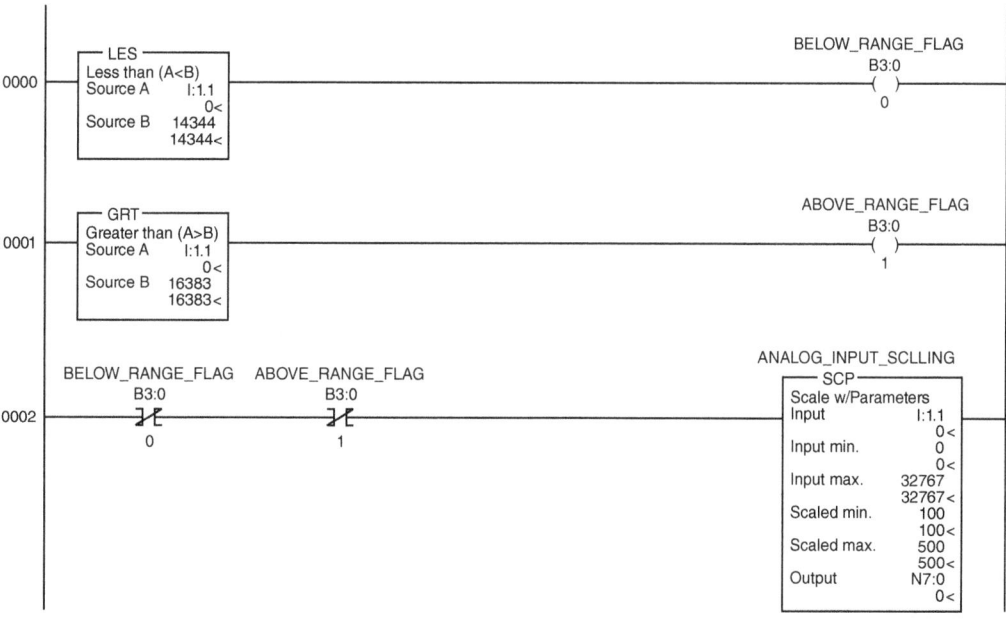

FIGURE 8.11   Example 8.1.5.3 ladder program.

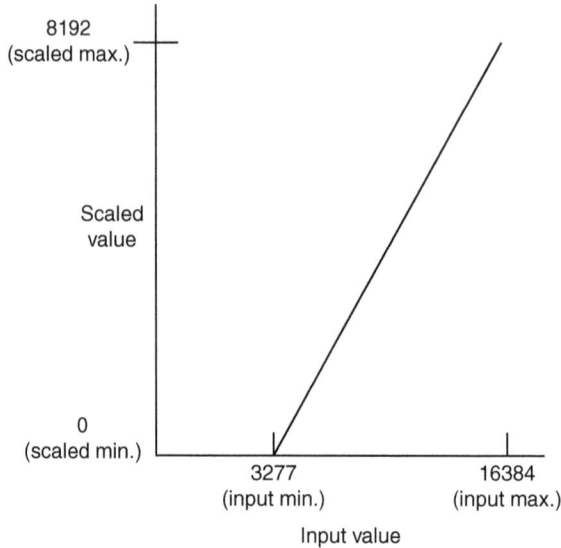

FIGURE **8.12** Output voltage input current relation.

The following equation expresses the linear relationship between the input current count value and the resulting scaled voltage output count value:

$$\textbf{Scaled value} = \textbf{input value} * \text{slope} + \text{offset}$$

where slope = (scaled max. − scaled min.)/(input max. − input min.) = 8192/13,107

$$\text{Offset} = \text{scaled min.} - (\text{input min.} * \text{slope}) = 0 - 3277 * (8192/13{,}107) = -2048$$

$$\textbf{Scaled value} = \textbf{input value} * (8192/13{,}017) - 2048$$

The scaling instruction available in the SLC 5/02 processor, which does not support floating point arithmetic, is used to realize an efficient program. The scaling instruction uses the same multiply, divide, and add algorithm, but it does so with a single "rate" value instead of the scaled range and input range values. The "rate" is defined as (scaled range/input range) × 10,000. The rate is 6250 (8192 * 10,000/13,107) for this programming example. Figure 8.13 shows the implemented ladder rungs realizing the stated specifications. The analog input count is assumed in I:1.0 and the analog output scaled count is sent to O:1.0.

8.2 PID Control Configuration and Programming

Proportional integral derivative closed-loop control is one of the most widely used process control technique. It reacts to process violations or deviation from desired behavior using a strategy based on knowledge about error values, the history of error, and the future forecast of errors. This section covers PID closed-loop control with practical aspects of its implementation in industrial automation.

8.2.1 Closed-Loop Control System

As stated earlier in Chap. 1, a control system is a collection of hardware and software designed to produce desired process behavior in real time. Figure 8.14 shows the

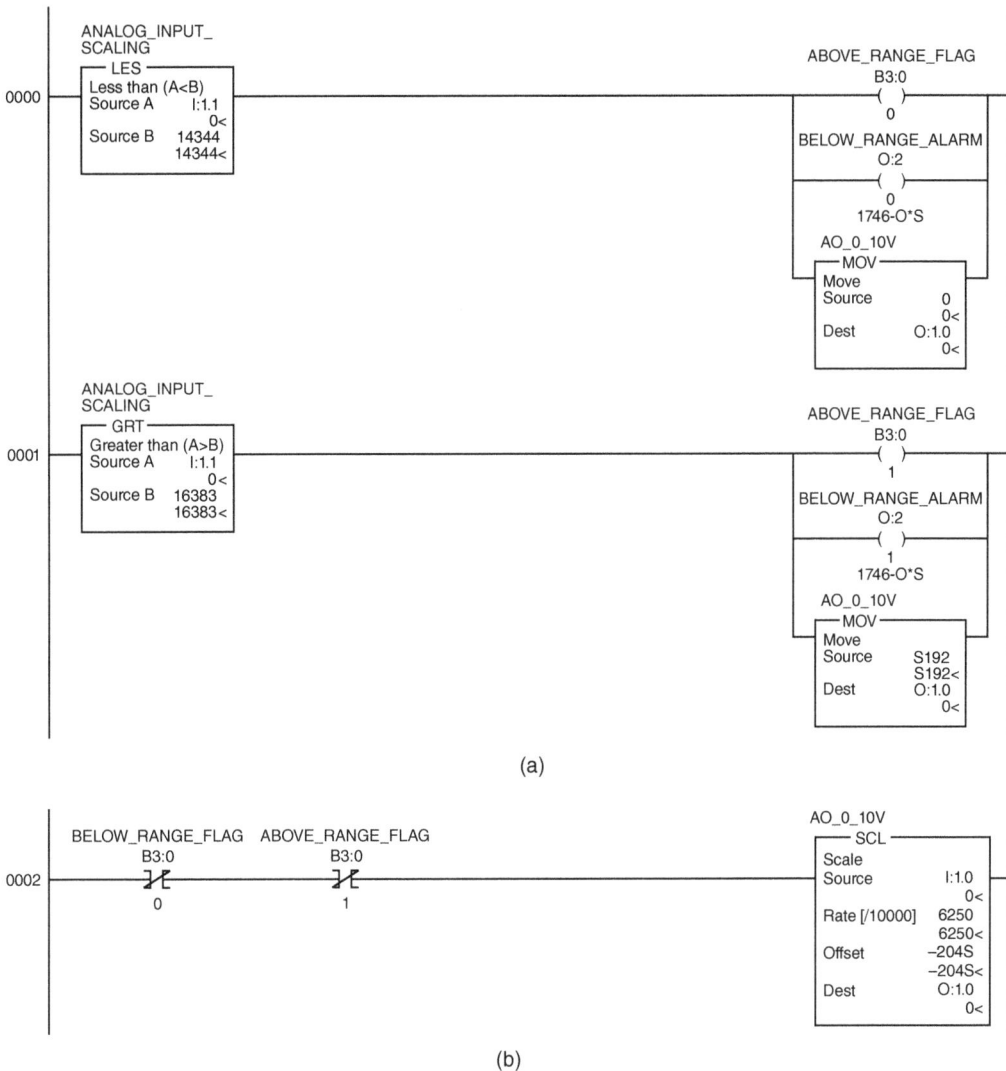

(a)

(b)

Figure 8.13   (a) Example 8.1.5.4 ladder program; (b) Example 8.1.5.4 ladder program.

functional block diagram for a single-variable, closed-loop process control. The control system loop includes the process, final control element, controller, error detector, and a measurement element. The following are the key variables and associated action in the closed loop:

- The set point is the user-defined desired value of the controlled variable, which is the process variable to be regulated.

- Measurement element output indicates the controlled variable status also called *process output variable* or *feedback signal*.

- The error is the difference between the set point and the controlled variable.

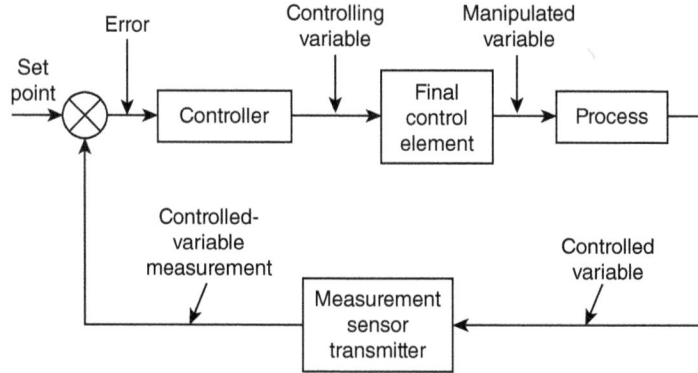

Figure 8.14 Single-variable closed-loop control.

- Output is an indication produced by the controller to initiate the action to be taken by the final control element to correct the error.

- Final control element is the actuator, which bring about the changes in the process and the selected controlled variable. The output of the final control element is called the *process controlling* or *manipulating variable*.

8.2.2 Control System Time Response

Time response of the control system can be determined based on the time characteristic of the process controlled variable value (X) following a step change of the controlling variable output value (Y). Most control systems are self-regulating, which means a new controlled variable equilibrium is reached immediately or shortly after a process change or disturbance. For example, the speed of an induction motor settles at the value that produces the required load torque. The motor will run at this equilibrium point under the same load conditions (for a rotating object, equal force and load lead to constant speed). A small increase in load due to disturbances will force the motor to decelerate (load greater than motor torque) and reach a lower speed point of equilibrium after short period of oscillations. Thus the speed of the induction motor is an excellent candidate for selection as controlled variable. Motor speed control is very common application in process control.

Figure 8.15 plots a process controlling variable output response to a step input change in the controlling variable for a possible controller implementation. Other responses are possible through a different controller design.

The time response can be determined using the controlled variable delay time (T_u), recovery time (T_g), and the final value (X_{max}). The variables are determined by applying a tangent line to the maximum value and the inflection point of the step response. In many systems, it is not possible to drive the process and record the response characteristic up to the maximum value because the process controlled variable cannot exceed specific limit. In this case, the rate of controlled variable rise is used to identify the system. The controllability of the system can be defined based on the ratio T_u/T_g as shown in Table 8.3.

A controlled system with dead and recovery times reacts to a step change in the controlling variable process input as shown in Fig. 8.16. T_d is the dead time, which is the

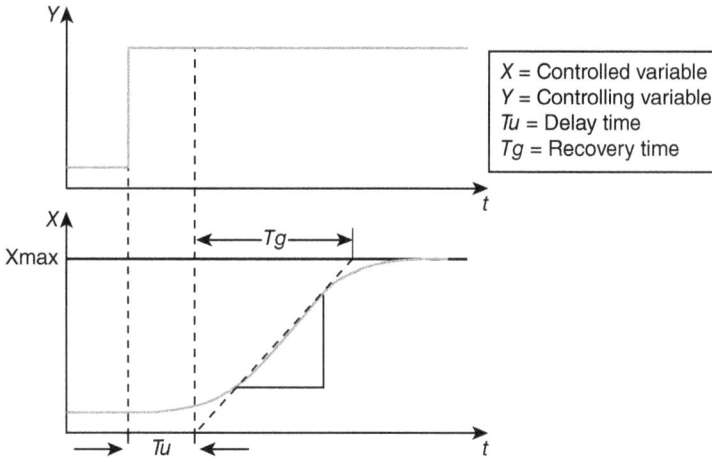

FIGURE **8.15** Step change responses.

Process Type	T_u/T_g	System Controllability
Slow reacting	< 0.1	Very good
Reacting	0.1 to 0.3	Can be controlled
Fast reacting	> 0.3	Difficult to control

TABLE **8.3** System Controllability

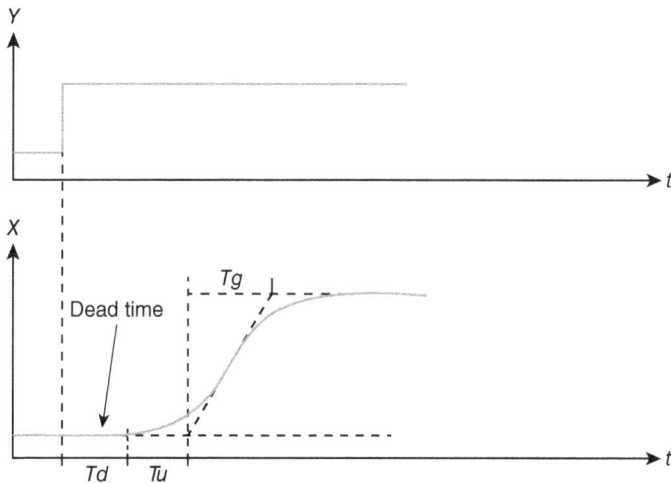

FIGURE **8.16** Control process with dead time response.

time interval from initiating a change in the process input signal (controlling variable) to the point where we observe the start of the process output (controlled variable) response. The controllability of a self-regulating controlled system with dead time is determined by the ratio of T_d/T_g. The dead time T_d must be a small fraction of the

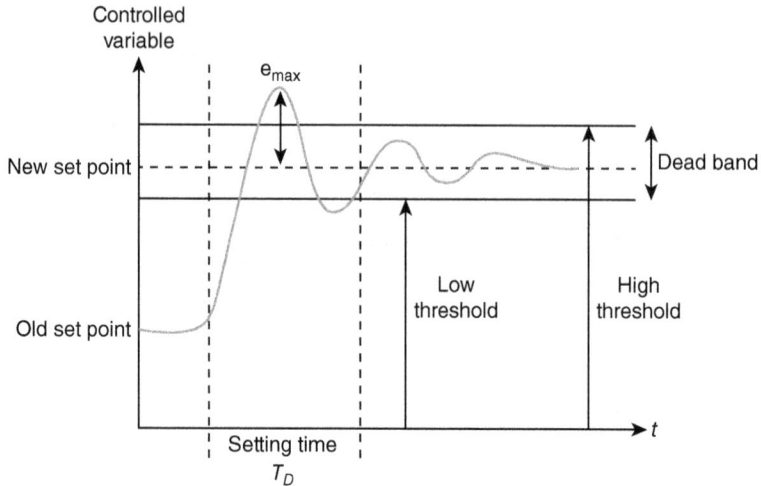

Figure 8.17   Typical oscillatory closed-loop control.

recovery time T_g for good controllability. Processes with large dead time are extremely difficult to control and include large boiler drums' temperature, cascaded tanks liquid volume, and irrigation canal downstream water level controls. In a typical temperature control application as shown in Fig. 8.17, the temperature inside a tank is selected as the controlled variable while the heater current represents the controlling variable of the closed-loop feedback controller.

Figure 8.17 shows an oscillatory control system response with no dead time. The controlled variable reacts immediately after a set point change reaching and then exceeding the new set point. Finally it goes through attenuating oscillations reaching steady-state value within the allowable dead band. The total time for the controlled variable to reach the new steady-state value is called *settling time* (T_D). In Fig. 8.16 response, $T_D = (T_a + T_g + T_u)$. Settling time is an important measure of the control system performance. Typical system behavior will show several and repeated oscillations before reaching steady-state conditions. The maximum error (e_{max}) in the controlled variable corresponds to the first overshoot response and is also a good measure of controller performance. In systems where oscillations are not permitted, different control strategy and controller design are used. This includes the use of on/off control, set point ramping to the final target using small increments, and slow-response tuned controller.

8.2.3   Control System Types

Controlled systems are classified based on their time response to step changes of the controller output value. We distinguish between the following controlled systems:

- Self-regulating controlled systems
  1. Proportional-action controlled systems
  2. First-order controlled systems
  3. Second-order delay element controlled systems

- Non–self-regulating controlled systems
- Controlled systems with dead time

Processes with no self-regulation or with excessive dead time are difficult to control and require special and more complex control techniques, which include feed forward, cascaded control, and dead time compensation. Those techniques are not covered in this book. The following is a brief coverage of the most common three self-regulating controlled systems:

1. **Proportional-action controlled systems:** In proportional-action-controlled systems, the controlled variable process value follows the controller output value (controlling variable) almost immediately. The ratio between the controlled variable (process output) and the controlling variable (process input) is defined as the proportional gain of the controller. Figure 8.18 shows a gate valve, which can be used to control steam flow in a piping system in a temperature control application. The valve opens by lifting a wedge out of the path of the steam. Gate valves are primarily used to permit or prevent the flow of liquid or gas. Specifically designed proportional action controlled gate valve can be used for regulating liquid flow. In this case, the opening of the valve represents the controller output (controlling variable) value and the corresponding liquid/gas flow rate or its effect, as in temperature or level control applications, is our process controlled variable value. The relation between the two variables is direct and almost immediate but often not linear.

Figure 8.19 shows the gate valve proportional-action controlled system dynamic behavior of the controlled variable (liquid temperature inside tank) after a step change in the manipulated variable (valve position), controlling variable. Ideal control action without any lag is not possible in practical systems. The characteristic curves clearly show that a proportional-action controlled system exhibits self-regulation since a new equilibrium is reached immediately after a step change. The following is the ideal equation relating the temperature (T) to the valve position (Y):

$$T = K_p * Y$$

K_p is the process proportional gain.

Figure 8.18   Industrial gate valve.

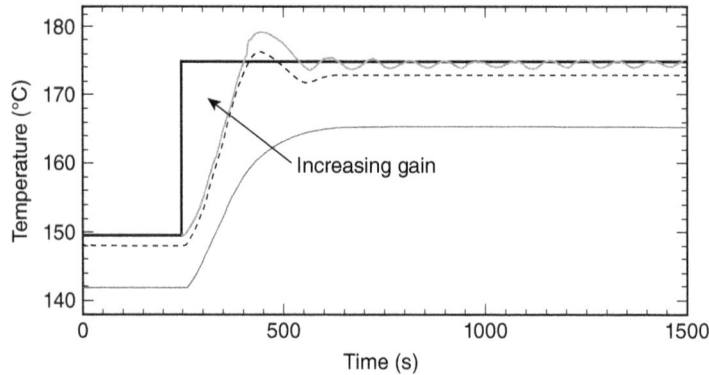

FIGURE 8.19 Proportional-action controlled system.

Small values for the proportional gain will produce slow response and large residual steady-state errors. Large proportional gain values can speed up the response but may also cause large temperature oscillations. Excessive proportional gains can cause large and unsustained oscillations, which can force a manual or automatic system shut down. Selecting and tuning the controller parameters including the proportional gain will be discussed later in this chapter.

2. **First-order controlled systems:** In a first-order system controlled system, the controlled variable value initially changes in proportion to the change of the controller output, the controlling variable value. The rate of change of the controlled variable value is reduced as a function of the time elapsed until process steady state is reached. This type of control is also known as a PT1 system. A water container that is heated with steam is an example of a first-order controlled system. In simple controllers, the time behaviors for the heating and cooling processes are assumed identical. Control is clearly more complex with different process cooling and heating time characteristics.

3. **Second-order delay element controlled systems:** In a second-order delay element controlled system, the process controlled value does not immediately follow a step change of the controller output value. The process controlled value initially increases in proportion to the positive rate of rise of the controller output value and then approaches the set point at a decreasing rate of rise. The controlled system shows what is known as a proportional response characteristic with second-order delay element. Pressure, flow rate, and temperature controls are all examples of this type, which is also known as a PT2 system.

8.2.4 Controllers Behavior

A controller provides the necessary feedback adjustments to the process in real time to bring about the desired changes in the selected controlled variable. A precise adaptation of the controller to the desired controlled variable time response depends on the precise settling of the control strategy, parameters tuning, and its ability to react adequately to set point and load disturbances. Controller's feedback can have a proportional action (P), proportional-derivative action (PD), proportional-integral action (PI), or

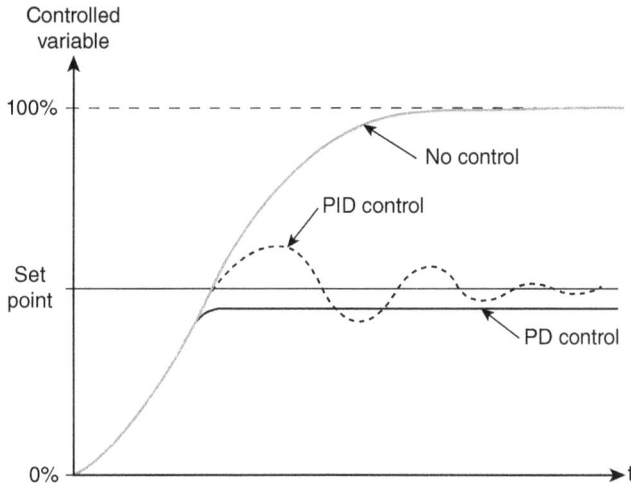

FIGURE 8.20 Closed-loop control types and performance.

proportional-integral-derivative action (PID). On/off control is a special flavor of the proportional control but with controller output either OFF (0%) or ON (100%). Controller actions are triggered due to deviations in the controlled variable behavior caused by process disturbances, load change, or the initiation of a new set point. The step response of the controller depends on its type. Figure 8.20 shows the step response of a P controller, PD controllers, and PID controllers. Proportional-integral (PI) control is the most common, which can help improve the control system response and reduce the final steady-state errors. A controller with derivative action is not appropriate if the control system has pulsing measured quantities as in the case of pressure or flow process control.

Let us briefly demonstrate how the proportional control mode works. The proportional mode is used to set the gain value of the controller. The setting for the proportional mode may be expressed as either: proportional gain or proportional band. Proportional gain (Kc) is the percentage change of the controller output relative to the percentage change in controller input, which is expressed as:

$$\text{Gain } (Kc) = \text{output change\%/input change\%}$$

Proportional band (PB) represents the same information and is the percentage of change of the controller input span which causes a 100% change in the controller output. The following equation relates gain and the PB:

$$\text{PB\%} = 100\%/\text{gain}$$

In a proportional only mode, the controller output change as a percentage of controller span/range is a function of the percentage change in error and the controller gain.

$$\text{Controller output change \%} = \text{error change \%} * \text{gain}$$

Error change is a percentage of set point.

The proportional control mode is simple but responds only to a change in error, thus it will not eliminate the residual error. Selection of the loop gain/PB is critical to the process behavior; high gain (%) may produce cycling (sustained oscillations), while low gain can result in a stable loop but with slow response. A simple tuning technique increases gain until the process cycles following a disturbance, then it reduces that gain by 50%.

Example 8.2.4.1 If the set point is suddenly changed 10% with a proportional band setting of 50%, the output will change as follows:

Controller output change % = input change % * gain

Gain = 100%/PB

Input change = 10%, PB = 50%, and gain = 100% /50% = 2

Controller output change = 10% * 2 = 20%

For a 12-psi controller span: controller output change = 0.2 * 12 psi = 2.4 psi

For a 16-mA controller span: con troller output change = 0.2 * 16 mA = 3.2 mA

8.2.5   Selection of the Suitable Controller Structures

To achieve optimum control results, select a controller structure that is suitable for the process and that can be adapted to the controlled system within specific limits. Table 8.4 provides an overview of suitable combinations of a controller structure and controlled system. Table 8.5 provides an overview of suitable combinations of a controller structure and physical quantity.

Controlled System		Controlled Structure			
		P	**PD**	**PI**	**PID**
	With dead time only	Unsuitable	Unsuitable	Suitable	Unsuitable
	PT1 with dead time	Unsuitable	Unsuitable	Well-suited	Well-suited
	PT2 with dead time	Unsuitable	Suited conditionally	Well-suited	Well-suited
	Higher order	Unsuitable	Unsuitable	Suited conditionally	Well-suited
	Not self-regulating	Well-suited	Well-suited	Well-suited	Well-suited

TABLE 8.4   Controlled Systems Structure Selection

Physical Quantity	Controller Structure			
	P	**PD**	**PI**	**PID**
	Sustained Control Deviation		**No Sustained Control Deviation**	
Temperature	For low performance requirements and proportional action controlled systems with $T_u/T_g < 0.1$	Well-suited	The most suitable controller structures for high-performance requirements (except for specially adapted special controllers)	
Pressure	Suitable, if the delay time is inconsiderable	Unsuitable	The most suitable controller structures for high-performance requirements (except for specially adapted special controllers)	
Flow rate	Unsuitable, because required GAIN range is usually too large	Unsuitable	Suitable, but integral action controller alone often better	Hardly required

TABLE 8.5 *Physical Quantity Control Structure Choices*

In the case of PI and PID controllers, which represent the most common types of controllers, controlled variable oscillations occur. Experience shows that extensive tuning of the initial parameters is always necessary. Level and temperature control inside very large container or a set of cascaded containers are examples of systems with dead time. Dead time represents the time that has to expire before a change can be measured at the system output. In systems with dead time, changes in the controlled variable due to the controller action are delayed by the amount of the dead time. A system with $T_u/T_g > 0.3$ is typically difficult to control. Several technical sources provide a best starting guess for the PID parameters based on known process models. Most PID parameters values recommendations are based on a 1/4 decay controlled-variable behavior, which means that consecutive over shoot oscillations maintain this decay ratio. Additional tuning of selected controller parameters must be performed during system checkout. Other tuning techniques, including experimental ones, are available for different and commonly desired process control behavior. Knowledge of the process transient behavior is essential to successful controller implementation. The tuning task is a team effort requiring input and contributions from the process domain experts.

8.3 PID Instruction

PID closed-loop control is a common method of process control in most industrial and commercial applications. This section covers the use of PID instruction blocks under the SLC-500 PLC system. Similar structures exist in most other PLC brands with slight changes in format but same basic operation.

Figure 8.21 shows the details of the PID instruction block. The PID block includes three parameter sets: one for the input process variable, one for the output control variable, and a 23-word control block. The PID instruction is normally placed on the rung as an output block without any conditional logic. During the programming/configuration of the PID block you must enter locations for the Control Block, Process Variable, and

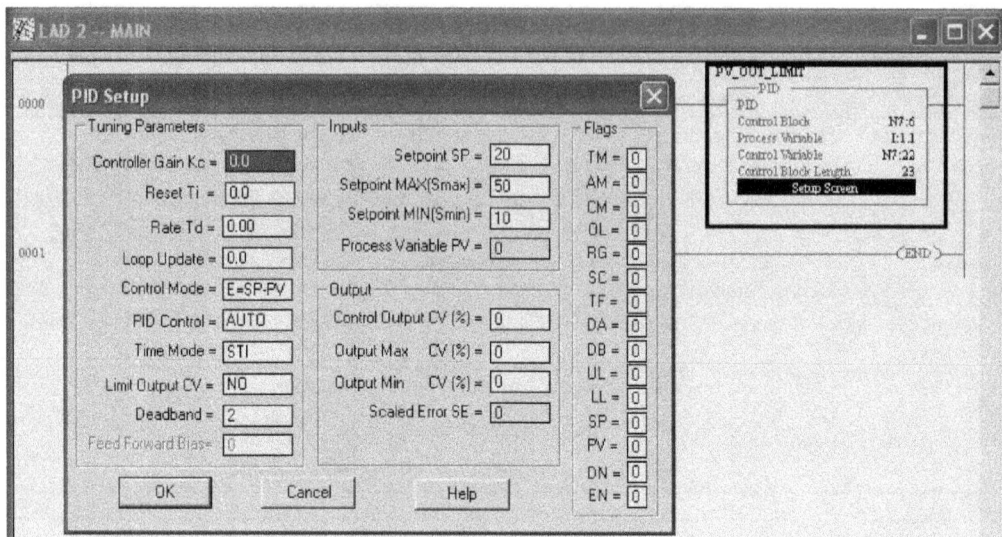

FIGURE 8.21 PID instruction block. (*Courtesy of the Allen Bradley RSLogix 500.*)

Control Variable immediately after the PID instruction is entered into the ladder logic.

The Process Variable (PV) is the address that stores the input variable. This address can be the location of the analog input word, where the value of the input address is stored. You can enter this value as an integer after scaling the input value in the range of 0 to 16,383.

The Control Variable (CV) stores the output of the PID instruction. Like the Process Variable, the CV's range is from 0 to 16,383. The upper range (16,383) is noted as the 100% or the "ON" position. You can scale the PID output range to fit your PID application specifications.

The Control Block is a 23-word block used by the PID instruction and configured by the user. It is typically assigned to contiguous words in the N7 file. Details are given in the next section.

8.3.1 SLC-500 PID Control Block

The Control Block file is the core of the PID programming because it stores the data required to perform its function. The file length is fixed at 23 words and it is entered as an integer file. For example, a Control Block entry of N7:6 will occupy elements N7:6 through N7:28. It is a good idea to use a unique data file to hold your PID control blocks, for example, the N10 file. Figure 8.22 shows the Control Block layout and associated assignments. The following is a brief description of the control block elements (words and word 0 bits); refer to the RSLogix 500 technical manual for more details:

Mode TM (word 0, bit 0): This function toggles the values between Timed and Selectable Timed Interrupts (STI) mode. Timed mode indicates that the PID executes its equation and updates its output by a specified value in the user specified "Loop Update" parameter (Word 13). In the STI mode, the PID instruction

	15	14	13	12	11	10	09	08	07	06	05	04	03	02	01	00
Word 0	EN		DN	PV	SP	LL	UL	DB	DA[1]	TF	SC	RG[1]	OL[2]	CM[2]	AM[2]	TM[2]
Word 1	PID sub error code (MSbyte)															
Word 2	Setpoint SP															
Word 3	Gain K_c															
Word 4	Reset T_i															
Word 5	Rate T_d															
Word 6	Feed forward/bias															
Word 7	Setpoint max (SMax)															
Word 8	Setpoint min (SMin)															
Word 9	Deadband															
Word 10	Internal use do not change															
Word 11	Output max															
Word 12	Output min															
Word 13	Loop update															
Word 14	Scaled process variable															
Word 15	Scaled error SE															
Word 16	Output CV% (0 to 100%)															
Word 17	MSW integral sum															
Word 18	LSW integral sum															
Word 19	Internal use do not change															
Word 20	Internal use do not change															
Word 21	Internal use do not change															
Word 22	Internal use do not change															

(1) Applies to the SLC 5/03 and higher processors, (2) You may alter the state of these values with your program

FIGURE 8.22 SLC500 PID control block structure. (*Courtesy of the Allen Bradley LogixPro 500.*)

should be placed in an STI interrupt subroutine. It then updates its output every time the STI subroutine is scanned. The STI time interval and the PID loop update rate must be the same for the equation to execute properly. Status register S:30 is used in conjunction with this mode timing.

Auto/Manual AM (word 0, bit 1): The **AM** function toggles between Auto and Manual mode. When the PID is in Auto mode the PID is controlling the output. When the PID is in Manual mode the user is controlling it. It is recommended that you tune your setup in manual mode. Once things are running smoothly you can switch to Auto mode.

Control CM (word 0, bit 2): This function toggles between the E = SP-PV (direct acting control) and E = PV-SP (reverse-acting control). In reverse acting control, the CV increases when the input Process Variable (PV) is smaller than the set point (SP). Direct acting is the opposite.

Output Limiting Enabled Bit OL (word 0, bit 03): This flag is set when you choose to limit the Control Variable (CV) parameter. This bit can be set or cleared in your ladder logic program.

SLC 5/03 and SLC 5/04 specific—Reset and Gain Range Enhancement Bit RG (word 0, bit 4): When this bit is set, it causes the Reset Minute/Repeat Value and the gain multiplier to be enhanced by a factor of 10 (Reset multiplier of 0.01 and Gain multiplier of 0.01).

Scale Set Point Flag SC (word 0, bit 05): This flag is cleared when scaling values are specified for the set point.

Loop Update Time Too Fast TF (word 0, bit 06): This bit is set in the PID algorithm if the Loop Update value you entered cannot be performed by the program (because of scan time limitations). If the bit is set, try to correct the problem by changing your Loop Update value to a slower value or move the PID instruction to an STI routine. The Reset and Rate Gain parameters will be in error if this bit is set.

Derivative (Rate) Action Bit DA (word 0, bit 07): When this bit is set, the derivative (rate) calculation is evaluated on the error instead of the PV. When this bit is cleared, the derivative calculation is performed on PV. This bit is only used by the SLC 5/03 and SLC 5/04 processors.

DB, Set When Error Is in DB (word 0, bit 08): This bit is set when the PV is within the 0 crossing dead band range.

Output Alarm, Upper Limit UL (word 0, bit 09): This bit is set when the calculated control output CV exceeds the upper CV limit.

Output Alarm, Lower Limit LL (word 0, bit 10): This flag is the same as the upper limit except the bit is set when the calculated control output CV exceeds the lower CV limit.

Set Point Out of Range SP (word 0, bit 11): The Set Point Out of Range bit is set when the SP exceeds the maximum scaled value or is less than the minimum scaled value.

Process Variable Out of Range PV (word 0, bit 12): This bit is set when the unscaled PV exceeds 16,383 or is less than zero.

PID Done DN (word 0, bit 13): This is set on scans when the PID algorithm is computed. It is computed at the Loop Update rate.

PID Enabled EN (word 0, bit 15): This bit is set when the rung that contains the PID instructions is enabled.

Set point SP (word 2): The set point is the control point of the process variable. This value can easily be changed in the ladder logic. While using PID control you will probably be adjusting this value often. Write the value in the third word in the control block (e.g., if your control block is N10:0, store the set point in N10:2). Without scaling, the range of this value is from 0 to 16,383. Otherwise the range is min scaled (word 8) to max scaled (word 7).

Gain K_c (word 3): This is the proportional gain. It ranges from 0.1 to 25.5 on the SLC 5/00 and 5/01s and 0 to 32,767 for the SLC 5/03s and 5/04s. A rule of thumb is to set the gain to one-half the value needed to cause the system to oscillate when the reset and rate terms are set to zero.

Reset T_i (word 4): This is the integral gain. The range for this gain ranges from 0.1 to 25.5 minutes per repeat. However, as in the proportional gain, the SLC 5/03s and 5/04s range from 0 to 32,767 minutes per repeat. A rule of thumb is to set the reset time equal to the natural period found in gain K_c.

Rate T_d (word 5): This is the derivative term. The range for this value is from 0.01 to 2.55 minutes. For the SLC 5/03s and 5/04s the range is from 0 to 32,767. It is recommended that this value be set to 1/8 of the integral time.

Feed Forward/Bias (word 6): Applications involving transport lags may require that a bias be added to the CV output in anticipation of a disturbance. This bias can be accomplished using the processor by writing a value to the Feed Forward/Bias element, the seventh element (word 6) in the control block file. The value you write is added to the output, allowing a feed forward action to take place. You may add a bias by writing a value between −16,383 and +16,383 to word 6 with your programming terminal or ladder program.

Maximum Scaled S_{max} (word 7): This parameter recognizes the set point value when it is in engineering units. The control input must be 16,383 (or 32,768 for the 5/03 and 5/04). The valid range is −16,383 to +16,383. The SLC 5/03s and 5/04s range is from −32,768 to +32,767.

Minimum Scaled S_{min} (word 8): This parameter is exactly like the S_{max} except the control input must be zero.

- Applies to the SLC 5/03 and higher processors.
- You may alter the state of these values with your program.

Dead Band DB (word 9): This parameter is always a positive value. The dead band's value ranges from above the set point value to below the set point value in which you enter. The dead band value is entered at the zero crossing point of the process variable (PV) and the set point. This implies that the dead band is only effective once the PV enters the dead band and has passed through the set point. The range of the dead band if it is scaled max (S_{max}) is 0, and if there is no scaling the range is from 0 to 16,383.

Loop Update (word 13): This is the time interval between PID calculations. The value is in 0.01 second intervals. A rule of thumb is to set the loop update time 5 to 10 times faster than the natural period of the load (this is determined by setting the reset and rate parameters to zero and then adjusting the gain until the output begins to oscillate). When in the STI mode, this value must equal the STI time interval value S:30. The valid range for this parameter is 1 to 2.55 seconds. *For the SLC 5/03 and 5/04, the range is from 0.01 to 10.24 seconds.*

Scaled Process PV (word 14): This parameter is for display only. This is the scaled value of the PV (the analog input). Without scaling the range of this parameter is 0 to 16,383. Otherwise, the range is minimum scaled (word 8) to maximum scaled (word 7).

Scaled Error (word 15): This parameter is also for display only. This is the scaled error, which is selected by the control mode parameter. The range is scaled maximum to scaled maximum, or 16,383 to −16,383 when no scaling exists.

Output CV% (word 16): This parameter displays the 0 to 16,383 CV output in terms of percentage (0 to 100%). If the Auto mode is picked with the F1 function key, this is for display purposes only. If you pick manual mode and you are using an APS data monitor, you can change the output percentage of the CV by adjusting it on the APS monitor. Writing to the CV using your program or nonintelligent programming device will not affect the CV. If you are using a non-APS device, you must write directly to the CV, which ranges from 0 to 16,383.

Integral Sum (word 17 and 18): A two-word location used to store the value of the calculated integral sum.

8.3.2 SLC-500 Tank Level PID Control

This section details the implementation of a PID control of tank level process using the Allen-Bradley RSLogix software. Description of all the steps needed to program and implement the desired tank level regulation is included along with the associated commands and the ladder software screens. PID control is a common and popular type of control used in process control and automation today. It is used as an output ladder instruction, which contain physical properties values such as temperature, pressure, liquid level, or flow rate. PID instructions are typically used to control closed-loop processes utilizing input and output signals interfaced to an appropriately configured analog modules. The analog input provides current measurements from the sensors for the physical entity we desire to regulate while the analog output is used to drive the final control actuator to bring about the needed regulations.

The PID block controls the level process by sending the correct output signal to the control valve stepper motor to force the desired change in tank level. The greater the error between the set point and the level process variable, the larger the output signal (larger valve opening), which is known as *direct-acting controller*. The PID algorithm or the controller output is:

$$Output = K_C \left[(E) + \frac{1}{Tt_I} \int (E)\, dt + T_D \cdot \frac{d(E)}{dt} \right] + \text{Feed Forward/Bias}$$

where K_C = proportional gain, a tuning parameter

$1/T_i$ = reset time, a tuning parameter

T_D = rate term, a tuning parameter

E = error; SP – PV

t = time or variable of integration; takes on values from time 0 to the present time t

Feed forward/Bias = controller output at zero error

An example of level PID control is shown in Fig. 8.23. The PID closed-loop control holds the level process variable close to a chosen set point. The process variable measurement is provided by a level sensor after sensory validation and fusion if redundant sensors are used. The controller output derives the tank outlet control valve. Figure 8.24 shows the closed-loop PID controller schematic.

A sensor measures the tank level and converts it into a voltage signal 0 to 10 V; 0 V corresponds to the level when the tank is empty (0 L) whereas 10 V is the level indication for a completely filled tank (1000 L). The sensor is connected to the second analog input channel of the SLC 5/03 PLC trainer, I:1.1. An SS1 selector switch (I:2/0) is used to initiate a sudden set point change in the tank level. This level is to be controlled alternatively for 100 L level SP (SS1 = 0) or 700 L level SP (SS1 = 1). The PID CV value needs

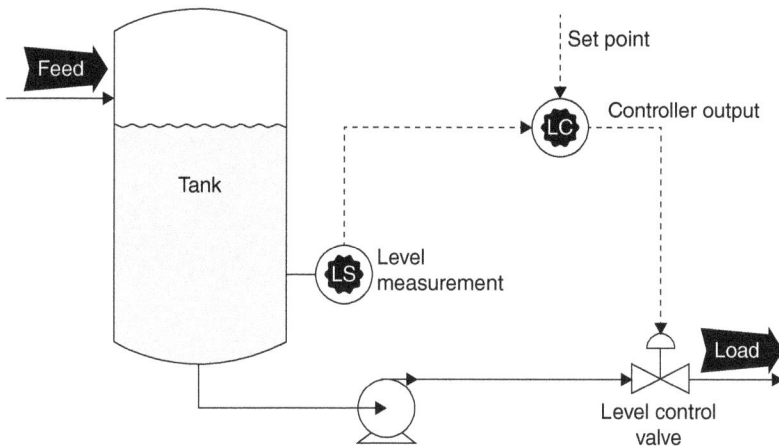

FIGURE 8.23   Level control PID closed-loop control.

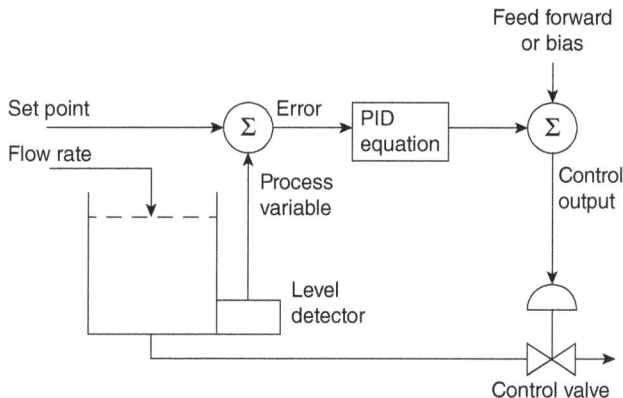

FIGURE 8.24   Level control PID closed-loop control schematic.

to be converted from the internal percentage value to the analog output count (0 to 16,383), which is used to drive the tank outlet valve. Figure 8.25 lists the ladder implementation for the PID tank level control. PI mode is used with auto update and direct-acting control setting in this implementation. Only the PID-related section of the code is shown, control block words must be initialized before program execution. PID tuning is briefly discussed in the next section.

PID closed control loop's implementation requires expert knowledge about the process operation and time behavior under varying disturbances and load changes. As with other automation tasks, it does demand collaboration and team work among project team members with different expert domain knowledge. Often the person implementing the PLC software lacks expertise with some aspects of the process behavior, as in electrical, chemical, mechanical, environmental, nuclear, manufacturing, and many others. Team work and effective project management are the key elements for the success of large process control and industrial automation projects.

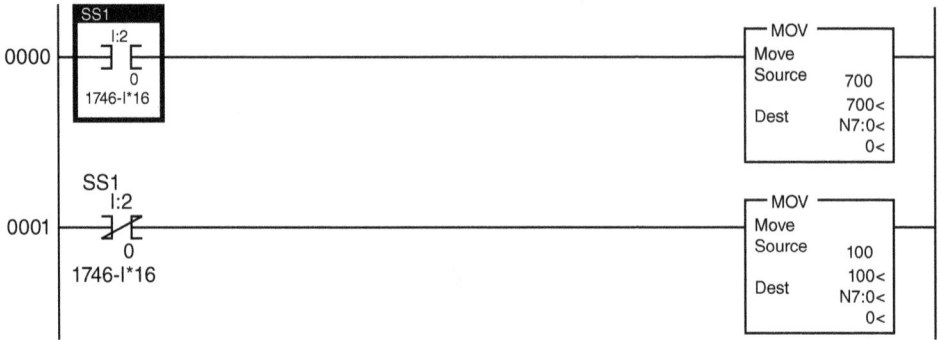

(a)

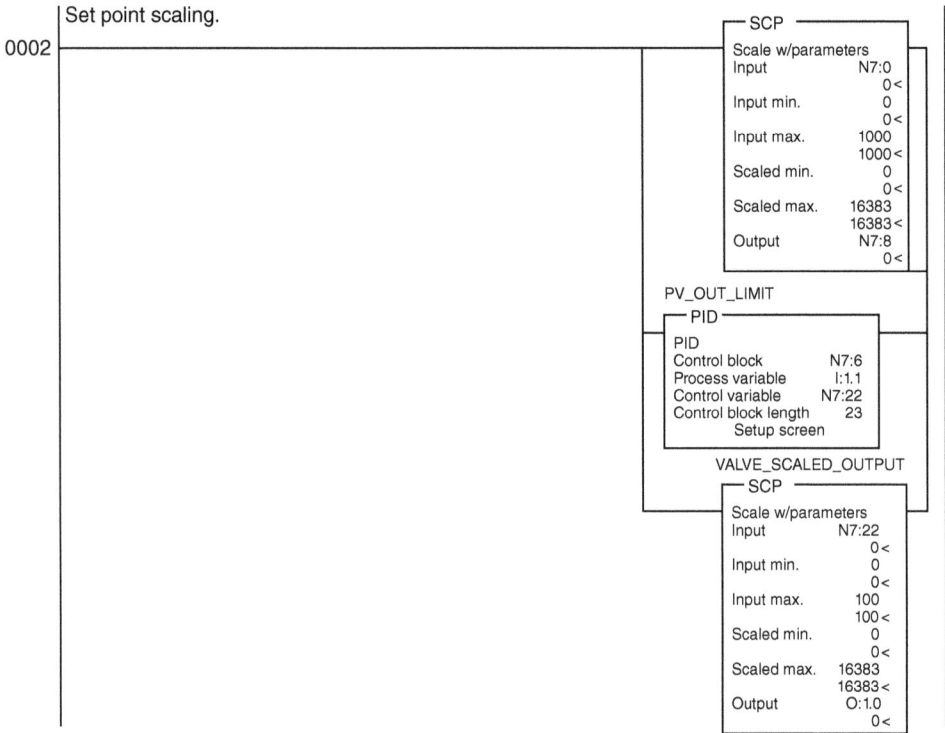

(b)

Figure 8.25   (a) Level control PID closed-loop control ladder program; (b) level control PID closed-loop control ladder program.

8.4   PID Tuning

Process control refers to the methods that are used to control the selected process variables when producing an end product. For example, factors such as the temperature of the materials, the pressure under which the materials are held, the proportion of one ingredient to another, the speed at which the ingredients are mixed, and for how long can significantly impact the quality of an end product. Control/automation of the production process is easily justified for the following key three reasons:

1. Increase efficiency by maintaining the process at a specific operating/reaction points. For example, control points might be the temperature and pressure at which a chemical reaction takes place. Accurate control of temperature and pressure in this case ensures process efficiency and save money required to produce the end product.

2. Reduce variability, which ensures a consistently high-quality product and eliminates the need for product padding to meet the required specifications.

3. Ensure safety; failing to control all of the relevant variables can produce a runaway process, which can be catastrophic.

PID closed-loop feedback controllers are widely used in industrial control systems. A PID controller calculates an error value as the difference between a set point and a measured process variable. The controller acts to minimize the error by changing the process through the regulation of the selected manipulated variable. The PID controller algorithm involves three constant parameters; the proportional (P), the integral (I), and the derivative (D) values. P is proportional to the *present* error, I is proportional to the accumulation of *past* errors, and D is proportional to the *current* rate of error change. The weighted sum of these three actions is used to adjust the process through a control element such as the setting of a control valve, the position of a damper, or the frequency supplied to an induction motor.

Many PID loops are configured to control a process variable, such as flow, through the regulation of a mechanical device, such as a valve. Mechanical maintenance can be a major cost and wear leads to control degradation in the form of either sticking of contacts or a delay in the mechanical response to an input signal. The rate of mechanical wear is mainly a function of how often a device is activated to make a change. Where wear is a significant concern, the PID loop must have an output dead band to reduce the frequency of activation of the output (change in valve position). This is accomplished by modifying the controller to hold its output steady if the change in the process variable and the corresponding error is within the defined dead band range. The process variable measurement must leave the dead band before the actual controller output will change.

Most control applications require the use of only one or two of the available PID control actions to provide the appropriate system behavior, which can be achieved by setting the unused parameters to zero. The most common PID controllers do not use the derivative term, since derivative action is sensitive to measurement noise, and thus are called *PI controllers*. The absence of the integral term will produce large residual errors at different operating points, which might be unacceptable in most applications. Most industrial PID controllers are implemented in PLCs or as a panel-mounted digital controller. Digital/PLC implementations have the advantages that they are relatively cheap and are flexible with respect to the software implementation of the PID algorithm.

8.4.1 PID Closed-Loop Tuning

The process of setting the PID parameters is known as the closed-loop tuning. This process includes the selection and adjustment of up to three control parameters: the proportional gain, the integral gain, and the derivative gain for the desired process control

Tuning Method	Advantages	Disadvantages
Manual tuning	Online technique, which does not require math or process modelling.	Requires a process domain expert and interdisciplinary collaboration.
Ziegler-Nichols	Online technique, which is a proven way for obtaining first good parameters estimate.	An aggressive tuning exposing process to excessive upsets and requires multiple trials.
Cohen-Coon	Offline technique based on an established and well-defined process model.	Works good only for first-order processes and requires math.
Software tools	Allows simulation before downloading, can be used offline or online, and produces consistent tuning.	Software and training cost. Also requires process modelling and math knowledge.

TABLE 8.6   Common PID Tuning Methods

response. Control systems have varying behavior and individual applications have different requirements. PID tuning is a difficult problem requiring excellent knowledge about the process dynamics/behavior, as each of the three parameters can be assigned one of infinite possible values. There are accordingly various methods for PID loop tuning; this section describes few commonly used traditional manual methods.

Designing and tuning a PID controller can be viewed as conceptually intuitive and simple process, but can be hard in practice, if multiple and often conflicting objectives such as short time duration transient and high stability are to be achieved. System performance can generally be improved by careful tuning the implemented PIDs. For most processes having a high degree of nonlinearity; PID parameters that work well at full-load conditions might not work during the process ramping up or down, which can be corrected by using different parameters for different load/operating conditions. The PID-controlled process can become unstable with incorrectly chosen parameters. Instability can be caused by excess gain, particularly in the presence of significant lag and dead time in the controlled process.

The choice of a tuning method depends mainly on two factors: the response time of the system and whether or not the loop can be taken offline for tuning. If the system can be taken offline, the best tuning method often involves subjecting the system to a step change in input, measuring the output as a function of time, and using the obtained response measurements to calculate the control parameters. If the system cannot be taken offline for tuning, the most effective online methods generally assumes good knowledge of a process model, then setting the PID parameters based on the dynamic model results. Manual tuning methods can be relatively inefficient if the loops have longer lag or excessive dead time. Table 8.6 summarizes the advantages/disadvantages for four commonly used methods for PID tuning.

8.4.2   Manual Tuning

If the system must remain online, one tuning method is to first set K_i $(1/T_i)$ and K_d (T_d) values to zero. Increase the K_p (K_c) until the output of the loop oscillates, then the K_p should be set to approximately half of that value for a quarter amplitude decay type response. Then increase K_i until any offset is corrected in sufficient time for the process. Notice that too much K_i may cause process instability. Finally, increase K_d, if required,

Tuning Parameter	Rise Time	Overshoot	Settling Time	Steady State Error	Stability
K_p	Decrease	Increase	Small change	Decrease	Degrade
K_i	Decrease	Increase	Increase	Eliminate	Degrade
K_d	Minor change	Decrease	Decrease	No effect	Improve if K_d small

TABLE 8.7   Effect of Tuning Parameters Increase on Response

until the loop is acceptably quick to reach its reference after a load disturbance. However, too much K_d will cause excessive response and overshoot. A fast PID loop tuning usually overshoots slightly to reach the set point more quickly; however, some systems cannot accept overshoot, in which case an over-damped, closed-loop system is required, which will require a K_p setting significantly less than half that of the K_p setting that was causing oscillation. Table 8.7 shows the effects of increasing a tuning parameter independently on the transient process behavior.

8.4.3   Ziegler-Nichols Method

The Ziegler-Nichols tuning method is a heuristic method of tuning a PID controller. It was developed by John G. Ziegler and Nathaniel B. Nichols. It is performed by setting the integral (I) and derivative (D) gains to zero. The proportional (P) gain K_p is then increased from zero until it reaches the ultimate gain K_u, at which the output of the control loop oscillates with a constant and sustained amplitude. K_u and the oscillation period T_u are used to set the P, I, and D gains depending on the type of controller used as shown in Table 8.8.

8.4.4   Cohen-Coon Tuning Method

Under Manual mode, wait until the process is at steady state. Next, introduce a step change in the process input (controller output). Based on the process output (process variable to be controlled), obtain an approximate first-order process with a time constant τ delayed by τ_{DEL} units from when the input step was introduced. The values of τ and τ_{DEL} can be obtained by first recording the following time instances as shown in Fig. 8.26:

t_0 = time when input step was initiated

t_2 = time when output is half the steady-state point occurs

t_3 = time when 63.2% of the steady-state output point occurs

Control Type	K_p	K_i	K_d
P	$0.5K_u$	–	–
PI	$0.45K_u$	$1.2K_p/T_u$	–
PD	$0.8K_u$	–	$K_pT_u/8$
Classic PID	$0.60K_u$	$2K_p/T_u$	$K_pT_u/8$
Some overshoot	$0.33K_u$	$2K_p/T_u$	$K_pT_u/3$
No overshoot	$0.2K_u$	K_p/T_u	$K_pT_u/3$

TABLE 8.8   Ziegler-Nichols Tuning Parameters

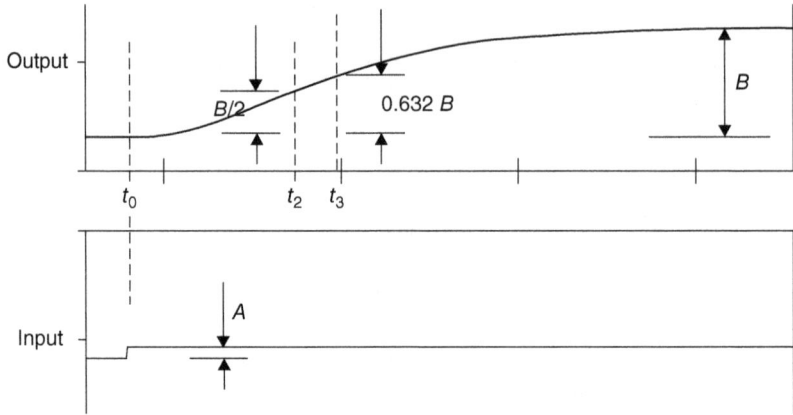

Figure 8.26 Cohen-Coon tuning measurements.

From the measurements produced by the step change test: t_0, t_2, t_3, A, and B, evaluate the following process parameters:

$$t_1 = [(t_2 - \ln(2) * t_3)/(1 - \ln(2)]$$
$$\tau = t_3 - t_1$$
$$\tau_{DEL} = t_1 - t_0$$
$$K = B/A$$

Based on the parameters K, τ and τ_{DEL}, Table 8.9 shows the formulas for calculating the PID controller parameters K_c, T_I, and T_D.

8.4.5 PID Tuning Software

Most modern industrial facilities no longer tune loops using the manual calculation methods, as in Manual and the Ziegler-Nichols tuning methods. Instead, PID tuning and loop optimization software are used to ensure consistent results. These software packages will gather the data, develop process models, and suggest optimal tuning. Some software packages can even develop tuning by gathering data from reference changes. In addition to being familiar with the software features and technical details, experience and knowledge of both the process dynamic behavior under load change or disturbances is a must. This normally requires a person with domain experience coupled with good knowledge of control systems and the associated math/modelling issues.

Control Type	K_c	T_I	T_d
P	(1/(K*r))*(1+r/3)		
PI	(1/(K*r))*(0.9+r/12)	[(30+3*r)/(9+20*r)]*τ_{DEL}	
PID	(1/(K*r))*(4/3+r/4)	[(32+6*r)/(13+8*r)]*τ_{DEL}	[4/(11+2*r)]*τ_{DEL}

Table 8.9 Cohen-Coon PID Tuning Parameters

Mathematical PID loop tuning induces an impulse in the system, and then uses the controlled system's frequency response to design the PID loop values. In loops with response times of several minutes, mathematical loop tuning is recommended, because trial and error can take days just to find a stable set of loop values. Optimal values are harder to find. Some digital loop controllers offer a self-tuning feature in which very small set point changes are sent to the process, allowing the controller itself to calculate optimal tuning values. Formulas are available to tune the loop according to different performance criteria. Many patented formulas are now embedded within PID tuning software and hardware modules. The software will model the dynamics of a process by applying a disturbance and then the calculation of the PID control parameters from the observed response.

PID controllers have limitations and can produce unacceptable performance if used alone. They also have difficulties in the presence of nonlinearity, noticeable signal noise, large lag, and extensive dead time. Interactions and interdependencies among different feedback closed loops can also be a cause of great challenge. Advanced control techniques can be used to overcome some of these challenges. This includes cascaded control loops, feed forward control, fuzzy logic control, and supervisory control. These advanced techniques are not the focus of this book but advanced readers can refer to the literature for more details.

8.4.6 Integral Windup and PI Controllers

A PI controller (proportional-integral controller) is a special case of the PID controller in which the derivative (D) of the error is not used. The controller output is given by:

$$K_p * E + K_I * \int E \cdot dt$$

where E is the error or deviation of actual measured value (PV) from the set point (SP):

$$E = SP - PV$$

The PI control is the most common type of closed-loop control. Unlike proportional control mode it can eliminate the steady-state residual errors while producing fast and acceptable response. One common problem resulting from the ideal PID implementations is integral windup. Following a large change in set point the integral term can accumulate an error larger than the maximal value for the regulation variable (windup), thus the system overshoots and continues to increase until this accumulated error is unwound. This problem can be addressed by:

- Disabling the integration until the PV has entered the controllable region.
- Preventing the integral term from accumulating above or below predetermined bounds.
- Back-calculating the integral term to constrain the regulator output within feasible bounds.

Another approach is the set point ramp up/down technique, which can be used to eliminate the integral windup effect. For example, a 100-point step change in SP can be achieved by requesting a SP of 20, 40, 60, 80, and 100. Every SP change takes place once the process variable achieves the dead band requirements for the previous SP. This approach simplifies the PID tuning procedure as it is easier to perform as a result of the smaller load/step changes in SP.

Chapter 8: Home Work Problems and Laboratory Projects

1. What are the actual decimal counts values used by the SLC-500 analog I/O modules? What does it mean in terms of analog-to-digital conversion resolution?

2. Describe one way to maximize the resolution of analog I/O signals. How does resolution affect analog signal errors?

3. Describe the function of the "scale" instruction used in analog I/O processing.

4. An analog input voltage signal in the range from 0 to 10 V is received in the standard form 0 to 27,648. The signal is connected to the PLC configured analog input module, perform the following:

    a. Show a rung to scale the input signal from (0–10 V) to a 12-bit resolution digital counts (0–4095).

    b. Show another rung to scale the signal to engineering units in the range from 50 to 350°F.

5. For problem 4, show a rung(s) to validate the input signal and force the equivalent standard digital count to be in the range 0 to 27,648.

6. Refer to problem 4, if the equivalent standard digital count is from 0 to 15,300. What is the equivalent normalize count? Assuming a normalize count of 0.6257234, what is the equivalent 12 bits resolution count?

7. Study the rung shown in the following figure and answer the following questions:

    a. What is the effect of the LIMIT instruction on the analog input count?

    b. Show another way of achieving the same LIMIT function.

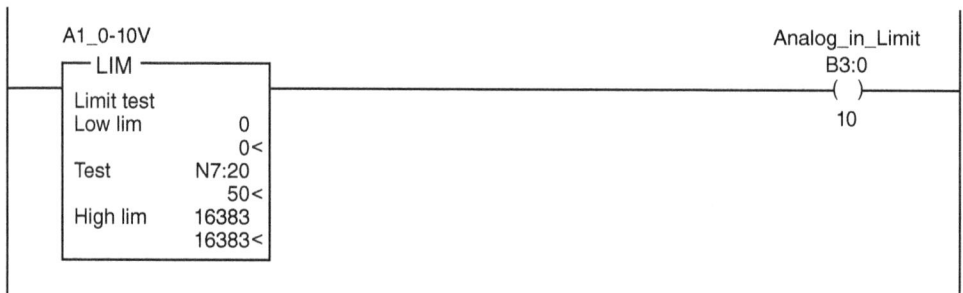

```
   A1_0-10V                                                    Analog_in_Limit
  ┌─ LIM ──────────┐                                                 B3:0
  │ Limit test     │                                                ─( )─
  │ Low lim      0  │                                                  10
  │            0<   │
  │ Test     N7:20  │
  │           50<   │
  │ High lim 16383  │
  │        16383<   │
  └────────────────┘
```

8. Define the following terms and give an example:

    a. A closed-loop control system
    b. An open-loop control system
    c. Single-variable control system
    d. Multivariable control system

9. Define the following terms and quantify their influence on control process behavior:

    a. Dead time
    b. Settling time
    c. Recovery time
    d. Delay time

10. Explain two methods used to determine the time response of a control system.

11. Draw a block diagram for a single-variable closed loop and describe the function of each block operation.

12. How is time response of a control system determined?

13. Self-regulated control systems discussed in this chapter are classified as three types as follows; briefly explain each type.

    a. Proportional-action controlled systems
    b. First-order controlled systems
    c. Second-order delay element controlled systems

14. An induction motor is a self-regulating device. The steady-state torque and speed is a linear relation in the operating range. As load increases, the motor torque increases and the speed decreases and a new steady-state operating state is reached. Draw the relation between torque and motor speed. Show how self-regulation works under a small motor load change?

15. What does non-self-regulating control system mean? Give an example.

16. Figure 8.27 shows the behavior of the controlled variable under transient condition. Answer the following:

    a. If the set point is 200°F and the dead band is ±4°F. Determine the maximum error in °F.
    b. What is the settling time? How much is it in this example?

17. Figure 8.28 diagram shows a closed-loop system for tank outlet flow control. Redraw the diagram to include supervisory control system and describe the advantages of the supervisory control.

18. Study the rungs shown in Fig. 8.29 and answer the following questions:

    a. What is the scaling equation for analog input I:1.1?
    b. Explain the function of the shown ladder program.

19. What is the concept of controllability? What is the effect of dead time on controllability? Show examples of variation of controllability.

20. What is the effect of dead time on the controller design? Which control techniques are more suitable for processes with large dead time?

Laboratory 8.1—Tank Level Sensors Measurement Processing and Monitoring

The objective of this laboratory is to provide hands-on knowledge of the analog programming used in commercial and industrial application. This lab demonstrates fundamental issues associated with analog I/O programming.

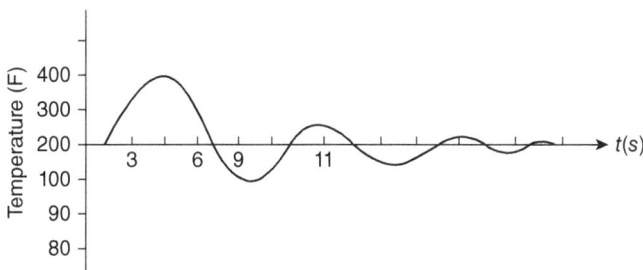

FIGURE 8.27 Problem 16 controlled variable.

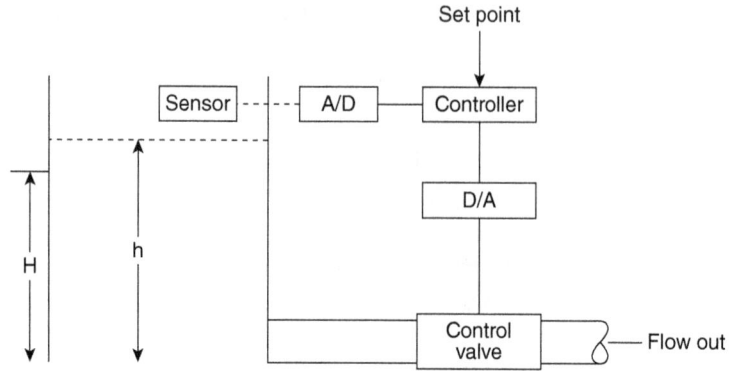

FIGURE 8.28 Problem 17 closed-loop control.

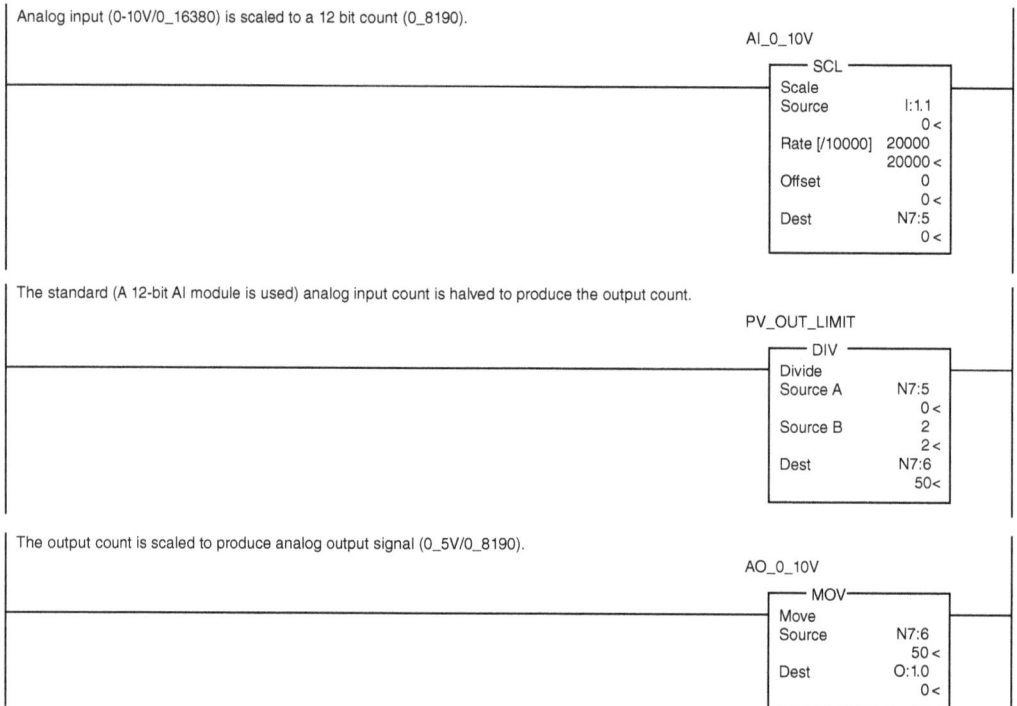

FIGURE 8.29 Problem 18 ladder rung.

Two analog signals are configured and used as tank-level measurements. Both are connected to the NIO4V composite analog module. Analog input 1 (I:1.0) measures tank level 1 and analog input 2 (I:1.1) measures tank level 2. The tank level has a range of 0 to 40 m. The two measurements are used to generate an average for the two tanks levels. The average is sent to analog output 0 (O:1.0). The corresponding scaled voltage value from a 0 to 10,000 mV is stored in an internal register (N7:11) to be displayed on the local panel meter and in engineering units on the HMI. The difference between the two analog input signals is sent to analog output 1 (O:1.1). The corresponding scaled voltage value, from 0 to 10,000 mV, is stored in an internal register (N7:12) to be displayed on the local panel meter and in engineering units on the HMI.

Specification:

1. Configure your analog input module for the two input level signals. An NOI4V composite module is assumed in our implementation but any other SLC-500 analog input module can be used. Partial ladder implementation is shown in Fig. 8.30 with adequate documentation.

2. The first two rungs in this laboratory read the two analog input voltage signals from 0 to 10 V in the standard count range from 0 to 16,383. The two counts are

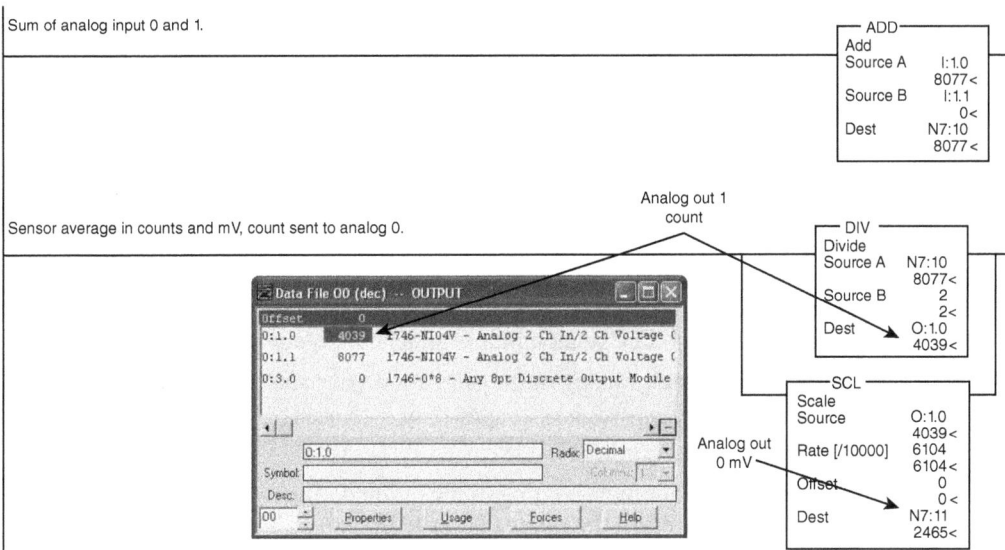

(a)

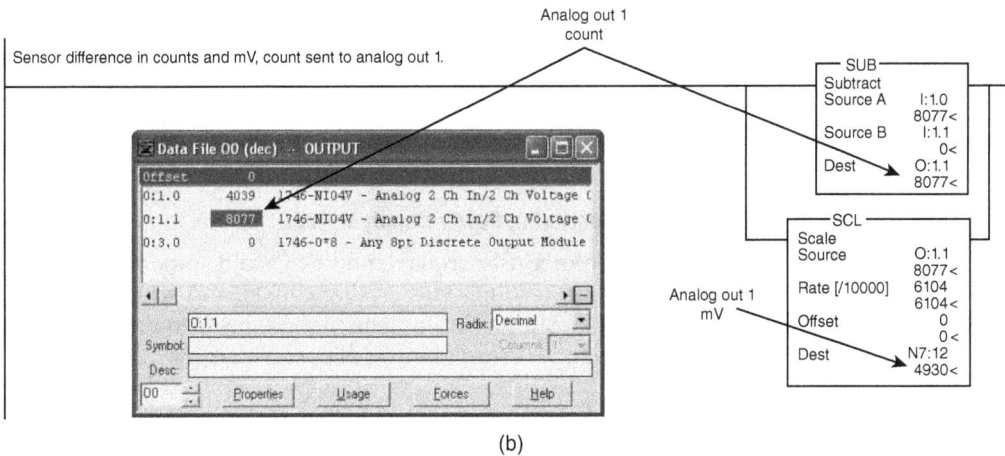

(b)

FIGURE 8.30 (a) 1 analog inputs average; (b) analog inputs difference.

averaged and sent to output 0. Next, the scaled average from 0 to 10,000 mV is stored for further display.

3. Repeat the previous step for the analog input count difference. The difference in count is sent to output 1 and the corresponding scaled mV value is stored for further use later in the program.

4. Configure the HMI screen for displaying both the average count and the difference between the two analog inputs in both count value and in mV.

5. Configure your panel or trainer to display the count and mV values.

6. Download your program to the PLC and perform the checkout.

Requirements:

- Configure the HMI or your panel to display tank level 1, tank level 2, and the average in engineering units.

- Program the laboratory rungs as specified earlier.

- Validate the analog input readings. Values above allowable range should be capped at the maximum level, while values below the range are set to zero.

- Document the program and conduct system checkout using the watch table method.

Implementation Steps

1. Configure the two PLC analog inputs attached to your analog module.

2. Configure the single analog output signal on your analog module.

3. Use two 10-V potentiometers to supply the analog input signals (0 to 10 V) to the input module.

4. Change the potentiometer setting in the defined range to represent a tank level in engineering units in the range of 0 to 40 m (0 to 10 V).

5. Configure a status page in the HMI to display the two tank levels, their average, and deviation in both counts and in meters as you change the potentiometer setting from the minimum to the maximum voltage (0 to 10 V).

6. Display the digital count in the 14 bits resolution (0 to 16,383) on the HMI status page.

7. Monitor the voltmeter readings and verify the analog signals scaling and measured value.

Laboratory 8.2—Validating and Monitoring Power Supply Voltage

This laboratory's objective is to enforce user understanding of analog modules configuration, input/output signal wiring, reading/interpreting counts, and scaling to engineering units. It is designed to be simple in implementation, just requiring a small –5/+5 and –12/+12 V power supply or breadboard with power supply.

Requirements:

- Configure two channels on your analog input module and wire to from the two variable power supply terminals: the ±5 V and the ±12 V.

- Change the power supply two settings and observe the PLC count readings and the corresponding mV scaled values as shown in the ladder rungs of Fig. 8.31.

- Add the necessary logic and the corresponding rungs to cap the input signals to the ranges of –5 to 5 V and –12 to 12 V.

- Configure the HMI to provide continuous display of all input values in counts and in volts.

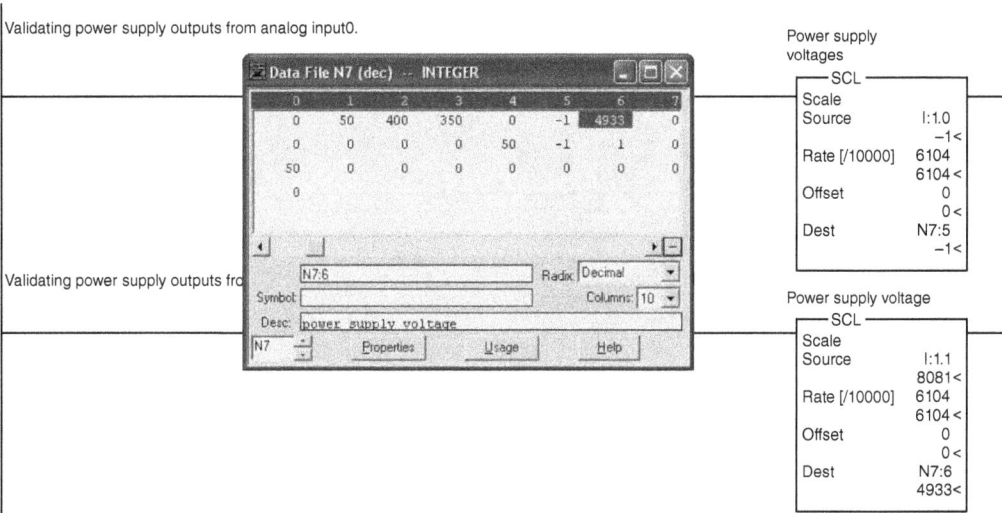

(a)

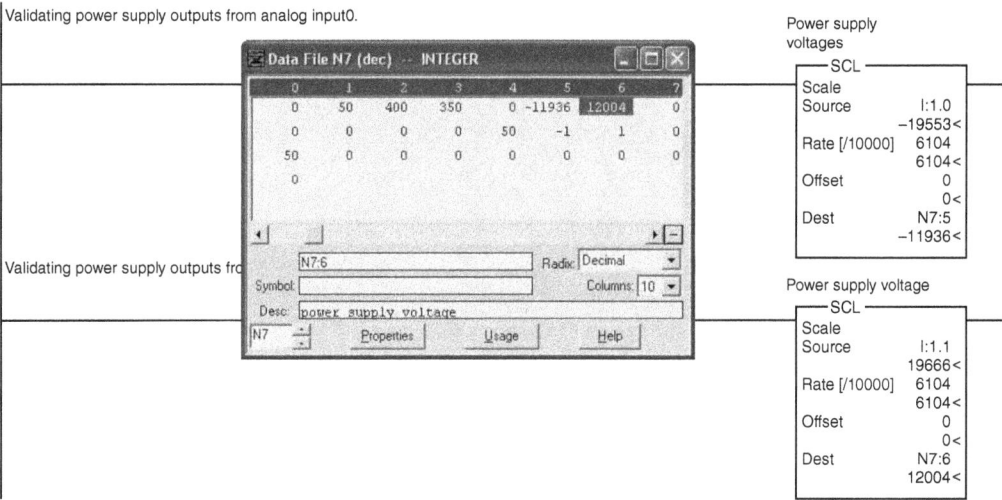

(b)

FIGURE 8.31   (a) Power supply ±5 V validation; (b) power supply ±12 V validation.

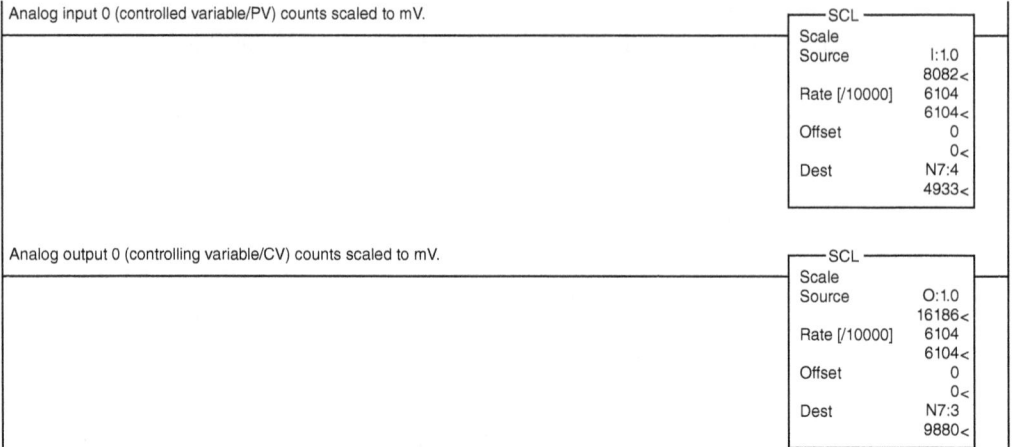

(a)

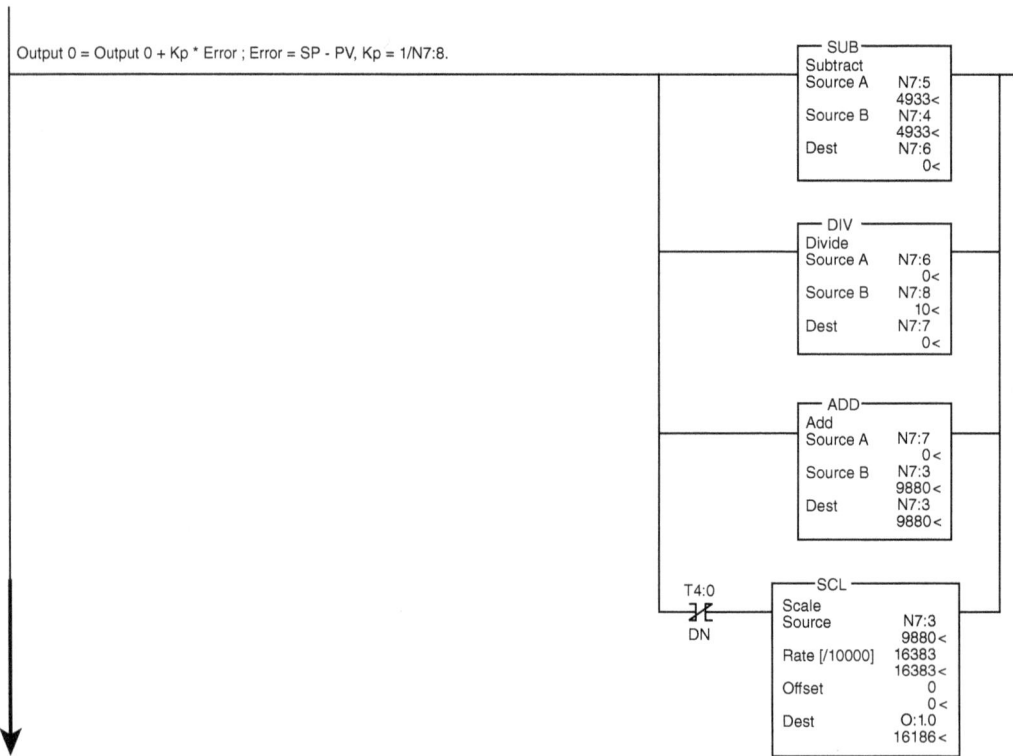

(b)

FIGURE 8.32 (*a*) PID control ladder; (*b*) PID control ladder; (*c*) PID control ladder.

Proportional controller timer (update rate of once per second).

```
  T4:0                                            ┌─ TON ─────────────┐
  ─┤/├─                                           │ Timer On Delay     ├─(EN)─
   DN                                             │ Timer        T4:0  │
                                                  │ Timer Base    1.0  ├─(DN)─
                                                  │ Preset         1<  │
                                                  │ Accum          0<  │
                                                  └────────────────────┘
```

Scale set point to mV (SP is half analog input 1).

```
                                                  ┌─ SCL ─────────────┐
                                                  │ Scale              │
                                                  │ Source       I:1.1 │
                                                  │             8081<   │
                                                  │ Rate [/10000] 6104  │
                                                  │             6104 <  │
                                                  │ Offset          0   │
                                                  │               0 <   │
                                                  │ Dest         N7:5   │
                                                  │            4933 <   │
                                                  └────────────────────┘
```

(c)

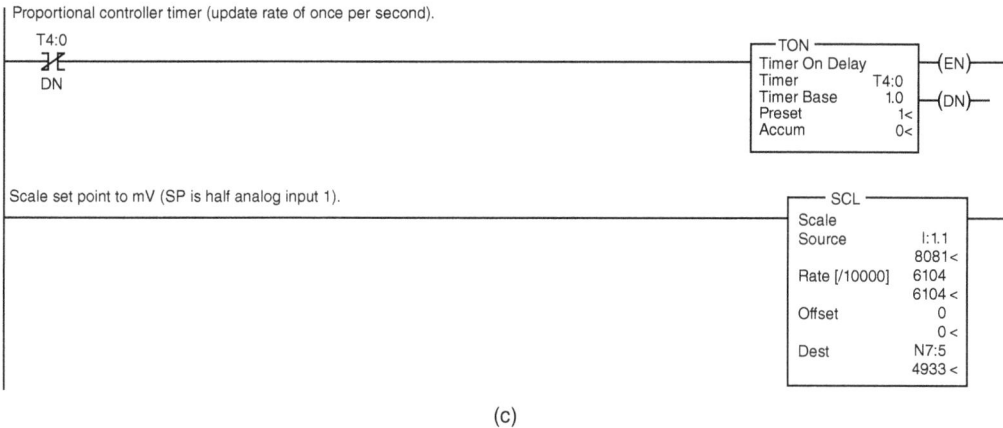

FIGURE 8.32 *(Continued)*

Laboratory 8.3—Simple Closed-Loop PID Control

This laboratory implements the operation of a PID controller in the proportional (P) mode. The PID instruction can be configured and used to implement other control modes as was shown in Chap. 8, Section 8.3. The objective is to get the user familiar with the elements and calculations involved along with the needed scaling operations between signals value, digital counts, and engineering units.

Specifications:

A sensor is used to measure a tank level and provide an input analog signal from 0 to 10,000 mV for the operation level range. The tank level (I:1.0) is to be controlled at a set point level, thus it is labeled as the "controlled variable" or as in PID notation the "Process Variable/PV." The set point (I:1.1) is provided as an analog input through an available module channel. The P controller gain inverse is provided as a constant stored in N7:8; the actual proportional gain is typically a fraction from 0 to 1. The level control will be achieved through the regulation of a valve, which receives its signal from the controller output (O:1.0). The controller output is labeled "Controlling Variable" or as in PID notation the "Control Variable / CV." All sampling and calculations are done every scan except the controller output, which in our implementation gets updated once every second.

Requirements:

- Study the ladder implementation shown in Fig. 8.32. Make sure the individual rung documentation is correct and make changes as needed.
- Configure the analog module/modules to accommodate the PV, SP, and CV PID used analog signals for the proportional control.
- Add the necessary configuration and logic to accommodate an HMI for displaying the PV and the CV. Also to facilitate the input of the SP from the remote HMI location.
- Download your program to the PLC processor and perform the checkout. Make changes in the proportional gain and observe the changes in response.
- Calculate the proportional gain using one of the online tuning methods discussed in Chap. 8, Sec. 8.4 and record your observations.

CHAPTER 9

Comprehensive Case Studies

This chapter is intended to be a cap stone project encompassing most of the concepts covered in this book. Case studies covered are part of two multi sites industrial projects implemented by the authors. The coverage is abbreviated to one site and is implemented using the AB SLC-500 PLC system.

NASA space rover lands on Mars, August 5, 2012.

Chapter Objectives

- Understand and document complete process description.
- Design the control system I/O and PLC memory map.
- Develop complete system logic diagrams from specifications.
- Implement ladder and HMI configuration, communication, programming, and checkout.

The two case studies selected for this chapter are intended to be a cap stone project encompassing most of the concepts covered in this book. The first project is part of a large, networked, multiple-site irrigation canal water control process that was implemented by the authors several years ago in the Delta region of the river Nile in Egypt. The second project deals with a common process control task in the wastewater treatment industry, which has to do with the pumping station control. The coverage in both projects is simplified to an abbreviated one site and is transformed to a newer implementation using the Allen-Bradley SLC-500 PLC system.

9.1   Irrigation Canal Water Level Control

Irrigation water is channeled through two motorized vertical gates from an upstream to a downstream agriculture area. Motorized gates, lower and upper pairs, are used to regulate the water flow and thus the downstream level. Fully closed vertical gates will produce continuous reduction in the downstream water level. Low downstream water level will not allow irrigation in the downstream area. Fully raised vertical gates will result in excessive irrigation in the downstream area and, thus, much water will be wasted. Adequate regulation of the vertical gates position can maintain desired downstream water levels at different times and thus support required irrigation cycles and preserve water resources at the same time. Therefore, coordination among all networked PLCs regulation sites is a must.

To simplify the case study, we will assume two vertical gates and only one PLC regulator site. Two identical constant speed motors derive the two vertical gates. The motors can move the gates up or down, but only one motor can run at any time. This is done to reduce the overall power requirements of the regulator site. Also, small variation in the gate's position is needed at any time due to the large downstream water level transient. The lowest position gate should be selected for a raise operation and the highest position gate must be selected for a lowering command. Both gates are equipped with position sensors, fully closed, and fully opened limit switches.

Two downstream level sensors provide redundant measurements at two points downstream locations. These sensors must be validated as they should provide similar readings consistent with previous measurements. We will assume two validated sensor input values for this project. Upstream flooding limit switch is used to provide the regulator an indication of possible flooding and thus command the system to raise both gates gradually using the same selection criteria until the flooding condition is removed or downstream level is close to the upstream level.

Each motor is equipped with an overload alarm switch, which is used to trigger any unusual conditions, such as over temperature or load. The motors provide an input discrete signal within 5 seconds from the start action, indicating if the motor is running or not. If a motor fails to start then the other motor is selected and an alarm is issued. Motors can also start by activating the push button (PB) located on the local panel if the AUTO/MAN switch is in manual (MAN) and the LOCAL/REMOTE switch is on LOCAL.

A selected motor is scheduled to run for 15 seconds. This is followed by an idle period of 10 minutes. These values are experimental values derived during the system checkout with major help from the manual system operator who was considered the domain expert for this application. No motor is allowed to run during the idle time. This is done to prevent repetitive activation of the motors during downstream water level transient. It is important to keep the gates at close positions to minimize the loading on both gates structure. An emergency shutdown switch (ESD) is available for the operator to shut down the system in addition to the START and the STOP PB switches. The ESD switch is physically wired in series with the PLC system MCR coil.

9.1.1   System I/O Map

The first step in the design of a PLC control application is the translation of the process specification to actual input/output (I/O) resources. This is known as the PLC I/O map. This important step lists all I/O tags, assigned PLC addresses, and description. Figure 9.1 lists the irrigation control system discrete inputs, whereas Fig. 9.2 shows the corresponding PLC input tags. Figures 9.3 and 9.4 repeat the same process for the discrete outputs. Notice that none of the real analog inputs/outputs for this control process is listed. We only limited our case study to ON/OFF control based on water level analog real-time measurements relative to user-defined set point for the downstream water level.

9.1.2   Logic Diagrams

Logic diagrams are recommended for sound documentation and as means of a transition to the actual ladder programming implementation. This step as demonstrated makes ladder programming much easier. The following logic diagram shows the

Tag Name	Address Number	Comments
AUTO	I:2/1	Auto/Manual Selector Switch
REMOTE	I:2/2	Remote/Local Selector Switch
VG1_ROL	I:2/3	VG1 Running On Line
VG2_ROL	I:2/4	VG2 Running On Line
VG1_LS_RAISED	I:2/5	VG1 Limit Switch fully Raised
VG2_LS_RAISED	I:2/6	VG2 Limit Switch fully Raised
ESD	I:2/7	Emergency Shutdown
VG1_LS_LOWERED	I:2/8	VG1 Limit Switch fully Lowered
VG2_LS_LOWERED	I:2/9	VG1 Limit Switch fully Lowered
_____	I:1.0	Vertical gate 1 analog position
_____	I:1.1	Vertical gate 2 analog position
_____	N7:1	Downstream Water Level 1
_____	N7:2	Downstream Water Level 2

FIGURE **9.1**   Irrigation system input.

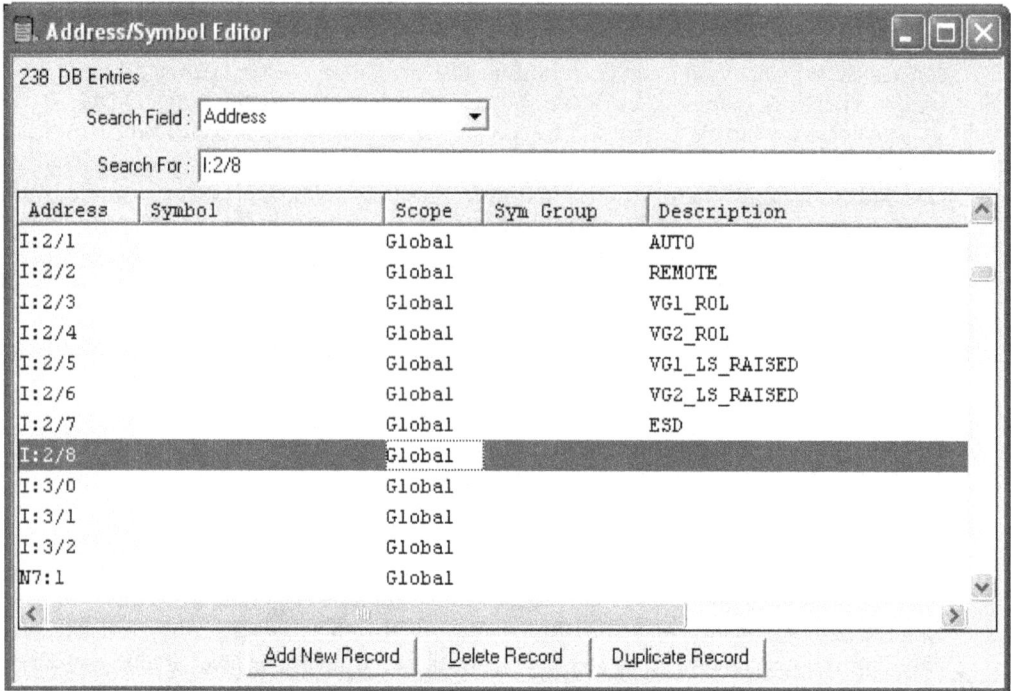

Figure 9.2 Irrigation-system PLC input tags.

Tag Name	Address Number	Comments
VG1_ Raised	0:3/0	Vertical gate1 raise one step
VG2_ Raised	0:3/1	Vertical gate2 raise one step
VG1_Lowered	0:3/2	Vertical gate1 lower one step
VG2_Lowered	0:3/3	Vertical gate2 Lower one step

Figure 9.3 Irrigation system outputs.

implementation for the vertical gate 1 raise operation. Other operations logic diagrams can be constructed in a similar manner for vertical gate 1 lowering operation and both the raise and lowering for VG2.

The logic diagram is shown in Fig. 9.5 and is segmented into three parts. Part (a) derives VG1 selection for a gate 1 "Control Up" raise operation. The "Control Up" raise command requires that the desired set point be greater than or equal to the downstream average water level, downstream average level is outside the dead band, and that VG1 is the next gate to go up. VG1 is next to go up if it is available, is not fully raised, VG2 is not next to go up, and VG1 has the lowest absolute level position. VG1 is available to go up only if the system is on Auto, VG1 motor is not running, VG1 motor did not fail to start, and the system is placed on REMOTE operation. The system is placed on LOCAL to allow motor start only from the field using the local panel START/STOP PB.

FIGURE 9.4 Irrigation-system PLC output tags.

Part (*b*) shows the logic for starting VG1 motor for a 15-second duration. This raised duration should cause increase in downstream water level, which is expected to stabilize after a 10-minute idle time. The system enters the idle state after every gate raise or lower operation. No gate movement is permitted during the idle duration. Part (*c*) details the motor-fail-to-start logic for VG1. Notice that other logic diagrams are needed to produce some of the indicated conditions in the logic diagram, as in the "VG1 LOWEST IN POSITION" shown in Fig. 9.5*a*. The logic diagrams for VG2 will produce and document some of the used conditions, as in the "VG2 NEXT UP" in Fig. 9.5*a*.

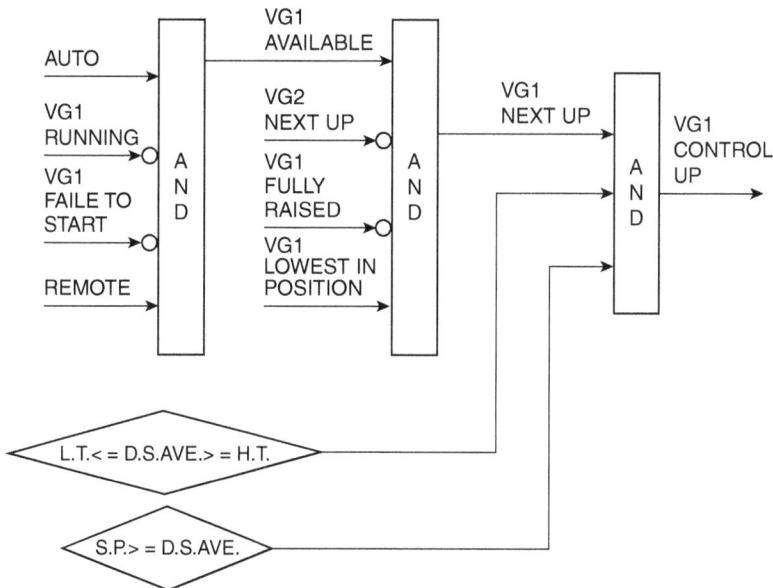

FIGURE 9.5 (*a*) Downstream level control logic diagrams; (*b*) downstream level control logic diagrams; (*c*) downstream level control logic diagrams.

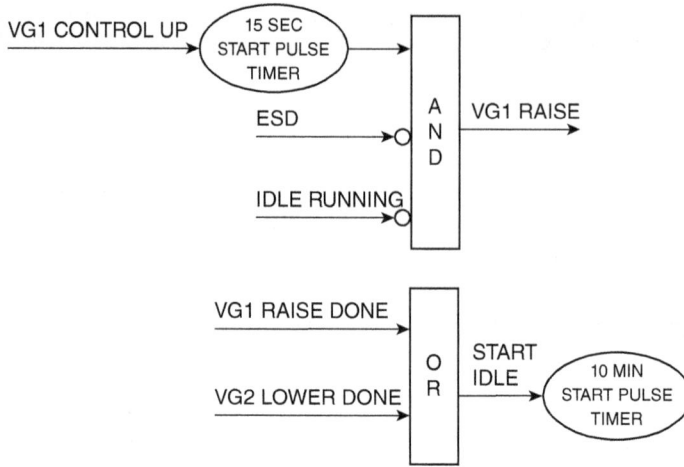

Irrigation project vertical gate up logic diagram (Part 2)

(b)

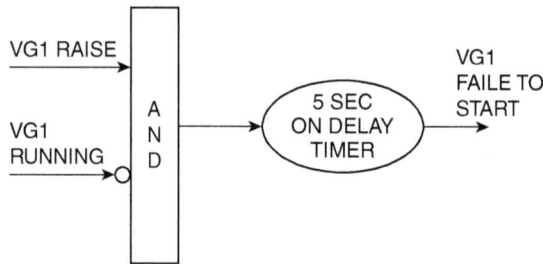

Irrigation project vertical gate up logic diagram (Part 3)

(c)

FIGURE 9.5 *(Continued)*

9.1.3 Automated System Building Blocks

The CPU supports the following types of subroutines that allow an efficient structure for the program:

- File 2 (main file) define the structure of the program.
- Subroutine (SUB3, SUB4, etc.) contain the program code that corresponds to specific tasks or combinations of parameters. Each subroutine file provides a set of input and output parameters for sharing data with the calling file.

Figure 9.6 shows the PLC subroutines (initialization, alarms, downstream average, set point validation, vertical gate 1 raise, vertical gate 2 raise, and vertical gate position). The exact ordering of the implemented subroutine is irrelevant since the scanning of the ladder code is continuous and repeats at high rate, at least three complete

(a)

Figure 9.6 (a) Irrigation system PLC subroutines (project view); (b) irrigation system PLC subroutines (ladder view).

scans per second. Critical tasks might require a definite ordering of some functional blocks. Most initialization tasks are executed only once on power up or system reset conditions. The figure shows the project view in part (a) and the ladder view in part (b).

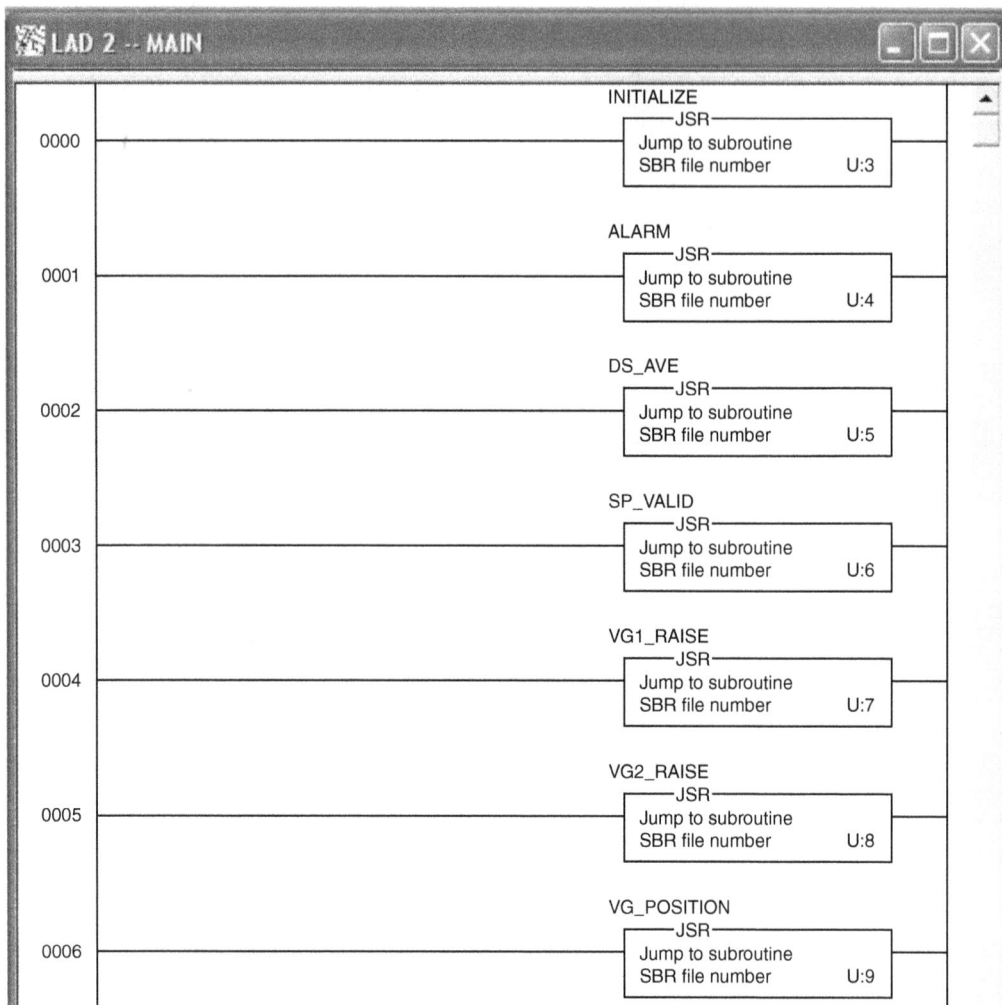

(b)

FIGURE 9.6   (Continued)

9.2   Irrigation Canal Ladder Implementation

The set point validation subroutine shown in Fig. 9.7 consists of three rungs: set point high limit/low limit validation, and dead band calculations. The following is a brief documentation of the rungs:

- Rung 1 compares the set point (N7:5) to the high limit (400 m), if greater than move the high limit into the set point.
- Rung 2 compares the set point (N7:5) to the low limit (50 m), if less than move the low limit into the set point.
- Rung 3 calculates the high/low threshold, assuming dead band (4 m).

Set point validation.

```
 ┌─────── GRT ───────┐                          ┌────── MOV ──────┐
 │ Greater than (A>B) │                          │ Move             │
 │ Source A     N7:5  │                          │ Source      400  │
 │               50 < │                          │            400 < │
 │ Source B     400   │                          │ Dest        N7:5 │
 │              400 < │                          │             50 < │
 └────────────────────┘                          └──────────────────┘
```

Set point validation.

```
 ┌─────── LES ───────┐                          ┌────── MOV ──────┐
 │ Less than (A<B)    │                          │ Move             │
 │ Source A     N7:5  │                          │ Source       50  │
 │               50 < │                          │             50 < │
 │ Source B      50   │                          │ Dest        N7:5 │
 │               50 < │                          │             50 < │
 └────────────────────┘                          └──────────────────┘
```

Calculate high threshold (N7:7) and low threshold (N7:6) around the dead band

```
                                          ┌─────── ADD ───────┐
                                          │ Add                │
                                          │ Source A     N7:5  │
                                          │               50 < │
                                          │ Source B       2   │
                                          │                2 < │
                                          │ Dest        N7:7   │
                                          │              52 <  │
                                          └────────────────────┘

                                          ┌─────── SUB ───────┐
                                          │ Subtract           │
                                          │ Source A     N7:5  │
                                          │               50 < │
                                          │ Source B       2   │
                                          │                2 < │
                                          │ Dest        N7:6   │
                                          │              48 <  │
                                          └────────────────────┘
```

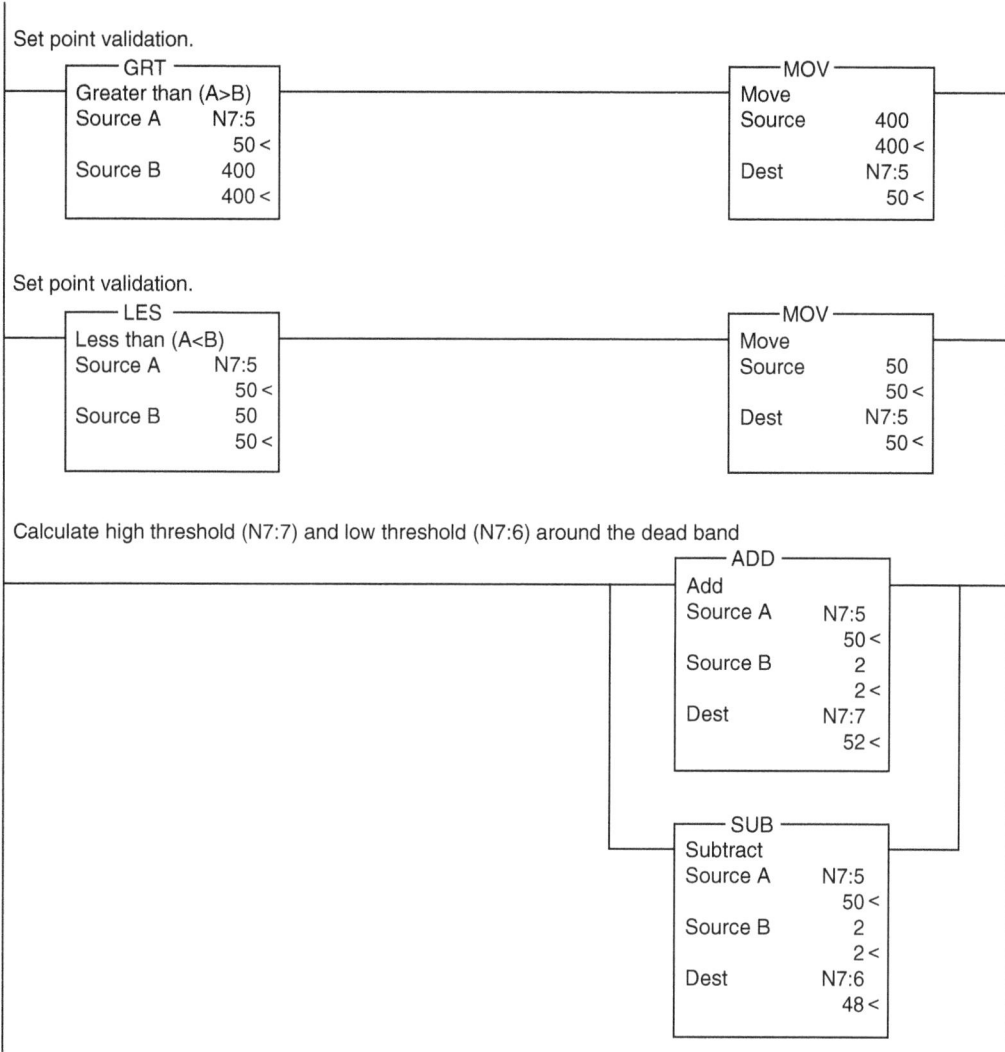

Figure 9.7   Set point validation rungs.

The initialization subroutine in Fig. 9.8 consists of one rung, which clears the accumulated values for VG1 raise one step, VG2 raise one step, VG1 fail to start, VG2 fail to start, and the dead-time timer as the system is placed in AUTO.

Alarm subroutine consists of five rungs, including Vertical Gate 1 Failed to Start, Vertical Gate 2 Failed to Start, Downstream 1 Failed, Downstream 2 Failed, and common alarm. The ladder diagram works as shown in Fig. 9.9. If VG1 or VG2 fails to start, or DS1 or DS2 fails, the common alarm will be set.

Figure 9.10 shows a ladder logic diagram for the Vertical Gate 2 Failed to Start alarm. This diagram assumes an examine-if-close I:2/1 (AUTO), O:3/1 tag name (VG2_RAISE), normally open contact I:2/3 tag name (VG1_ROL), ON-delay timer (TON)

```
First pass                                                    VG1_RAISE_TIMER
S:1                                                                  T4:0
──┤ ├──────────────────────────────────────────────────────────────( RES )──┐
  15                                                                         │
                                                              VG2_RAISE_TIMER │
                                                                    T4:1      │
                                                              ──────( RES )───┤
                                                                             │
                                                              VG1_FTS_TIMER   │
                                                                    T4:2      │
                                                              ──────( RES )───┤
                                                                             │
                                                              VG2_FTS_TIMER   │
                                                                    T4:3      │
                                                              ──────( RES )───┤
                                                                             │
                                                              DEAD_TIME_TIMER │
                                                                    T4:4      │
                                                              ──────( RES )───┘
```

FIGURE 9.8 Initialization rung.

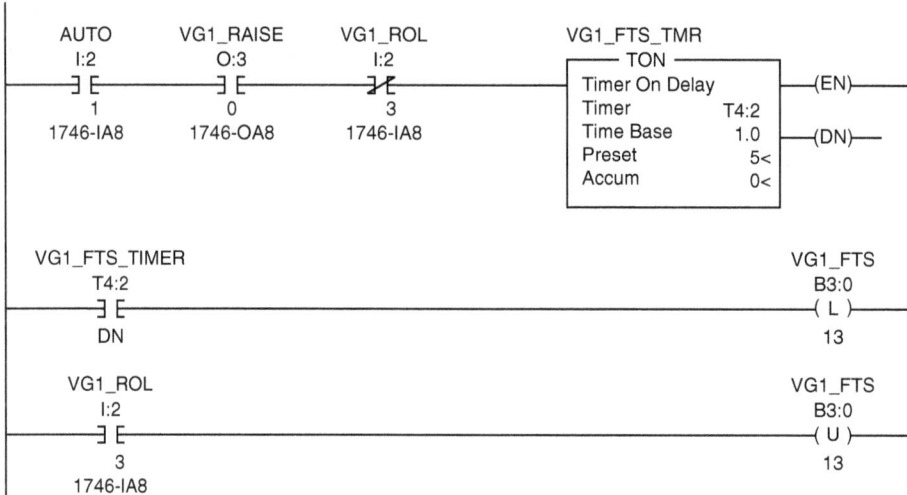

```
  AUTO          VG1_RAISE      VG1_ROL          VG1_FTS_TMR
  I:2              O:3           I:2        ┌──────── TON ────────┐
──┤ ├────────────┤ ├───────────┤/├─────────┤ Timer On Delay      ├──(EN)──
   1              0             3          │ Timer        T4:2    │
 1746-IA8      1746-OA8      1746-IA8      │ Time Base    1.0     ├──(DN)──
                                           │ Preset         5<    │
                                           │ Accum          0<    │
                                           └──────────────────────┘

  VG1_FTS_TIMER                                               VG1_FTS
     T4:2                                                      B3:0
──────┤ ├─────────────────────────────────────────────────────( L )──
      DN                                                        13

  VG1_ROL                                                     VG1_FTS
    I:2                                                        B3:0
──────┤ ├─────────────────────────────────────────────────────( U )──
      3                                                         13
   1746-IA8
```

FIGURE 9.9 Vertical Gate 1 Failed to Start rung.

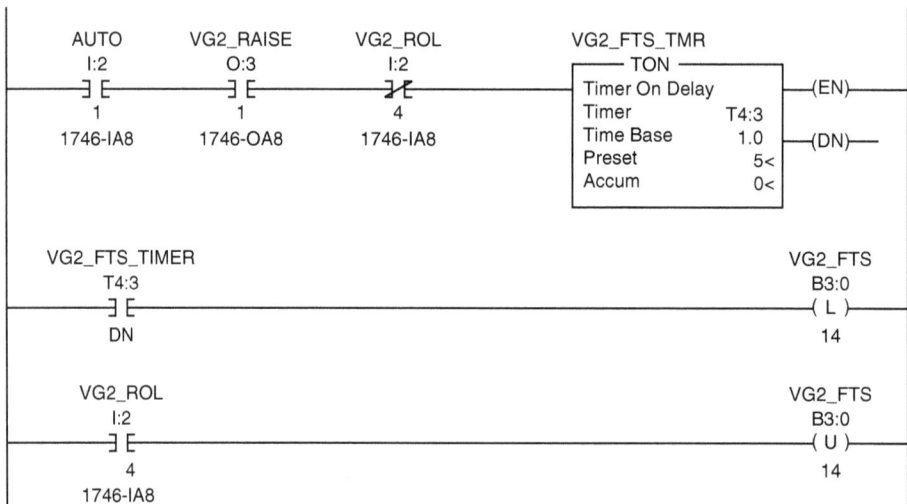

```
  AUTO          VG2_RAISE      VG2_ROL          VG2_FTS_TMR
  I:2              O:3           I:2        ┌──────── TON ────────┐
──┤ ├────────────┤ ├───────────┤/├─────────┤ Timer On Delay      ├──(EN)──
   1              1             4          │ Timer        T4:3    │
 1746-IA8      1746-OA8      1746-IA8      │ Time Base    1.0     ├──(DN)──
                                           │ Preset         5<    │
                                           │ Accum          0<    │
                                           └──────────────────────┘

  VG2_FTS_TIMER                                               VG2_FTS
     T4:3                                                      B3:0
──────┤ ├─────────────────────────────────────────────────────( L )──
      DN                                                        14

  VG2_ROL                                                     VG2_FTS
    I:2                                                        B3:0
──────┤ ├─────────────────────────────────────────────────────( U )──
      4                                                         14
   1746-IA8
```

FIGURE 9.10 Vertical Gate 2 Failed to Start rung.

336

with a 5-second preset time, and B3/13 output coil tag name (VG1_FTS). The ladder diagram works as follow:

- During the first scan as the MAN/AUTO switch is placed in AUTO, VG1 raise output is set while I:2/4 tag name (VG2_ROL) is OFF. The power flows to the timer and the timer starts timing.
- It takes up to 5 seconds to receive the associated motor running online contact from the initiation of the motor start command.
- If motor running online signal is not received within 5 seconds, the Fail-to-Start coil (VG2_FTS) is energized indicating a motor failure.

Figure 9.11 shows a ladder logic diagram for the Downstream 1 Fail alarm. This diagram assumes greater than or equal and less than or equal instructions and output coil B3/11 tag name (DS1_Fail). The ladder diagram works as follows: If downstream 1 is greater than or equal to downstream average high threshold or less than or equal to downstream average low threshold, then power flow to the output coil and B3/11 is set. This is an indication of level sensor 1 failure/alarm.

Figure 9.12 shows a ladder logic diagram for the Downstream 2 Fail alarm. This diagram assumes greater than or equal and less than or equal instructions and output coil B3/12 tag name (DS2_Fail). The ladder diagram work as follows: If downstream 2 is greater than or equal to downstream average high threshold or less than or equal to downstream average low threshold, then power flow to the output coil and B3/12 is set. This is an indication of level sensor 2 failure.

Figure 9.13 shows a ladder logic diagram for the common alarm. This diagram uses examine-if-close B3:0/13 tag name (VG1_FTS); examine-if-close contact B3:0/14 tag name (VG2_FTS), B3:0/11 tag name (DS1_FAIL), B3:0/12 tag name (DS2_FAIL), and output coil B3:0/10 tag name (COMMON_ALARM). The ladder diagram works as follows: The rung common alarm becomes true if the vertical gate 1 failed to start, or

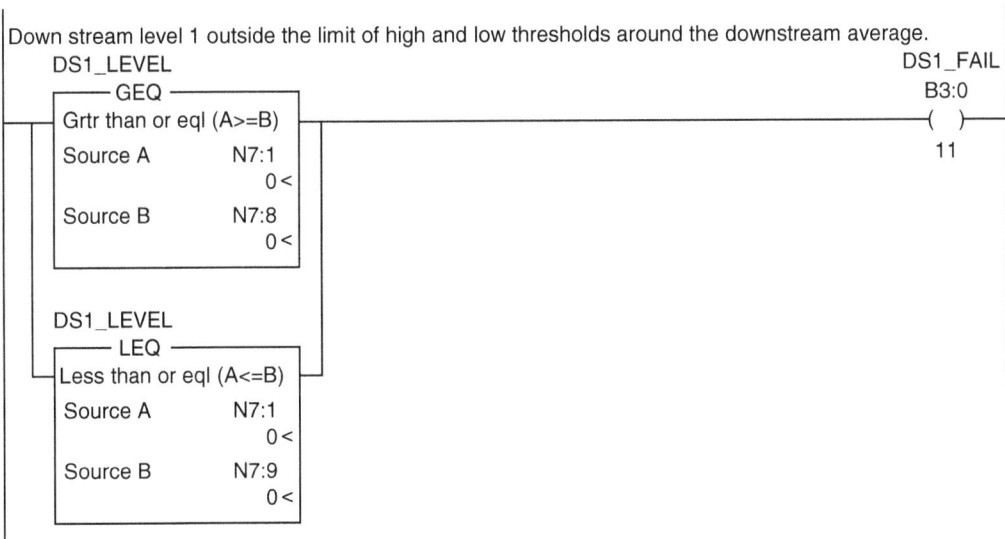

FIGURE 9.11 Downstream 1 Transmitter Fail rung.

Down stream water level 2 outside the limit of high and low thresholds around the downstream average.

```
DS2_LEVEL                                                              DS2_FAIL
    ┌──── GEQ ──────────┐                                                B3:0
 ┌──┤ Grtr than or eql (A>=B) ├────────────────────────────────────────( )──
 │  │ Source A      N7:2  │                                                12
 │  │                 0<  │
 │  │ Source B      N7:8  │
 │  │                 0<  │
 │  └───────────────────┘
 │
 │  DS2_LEVEL
 │  ┌──── LEQ ──────────┐
 └──┤ Less than or eql (A<=B) ├──┘
    │ Source A      N7:2  │
    │                 0<  │
    │ Source B      N7:9  │
    │                 0<  │
    └───────────────────┘
```

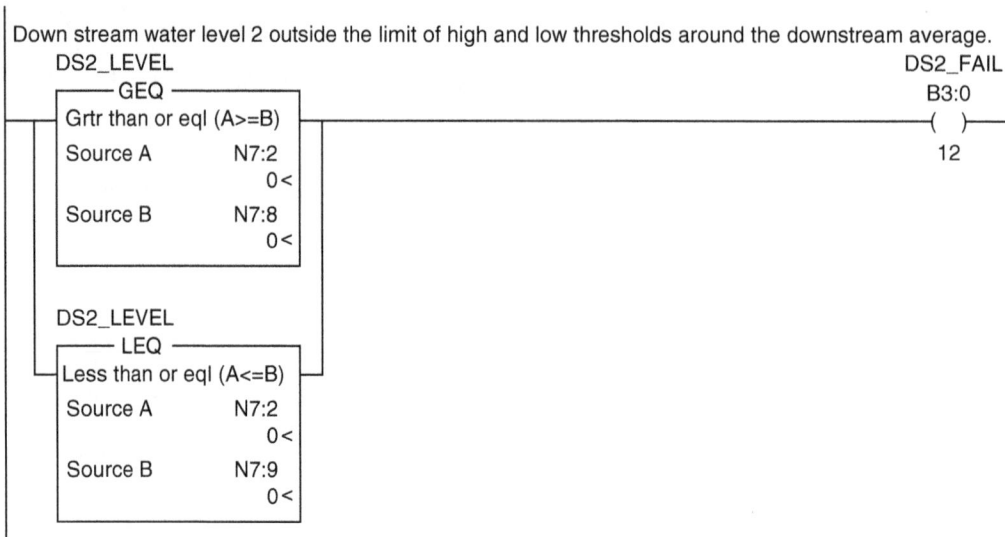

FIGURE **9.12** Downstream 2 Transmitter Fail rung.

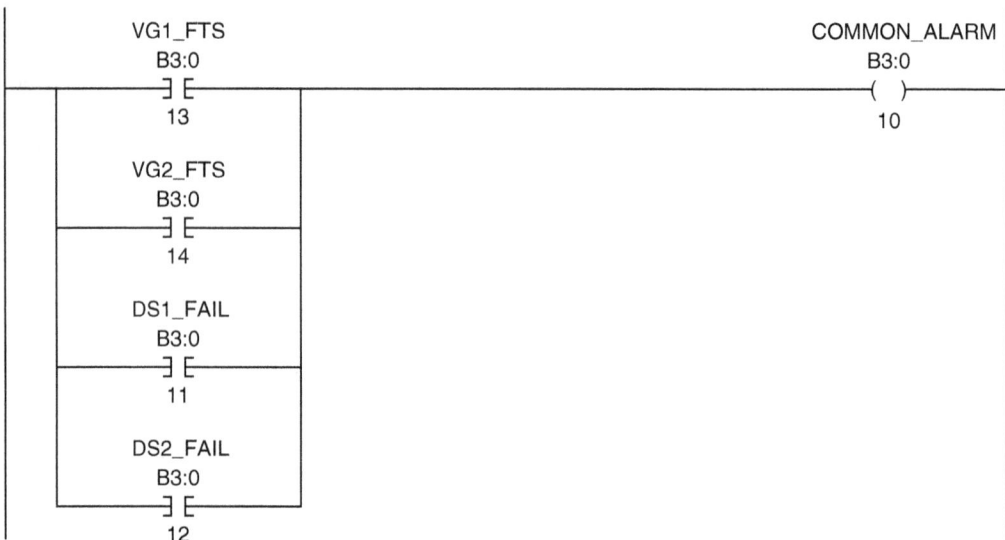

```
     VG1_FTS                                              COMMON_ALARM
      B3:0                                                     B3:0
 ┌─────┤ ├──────────────────────────────────────────────────( )──
 │       13                                                    10
 │    VG2_FTS
 ├─────┤ ├──
 │       14
 │    DS1_FAIL
 ├─────┤ ├──
 │       11
 │    DS2_FAIL
 └─────┤ ├──
         12
```

FIGURE **9.13** Common alarm rung.

vertical gate 2 failed to start, downstream 1 fail, or downstream 2 fail. The power flows to the output coil common alarm.

Downstream average subroutine consists of three rungs: DS average, downstream 1 level sensor validation, and downstream 2 level sensor validation. Figure 9.14 shows a ladder logic diagram for the downstream average. This diagram assumes examine-if-open contact B3:0/13 tag name (VG1_FTS); examine-if-open contact B3:0/14 tag name (VG2_FTS), examine-if-open contact B3:0/11 tag name (DS1_FAIL), examine-if-open

Calculate down stream average.

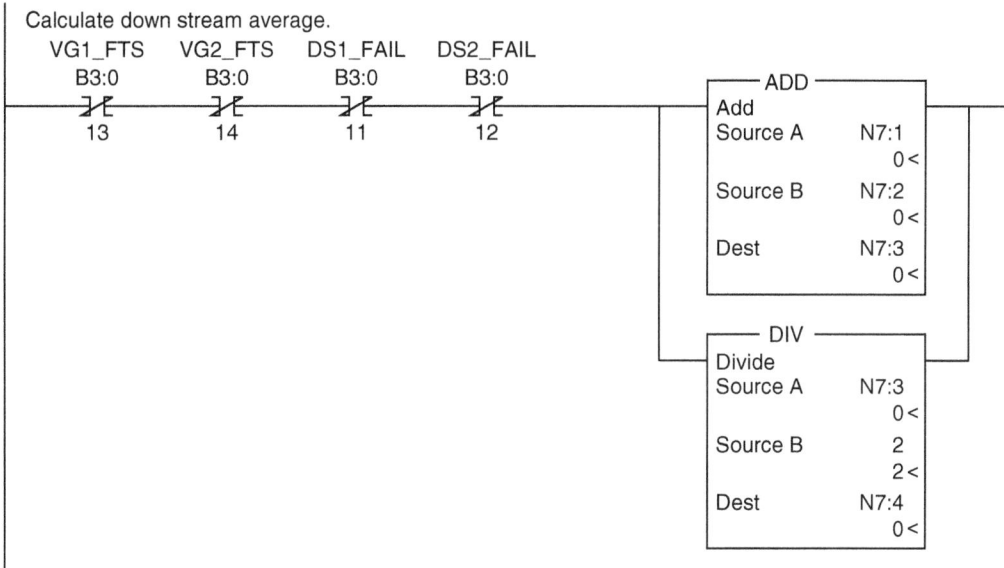

FIGURE 9.14   Downstream average calculations.

contact B3:0/12 tag name (DS2_FAIL), ADD instruction, and DIV instruction. The ladder diagram works as follows: If vertical gate 1 failed to start is OFF, vertical gate 2 failed to start is OFF, and downstream 1 and downstream 2 fail is OFF, power flows to the ADD and DIV instructions and the instructions are executed. The ADD instruction adds the value of N7:1(DS1_LEVEL) to the value of N7:2 (DS2_LEVEL) and outputs the sum at the N7:3 (DS_SUM). The DIV instruction divides (DS_SUM) by 2 and outputs the quotient at N7:4 (DS_AVE_LEVEL).

Figure 9.15 shows a ladder logic diagram for the downstream 1 level sensor validation. This diagram assumes examine-if-close contact B3:0/12 tag name (DS2_FAIL) and a MOV instruction. The rung initially during the first scan as downstream 2 fail is ON; the MOV instruction is executed and it copies the content of N7:1 (DS1_LEVEL) into N7:4 (DS_AVERAGE).

Figure 9.16 shows a ladder logic diagram for the downstream 2 level sensor validation. This diagram assumes examine-if-close contact B3:0/11 tag name (DS1_FAIL) and a MOV instruction. The rung initially during the first scan as downstream 1 fail is ON;

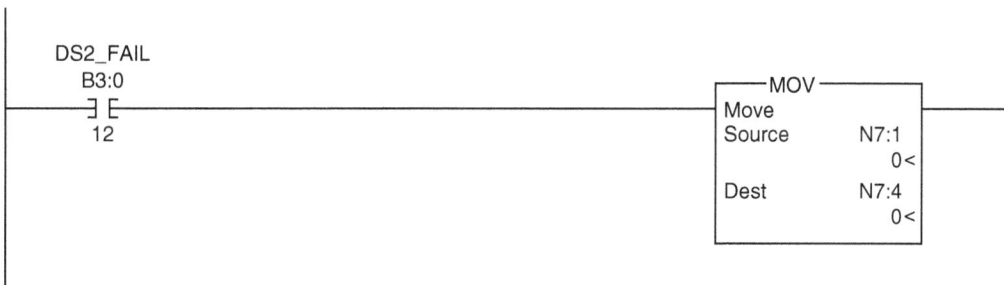

FIGURE 9.15   Downstream average level update using sensor 1.

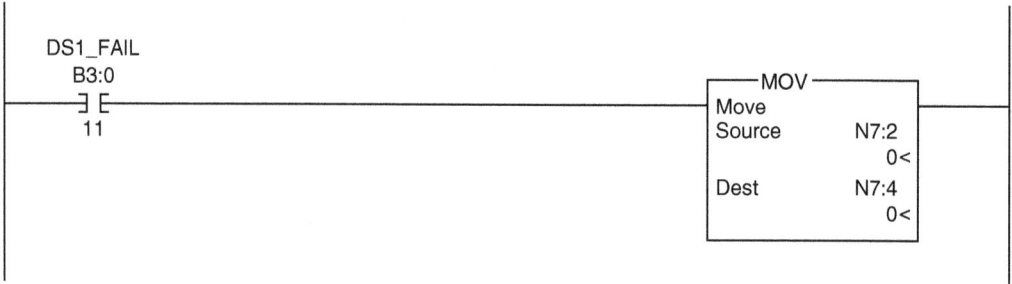

FIGURE **9.16**   Downstream average level update using sensor 2.

the MOV instruction is executed and it copies the content of N7:2 (DS2_LEVEL) into N7:4 (DS_AVERAGE).

Vertical gate position subroutine consists of four rungs: (1) vertical gate 1 highest in position, (2) vertical gate 1 lowest in position, (3) vertical gate 2 highest in position, and (4) vertical gate 2 lowest in position, as shown in Fig. 9.17. The Greater Than or Equal instruction (GRT) compares I:1.0 (VG1 position) to I:1.1 (VG2 position). Once AUTO/MAN switch is placed in AUTO, if VG1 position is higher than or equal to VG2 position, then the output B3:0/1 tag name (VG1_HIGHEST_IN_POS) is set. Notice that if the comparison used is only greater than and the two vertical gates are having equal position, then no gate will be selected to move.

Figure 9.18 compares VG1 position with VG2 position. When the MAN/AUTO switch is placed in AUTO and the VG1 position is less than or equal to the VG2 position, then the output coil B3:2/0 tag name VG1_LOWEST_POS is set. Notice that if the

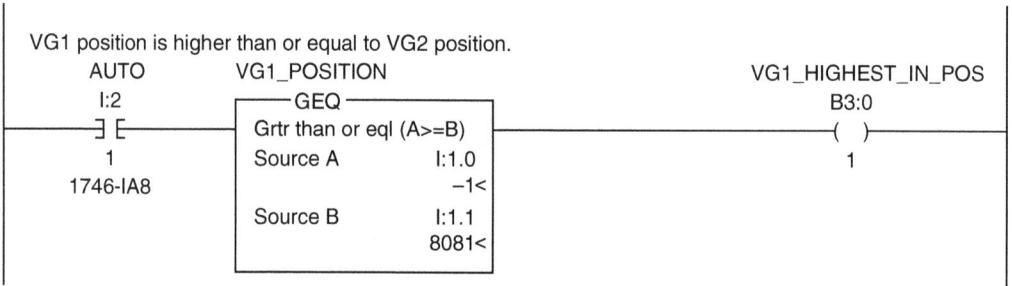

FIGURE **9.17**   Vertical gate 1 highest position rung.

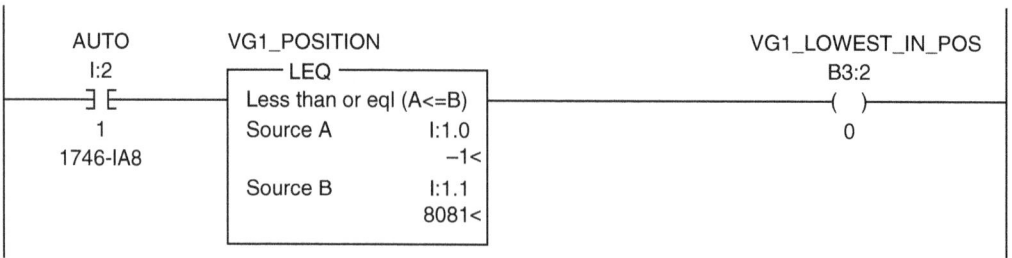

FIGURE **9.18**   Vertical gate 1 lowest position rung.

comparison used is only less than and the two vertical gates have equal positions, then no gate will be selected or moved.

Figure 9.19 compares I:1.1 (VG2 position) to I:1.0 (VG1 position). When AUTO/MAN switch is placed in AUTO and VG2 position is greater than or equal to VG1 position, then the output B3:0/2 tag name (VG2_HIGHEST_IN_POS) is set. Notice that if the comparison used is only greater than and the two vertical gates are having equal position, then no gate will be selected to move.

Figure 9.20 compares VG2 position with VG1 position. When the MAN/AUTO switch is placed in AUTO, the VG2 position is less than or equal to the VG1 position, then output coil B3:2/1 tag name VG2_LOWEST_POS is set. Notice that if the comparison used is only less than and the two vertical gates have equal positions, no gate will be selected or moved.

Vertical gate 1 raise subroutine consists of seven rungs: (1) vertical gate 1 available, (2) next up, (3) control up, (4) vertical gate 1 raise one step timer, (5) vertical gate 1 dead time enable, (6) dead-time timer, and (7) vertical gate 1 raise one step. As shown in Fig. 9.21, if AUTO/MAN switch is in AUTO, vertical gate 1 not running, not failed to start, and LOCAL/REMOTE switch in remote, then vertical gate 1 is available (VG1_AVE).

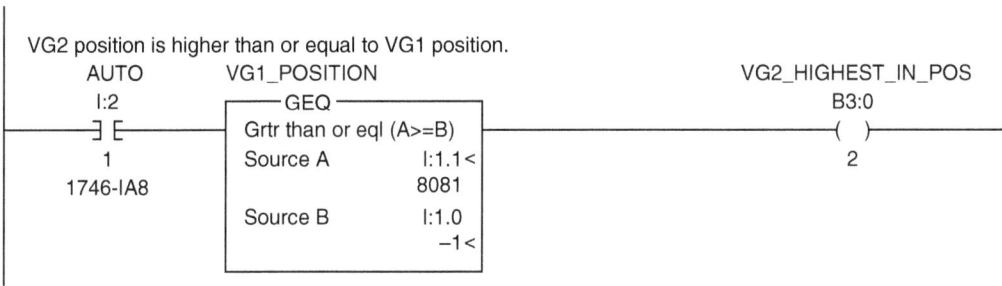

FIGURE 9.19 Vertical gate 2 highest position rung.

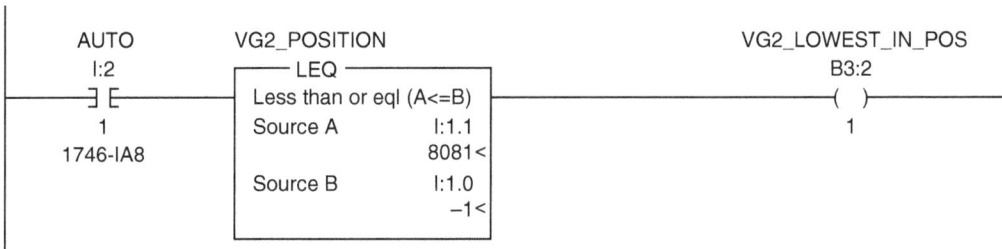

FIGURE 9.20 Vertical gate 2 lowest position rung.

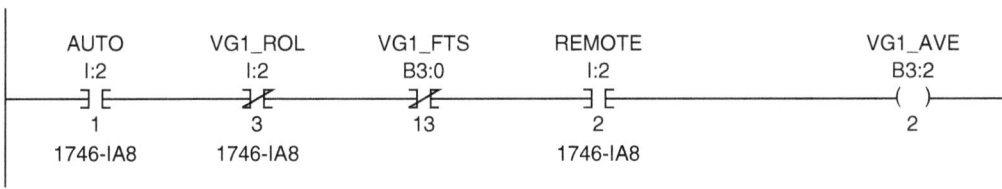

FIGURE 9.21 Vertical gate 1 available rung.

As shown in Fig. 9.22, if VG1 is available, lowest in position, not fully raised, and VG2 is not next to go up, then VG1 is next to go up.

As shown in Fig. 9.23, if vertical gate 1 is next up, set point is outside the dead band, and set point is greater than downstream average level, then CONTROL_UP is set. The next-to-go-up gate action is latched due to possible transient in water level measurements.

As shown in Fig. 9.24, if vertical gate 1 control up is set as a pulse, then T4:0 will start timing and timer timing bit (T4:0/TT) will be set latching around B3:0/3 keeping the timer running for 15 seconds.

As shown in Fig. 9.25, while T4:0/TT is ON , ESD is OFF, and dead-time timer is not running, vertical gate 1 is ON raising the gate one step, which is equivalent to 15 seconds

As shown in Fig. 9.26 first rung, if vertical gate 1 raises transition from high to low, B3:1/0 will be ON for one scan due to the one shot (VG1_DEAD_TIME_OSR). For the second rung if B3:1/0 or B3:1/3 is active, timer (T4:4) will be set and the timer timing bit T4:4/TT will latch around (B3:1/0 or B3:1/3), which causes T4:4 (dead-time timer) to run for 15 minutes.

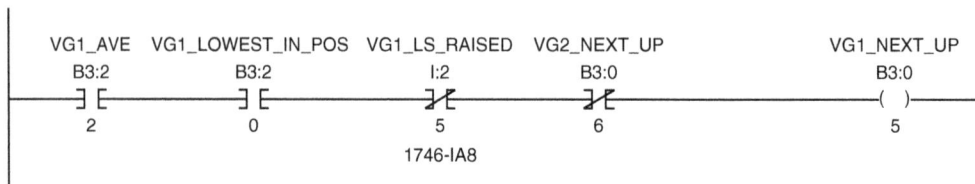

FIGURE 9.22 Vertical gate 1 next-to-go-up rung.

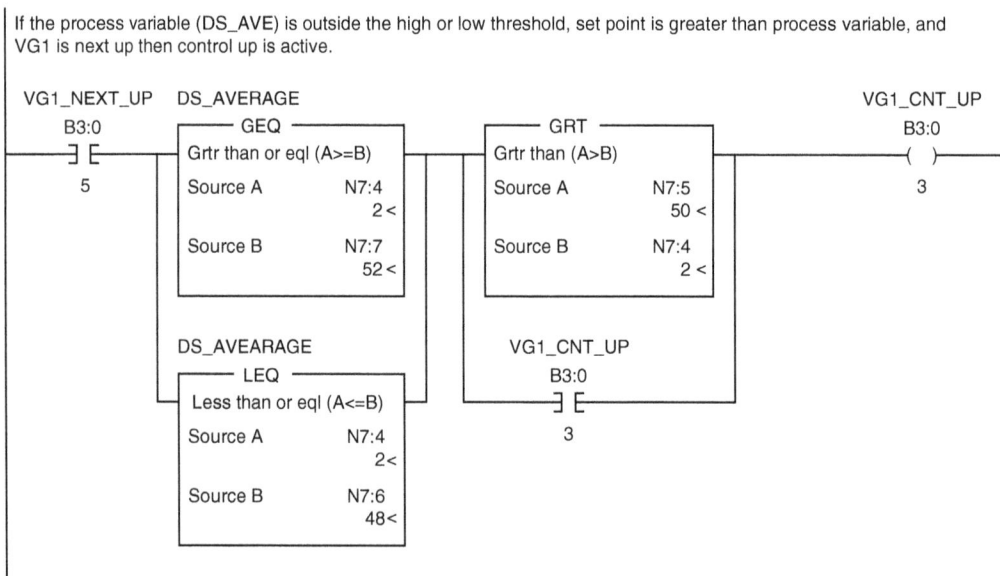

FIGURE 9.23 Vertical gate 1 up control logic.

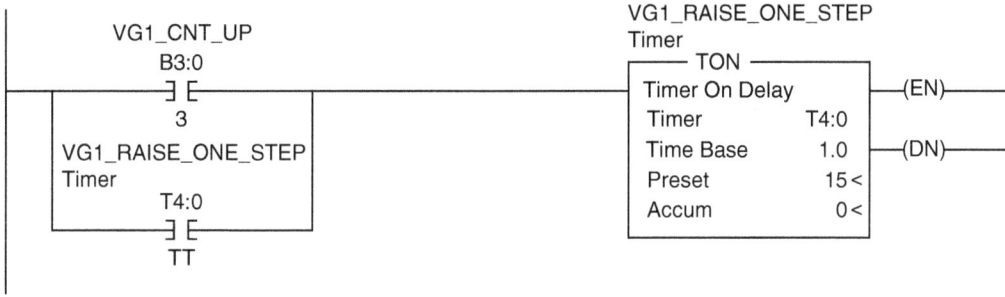

FIGURE 9.24 Vertical gate 1 raise one-step timer.

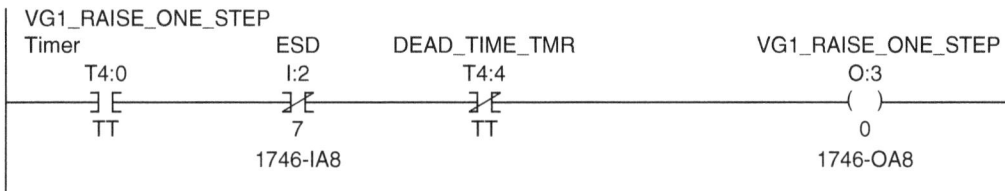

FIGURE 9.25 Vertical gate 1 raise one-step.

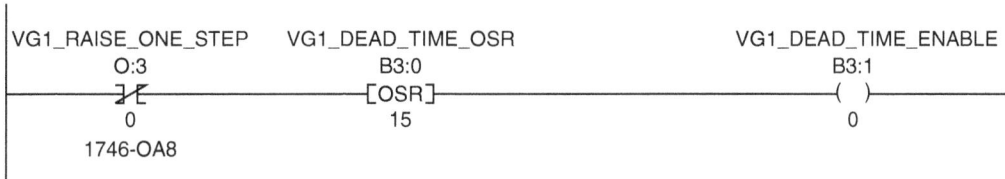

(a)

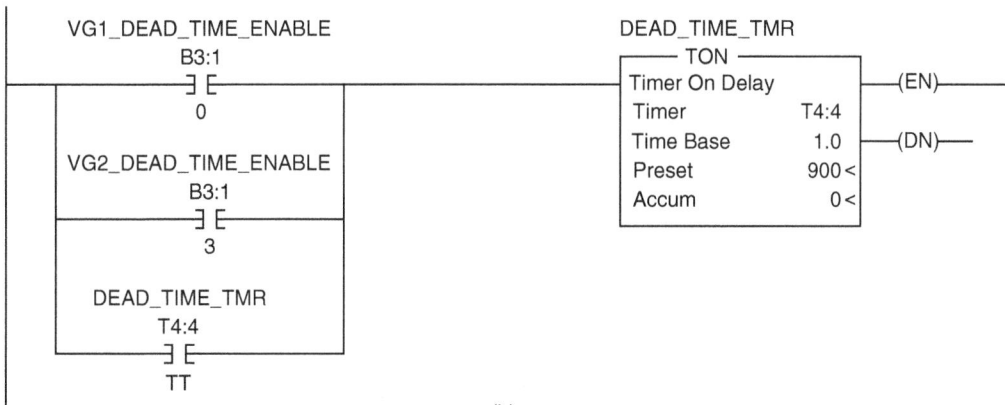

(b)

FIGURE 9.26 (a) Vertical gate 1 dead timer enable; (b) vertical gates dead timer.

As shown in Fig. 9.27, while T4:0/TT is active, ESD is OFF, and dead-time timer is not running, vertical gate1 is on raising the gate one step, which is equivalent to 15 seconds.

Vertical gate 2 raise subroutine consists of six rungs: (1) vertical gate 2 available, (2) next up, (3) control up, (4) VG2_RAISE_ONE_STEP timer, (5) VG2_RAISE_ONE_STEP, and (6) vertical gate 2 dead time enable. As shown in Fig. 9.28, if AUTO/MAN switch is in AUTO, vertical gate 2 not running, not failed to start, and LOCAL/REMOTE switch is in Remote, then vertical gate 2 is available.

As shown in Fig. 9.29; if vertical gate 2 is available, lowest in position, not fully raised, and vertical gate 1 is not next to go up, then vertical gate 2 is next to go up.

As shown in Fig. 9.30, if vertical gate 2 is next up, set point is outside the dead band, and set point is greater than downstream average level, then CONTROL_UP is set.

As shown in Fig. 9.31, first rung, if vertical gate 2 control up is set as a pulse, T4:1 will start timing, and timer timing bit (T4:1/TT) will be set latching around B3:0/4 keeping the timer running for 15 seconds. Once a gate is selected, it will be moved up for the entire 15 seconds unless it reaches the top position. For the second rung, while T4:1/TT is active, ESD is OFF, and dead-time timer is not running, vertical gate 2 is active raising the gate 2 one step, which is equivalent to 15 seconds.

As shown in Fig. 9.32, if vertical gate 2 raises transition from high to low, B3:1/3 will be ON for one scan and will latch around the dead-time timer, as was shown in Fig. 9.26.

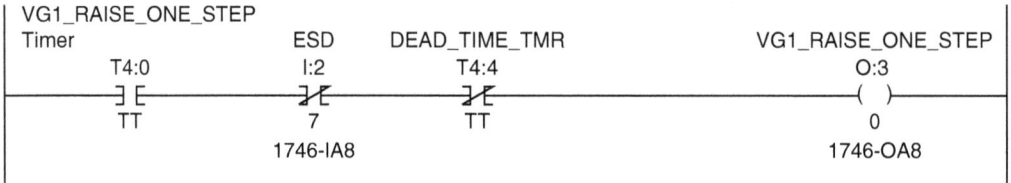

Figure 9.27 Vertical gate 1 raise.

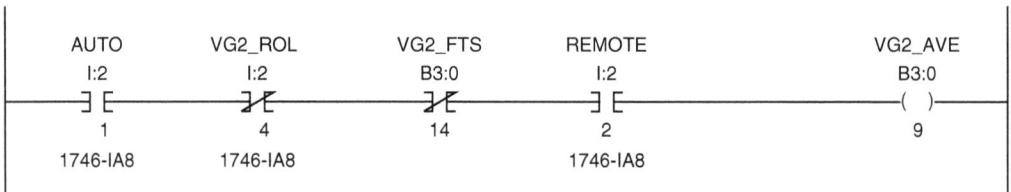

Figure 9.28 Vertical gate 2 available logic.

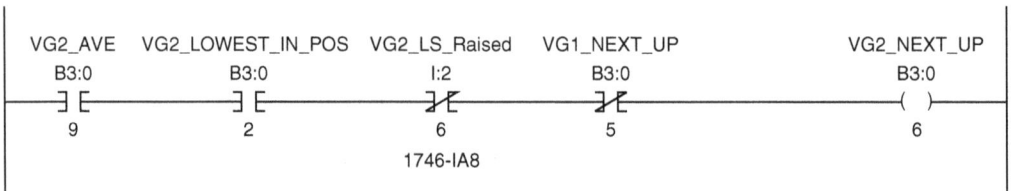

Figure 9.29 Vertical gate 2 next-to-go-up rung.

If the process variable (DS_AVE) is outside the high or low threshold, set point is greater than process variable, and VG2 is next up then control up is active.

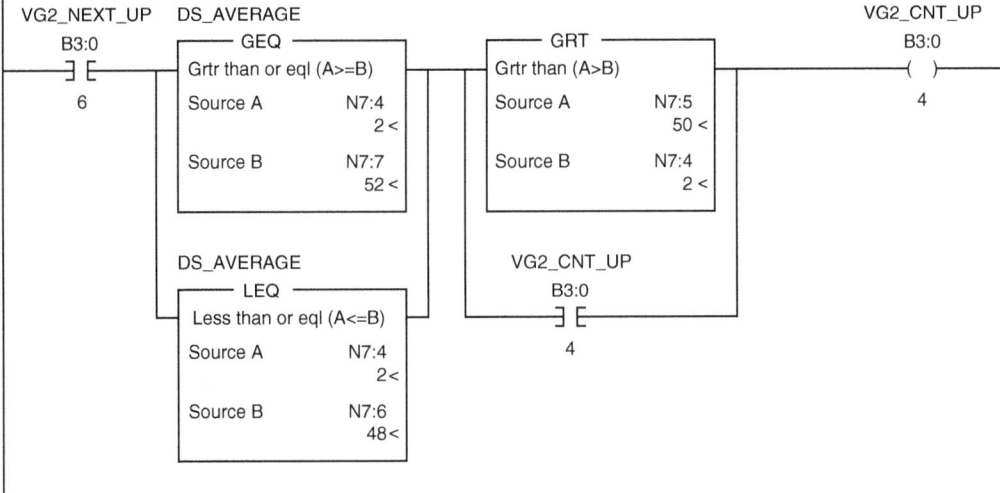

FIGURE **9.30** Vertical gate 2 up control logic.

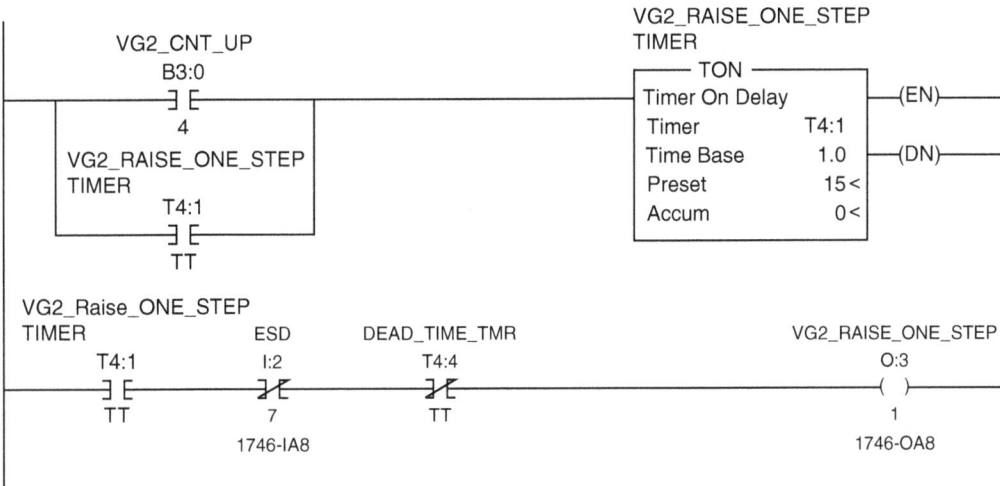

FIGURE **9.31** Vertical gate 2 raise one step.

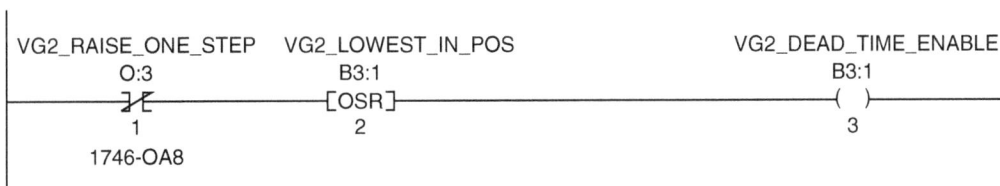

FIGURE **9.32** Vertical gate 1 dead timer.

9.3 Wet Wells Pump Station Control

High flow rate storm rain water is channeled to two large wet wells: the east wet well and the west wet well. The water is pumped to the river from the two connected wells at a constant rate using a predefined process sequence control. Two motor-derived, constant-speed immersed pumps are used: one in the east wet well and the other in the west wet well. Each pump is equipped with an overload alarm switch, which is used to trigger any unusual conditions such as over temperature or overload. The motors provide an input discrete signal indicating if the motor is running or not. The motors can also start by activating the PB located on the local panel if the AUTO/MAN switch is in Manual.

Three float switches are used to provide an accurate indication of the water level at three prespecified critical east/west wet well. The Low Level Float switch triggers the stopping of the running pump. The High Level Float switch triggers the starting of the scheduled pump. If the scheduled pump fails to start within 5 seconds, the second pump is selected and started. An alarm must be issued to alert the operator of any motor failure. The Very High Level Float switch triggers the starting of both pumps. If either of the two pumps fails to start, the corresponding alarm is activated by the control.

Pumps are scheduled to run according to an operator-predefined calendar. This input is expected in hours of accumulated total pump run time. The two pumps must alternate while the water level is below the very high level and above the low level. The two pumps run at levels above the very high level and cascaded timers are not altered during this condition.

9.3.1 System I/O Map

The first step in the design of a PLC control application is the translation of the process specification to actual I/O resources. This is known as the PLC I/O map. This important step lists all I/O tags, assigned PLC addresses, and description. Figure 9.33 lists the wet well pump control system discrete inputs and the corresponding PLC input tags. Figure 9.34 shows the same process for the PLC symbol editor screen. Figures 9.35 and 9.36 repeat the same process for the control system discrete outputs. Notice that none of the real analog inputs/outputs for this control process is listed. We only limited our case study to ON/OFF control based on water level simulated analog real-time measurements relative to user-defined set point for the wet well water levels.

Tag Name	Address Number	Comments
OFF_FLOAT	I:2/0	Off Float Switch
ON_FLOAT	I:2/1	On Float Switch
OVERRIDE_FLOAT	I:2/2	Override Float Switch
E_ROL	I:2/3	East Pump Running
W_ROL	I:2/4	West Pump Running
AUTO	I:2/5	Auto/Manual Selector

FIGURE 9.33 Pump station system input.

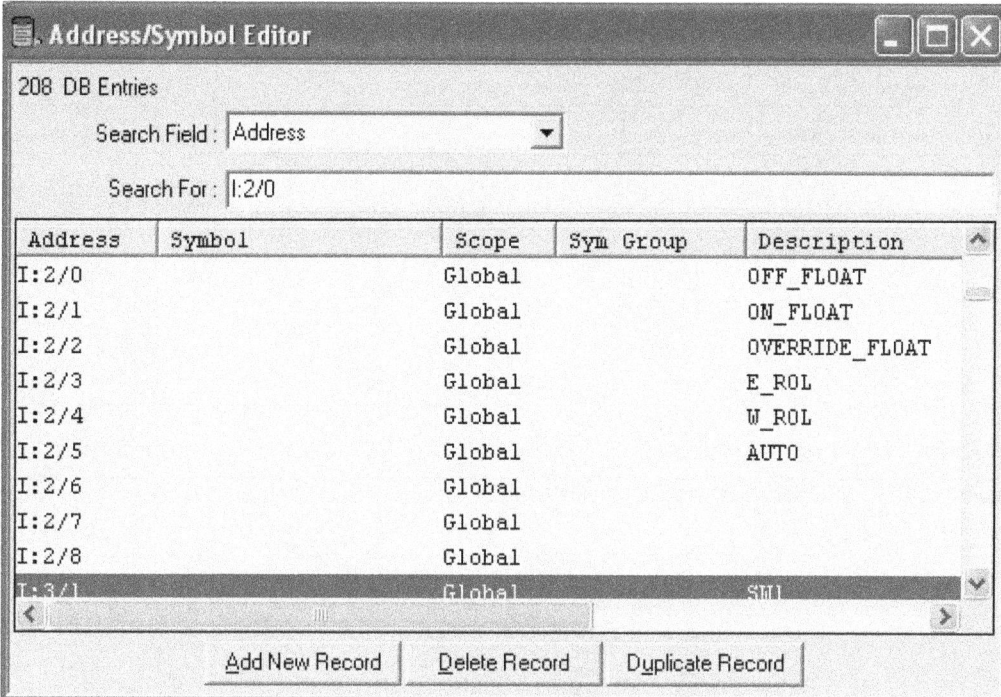

FIGURE 9.34 Pump station system PLC input tags.

Tag Name	Address Number	Comments
E_PUMP	O:3/0	East Pump
W_PUMP	O:3/1	West Pump
E_FTS	O:3/2	East Pump Fail To Start
W_FTS	O:3/3	West Pump Fail To Start
COMMON_ALARM	O:3/4	Common Alarm

FIGURE 9.35 Pump station system outputs.

9.3.2 Automated System Building Blocks

The CPU supports the following types of subroutines that allow an efficient structure for the program:

- File 2 (main file) defines the structure of the program.
- Subroutine (SUB3...) contains the program code that corresponds to specific tasks or combinations of parameters. Each subroutine file provides a set of input and output parameters for sharing data with the calling file.

Figure 9.37 shows the subroutines designed and implemented for the wet well pumping station control in the project view. Figure 9.38 shows the same in the PLC ladder view.

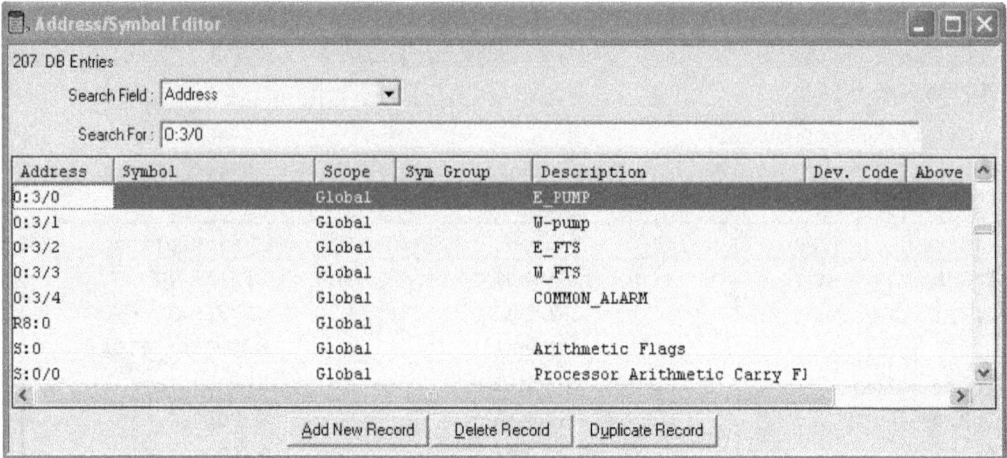

FIGURE **9.36**   Pump station PLC system outputs tags.

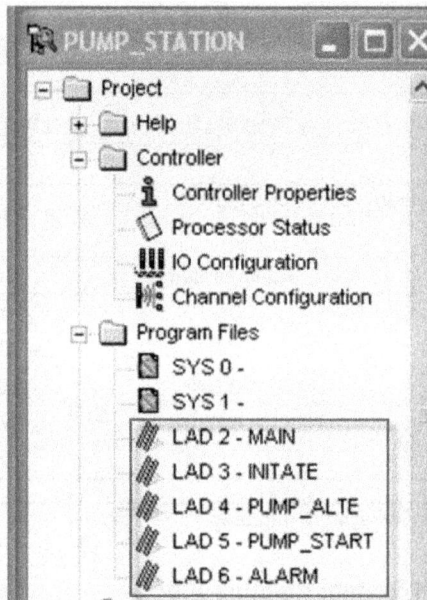

FIGURE **9.37**   Pump station system PLC subroutines (project view).

9.4   Pumping Station Ladder Implementation

The initialization block "INITIATE" is shown in Fig. 9.39. A one shot causes this rung to execute once when selector switch AUTO/MAN switch is in AUTO.

9.4.1   Pump Alarms

The pump alarm subroutine includes eight rungs (Figs. 9.40 through 9.42). One common alarm is dedicated for the east wet well and the other for the west wet well. A

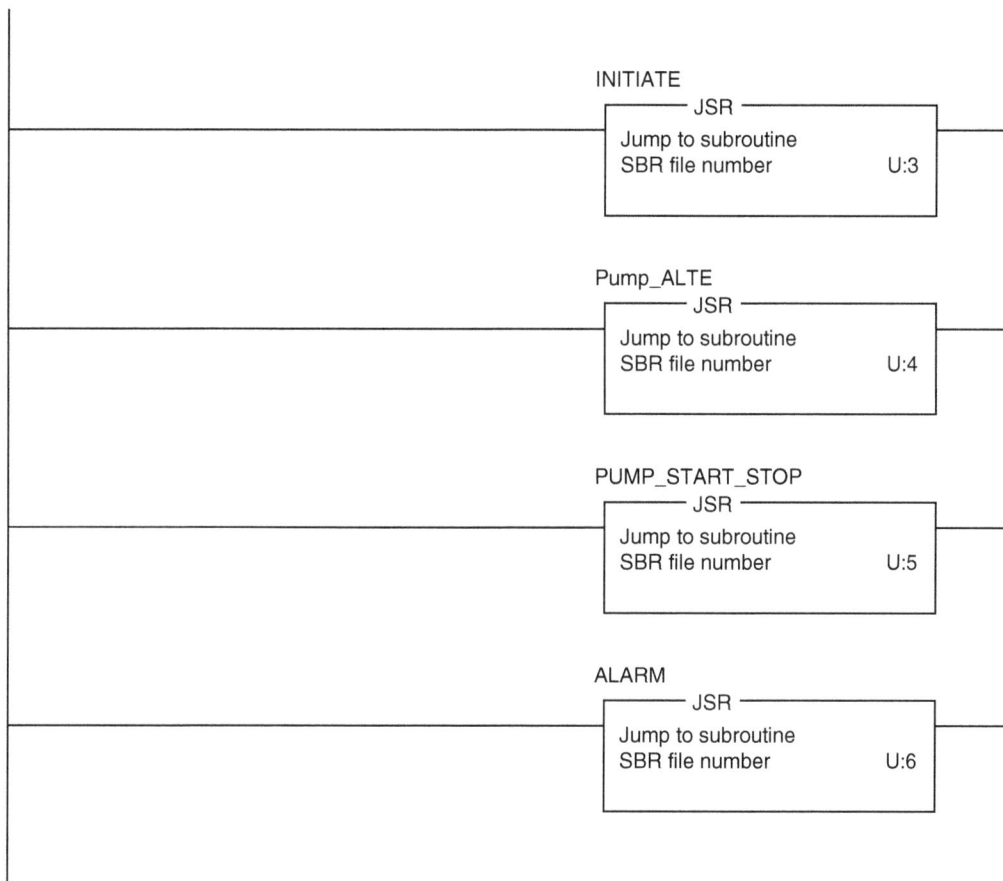

FIGURE 9.38   Pump station PLC subroutines (ladder view).

common alarm is triggered from east pump motor fail to start, west pump motor fail to start, or emergency shutdown.

- The east pump fail to start is ON, if motor running input is not received within 5 seconds from the initiation of the rung command.

- An alarm is issued once the selected pump fails to start. The operator is expected to attend to this failure and clear the cause in order to allow and enforce the pump alteration calendar. Having a situation where both pumps failing to start can constitute an emergency condition and must be eliminated. A third stand by pump and the manual control system can eliminate this problem. The latch for the fail to start is used because the same condition is used to allow the pump output.

- West pump fail to start is ON if the motor running input is not received within 5 seconds from the west pump motor run output command. The latch for the fail to start is used because the same condition is used to allow the pump output.

- The common alarm goes ON if either of the east pump fails to start or the west pump fails to start or emergency shutdown (ESD) is triggered.

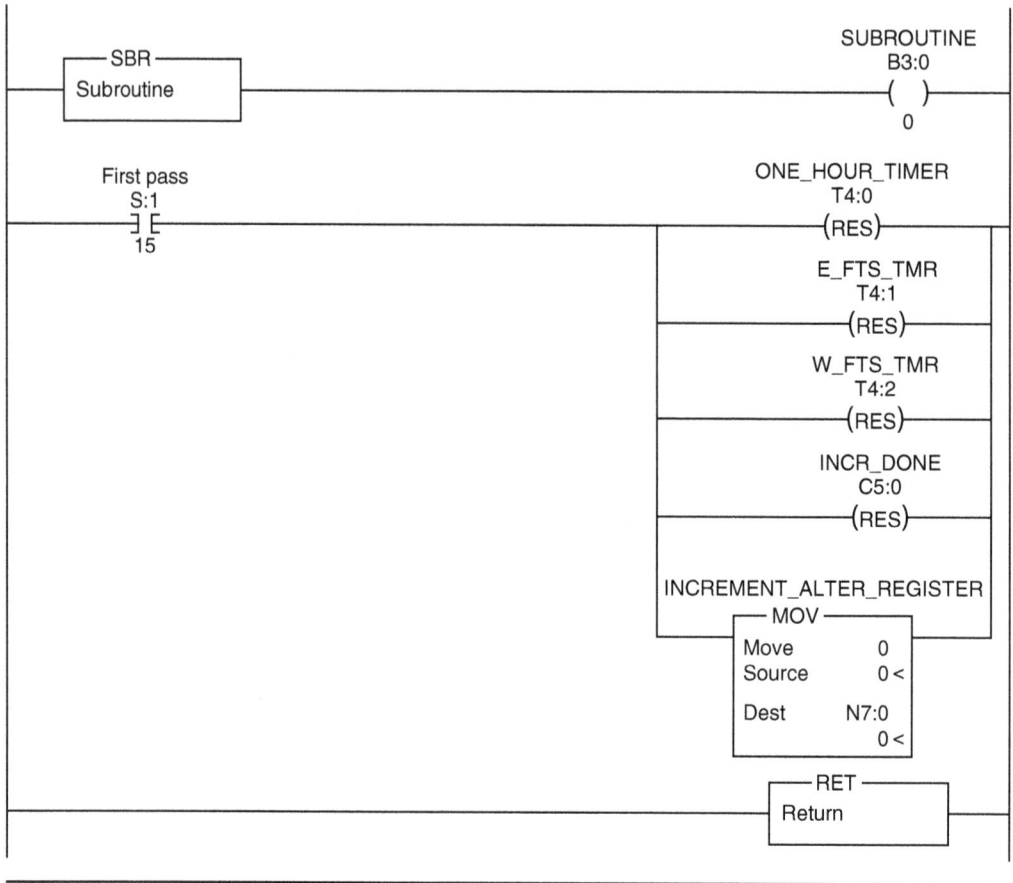

FIGURE **9.39**   Initialization subroutine rungs.

- The two pumps alteration follows a defined calendar based on scheduled run time in hours (simulated by 60 seconds). An ON-delay retentive timer shown in Fig. 9.43 is configured for a 1-hour preset value (T4:0.ACC). The done bit of this timer is used to trigger an up-counter, which is configured to implement the desired pump schedule calendar Fig. 9.44.

- The pump calendar is initialized by the operator measured in whole hours, which indicates the time intervals for the two pumps alterations. The up-counter shown in Fig. 9.44 is used to keep track of the accumulated pump run time in hours. The counter is incremented every hour of operation.

- A memory word (N7:0), as shown in Fig. 9.45, is used to select one of the two pumps to run and is named *increment register* (INCRM).

- The increment (INCRM) register will increment every time the user calendar expires. The even values of this register (N7:0/0 is false) will be used to select and start the east pump. The odd values of this register (N7:0/0 is true) will be used to select and start the west pump. Figures 9.46 and 9.47 implement this logic.

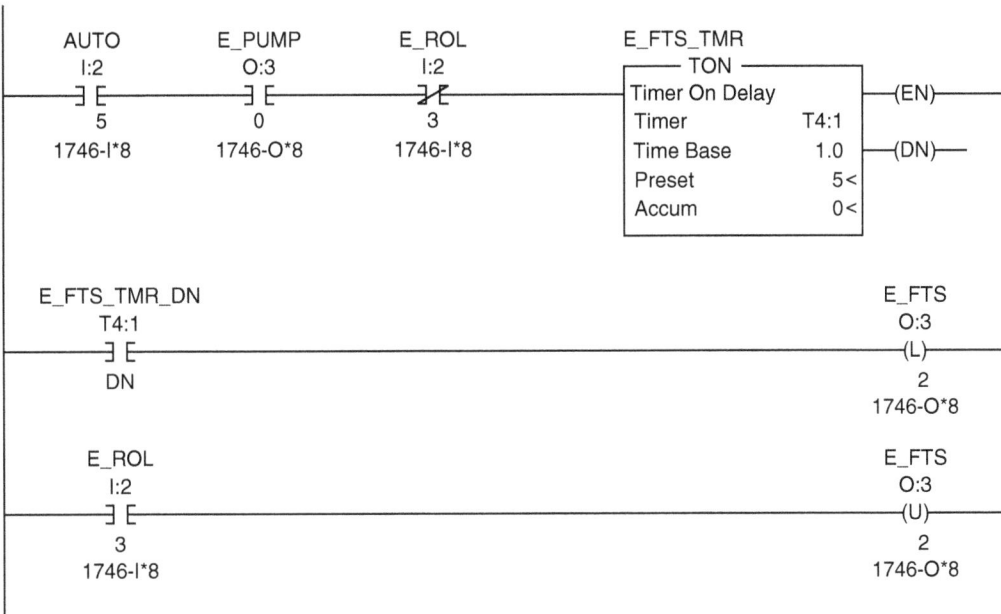

FIGURE 9.40 East pump failed to start rung.

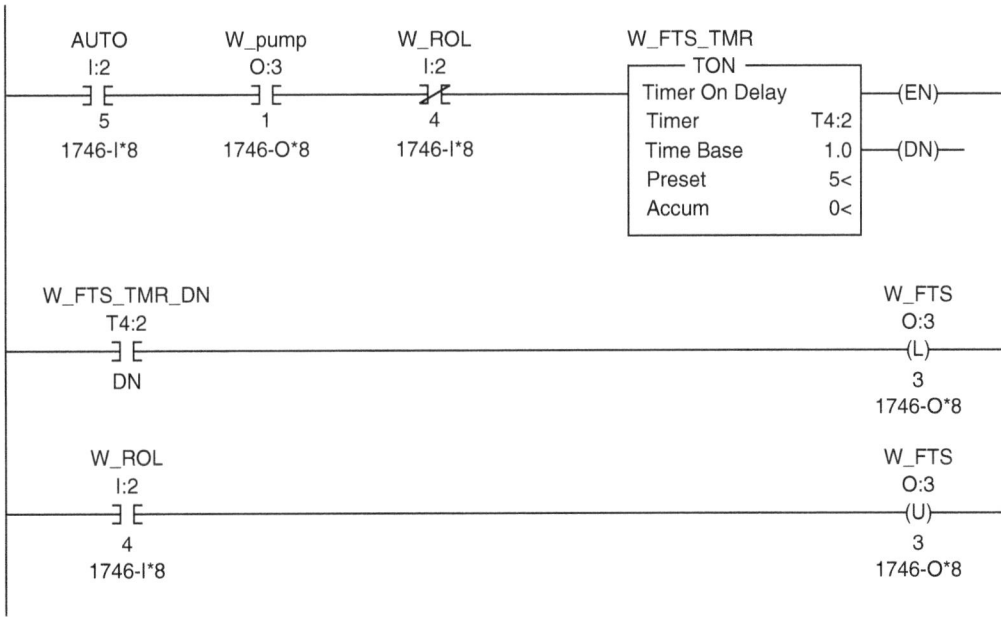

FIGURE 9.41 West pump failed to start rung.

```
        E_FTS                                    COMMON_ALARM
        O:3                                           O:3
       ─┤ ├─┬─────────────────────────────────────────( )──
         2  │                                            4
      1746-O*8                                       1746-O*8

        W_FTS                                      
        O:3                                        
       ─┤ ├─┘
         3
      1746-O*8
```

FIGURE 9.42 Common alarm rung.

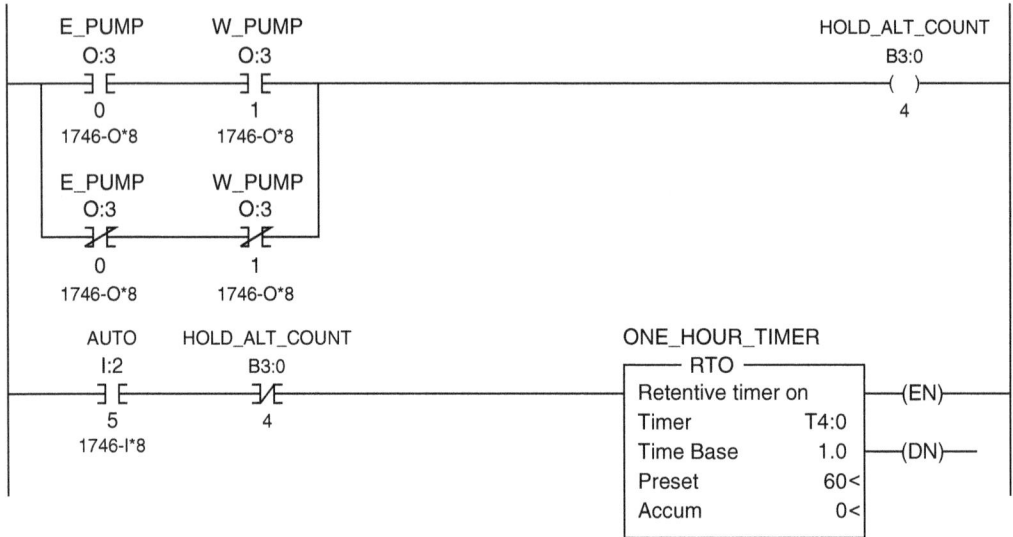

```
     E_PUMP       W_PUMP                          HOLD_ALT_COUNT
      O:3          O:3                                 B3:0
    ──┤ ├──┬──────┤ ├──┬──────────────────────────────( )──
       0   │       1   │                                4
    1746-O*8      1746-O*8

     E_PUMP       W_PUMP
      O:3          O:3
    ──┤/├──┴──────┤/├──┘
       0            1
    1746-O*8      1746-O*8

     AUTO      HOLD_ALT_COUNT        ┌─── ONE_HOUR_TIMER ──┐
      I:2          B3:0              │       RTO            │
    ──┤ ├──────────┤/├──────────────┤ Retentive timer on   ├──(EN)──
       5            4                │ Timer         T4:0   │
    1746-I*8                         │ Time Base     1.0    ├──(DN)──
                                     │ Preset        60<    │
                                     │ Accum          0<    │
                                     └──────────────────────┘
```

FIGURE 9.43 Pump station 1-hour timer rungs.

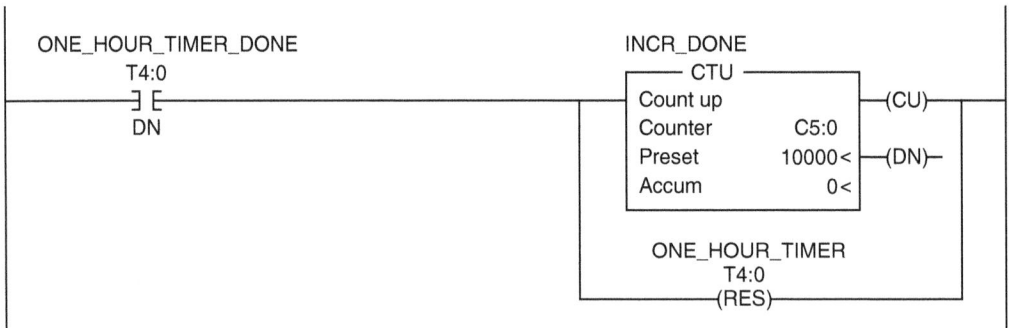

```
   ONE_HOUR_TIMER_DONE                  ┌──── INCR_DONE ──────┐
        T4:0                            │        CTU           │
      ──┤ ├───────────────────────────┬┤ Count up             ├──(CU)──
         DN                           ││ Counter       C5:0   │
                                      ││ Preset     10000<    ├──(DN)──
                                      ││ Accum          0<    │
                                      │└──────────────────────┘
                                      │
                                      │   ONE_HOUR_TIMER
                                      │       T4:0
                                      └──────(RES)───────
```

FIGURE 9.44 Pump station counter rung.

INCR_DONE   INCR_ONE_SHOT   INCREMENT_ALTER_REGISTER
        ┌─── GEQ ───┐   B3:0   ┌─── ADD ───┐
        Grtr than or eql (A>=B)   ─[OSR]─────────────────────   Add
                                        5   Source A   N7:0
        Source A   C5:0.ACC   0<
                          0<
                                                                          Source B   1
        Source B   N7:10   1<
                          0<
                                                                          Dest   N7:0
                                                                                             0<

                                                                                  INCR_DONE
                                                                                    C5:0
                                                                                  ─(RES)─

                                                                                  ONE_HOUR_TIMER
                                                                                    T4:0
                                                                                  ─(RES)─

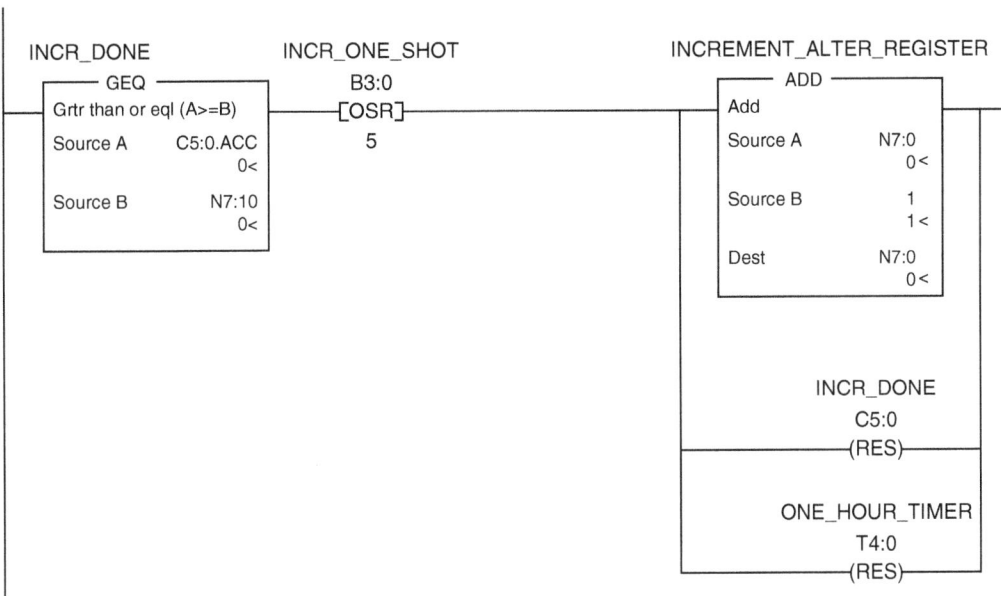

FIGURE 9.45 Pump station ADD rung.

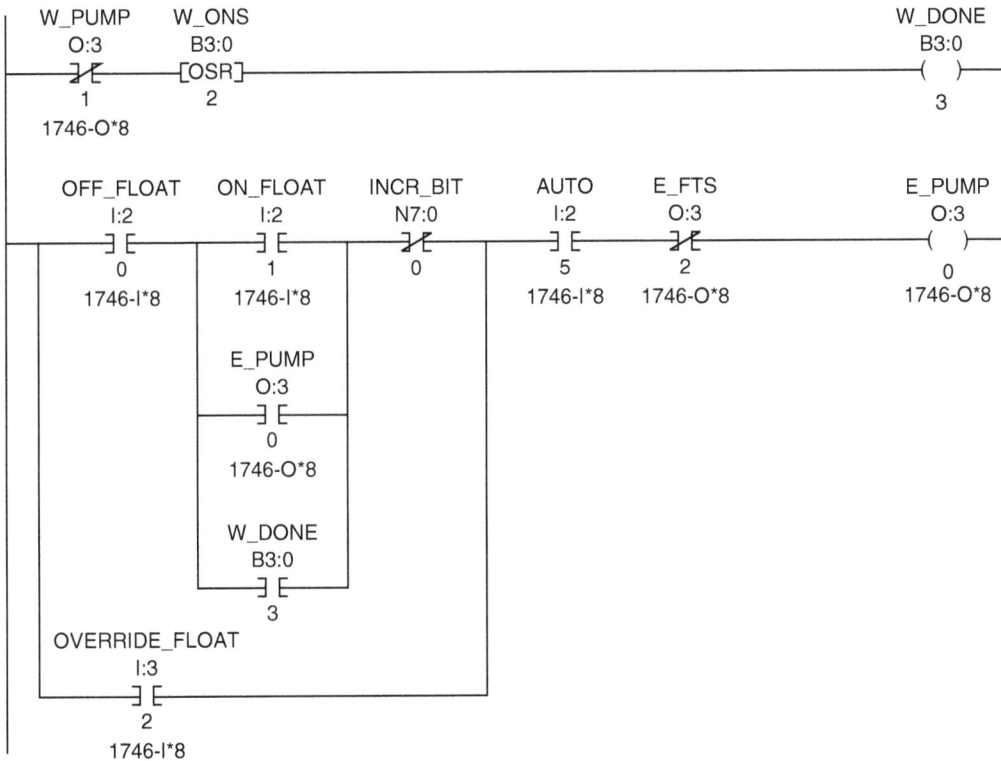

W_PUMP   W_ONS   W_DONE
  O:3   B3:0   B3:0
 ─] [──   ─[OSR]──   ─( )─
   1   2   3
1746-O*8

    OFF_FLOAT   ON_FLOAT   INCR_BIT   AUTO   E_FTS   E_PUMP
      I:2   I:2   N7:0   I:2   O:3   O:3
    ─] [──   ─] [──   ─]/[──   ─] [──   ─]/[──   ─( )─
      0   1   0   5   2   0
   1746-I*8   1746-I*8   1746-I*8   1746-O*8   1746-O*8

                E_PUMP
                 O:3
               ─] [──
                 0
              1746-O*8

                W_DONE
                 B3:0
               ─] [──
                 3

   OVERRIDE_FLOAT
       I:3
     ─] [──
       2
   1746-I*8

FIGURE 9.46 East pump rung.

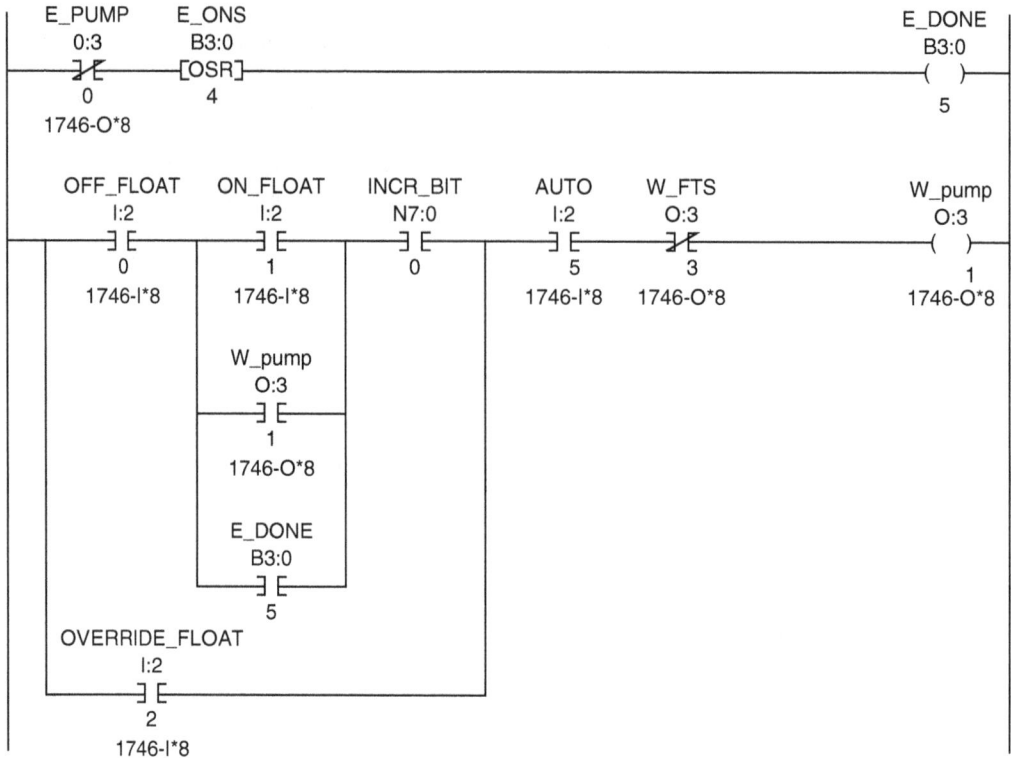

Figure 9.47   West pump rungs.

- If system is placed in AUTO and the wet well water level exceeds the high limit, the scheduled pump will be selected and run. If the selected pump fails to start, an alarm is issued and the other pump is selected and started. If the water level exceeds the high high limit, both pumps are selected and started regardless of the defined calendar.

Chapter 9: Home Work Problems and Laboratory Project

Laboratory 9.1—Conveyor System Speed Control Capstone Project

The conveyor system shown in Fig. 9.48 is a six-station, cellular-module flexible manufacturing system (FMS). One belt provides closed circular motion of pallets between stations, while the second provides straight motion parallel to the long leg of the inner belt. Two transient belts allow for pallet transfer between the two conveyors. A detailed description of the conveyor system I/O wiring, the graphical user interface, and each station operation is given in the next sections. A schematic drawing of the conveyor system, including the two belts and the six stations, is shown in Fig. 9.48.

The conveyor system was configured to provide the FMS platform for this project. It will also be used to simulate and test the control algorithms and implementations of

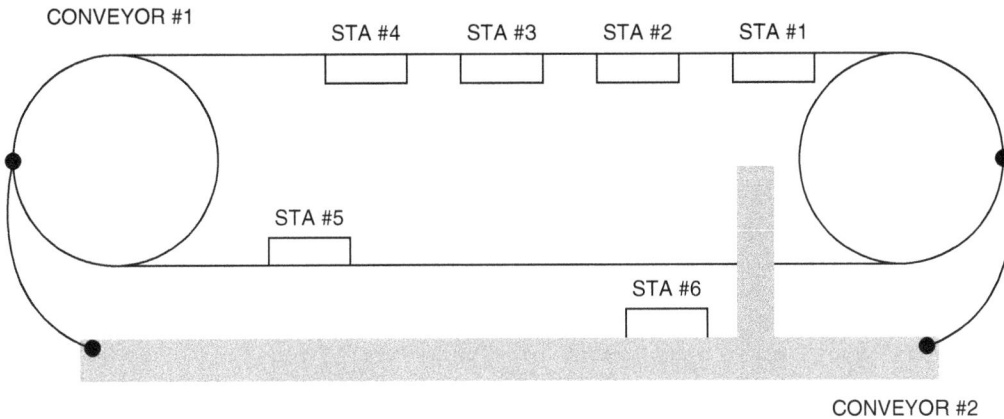

FIGURE **9.48** Circular six-station conveyor system.

the operation, which are described next. The following is a brief description of the conveyor system work stations and their associated control implementations:

Station 1, Loading Cell: This station is equipped with two solenoids: A and C. With no power applied, a pallet can enter the station and lock into position. When the two solenoids receive power, they switch position and prevent additional pallets from entering the station.

Once a pallet is detected in position at station 1, a start pulse is output to the loading robot. The robot sends a "done" pulse to the PLC upon completion of part loading. For the study, a 5-second timer is used to simulate the loading task. The timer timing bit is used as the Go command to the robot. The timer done bit is used as the done acknowledgment from the robot, which moves the pallet to station 2.

Two alarms are associated with station 1. The station 1 "loading jam" alarm is activated whenever the done from robot is active and the pallet remains in position. The station 1 "task time out" alarm becomes active whenever the Go command is not acknowledged by the loading robot in 5 seconds.

Station 2, Production Feed Rate Regulation: This station is equipped with a pallet advance solenoid. When this solenoid receives power and energizes, it activates and advances the pallet to station 3. The PLC receives input signals from two limit switches indicating the advance solenoid positions (forward or backward).

When the pallet is in position, the back limit switch is ON, the station 2 output is clear, and the station 3 prestop is inactive. The advance solenoid will energize and move the pallet to station 3. Otherwise, when station 1 output is clear, the advance solenoid is de-energized and retracted.

One alarm associated with station 2 is the "production rate" jam alarm. This alarm is activated by a malfunction in the advance solenoid two-limit switches.

Station 3, Production/Assembly Task: This station is equipped with a pre-stop solenoid that can be energized to prevent the pallet from entering the station. A stop solenoid can be energized to stop the pallet and a locator clamp solenoid is used to clamp the pallet in position. Station 3 has a "raise" solenoid that is used to get the part into the assembly cell and a "lower" solenoid that is used to lower the pallet on the conveyor belt. Five digital inputs are interfaced to the PLC: (1) pallet approaching station, (2) pallet in station, (3) station output clear, (4) up-reed, and (5) down-reed light switches.

When the pallet is in position, the "stop pallet" solenoid is activated, the "clamp" solenoid is energized, and the "raise" solenoid is powered to raise the pallet. After a 5-second timer simulating the assembly operation interface, as described in station 1, the "lower" solenoid energizes. Once the pallet is completely lowered, the clamp locator is released, the stop solenoid is de-energized, the pallet moves to the next station, and the pre-stop solenoid is de-energized to allow the next pallet to enter the station.

Two alarms are associated with this station: cycle-up failure and cycle-lower failure. At station 3 cycle-up failure, alarm is activated whenever the raise solenoid is on the up position and the reed switch remains OFF for 1 second. The cycle-down failure alarm works the same way for the lower operation.

Station 4, Testing and Shorting: This station is equipped with four solenoids: raise, lower, pre-stop, and stop. The functions of the solenoids are similar to those explained for station 3. The station also has six switches inputs: (1) pallet at station, (2) station up, (3) station at middle position, (4) station down, (5) pallet at output, and (6) output clear.

Station 4 can reject a pallet to a transfer conveyor area (continuation of conveyor 1) or move the pallet to conveyor 2. The reject command can be initiated from the user interface by the operator. The station has three positions: up, middle, and down. When both lower and raise solenoids de-energized, the station is in the middle position. When the pallet is in position, the station is in the middle position. A reject signal is initiated from the user interface when station 6 "request to push" command is not activated. The raise solenoid is energized to raise the pallet and move it to the transfer conveyor (pallet rejected).

Station 4 has two alarms, "cycle-up" failure and "cycle-down" failure. The operation of these alarms is exactly the same as for station 3.

Station 5, Production Output Rate Regulation: This station is equipped with two solenoids: "lane stop" and "wipe off" solenoids. It also has two inputs: "lane clear" and "feed lane full."

If the operator requests a station 4 reject or a station 6 push, the "wipe off" solenoid will energize and allow the pallet to move to conveyor 2. Otherwise the pallet continues on conveyor 1. No alarms are assigned to this station.

Station 6, Unloading Cell: This station is equipped with four solenoids: advance pusher, pre-stop, pusher stop, and lane stop. The PLC receives six switches inputs from this station: (1) pusher forward, (2) pusher backward, (3) pallet present last, (4) pallet present right, (5) lane clear to push, and (6) output clear.

The operator can request to push from the user interface/HMI. When a pallet is present left or right, a request to push is made from the HMI. If a request to reject from station 4 is inactive and station 6 lane is clear to push, the lane stop solenoid will stop the pallet and pre-stop solenoid will prevent incoming pallets from entering the station. The advanced pusher solenoid will energize and unload the pallet to the transfer conveyor area. When the pusher is back, all solenoid reverse their action, the next pallet enters the station, and the push cycle is repeated.

Two alarms are associated with this station: advance pusher forward and backward failure. The alarms operation is similar to those of station 4. Each of the above six stations has a separate Enable/Disable switch, which can be used to bypass the station. This feature allows for flexible manufacturing activities. Not only we can skip certain manufacturing operations, but the entire station can be replaced, modified, or removed. The entire system is equipped with an ESD switch. The ESD switch is wired in series with the MCR coil. As with other systems, all hardware modules and the two conveyor motors are provided power through the ESD switch.

Conveyor System Input/Output Listings

The next two tables provide detailed listing of all inputs and outputs used in the conveyor system control.

Conveyor System Inputs

Device Name	Address	Description
STA #1PIP LS	I:2/0	120-V ac signal limit switch to indicate pallet in position
STA #1 OUTCLR SEN	I:2/1	120-V ac sensor to indicate station #1 output is clear
STA #1 ACTIVE SEL SW	I:2/2	120-V ac on/off selector switch to enable/disable station #1 function
STA #2 CYL FWD LS	I:2/3	120-V ac limit switch to indicate the advance solenoid (SOL) forward position
STA #2 CYL BAC LS	I:2/4	120-V ac limit switch to indicate advance SOL backward position
STA #2 PIP LS	I:2/5	120-V ac sensor to indicate that STA #2 pallet in position
STA #2 OUT CLR	I:2/6	120-V ac sensor to indicate STA #2 output is clear
STA #2 ACTIVE SEL SW	I:2/7	120-V ac on/off selector switch to enable/disable station #2
STA #3 STA UP REED SW	I:3/0	STA #3 120-V ac sensor to indicate raise SOL upper position ac sensor to indicate raise SOL upper position
STA #3 STA DOWN REED SW	I:3/1	STA #3 120-V ac sensor to indicate raise SOL down position
STA #3 PAL AT IN	I:3/2	STA #3 120-V ac sensor to indicate pallet at input
STA #3 PAL AT IN STA PROX	I:3/3	120-V ac sensor to indicate pallet in STA #3
STA #3 OUT CLR SW	I:3/4	120-V ac limit switch to indicate STA #3 output is clear
STA #3 ACIVE SEL SW	I:3/5	120-V ac on/off selector switch to enable/disable STA #3
STA #4 UP LS	I:3/6	120-V ac limit switch to indicate STA #4 upper position
STA #4 MID LS	I:3/7	120-V ac limit switch to indicate STA #4 mid position
STA #4 DOWN LS	I:4/0	120-V ac limit switch to indicate STA #4 down position
STA #4 PAL AT IN PROX	I:4/1	120-V ac sensor to indicate STA #4 pallet at input position
STA #4 PAL AT OUT PROX	I:4/2	120-V ac sensor to indicate STA #4 pallet at input position

Device Name	Address	Description
STA #4 OUT CLR	I:4/3	120-V ac sensor to indicate STA #4 output is clear
STA #4 TRANS AREA CLR	I:4/4	120-V ac sensor to indicate STA #4 transition area is clear
STA #4 ACTIVE SEL SW	I:4/5	120-V ac on/off selector switch to enable/disable station #4
STA #5 LANE CLR	I:4/6	120-V ac fiber sensor to indicate lane is clear
STA #5 FEED LANE FULL	I:4/7	120-V ac sensor to indicate STA #5 lane is full
STA #5 ACTIVE SEL SW	I:5/0	120-V ac on/off selector switch to enable/disable station #5
STA #6 PUSHER BACK LS	I:5/1	120-V ac limit switch to indicate pusher back position
STA #6 PUSHER FWD LS	I:5/2	120-V ac limit switch to indicate pusher forward position
STA #6 PALLET PRES PX	I:5/3	120-V ac proximity switch to indicate pallet press left position
STA #6 LANE CLRTO PUSH	I:4/4	120-V ac sensor to indicate STA #6 lane is clear
STA #6 OUT CLR	I:4/5	120-V ac sensor to indicate STA #6 output is clear
STA #6 ACTIVE SEL SW	I:4/6	120-V ac on/off selector switch to enable/disable station #6
STA #7 ZONE CLR FIBER	I:4/7	120-V ac fiber switch to TA #7 zone clear
CON V #1 STOP MTR	I:5/0	120-V ac PB to stop conveyor #1 motor
CON V #1 START MTR	I:5/1	120-V ac PM to start conveyor #1 motor
CON V #2 STOP MTR	I:5/2	120-V ac PB to stop conveyor #2 motor
CON V #2 START MTR	I:5/3	120-V ac PB to start conveyor #2 motor
STA #6 PIP	I:5/4	120-V ac indicate pallet in position right
CONV#1 SPEED	I:5/6	Conveyor #1 0- to 10-V dc analog input signal
CONV#2 SPEED	I:5/7	Conveyor #2 0- to 10-V dc analog input signal

Conveyor System Outputs

Device Name	Address	Description
STA #1 CYL A	O:3/0	120-V ac signal to STA #A solenoid A
STA #1 CYL B	O:3/1	120-V ac signal to STA #B solenoid A
STA #1 CYL C	O:3/2	120-V ac signal to STA #C solenoid A
STA #2 ADV SOL	O:3/3	120-V ac signal to STA #2 advance solenoid
STA #3 RAISE STA SOL	O:3/4	120-V ac signal to STA #3 to raise solenoid
STA #3 LOWER STA SOL	O:3/5	120-V ac signal to STA #3 to lower solenoid
STA #3 PRE-STOP SOL	O:3/6	120-V ac signal to STA #3 pre-stop solenoid
STA #3 STA STOP SOL	O:3/7	120-V ac signal to STA #3 start/stop solenoid
STA #3 LOCATOR CLMP SOL	O:4/0	120-V ac signal to STA #3 to raise solenoid
STA #4 RAISE STA SOL	O:4/1	120-V ac signal to STA #4 raise station solenoid

Device Name	Address	Description
STA #4 LOWER STA SOL	O:4/2	120-V ac signal to STA #4 lower solenoid pallet at input
STA #4 PRE-STOP SOL	O:4/3	120-V ac signal to STA #4 pre-stop solenoid
STA #4 LANE STOP SOL	O:4/4	120-V ac signal to STA #4 lane stop solenoid
STA #5 LANE STOP SOL	O:4/5	120-V ac signal to STA #5 lane stop solenoid
STA #5 WIP OFF SOL	O:4/6	120-V ac signal to STA #5 lane stop solenoid
STA #6 ADANCE PUSHER SOL	O:4/7	120-V ac signal to STA #4 wipe off solenoid
PRE-STOPS SOL	O:5/0	To STA #6 pre-stop solenoid
STA #6 PUSHER STOP SOL	O:5/1	120-V ac signal to STA #6 pusher stop solenoid
STA #6 LANE STOP SOL	O:5/2	120-V ac signal to STA #6 lane stop solenoid
STA #7 ALLOW AIR LOGIC SOL	O:5/3	120-V ac signal to STA #7 allow air logic solenoid
STA #1 ALLOW	O:5/4	120-V ac to conv #1 motor
STA #2 ALLOW	O:5/5	120-V ac to conv #2 motor
CONV #1 SPEED	N7:4	Analog output signal to ac drive for conveyor #1 motor
CONV #2 SPEED	N7:30	Analog output signal to ac drive for conveyor #2 motor

Graphical User Interface

Eight graphics screens should be implemented for the control of the conveyor system. In addition to the directory page, seven pages should be provided in the HMI interface. These are as follows: system overview, system status, control and trending, alarm graphics, conveyor 1 control, conveyor 2 control, and alarms summery page. The next subsections briefly describe each of the eight HMI pages. Specific details of the user interface implementation will depend on the HMI used, for example, color or monochrome, and its capabilities.

Directory page: This screen lists the seven available options and provides for the selection of individual pages. The pressing of a function key will take the user to selected screen.

Overview page: This page shows the two conveyor belts, station Enable/Disable switch status, station task busy/not busy status, motor 1 and motor 2 running/not running status, and color-coded status for assign station. The color code provides red for station enabled, green for station disable, red for conveyor running, and green for conveyor not running. No user entry is assigned for this page.

Control and trend: This page provides the status of the six stations used in the control system: Enable/Disable, alarms for stations and conveyors, Acknowledge/Reset of alarms, and associated color-code definition. It also provides for user input commands: station 4 reject/not reject, station 6 push/not push, conveyors 1 and 2 speed, and the trending of the two conveyor speeds, up or down. Three modes of control are provided from this page. In Manual mode, the operator can demand changes in conveyor speed by entering either percentage speed. The automatic mode is designed to allow users to regulate the conveyor speed using the PID control. The user initiates such control by entering the desired conveyor speed set point.

Status page: This page displays status information of the system in numeric or symbolic forms. Motor speeds are displayed in percentage value (%).

Alarm page: This page provides detailed listing and a description of all system alarms. The control and trend page provides for an annunciation of an alarm without any details.

Conveyor 1 control: This page displays the status of the controller for the conveyor system, only for conveyor 1.

Conveyor 2 control: This page displays the status of controller for the conveyor system, only conveyor 2.

Alarm Summary page: The alarm summary page displays all alarm transactions including current and past alarms, and indicates whether or not they have been acknowledged.

Requirements:
Use the aforementioned specifications to perform the following:

- Document and implement layer 1 of the conveyor system speed control. Present the implementation with your project group to the course/technical training instructor. Implement recommended modifications and secure a satisfactory layer 1 review.

- Proceed with the design of layer 2, perform the design review presentation, implement required changes, and secure final approval for the layer 2 document.

- Perform the final implementation and checkout as detailed in layer 3 discussions and coverage. This includes the ladder logic and HMI implementations along with the communication/networks configuration and the overall system debugging/checkout. Prepare for final presentation and system operation demonstration.

- Consider valid inputs and recommendations obtained during the final presentation and finalize your implementation. Document all tasks and write the final project technical report and manual.

Odd-Numbered Home Work Problem Solutions

Chapter 1 Home Work Solutions

1. a. A set point variable is the desired value for a "controlled variable" that is set through the operator, the master controller, or a networked PC/HMI.

   b. A controlled variable is the process variable that must be maintained precisely within a dead band around the set point. Typically, the variable is chosen to represent the state of the system and is named the "controlled variable." Examples of controlled variables are temperature, pressure, speed, flow rate, chemical concentration, density, etc.

   c. A manipulated variable is the selected processs variable used to regulate and maintain the controlled variable near the set point value. The variable chosen to control the system's state is termed the "manipulated variable." It is also called "controlling variable." Examples of manipulated variables are fuel/liquid flow, steam feed, variable frequency drive, heater current, etc.

   d. Direct acting control: the controller output increases in order to increase the process controlled variable.

   e. Reverse acting control: the controller output decreases in order to increase the process controlled variable.

3. An open-loop controller is a type of controller that computes its input into the process system using only the current state. It is also called a nonfeedback controller.

5. The following are four advantages of PLC control over hardwired relay control:

   - PLCs: Changes in logic can easily be implemented in software.
     Control relays: Changes require more complex hardware modifications.

   - PLCs: Easier to customize and download software.
     Control relays: Requires construction of new control panels.

   - PLCs: Less maintenance and faster speed of control.
     Control relays: Mechanical parts require more maintenance and reduce speed of control.

- PLCs: PLCs are more robust and redundancy is available. Control relays: Less reliable because of the use of individual components.

7. The following three primary steps are used in implementing single-variable closed-loop control:

   1. Select the PV (controlled variable), which must be easier to measure and regulate. The controlled variable should exhibit self-regulation under small disturbances.

   2. Select the right sensor and final control element.

   3. Design and implement the control strategy.

9. An oven set point = 220°C, measured value = 200°C, range = 200 to 250°C. The following are the calculated values:

   a. Absolute error = 20°C

   b. Error as percent of SP = (20/210)*100 = 9.524%

   c. Error as percent of range = (20/50)*100 = 40%

11. The following are the required explanations:

   a. The function of a process controller is to optimize the process behavior by maintaining the process variable near the desired values.

   b. The final control element furnishes the necessary power amplification between the low energy levels of the controllers and the much higher energy levels of the process as needed to perform their function.

   c. The objectives of a control system is to increase productivity and quality of the final products, which can be divided into three groups:

      - Economic incentive
      - Safety
      - Equipment protection

13. The circuit shown is wired wrong as you can stop the motor manually while the selector switch is in AUTO. Check Fig. 1.24 of the text book for more details.

15.

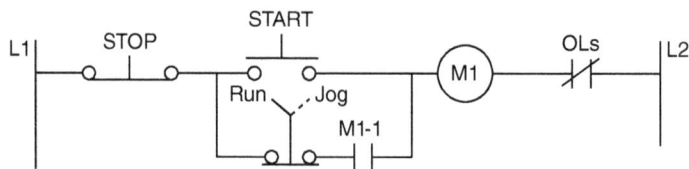

17. Replace the normally open PB Start and Stop switches using a three-way selector switch. The switch should have a Start, Stop, and OFF positions. This arrangement will mechanically interlock the Start (latch) and the Stop (Unlatch) actions.

19. The overloads of the magnetic starter should be moved outside the reversing loops between the power source terminals and the motor terminals.

Chapter 2 Home Work Solutions

1. The advantages of using a PLC in industrial automation include safety, continuous quality improvements, long-term cost reductions, flexible manufacturing, scalability, and easier maintainability. Also, as shown in the following table they eliminate control relays, which provide additional advantages.

PLCs	Control Relays
Changes in logic can easily be implemented in software.	Changes require more complex hardware modifications.
Easier to customize and download software.	Requires construction of new control panels.
New I/O modules, expansion chassis, HMI's, and software patches can be added. Networked control systems can be utilized.	Expansion is possible but at a higher cost.

3. The function of a PLC digital input module and digital output module are described as follows:

   a. A digital input module performs four main tasks: senses the presence of an input signal, maps the input signal which is typically a 120 V ac or 24 V dc to a low dc voltage, isolates the input signal from the mapped output signal, and produces a dc signal to be sensed by the PLC processor (CPU) during the input scan cycle.

   b. The output module switches the hot point (L1) using a TRIAC switch to the output coil terminal as it receives command from the PLC CPU during logic scanning.

5. The following are the definitions of the three terms:

   a. Program scan is the time it takes to solve the entire program logic in one cycle.

   b. An address is a unique designation of the instruction and its associated parameters used in the program.

   c. An instruction is the implementation of certain steps or a function to be executed in the program.

7. a. The main file (file 2) coordinates the execution/scanning sequence of the entire program by calling available subroutines, which exist in the remaining files.

   b. Nesting subroutines takes place when calling subroutines from within the executing subroutine.

9. The function of the power supply used in a PLC is as follows:

   The PLC power supply main function is to provide the appropriate voltage levels and drive the needed currents for the processor module, all deployed I/O modules, signal, and communication modules.

11. Output coil instruction is set to state "1" if rung operation logic (ROL) is "1." If ROL is "0," the output coil is set "0." The latch output instruction is set to state "1" if ROL is "1." If the same ROL is "0," the output will not change.

13. The conditions for MOTOR M to turn on are PB is not pressed, START PB is pressed, and LS1 is closed. Examine the following line diagram shown and verify these conditions.

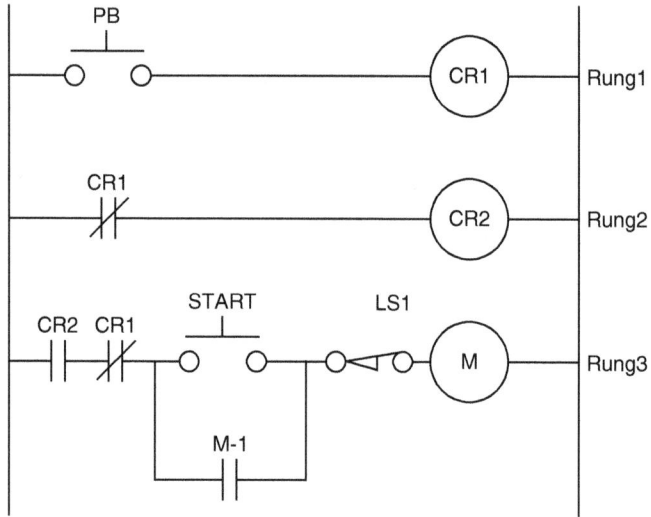

PB
CR1 — Rung1

CR1
CR2 — Rung2

CR2 CR1   START   LS1
M — Rung3
M-1

15. $SV = (SW1 + SW2)(SW3);$

   SV is a solenoid valve.

   SW1, SW2, and SW3 are three ON/OFF switches.

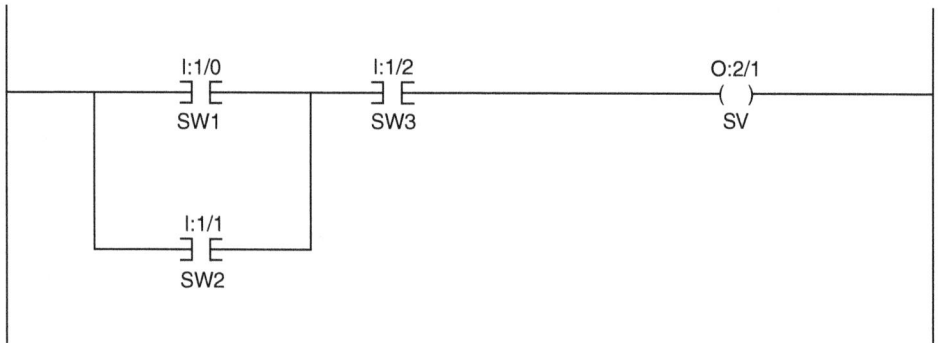

I:1/0   I:1/2   O:2/1
SW1   SW3   SV

I:1/1
SW2

17.

Parameter	Address Name	Value
Source A	N7:1	0101 0101 0101 0101
Source B	N7:2	0101 0001 1010 1011
Dest.	N7:3	0101 0001 0000 0001

19.

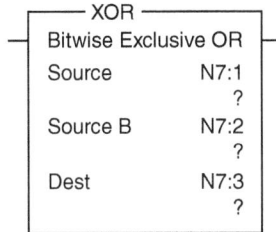

Address	Value
N7:1	01010101 11010101
N7:2	01010001 10101011
Dest	00000100 01111110

21.

a.

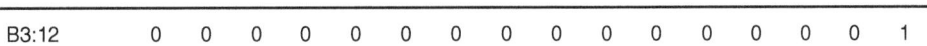

B3:12   0   0   0   0   0   0   0   0   0   0   0   0   0   0   0   1

b.

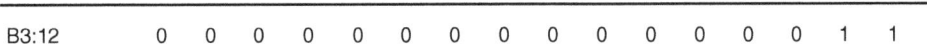

B3:12   0   0   0   0   0   0   0   0   0   0   0   0   0   0   1   1

23.

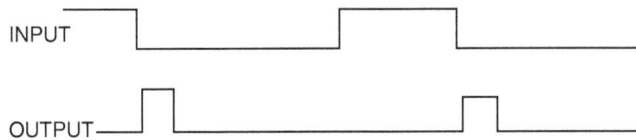

When input transitions from high to low, the output stays high for only one scan cycle.

25.

27. The instruction is only executed if the preceding logic for the same rung is true (power flows to the L coil), then (L) is activated. When preceding rung input is false, then L maintains the active status. L remains active until unlatch action is executed.

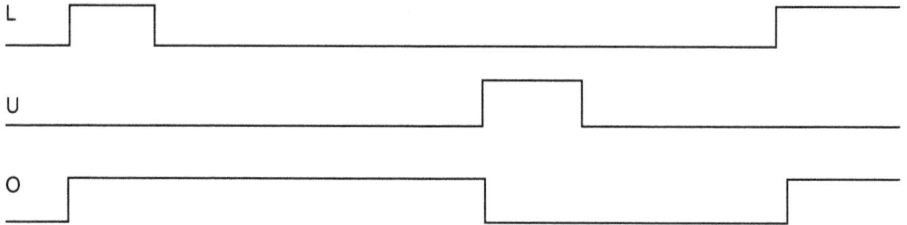

29. N7:2 = 4 and N7:3 = 1111 1111 1111 1011 Bin. = −5 dec.

31.

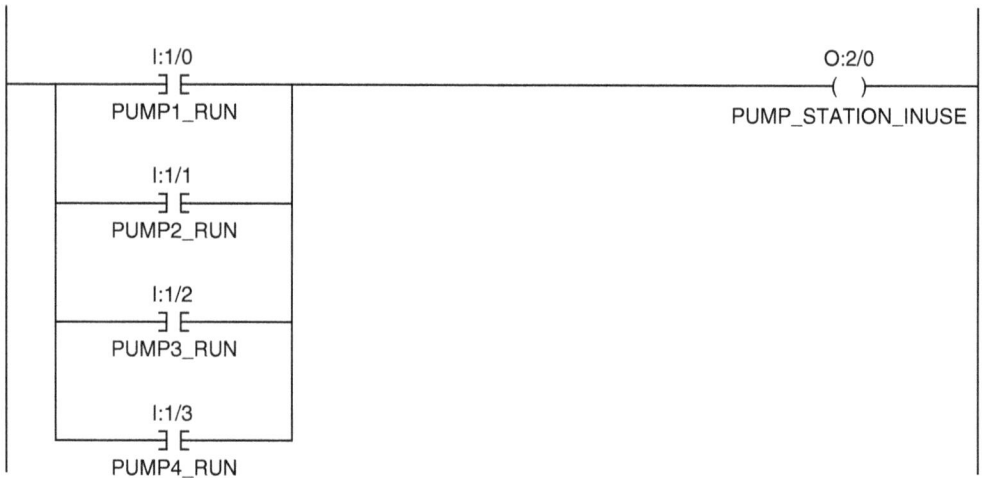

Chapter 3 Home Work Solutions

1. The TON is a non-retentive timer (when input to the timer is off the accumulated value resets to zero), while RTO is retentive timer (when the input to the timer loose power accumulated value is latched). A rest instruction is required in order to reset the accumulated value to zero.

3.

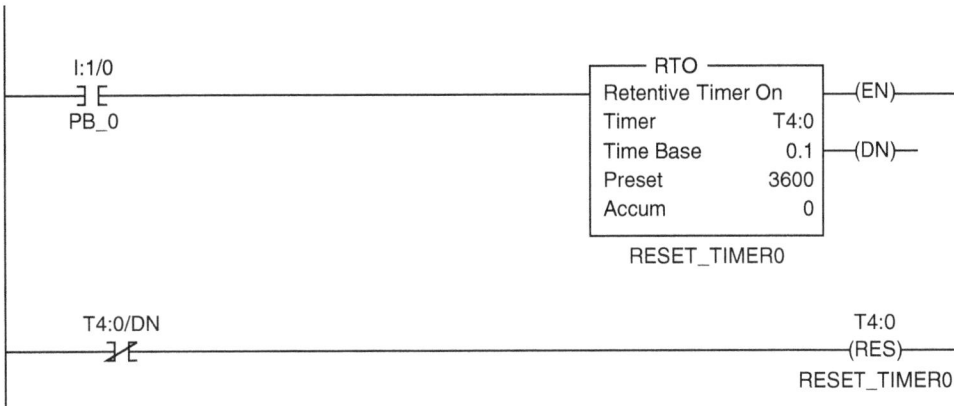

```
     I:1/0                                 ┌──── RTO ─────┐
 ───┤ ├───────────────────────────────────┤ Retentive Timer On ├──(EN)──
     PB_0                                  │ Timer      T4:0     │
                                           │ Time Base   0.1     ├──(DN)──
                                           │ Preset     3600     │
                                           │ Accum         0     │
                                           └──────────────────────┘
                                                RESET_TIMER0

     T4:0/DN                                                   T4:0
 ───┤/├────────────────────────────────────────────────────────(RES)──
                                                            RESET_TIMER0
```

The XIO (T4:0/DN) instruction in the second rung should be replaced by an XIC instruction.

5. a. M1 is on, M2 is OFF, SOL1 is OFF, SOL2 is ON

   b. M1 is OFF, M2 is ON, SOL1 is ON, and SOL2 is OFF

   c. M1 is OFF, M2 is ON, SOL1 is ON, and SOL2 is OFF

7. EN = ON

   TT = ON

   DN = OFF

9.

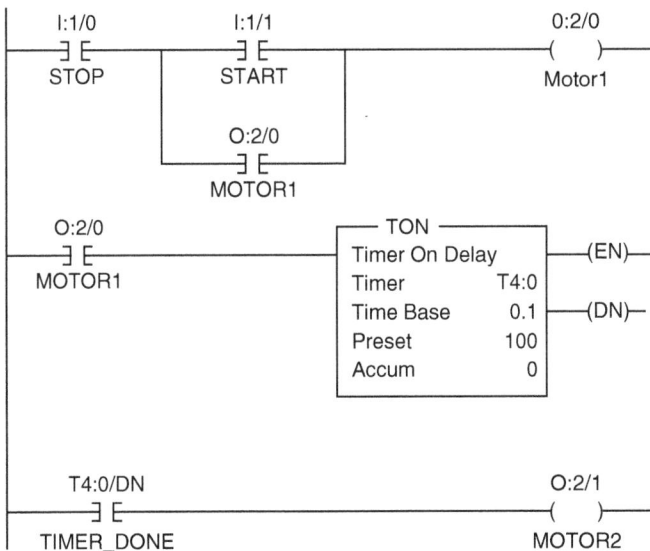

```
     I:1/0          I:1/1                      O:2/0
 ───┤ ├────────┬───┤ ├───────────────────────( )──
     STOP      │    START                     Motor1
               │
               │    O:2/0
               └───┤ ├───
                    MOTOR1

     O:2/0                    ┌──── TON ─────┐
 ───┤ ├─────────────────────┤ Timer On Delay ├──(EN)──
     MOTOR1                  │ Timer     T4:0 │
                             │ Time Base  0.1 ├──(DN)──
                             │ Preset     100 │
                             │ Accum        0 │
                             └────────────────┘

     T4:0/DN                              O:2/1
 ───┤ ├─────────────────────────────────( )──
     TIMER_DONE                          MOTOR2
```

11. a. ON and OFF, respectively.

b. Motor 1 stays on. Only the RTO reset will clear the DN bit.

c. Retain its last status.

d. ON, OFF, and ON, respectively after 40 seconds from the activation of PB1.

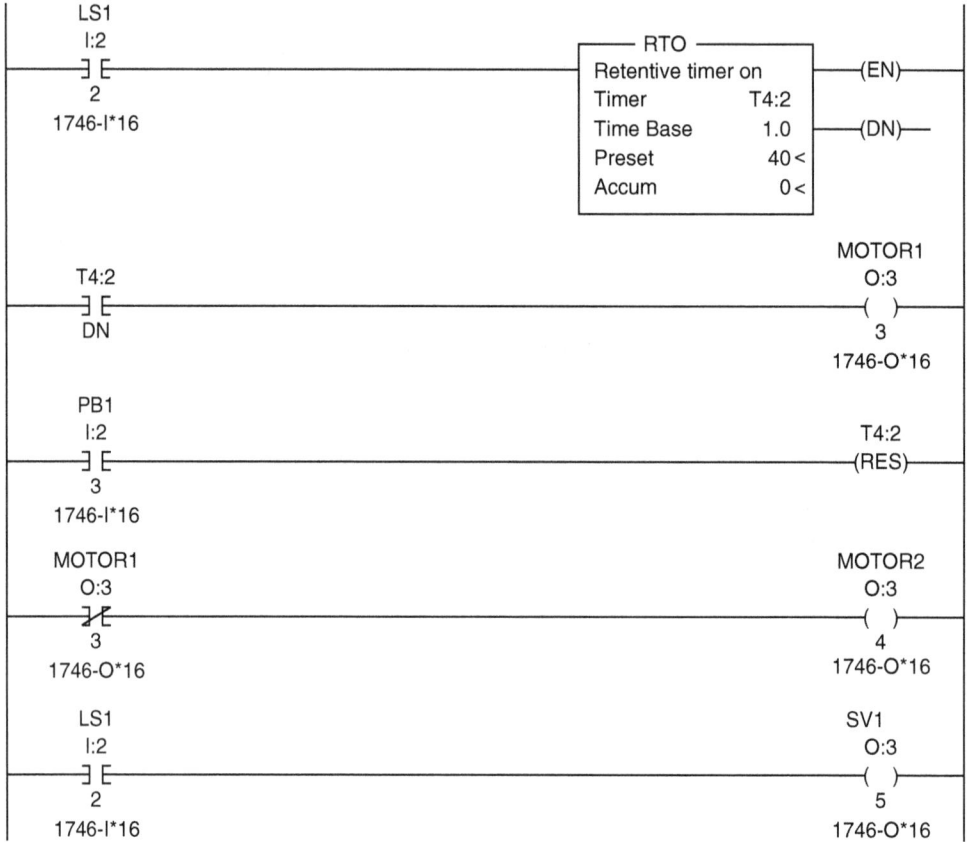

```
   LS1
   I:2                                    ┌──────── RTO ────────┐
────┤ ├──────────────────────────────────│ Retentive timer on  ├──(EN)──
    2                                     │ Timer          T4:2 │
 1746-I*16                                │ Time Base       1.0 ├──(DN)──
                                          │ Preset         40 < │
                                          │ Accum           0 < │
                                          └─────────────────────┘

                                                              MOTOR1
   T4:2                                                        O:3
────┤ ├───────────────────────────────────────────────────────( )──────
    DN                                                           3
                                                             1746-O*16

   PB1
   I:2                                                         T4:2
────┤ ├───────────────────────────────────────────────────────(RES)────
    3
 1746-I*16

 MOTOR1                                                       MOTOR2
   O:3                                                         O:3
────┤/├───────────────────────────────────────────────────────( )──────
    3                                                            4
 1746-O*16                                                   1746-O*16

   LS1                                                         SV1
   I:2                                                         O:3
────┤ ├───────────────────────────────────────────────────────( )──────
    2                                                            5
 1746-I*16                                                   1746-O*16
```

13.

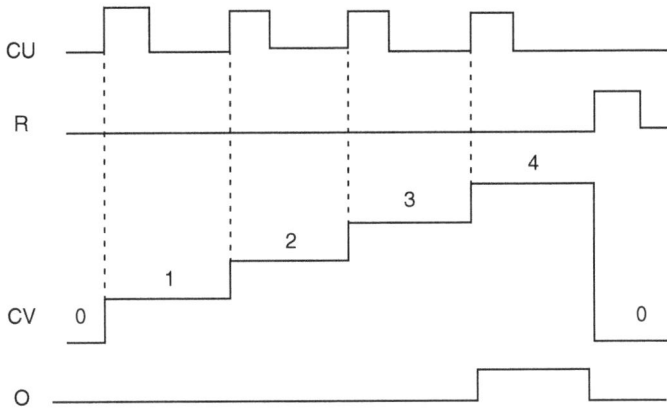

15. a. START PB should be pushed while the counter accumulated value is less than the preset value.

b. Accumulated value is larger or equal than the preset value.

c. Counter accumulated value is 0.

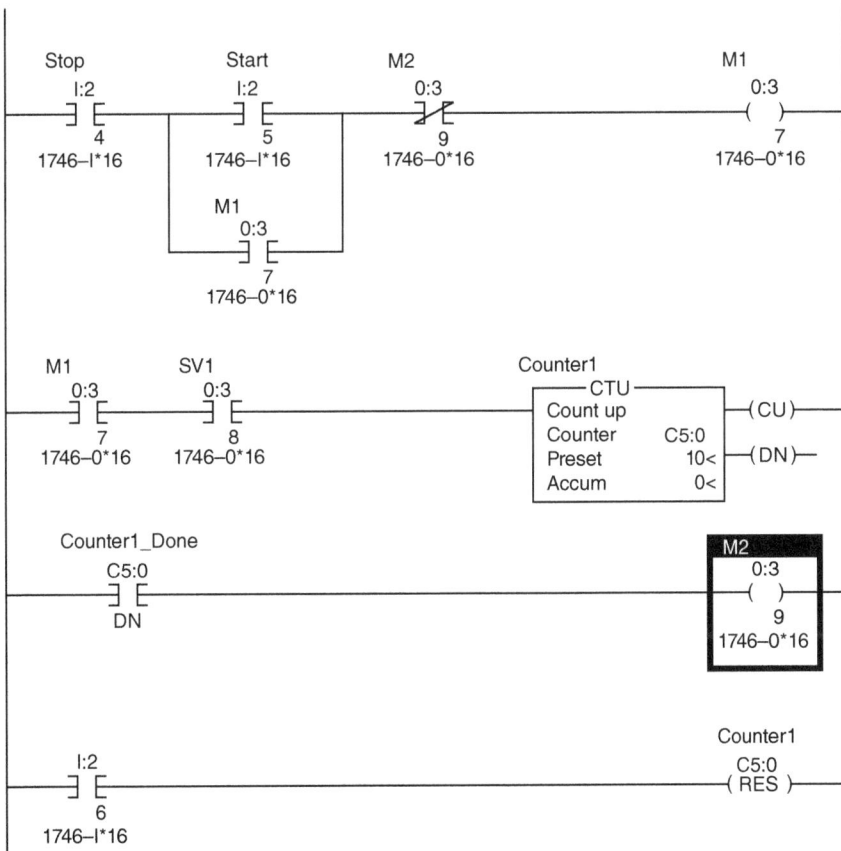

17. a. ON, ON, and OFF, respectively.

    b. ON, OFF, and ON, respectively.

19. a. Zero

    b. Three

    c. Ten

    d. When the counter accumulated value is greater or equal to the preset value.

21. The rungs shown in Figure 3.43 were programmed to indicte Pump fail to start alrm (Output O:2/7) when system is in AUTO , PUMP output is enabled, and the pump running input is not received in 5 seconds. When the rung was simulated the alarm was OFF under the above conditions. Explain why and fix the problem?

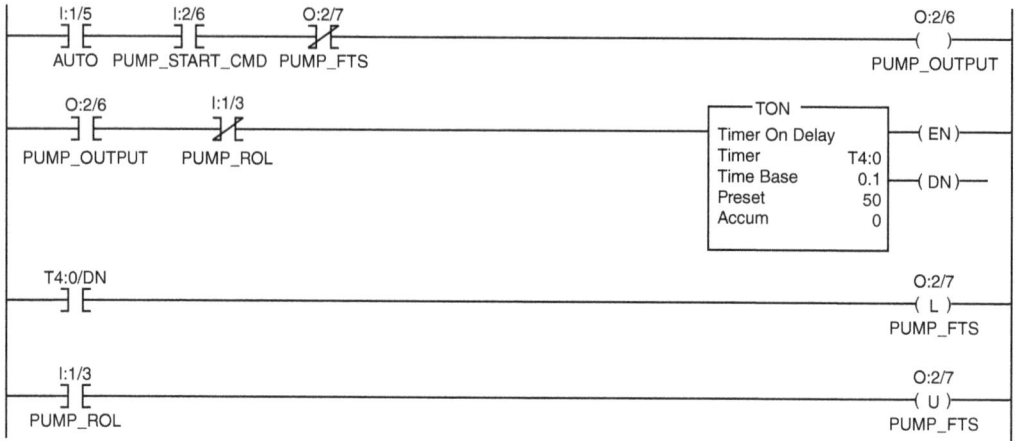

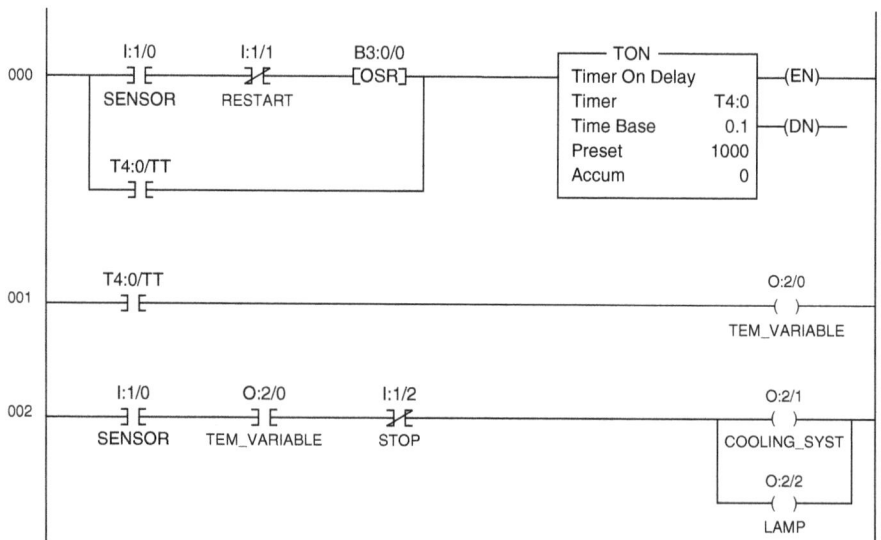

23.

```
      I:1/0                                    ┌──── CTU ────┐
    ──┤ ├──                                    │ Contact up  │──(CU)──
     CAR_ENTERING                              │ Counter  C5:0 │
                                               │ Preset    20 │──(DN)──
                                               │ Accum      0 │
                                               └─────────────┘

      C5:0/DN                                                  O:2/0
    ──┤ ├──                                                   ──( )──
                                                             PARKING_FULL

      I:1/1                                    ┌──── CTD ────┐
    ──┤ ├──                                    │ Contact down│──(CD)──
     CAR_EXITING                               │ Counter  C5:0 │
                                               │ Preset    20 │──(DN)──
                                               │ Accum      0 │
                                               └─────────────┘

      C5:0/DN                                                  O:2/1
    ──┤/├──                                                   ──( )──
                                                          PARKING_NOT_FULL

      I:1/2                                    ┌──── MOV ────┐
    ──┤ ├──                                    │ Move        │──(EN)──
      LOAD                                     │ Source  C5:0.PRE │
                                               │           ?  │
                                               │ Dest   C5:0.ACC │
                                               │           ?  │
                                               └─────────────┘

      I:1/3                                                    C5:0
    ──┤ ├──                                                   ──(RES)──
      RESET
```

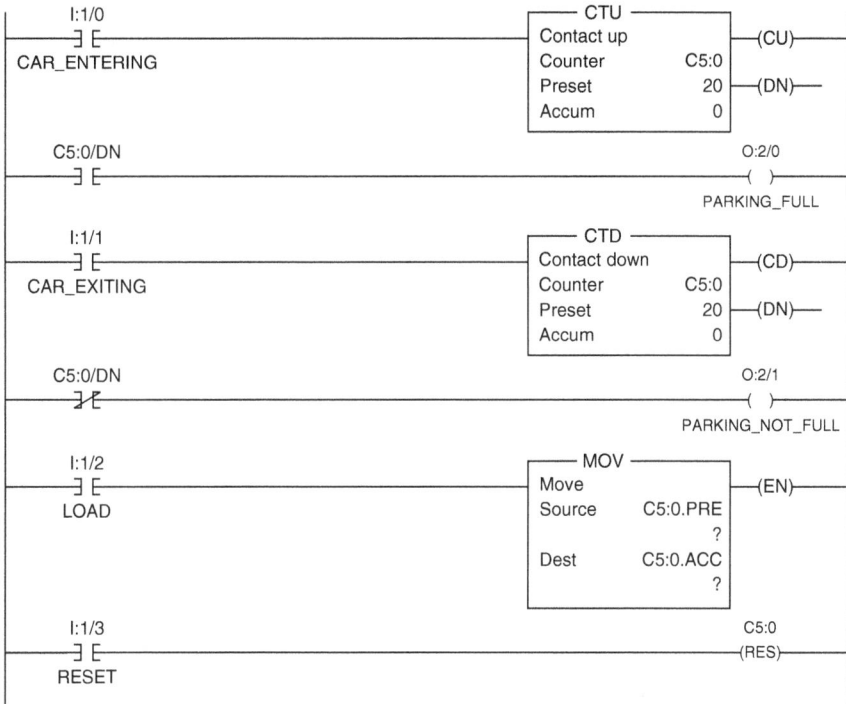

25. PL1 OFF and PL2 ON.

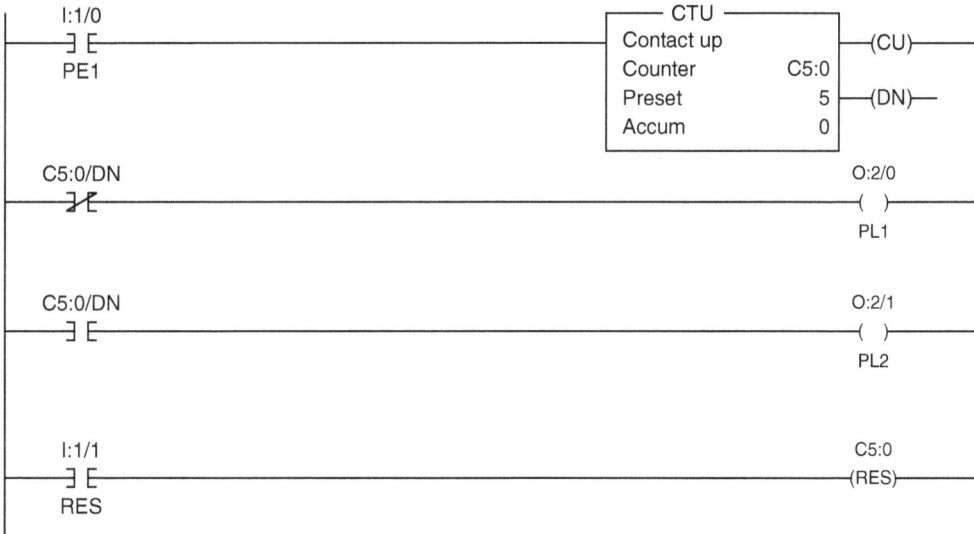

```
      I:1/0                                    ┌──── CTU ────┐
    ──┤ ├──                                    │ Contact up  │──(CU)──
      PE1                                      │ Counter  C5:0 │
                                               │ Preset     5 │──(DN)──
                                               │ Accum      0 │
                                               └─────────────┘

      C5:0/DN                                                  O:2/0
    ──┤/├──                                                   ──( )──
                                                               PL1

      C5:0/DN                                                  O:2/1
    ──┤ ├──                                                   ──( )──
                                                               PL2

      I:1/1                                                    C5:0
    ──┤ ├──                                                   ──(RES)──
      RES
```

Chapter 4 Home Work Solutions

1. File 3 is used for timers (three words per timer element). File 7 is used for integers (one word per integer).

3.

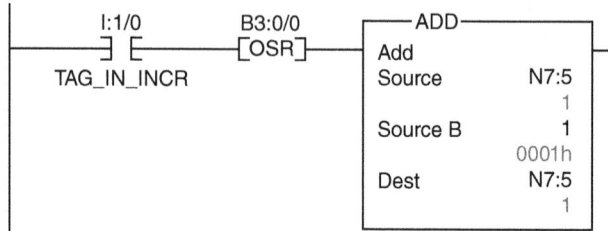

5.

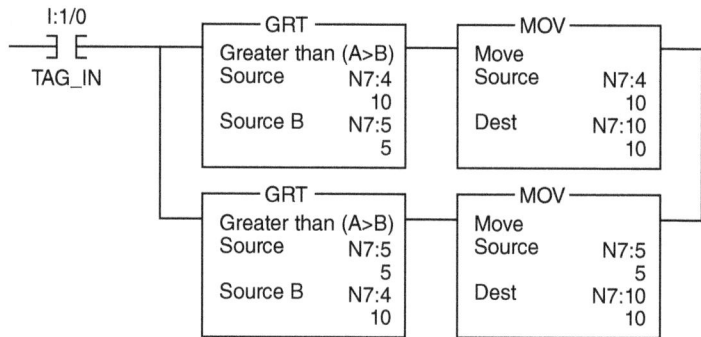

7. a. N7:0 will contain 3.

   b. N7:0 will repeatedly overflow.

9. a. B3:4 = −5 decimal = FFFB Hex.

   b. B3:4 = −5 decimal = FFFB Hex.

   c. B3:4 = −4 decimal = FFFC Hex.

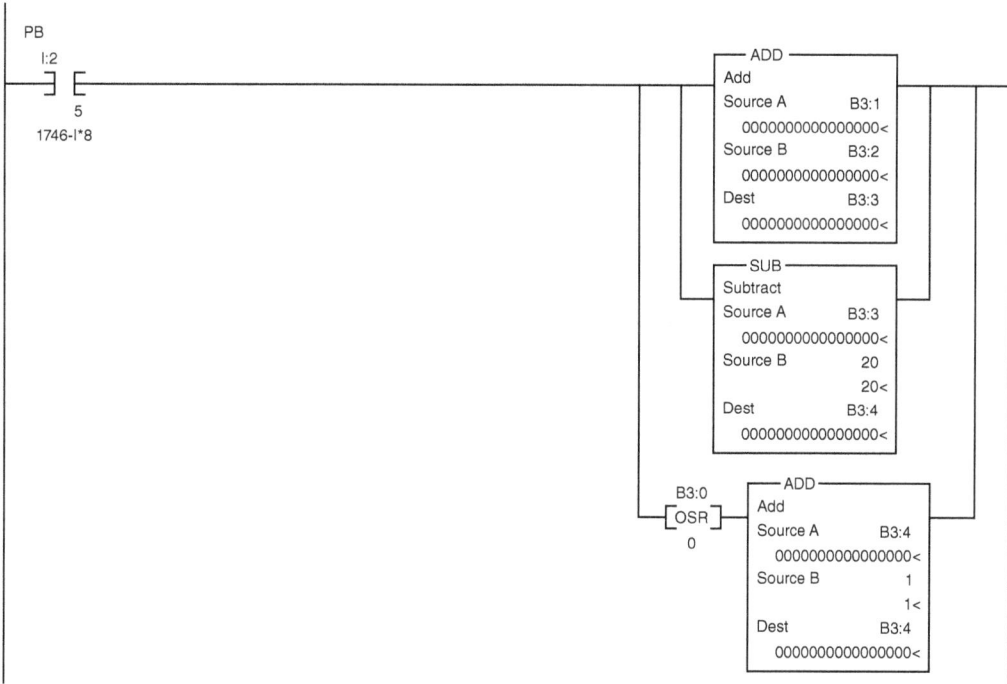

11.

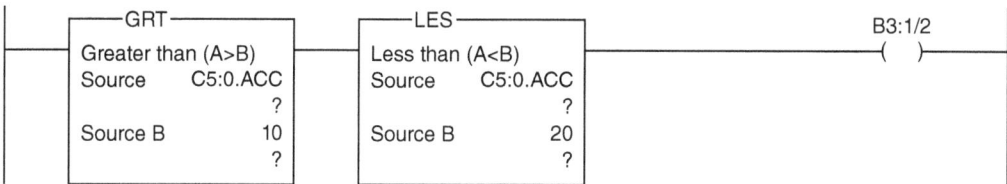

13. a. Push button

   b. If counter ACC is greater than 200 and no fault exist.

   c. If the counter is done or the START PB is pressed.

15.

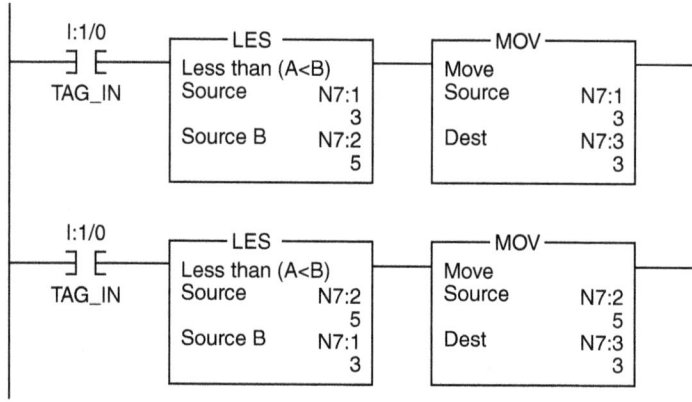

17.

19.

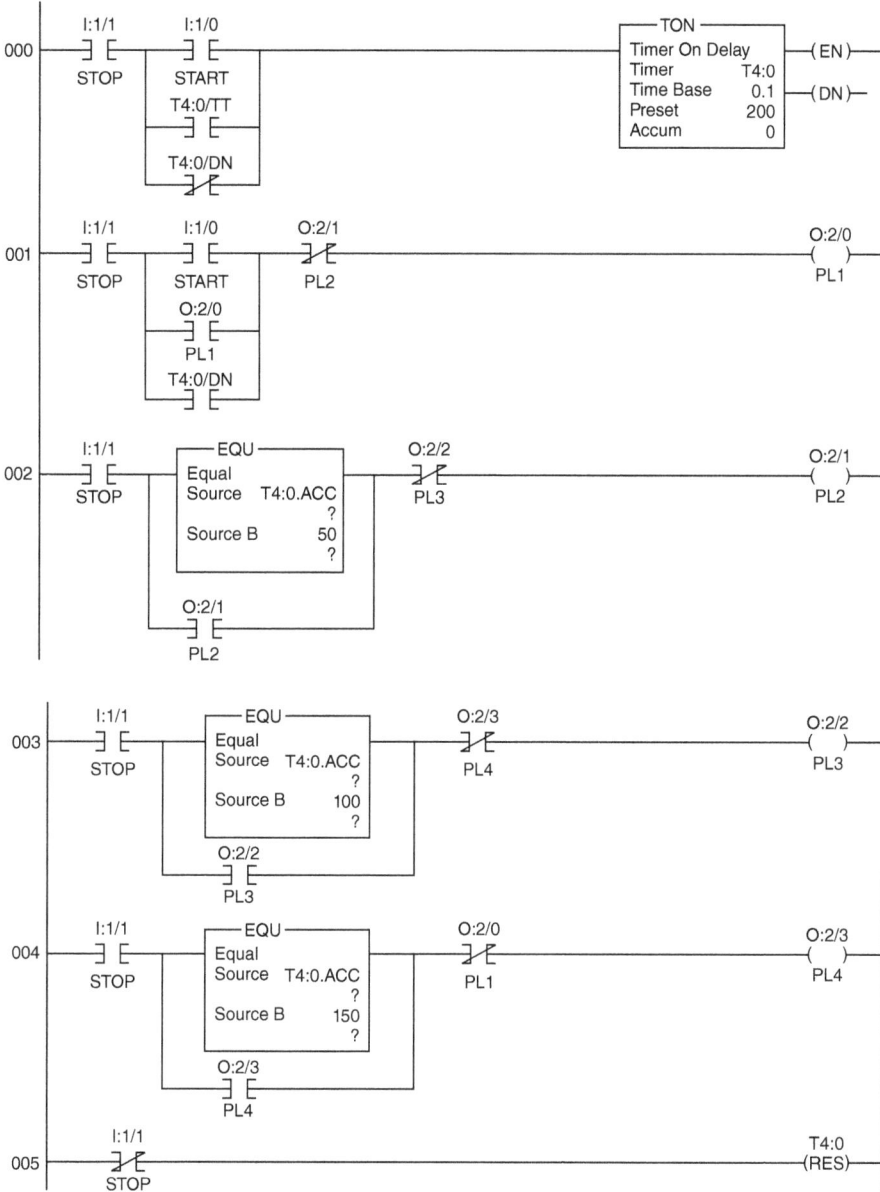

21. Motor 1 true when I:1/0 is on and (10 <= N7:10 <= 40)

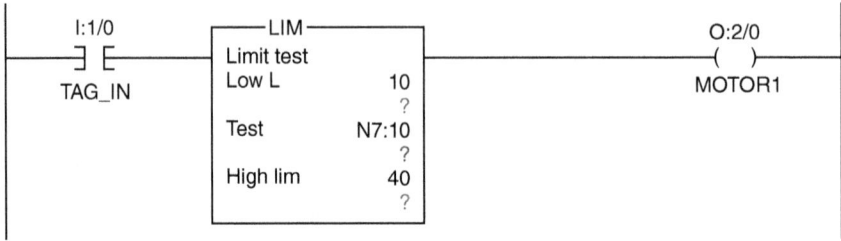

23.

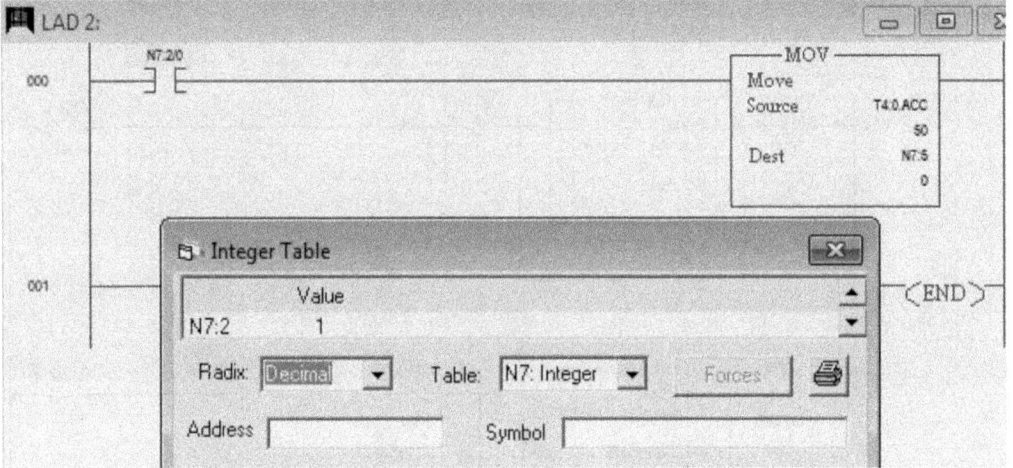

25.

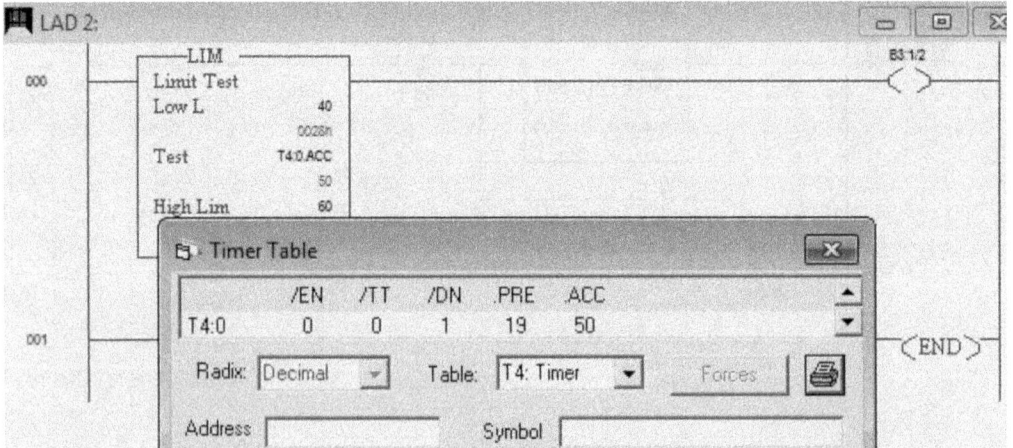

27.

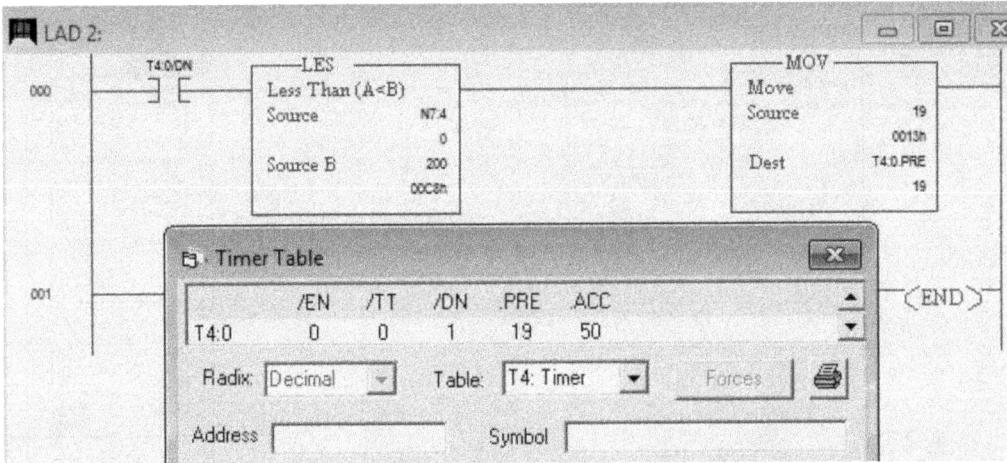

29.

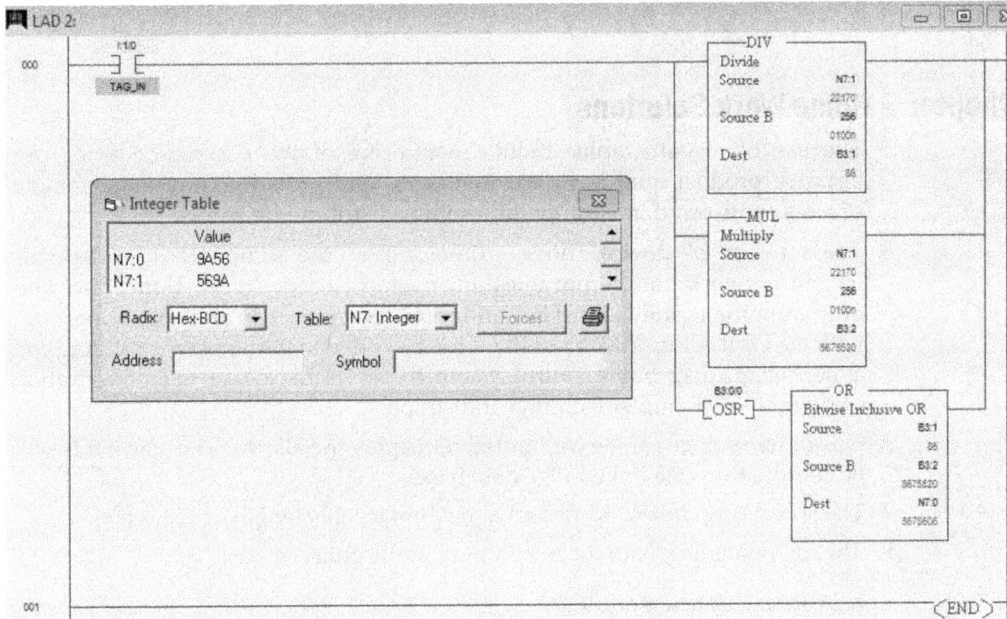

31. A one-rung implementation of the swap operation is shown below. The word to be converted is assumed in N7:1 and the swapped answer is stored in the same word.

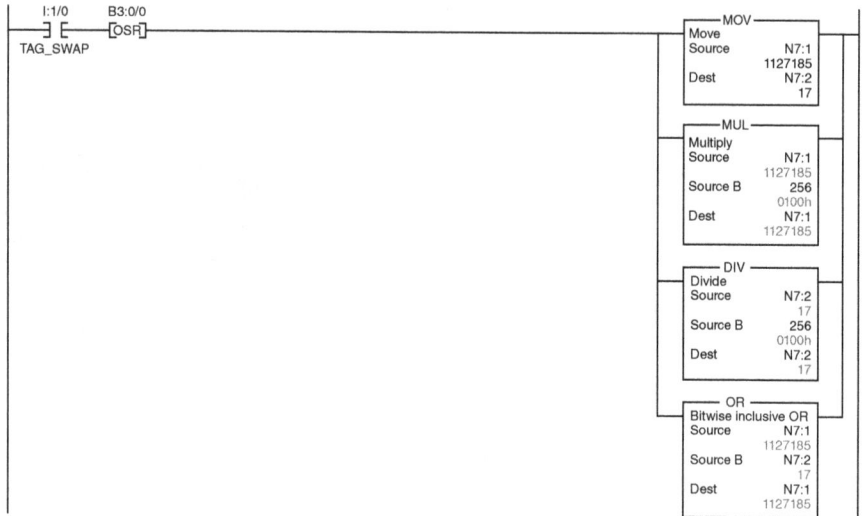

```
  I:1/0        B3:0/0                                          ┌─── MOV ──────────────┐
──┤ ├─────────[OSR]───────────────────────────────────────────┤ Move                  │
  TAG_SWAP                                                      │ Source       N7:1     │
                                                                │            1127185    │
                                                                │ Dest         N7:2     │
                                                                │                 17    │
                                                                └───────────────────────┘
                                                                ┌─── MUL ──────────────┐
                                                                │ Multiply              │
                                                                │ Source       N7:1     │
                                                                │            1127185    │
                                                                │ Source B      256     │
                                                                │            0100h      │
                                                                │ Dest         N7:1     │
                                                                │            1127185    │
                                                                └───────────────────────┘
                                                                ┌─── DIV ──────────────┐
                                                                │ Divide                │
                                                                │ Source       N7:2     │
                                                                │                 17    │
                                                                │ Source B      256     │
                                                                │            0100h      │
                                                                │ Dest         N7:2     │
                                                                │                 17    │
                                                                └───────────────────────┘
                                                                ┌─── OR ───────────────┐
                                                                │ Bitwise inclusive OR  │
                                                                │ Source       N7:1     │
                                                                │            1127185    │
                                                                │ Source B     N7:2     │
                                                                │                 17    │
                                                                │ Dest         N7:1     │
                                                                │            1127185    │
                                                                └───────────────────────┘
```

Chapter 5 Home Work Solutions

1. The use of visual graphic online monitoring of the process behavior can improve product quality, production rates, and provides convenient remote process control and integration of complex distributed systems.

3. The number of devices/nodes connected to the same network depends on the network and communication protocol used. At least one PLC and programming terminal must be configured for connection and communication with the HMI. Other HMIs and PLCs can be added to the network configuration as needed in a distributed control system. It is advisable to keep the configuration small for ease of implementation and support.

5. One or more page can be configured to display the alarms. The alarm bits will be coming from the linked PLC data base.

7. The speed is up to 19.2 Kbps and a maximum cable length of 15 Meters.

9. The following tasks cannot be executed in the online mode:

   a. Add or delete a module.

   b. Modify the I/O channel application or specific function association.

   c. Add predefined function blocks.

11. We will use an HMI for a tank level volume control, which is briefly described below:

A cylindrical tank has an area of 10 m². This application maintains a liquid volume of 10 to 50 m³ inside the tank at all times. Assume that the tank is equipped with two solenoid valves; fill (SV1) and drain (SV2). The tank level sensor is simulated by a count up (CTU) and a count down (CTD) counters. If the tank volume is greater than or equal to 50 m³, drain the tank by activating SV2. If the tank volume is less than or equal to 10 m³, fill the tank by activating SV1. Report the volume values in a tag labeled "Tank_Volume."

Two circle objects can be configured to indicate the fill and drain actions. Two additional circles can be configured to report the status of the solenoid valves; SV1 and SV2. Configure a dynamic text box to report the tank liquid volume. Two buttons can be configured to allow the operator to remotely initiate the fill or the drain action.

13. What is the purpose of Who Active utility in the RSLinx "RSWho"?

RSWho is one of RSLinx main windows, which shows networks and devices in a style similar to MS Windows. A variety of integrated configuration and monitoring tools are accessible from the right mouse button. The left pane of RSWho is the tree control, which shows networks and devices. The right pane is the list control, which shows all members of a collection. A collection is a network, or a device that is a bridge.

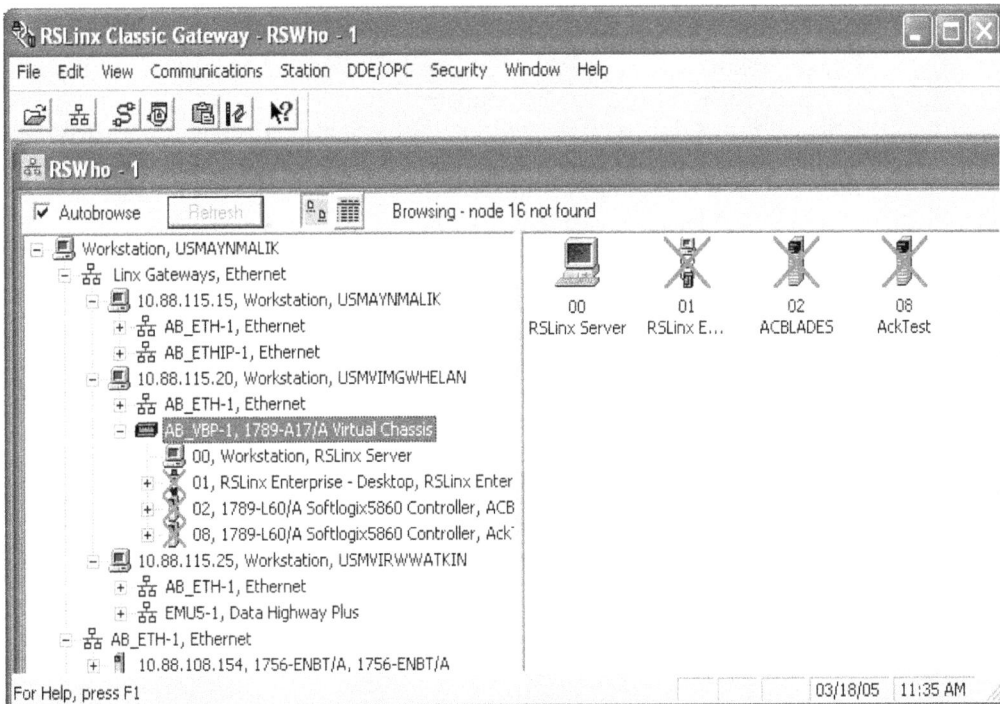

15. A typical example of a distributed control system is detailed in the case study (irrigation canal control) of Chap. 9. Seven regulators are used to control the downstream water level in an irrigation canal sections. Each regulator uses a local PLC, HMI, and panels. The local PLCs in the seven sites communicate status information to a central HMI and can receive commands from the central HMI. The central HMI can enforce supervisory global control over all sites.

17. The following is a brief list of supporting arguments:

    • Simplify the design and implementation process.
    • The use of shared configuration shortens the support requirements.
    • Selective configuration and downloading can easily achieved.
    • Less errors and better overall system monitoring and enhancements.
    • Improved scalability and improvements implementations.
    • Higher and more reliable service performance.
    • Easier data synchronization through supervisory central control.

19. A Local/Remote switch is typically used to interlock the two sources of commands. HMI commands are not active during the time the switch is placed in the Local position. Status information is reported in either of the two cases. Also user's privileges are assigned and authentication is enforced before the initiation of process commands.

21. As the number of networked Ethernet nodes increases, collisions and potential interference also increase. In addition to the limit on the number of nodes, there is also a limit on the proximity of adjacent nodes. Collisions will cause retransmission, which can be counterproductive in large real-time critical systems. Errors due to interference or noise will also cause retransmission and can affect real-time reporting and commands. The theoretical Ethernet segment node limit of 1024 is impractical and extremely higher than the actual number of nodes used in a single control system.

23. One possible scenario is to allow the user to enter a four digit BCD set value for the maximum number of scrapped bottles, which will be used by the logic as the limit for shutting down the conveyor system.

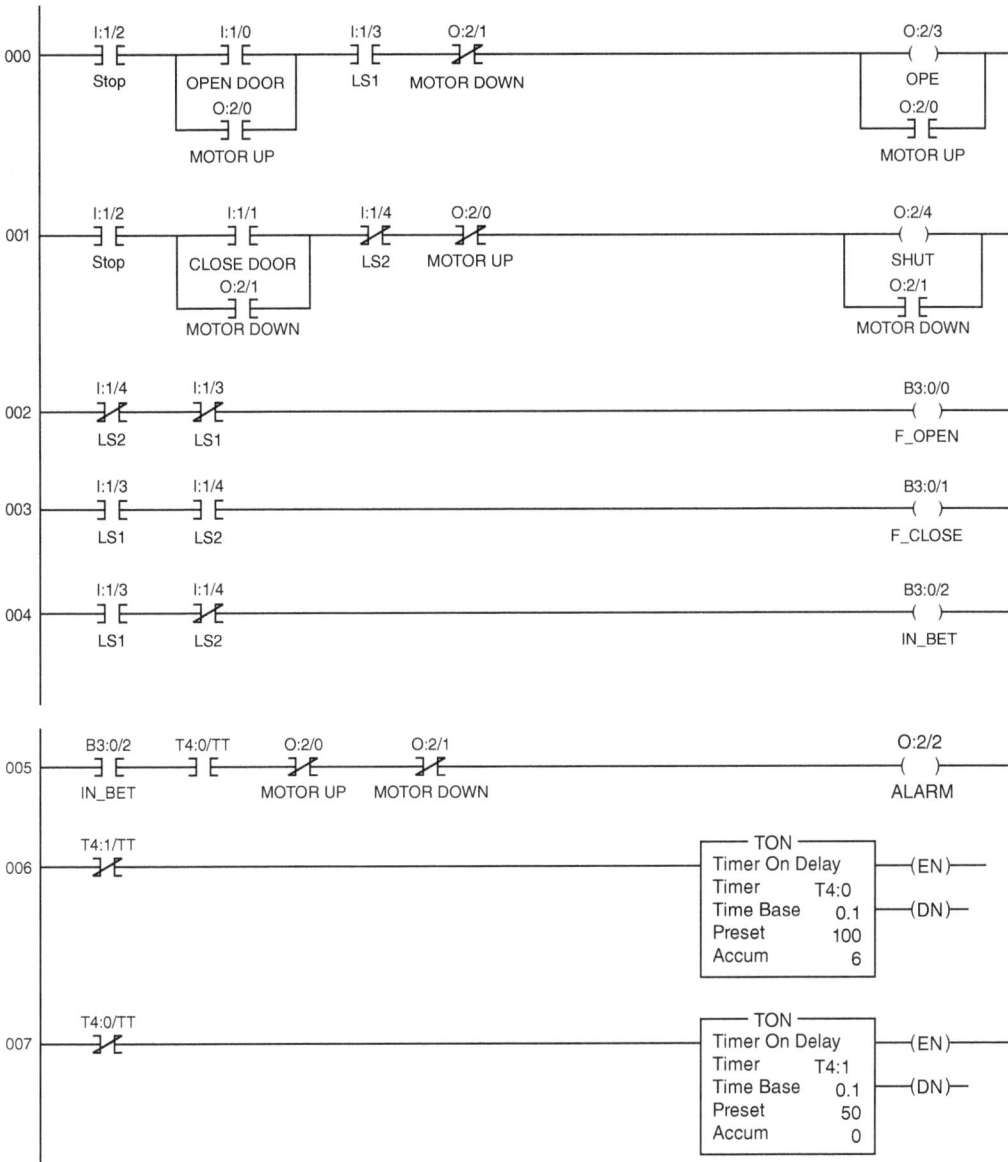

25.

Rung	Description

000
I:1/2 Stop — I:1/0 OPEN DOOR / O:2/0 MOTOR UP — I:1/3 LS1 — O:2/1 MOTOR DOWN — O:2/3 OPE / O:2/0 MOTOR UP

001
I:1/2 Stop — I:1/1 CLOSE DOOR / O:2/1 MOTOR DOWN — I:1/4 LS2 — O:2/0 MOTOR UP — O:2/4 SHUT / O:2/1 MOTOR DOWN

002
I:1/4 LS2 — I:1/3 LS1 — B3:0/0 F_OPEN

003
I:1/3 LS1 — I:1/4 LS2 — B3:0/1 F_CLOSE

004
I:1/3 LS1 — I:1/4 LS2 — B3:0/2 IN_BET

005
B3:0/2 IN_BET — T4:0/TT — O:2/0 MOTOR UP — O:2/1 MOTOR DOWN — O:2/2 ALARM

006
T4:1/TT —
TON
Timer On Delay — (EN)
Timer T4:0
Time Base 0.1 — (DN)
Preset 100
Accum 6

007
T4:0/TT —
TON
Timer On Delay — (EN)
Timer T4:1
Time Base 0.1 — (DN)
Preset 50
Accum 0

27.

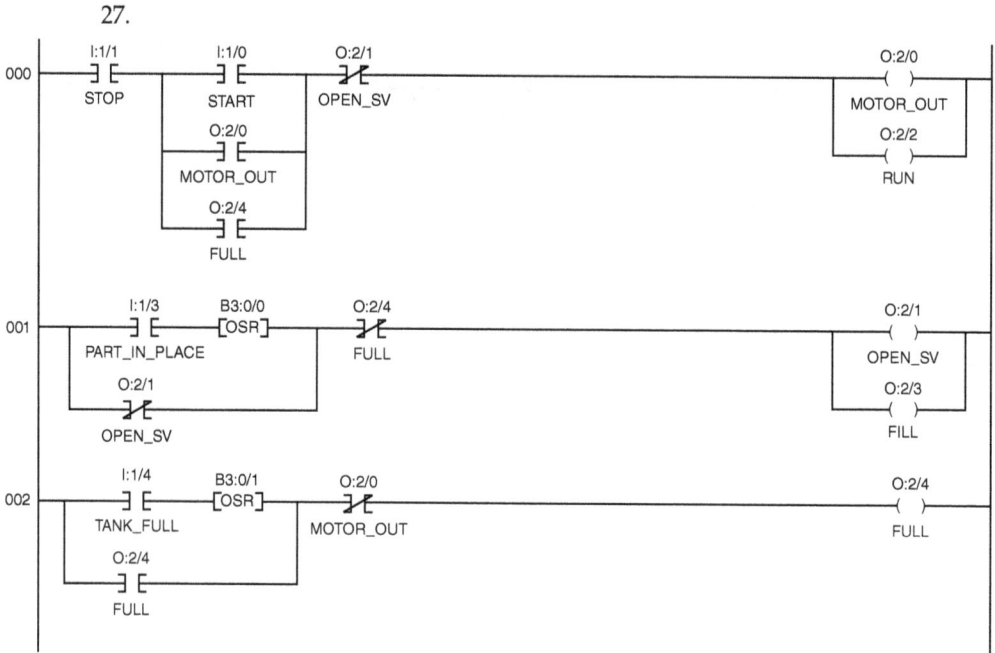

29. The below Table shows the status of the two limit switches as designed in the LogixPro simulator:

LS1	LS2	Door Status
0	0	Fully Open
1	0	In Between
0	1	Malfunction
1	1	Fully Shut

Assuming that the status of the two limit switches are reversed, the table below shows the new switches status:

LS1	LS2	Door Status
0	0	Fully Shut
1	0	Malfunction
0	1	In Between
1	1	Fully Open

31.

Rung 000:
I:1/1	I:1/0	O:2/1	B3:0/5		O:2/0
─┤ ├─	─┤ ├─	─┤/├─	─┤/├─		─( )─
STOP	START	SOL VALVE			MOTOR OUT
	O:2/0				O:2/2
	─┤ ├─				─( )─
	MOTOR OUT				RUN
	O:2/4				
	─┤ ├─				
	FULL				

Rung 001:
I:1/1	I:1/3	B3:0/0	O:2/4	O:2/1
─┤ ├─	─┤ ├─	─[OSR]─	─┤/├─	─( )─
STOP	BOX IN PLACE		FULL	SOL VALCE
	O:2/1			O:2/3
	─┤ ├─			─( )─
	SOL VALVE			FILL

Rung 002:
I:1/4	B3:0/1	O:2/0	O:2/4
─┤ ├─	─[OSR]─	─┤/├─	─( )─
LVL SENSOR		MOTOR OUT	FULL
O:2/4			
─┤ ├─			
FULL			

Rung 003:
I:1/1	O:2/1	B3:0/5
─┤ ├─	─┤ ├─	─(L)─
STOP	SOL VALVE	

Rung 004:
I:1/0	B3:0/5
─┤ ├─	─(U)─
START	

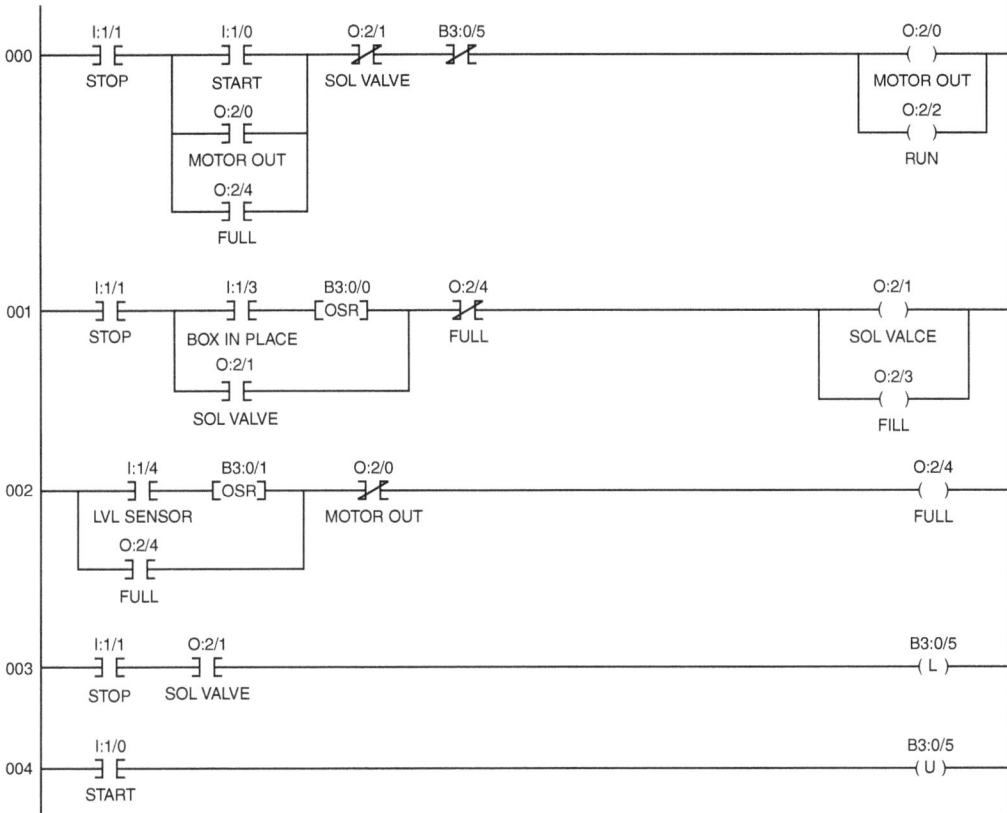

Chapter 6 Home Work Solutions

1. • Control system requirements
   • Start and stop procedures
   • Disturbances/alarms handling strategies

   Process constrains include an upper bound on all tasks completion time, precedence among tasks, and physical capability limitation. Additional constrains include task start time, interlocks, and timing/counting sequence.

3. 4 to 20 mA and 0 to 10 V are the most common and standard analog signals used in PLC control systems. A typical level control process will employ the measured level as an analog input signal and the output to the stepper motor driving the supply liquid valve as an analog output signal. Either of the two signals are typically a 4- to 20-mA current or 0- to 10-V signals.

5. • Individual tags and I/O addresses

   • Individual PLC data types

   • Individual blocks and tasks

   • Startup and shutdown procedures

7. The three modes used in testing the user program are PROG, REM, and RUN.

9. No. It is impossible to account for all possible future scenarios in a large process during system check out because of the fact that the process will not run under few rare specific situations during checkout to uncover those faults. HMIs provide visual indication of the process behavior, which can greatly help in refining the control strategy and achieve continuous quality improvements.

11. A normally closed contact emergency Stop push button is wired in series with the MCR coil to de-energize the coil in case of an emergency and disconnect the power to all I/O modules (CPU still receive power and its LEDs provide status information.

13. The standards are designed in order to protect personal/equipment in case of power outage. A device undesired turn ON or OFF can take place when power is restored if selector switches are used.

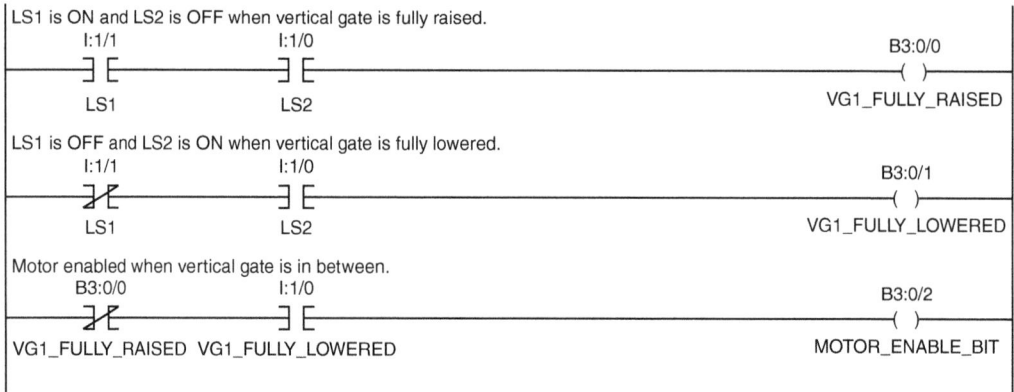

15.

LS1 is ON and LS2 is OFF when vertical gate is fully raised.

```
       I:1/1                    I:1/0                                      B3:0/0
      ─] [──────────────────────] [──────────────────────────────────────( )──
       LS1                      LS2                                  VG1_FULLY_RAISED
```

LS1 is OFF and LS2 is ON when vertical gate is fully lowered.

```
       I:1/1                    I:1/0                                      B3:0/1
      ─]/[──────────────────────] [──────────────────────────────────────( )──
       LS1                      LS2                                  VG1_FULLY_LOWERED
```

Motor enabled when vertical gate is in between.

```
       B3:0/0                   I:1/0                                      B3:0/2
      ─]/[──────────────────────] [──────────────────────────────────────( )──
 VG1_FULLY_RAISED  VG1_FULLY_LOWERED                               MOTOR_ENABLE_BIT
```

17. PL1 = ON, PL2 = OFF, and PL3 = ON

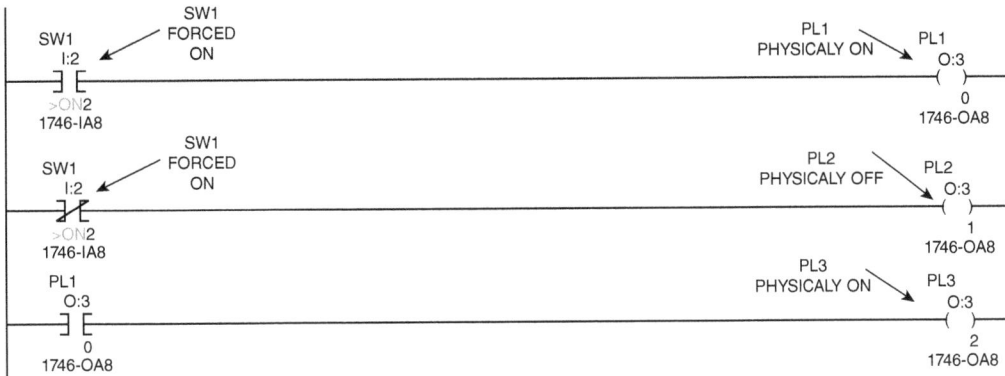

```
      SW1
SW1   FORCED                                    PL1          PL1
I:2   ON                                        PHYSICALY ON  O:3
─┤ ├─                                                        ─( )─
 >ON2                                                          0
1746-IA8                                                    1746-OA8

      SW1
SW1   FORCED                                    PL2          PL2
I:2   ON                                        PHYSICALY OFF O:3
─┤/├─                                                        ─( )─
 >ON2                                                          1
1746-IA8                                                    1746-OA8

PL1                                             PL3          PL3
O:3                                             PHYSICALY ON  O:3
─┤ ├─                                                        ─( )─
 0                                                            2
1746-OA8                                                   1746-OA8
```

19. PL2 is ON, PL3 is OFF, and PL4 is OFF

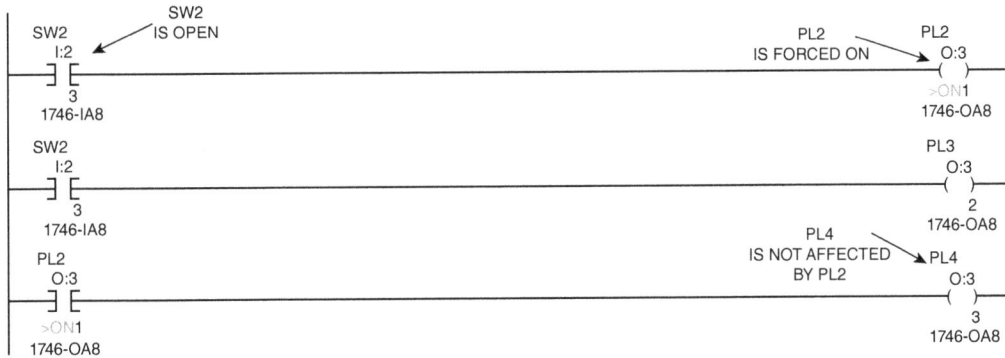

```
      SW2
SW2   IS OPEN                                   PL2          PL2
I:2                                             IS FORCED ON  O:3
─┤ ├─                                                        ─( )─
 3                                                           >ON1
1746-IA8                                                    1746-OA8

SW2                                                          PL3
I:2                                                          O:3
─┤ ├─                                                        ─( )─
 3                                                            2
1746-IA8                                                    1746-OA8
                                                PL4
PL2                                             IS NOT AFFECTED  PL4
O:3                                             BY PL2        O:3
─┤ ├─                                                        ─( )─
 >ON1                                                         3
1746-OA8                                                    1746-OA8
```

21. The program will not function correctly. Two solutions can be implemented; replace the analog input modules with standard hardware or perform the scaling over received analog input in the program software implementation. Most systems use standard analog input modules and occasionally perform the required input signal calibration/recalibration.

Chapter 7 Home Work Solutions

1. a. In computer process control, signal conditioning is used to adjust the measured signal to interface properly with the analog-to-digital conversion hardware.

   b. The transmitter has the function of propagating measurement information from the field site to the control room where the control action is to occur. Usually pneumatic or electronic signals are used.

   c. A multiplexer (or MUX) is a device that selects one of several analog or digital input signals and forwards the selected input into a single output line.

d. An electronic device that makes possible the transmission of data to or from a computer via telephone or other communication lines.

e. In an analog-to-digital conversion, the difference between the actual analog value and the equivalent analog value of the quantized digital value is called quantization error or quantization noise.

3. Answer is c.

5. "Seebeck effect" refers to converting temperature differences directly into electrical signal. It was discovered by a German physicist Thomas Johann Seebeck in 1821.

7. The following are the most common two standard analog signals:

$$4 \text{ to } 20 \text{ mA and } 0 \text{ to } 10 \text{ V dc}$$

9. More than one variable is selected and regulated in order to control a certain process.

11. a. Data logging implies recording the real-time data generated by an input device into computer memories or secondary storage.

b. Digital process control refers to the automated control of a process using digital computers or PLCs. Process control is used extensively in oil refining, chemical processing, electrical generation, and the food and beverage industries, where the creation of a product is based on series of processes being applied to raw materials.

c. Supervisory control is a general term for control of several individual controllers or control loops. It can also automatically generate the set point for the individual controllers in real time.

d. Analog-to-digital converter is part of the input module that converts analog input signal to a digital count.

e. Digital-to-analog converter is part of the output module that converts digital count to an analog signal.

13. The mode of control for which the amount of output is determined by the rate of change of the error is labeled as (on/off, proportional, derivative, or integral) control.

Answer is derivative control.

15. Answer is b.

17. a. $(300 - 50)/(5 - 0) = 50°\text{F}/\text{V}.$

b. $(300 - 50)/4095 = 0.06105°\text{F}/\text{bit}$

c. $100/0.06105 = 1638$

d. $5/4095 = 0.001221$

e. Average quantization error $= (0.001221/2) = 6.105006*10^{-6}$

19. In general, increasing the dead band causes the controller to turn the heater ON/OFF less often, while decreasing the dead band turns the heater more frequently.

21. $$\text{Proportional band} * \text{proportional gain} = 1$$
$$\text{Proportional band} = (1/\text{proportional gain}) = 1.33333$$

23. a. $$(350 - 40)/16 = 19.375°C/mA$$

   b. $$(350 - 40)/1023 = 0.3030°C/bit$$

   c. $$y = 0.303 * x + 40$$
   $$x = (y - 40)/0.303 = (159 - 40)/0.303 = 393$$

   d. $$\text{High threshold} = (SP + 0.5 * DB) = 252$$
   $$\text{Low threshold} = (SP - 0.5 * DB) = 248$$

   e. $$\text{Average quantization error} = 0.0303/2 = 0.01516°C$$

Chapter 8 Home Work Solutions

1. The maximum decimal count range of the SLC 500 analog modules is from −32768 to +32767 (as shown in Chap. 8, Fig. 8.5). The count corresponds to a current range of −20 to +20 mA or −20 to +20 volts. The analog input signal in the range from the minimum to the maximum value converts to the corresponding digital counts regardless of the module resolution actually used.

3. The "scale" instruction scales the value at the scale source input by mapping it using a linear equation defined by the line slope (rate/4000) and an offset. The two instruction parameters are calculated using the input and the output defined ranges.

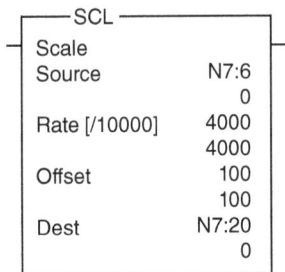

```
┌────────SCL────────┐
│  Scale            │
┤  Source      N7:6 ├
│                 0 │
│  Rate [/10000]  4000 │
│                 4000 │
│  Offset        100 │
│                100 │
│  Dest        N7:20 │
│                  0 │
└───────────────────┘
```

5.

```
       ANALOG_INPUT_
       SCALLING                                              BELOW_RANGE_FLAG
        ┌─ LES ──────────┐                                        B3:0
0000    │ Less than (A<B) │                                      ─( )─
        │ Source A   I:11 │                                        0
        │             0<  │                                  BELOW_RANGE_ALARM
        │ Source B  14344 │                                        O:2
        │           14344<│                                      ─( )─
        └─────────────────┘                                        0
                                                                1746-O*S
       ANALOG_INPUT_
       SCALLING                                              ABOVE_RANGE_FLAG
        ┌─ GRI ──────────────┐                                    B3:0
0001    │ Greater than (A>B) │                                   ─( )─
        │ Source A   I:11    │                                     1
        │             0<     │                                ABOVE_RANGE_ALARM
        │ Source B  16383    │                                     O:2
        │           16383<   │                                   ─( )─
        └────────────────────┘                                     1
                                                                1746-O*8

                                                          DIVIDE_RESULT_BY_
        BELOW_RANGE_FLAG   ABOVE_RANGE_FLAG                 INPUT_RANGE
            B3:0               B3:0                         ┌─ SCL ────────────┐
0002       ─┤/├─              ─┤/├─                         │ Scale            │
             0                  1                           │ Source     I:11  │
                                                            │             0<   │
                                                            │ Rate [/10000] 120│
                                                            │            120<  │
                                                            │ Offset      100  │
                                                            │            100<  │
                                                            │ Dest       N7:0  │
                                                            │             0<   │
                                                            └──────────────────┘
```

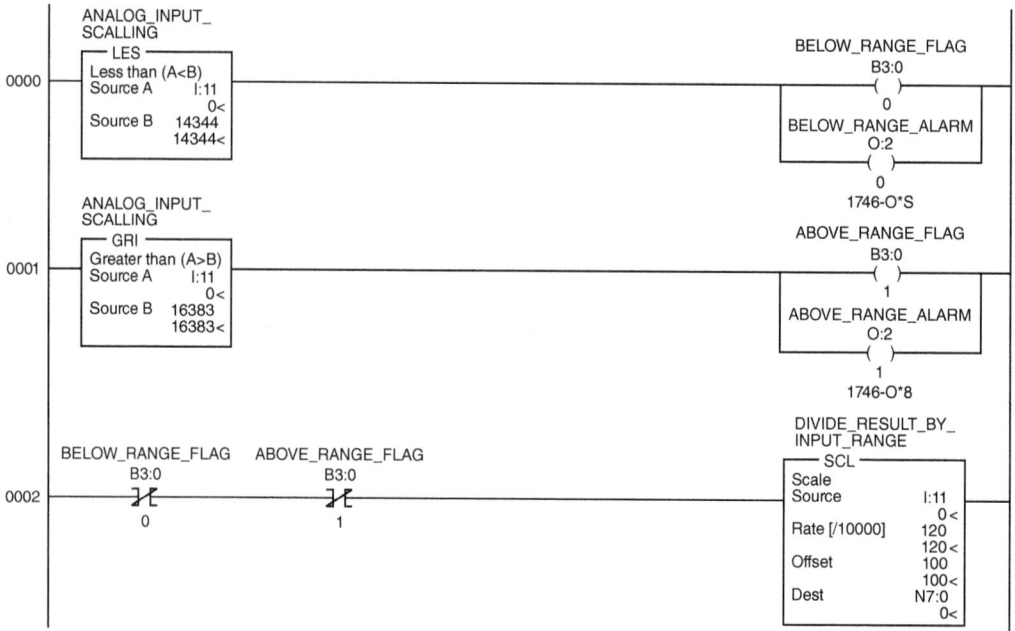

7. a. The limit instruction will validate the input analog signal count to be within the defined range, low limit to high limit. The output coil/bit B3:0/10 of the rung is true when the input signal value is valid.

b. The limit instruction can be replaced by using compare instructions to check if the signal count is between 0 (low limit) and 16,383 (high limit).

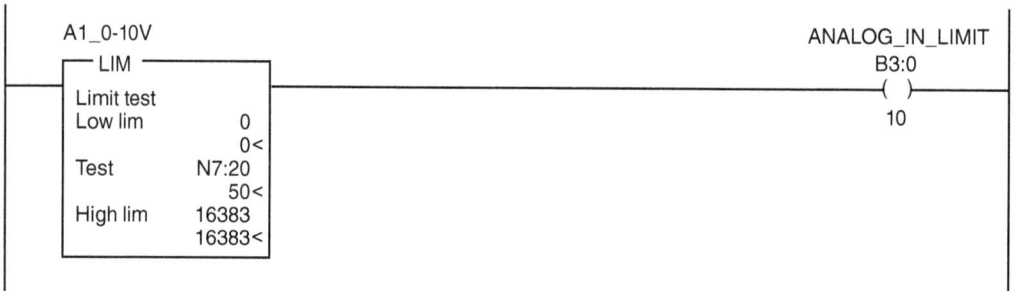

```
     A1_0-10V                                               ANALOG_IN_LIMIT
      ┌─ LIM ──────────┐                                         B3:0
      │ Limit test      │                                       ─( )─
      │ Low lim    0    │                                         10
      │             0<  │
      │ Test     N7:20  │
      │            50<  │
      │ High lim 16383  │
      │          16383< │
      └─────────────────┘
```

9. a. Dead time is the delay from when a controller output (CO) signal is issued until when the measured process variable (PV) first begins to respond.

b. Settling time is the time elapsed from the application of an ideal instantaneous step input to the time at which the output has entered and remained within a specified error band.

c. Recovery time is the duration of linear rise of the controlled variable after a process disturbance; please refer to Figs. 8.14 and 8.15 in Chap. 8.

d. Delay time is duration of initial small rise of the controlled variable after a process disturbance; please refer to Figs. 8.15 and 8.16 in Chap. 8.

11.

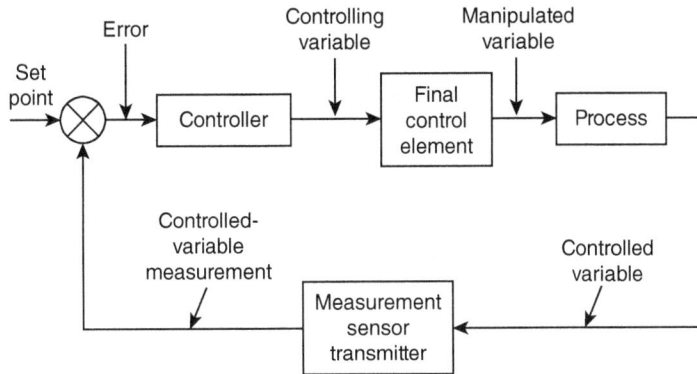

   a. Error is the difference between the set point and the controlled variable.

   b. The controller manipulates the error using the PID or other algorithms and sends the optimum controlling variable value to the final control element.

   c. Process refers to an interacting set of actions/operations that lead to the manufacture or development of some product.

   d. Measurement is the sensor/transmitter that converts a physical quantity into a signal suitable to be interfaced to the next stage.

13. a. In proportional-action controlled systems, the controlled variable process value follows the controller output value (controlling variable) almost immediately.

   b. In a first-order controlled system, the controlled variable value initially changes in proportion to the change of the controller output (the controlling variable value). The rate of change of the Controlled variable value is reduced as a function of the time elapsed until process steady state is reached.

   c. In a second-order delay element controlled system, the process controlled value does not immediately follow a step change of the controller output value. The Process Controlled value initially increases in proportion to the positive rate of rise of the controller Output value and then approaches the set point at a decreasing rate of rise.

15. The principal characteristic that makes a process self-regulating is that it naturally seeks a steady-state operating level if the controller output and disturbance variables are held constant for a sufficient period of time.

   Example: Cruse control of a car is a self-regulating process. If we keep the fuel flow to the engine constant while traveling on flat ground on a windless day, the car will settle out at some constant speed. If we increase the fuel flow rate by a fixed amount, the car will accelerate and then steady out at a different constant speed.

17.

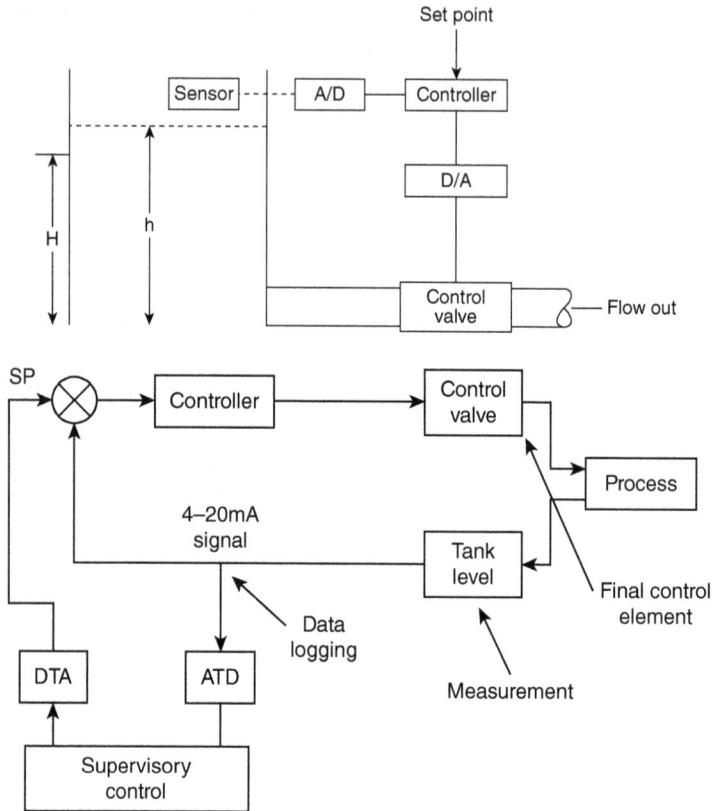

Supervisory control benefitted and made use of huge advancements in technologies, including universal standards, digital hardware, real-time operating systems, communication and networking, human machine interfaces, remote sensing, sensory fusion, redundancy, safety tools, and the wide spread of open system architectures. Some of the world largest international chemical and petroleum corporation own and operate the largest global computer networks. Every process controller, data acquisition system, human machine interface, actuators, sensors, and communication devices has to be accessible from any location in the world with proper authorization. This level of integration allows for highest level of coordination among all subsystems and thus producing optimum system control.

19. The concept of controllability denotes the ability to move a system around in its entire configuration space using only certain admissible manipulations. The exact definition varies slightly within the framework or the type of models applied.

   The following are examples of variations of controllability notions that have been introduced in the text book.

   - State controllability
   - Output controllability
   - Controllability in the behavioral framework

Glossary

A

Accumulated value   The number of counted events or timed intervals from a reference time or event mark.

Accuracy   Closeness between the value indicated by a measuring instrument and the corresponding true value. Sensor accuracy is based on U.S. NIST (NBS) standards.

Actual value   The present value of a variable.

Actuator   A device connected to an output module, which is able to regulate the behavior of a controlled process variable.

Address   A numerical identifier for a controller when used in computer communications.

Alarm (process alarm)   Warns when process values exceed the process alarm setting, a fixed value independent of set point.

Algorithm   A procedure for problem solving that is characterized by its time and space complexity as function of the problem size.

Alternating current (ac) I/O   An I/O module that converts ac signals originating or received by a field device to appropriate PLC processor logic level signal.

Analog device   An apparatus that measures or generates continuous signal, which is typically in voltage or current. Its accuracy depends on the resolution of the apparatus.

Analog input module   An input circuit that uses analog-to-digital converter to map measured continuous signal to a digital count that can be used by the processor.

Analog output module   An output circuit that uses digital-to-analog converter to map processor-generated digital count to continuous signal that can be used by the connected analog device.

Analog signal   A continuous signal that changes smoothly in a defined range rather than changing suddenly between discrete levels as in digital signals.

Analog to digital (ATD or A/D)   Electronic hardware that converts an *analog* signal like voltage, electric current, temperature, or pressure into a *digital* number that a computer can process and interpret.

Analog transmission   The transmission of data is done as a continuous signal, as opposed to an ON/OFF digital signal.

Anti-reset windup   This feature of the PID controller prevents the integral component (reset action) from operating when the controlled variable is outside the proportional band.

B

Big-endian   Most common format in data networking and the Internet protocol suite; such as IPv4, IPv6, TCP, and UDP. Stores the most significant byte at the lower memory address with the least significant byte being stored at the highest memory address.

Binary   A module/base 2 number system used in digital computers to represent information, data, and program code.

Binary bit   The smallest unit of information used in the module 2 numbering system. It is also known as a *binary digit*. It can carry either 0 or a 1 value.

Block diagram   A method used to represent the functional blocks and decision points of a system or an algorithm.

Boolean   A data type that stores the state of a single bit to either 0 (OFF) or 1 (ON).

Boolean algebra   A mathematical system used to express logic functions and associated logic operations.

Branch   A parallel logic path within a ladder logic network or rung.

Byte   A group of 8 adjacent bits categorized as 1 unit during the processor's three operations: fetch, read, or write cycles.

C

Calibration offset   An adjustment to eliminate the difference between the indicated value and the actual value.

Cascade control   Control in which the output of one controller is the set point for another.

Celsius   Formerly known as *centigrade*. A commonly used temperature scale, which assumes that water freezes at 0°C and boils at 100°C.

Central processing unit (CPU)   The very large integrated circuit that controls all PLC activities. It also manages all hardware and software resources/interfaces.

Clock   A circuit that generates timing pulses, which are used to synchronize the PLC, computer, or digital system operations.

Closed loop   A control system that utilizes feedback from a process to regulate its behavior.

Code   A set of programmed instructions defined for a given processor.

Coil   A presentation of the PLC output, which when energized, changes the status of the associated contacts.

Contact   A presentation of the PLC input or logic condition, which controls the propagation of current through the ladder logic network or rung.

Control cycle   The rate at which the output signal is updated.

Control mode   The output form or type of control action used by a controller to control a process: on/off, PI, PID, or manual.

Control relay   A relay used to control a task or a sequence of events.

Controlled variable   The process variable that is regulated in time and kept near a defined set point or desired value.

Cycle   A sequence of operations repeated regularly to complete or regulate a task.

D

Dead band   A region selected around the set point where control is withheld.

Dead time   The amount of time that it takes for the controlled variable to *start* changing after the controlling variable changes.

Debug   The process of locating and correcting errors in software, hardware interconnections, configurations, and communication.

Derivative   The rate of change of a process variable in time.

Deviation   The difference between the value of the controlled variable and the set point.

Diagnostic   Detection and isolation of errors and device faults.

Digital device   A device that accepts and processes discrete electric signals.

Digital signal   A signal that changes suddenly between certain levels, typically 0 (OFF) and 1 (ON).

Digital to analog (DTA or D/A)   An electric circuit that converts binary count to continuous analog signal.

Direct action   An output control action in which an increase in the controlling variable causes an increase in the controlled variable.

Distributed control   A method of control, which divide a large system into a number of integrated subsystems.

Done bit   A bit that is set to 1 oz, the instruction completes its task as in reaching a predefined preset value.

Down-counter   A counter that starts at a preset value and counts down to 0.

Download   Loading data (configuration or programs) from one computer to another networked intelligent device as HMIs and PLCs.

E

Edit   The process of ladder logic program modification to eliminate errors, change system operation, or optimize code.

Emergency shutdown relay (ESD)   A relay-associated contact used to inhibit electric power flow to control system PLC in an emergency situation.

Energize The physical connection of electric power to a device to initiate an action as in running a motor.

Endianness Refers to the order of the bytes, comprising a digital word, in computer memory.

Error signal A signal proportional to the difference between the actual process variable and its desired value.

Error The error is defined as set point minus the controlled variable.

Ethernet A common network protocol that uses Carrier Sense Multiple access Collision Detect (CSM/CD) access control.

Exclusive OR A logic element producing a true output only when its two inputs are different values.

Execution time The total time required for the completion of a program, an instruction, or a sequence of instructions.

F

Fahrenheit The temperature scale that sets the freezing point of water at 32°F and its boiling point at 212°F. The formula for conversion to Celsius is °C = 5/9* (°F − 32).

False A disabling logic state resulting from unsatisfied condition.

Fault A malfunction that interferes with normal system operation.

Feedback A correcting signal fed back to the controller to produce desired regulation of the process.

Floating point (real) Numbering representation for integers, fractions, or a mixed real number.

Flowchart A graphical representation of the steps, events, sequences, calculations, logic, and decisions used in an algorithm.

Force function An instruction that allows the operator to override the processor logic and control the state of a device.

Function keys Graphical keys on a computer/HMI or hard keys on a panel/keyboard that can be programmed according to a predefined operator user interface protocol.

G

Gate A logic circuit with one or more inputs and one output. The output is a function of all gate inputs.

Ground A conducting connection between a circuit or equipment chassis and the earth ground.

H

Hardwired The wire interconnection of electrical and electronic components/elements of a control system.

Hexadecimal A base 16 numbering system that uses the digits 0 to 9 and A to F to represent decimal values from 0 to 15.

HMI   Human Machine Interface.

Hysteresis   A band of change in controlled variable around the set point requiring no control action.

I

I/O   Input/output—either analog or discrete.

I/O address   A unique number assigned to each input/output device that corresponds to the device's location in the rack enclosure.

I/O module   A plug-in assembly, which contains two or more identical input or output circuits that provide the connection between a processor and the connected devices.

I/O scan time   The time required to update all local and remote I/O.

I/O update scan   The process of revising the bits in a PLC's I/O tables based on the latest results from reading the inputs and processing the outputs according to the control program.

Image table   A PLC-dedicated memory area for input/output data. Ones and zeroes represent I/O ON and OFF conditions, respectively.

Input   Process variable information that is supplied by the instrument.

Input scaling   The ability to scale input readings to the engineering units of the process variable.

Input type   The signal type that is connected to an input, such as thermocouple and switches.

Input/output address   A unique identifier assigned to each input and output.

Input/output scan time   The time required by the processor to monitor all inputs and control all outputs.

Input/output update   The continuous process of updating the entire input and output tables based on the latest scan results.

Integer   A positive or negative whole number.

Integral   The function in a PI or PID controller that adjusts the controlled variable based on its error history.

Interlock   A mechanism for preventing one output device or input action from being on while another is on.

Inverter   A digital circuit that performs inversion.

IP address   A unique Internet protocol address assigned to every networked device.

J

Jump instruction   A conditional or unconditional deviation from the sequential execution flow of a program.

Junction   The point where two dissimilar metal conductors join to form a thermocouple.

K

K   A measure used to express memory size in bits, bytes, or words. 1 K is equivalent to a decimal 1024.

Kilo (k)   A prefix used to quantify measurements in 1000 times.

L

Label   A name given to a membership function.

Ladder diagram   A graphical set of instructions that implements relay ladder functions.

Ladder logic   An industry standard for representing relay logic control systems.

Ladder relay instructions   Computer codes that implement relay coils and contacts and their corresponding functions in a PLC.

Latch instruction   The first half of an instruction pair, latch and unlatch. The PLC latch instruction energizes a specific output or internal coil until de-energized by the matching unlatch instruction.

Least significant bit (LSB)   The bit that carries the least weight in a byte or an arbitrary size word.

Limit switch   An electric switch, which is activated through motion or contact.

Little-endian   Stores the least significant byte at the lower memory address with the most significant byte being stored at the highest memory address. Popular for microprocessors, in part due to significant influence on microprocessor designs by Intel Corporation.

Load   The demand on a process, expressed in power, current, resistance, or torque.

Local area network (LAN)   A system of hardware and software that allows smart devices to communicate and exchange data within a small geographical area.

Logic diagram   A drawing that uses interconnected AND, OR, and NOT logic symbols to graphically describe a system's operation or control.

Logic level   The voltage magnitude associated with the 0 and 1 states in a digital binary system.

Loop tuning   The process of determining the proportional, integral, and derivative constants that will allow a PID controller to perform optimally.

M

Manual mode   The user observes the process and sets the output.

Master control relay (MCR)   A hardwired or software relay instruction that will de-energize its associated I/O devices when the instruction is de-energized.

Master rack   The enclosure containing the CPU or processor module.

Memory map   A diagram showing a system's memory addresses, as well as which programs and data are assigned to each section of memory.

Memory   The part of a programmable controller that stores data, instructions, and the control program.

Microprocessor   A large-scale integrated central processing unit on one chip.

Microsecond   One millionth of a second.

Millisecond   One thousandth of a second.

Modes of control   In Auto mode the controller calculates the output based on the error behavior. In Manual mode, the user sets the output. In remote, the controller is actually in auto but gets its command from another controller.

Module   An interchangeable, plug-in item containing electronic components.

Most significant bit (MSB)   The bit representing the greatest value of a nibble, byte, or word.

Most significant digit (MSD)   The digit representing the greatest value of a byte or word.

Motor control center   A control system utilizing variable speed drives by regulating voltage, current, or the frequency supplied to the motors.

Motor starter   A special relay designed to provide power to motors. It has a contactor and an overload relays connected in series. The contactor relay de-energizes under overload conditions.

Move instruction   A PLC instruction that moves data from one location to another.

N

Nested branches   A branch that begins and ends within another branch in a ladder logic rung.

Network   A set of devices or computers connected through a communication medium.

Nibble   A group of 4 bits.

Node   A station, such as a personal computer, a PLC, or an HMI that is connected to a network and can communicate through the network.

Noise   Unwanted electrical signals that usually produce signal interference in sensors and sensor circuits.

Nonvolatile memory   A memory designed to retain its memory while the power source is turned OFF.

NOR   A logic gate that produces a 1 (ON) output only when both of its two inputs are held at the 0 (OFF) state.

Normally closed contact (NC)   A contact that is conductive only when its operating coil is not energized.

Normally open contact (NO)   A contact that is conductive only when its operating coil is energized.

NOT   A logic operation that produces the inverse of its input.

O

OFF delay timer   An electromechanical relay with contacts that change state after a predetermined time from the point where power is removed from its coil.

Offline programming   The safest mode to develop PLC programs where the operation of the processor is stopped and all output devices are switched OFF.

ON delay timer   An electromechanical relay with contacts that change state after a predetermined time from the point where its coil is energized.

On/off control   A method of control in which the controller acts as a switch, turning the final control element either ON or OFF.

Online programming   The ability of a PLC processor and a programming terminal to make program changes while the processor is running.

Online   The state of being in continuous communication with the PLC processor.

Open loop   A control system that does not receive process feedback.

Open loop control   A control system with no feedback signal from sensors or auto corrections.

Optical isolation   Two electronic networks that are connected through a light-emitting diode and a photoelectric receiver. There is no electrical continuity between the two networks.

OR   A logic operation that produces a 0 (OFF) output only when both inputs are held at 0 (OFF).

Output image table   The area of a PLC's memory where information about the status of output devices is stored.

Output   Process variable output that is supplied by the PLC.

Output type   The form of PLC output such as relay, voltage pulse, or analog.

Over-damped response   A controlled variable response that overshoots the set point and then slowly settles back to it.

Overflow   Exceeding the defined range for a number representation, a timer, or a counter preset value.

Overload   A loading condition exceeding the value a device is designed to handle.

Overload relay   A special type relay designed to protect devices like motors by disrupting power flow under excessive overload conditions.

Overshoot   The amount by which a controlled variable exceeds the set point before reaching steady state.

P

Parallel circuit   A circuit where two or more connected components or contacts are connected to the same terminals

PB   The full span of proportional controller action, $PB = 100/gain$.

PC   Personal computer.

Peripherals   External devices, such as printers, disk drives, recorders, or other devices that are connected to a PLC.

PID control   A three-mode control action in which the controller has proportioning, integral (reset), and derivative (rate) action.

PLC   Programmable logic controller.

Potentiometer   A simple transducer that measures displacement based on resistance changes due to the movement of a wiper arm.

Power supply   The unit that supplies the necessary voltage and current to all system's circuitry.

Precision   The closeness of a setting, indication, calibration, or control to the actual value of the quantity being measured, usually expressed as a percentage of full scale.

Preset value   The number of time intervals or events to be counted.

Pressure switch   A switch that is activated at a prespecified pressure.

Process   A continuous chemical or manufacturing operation.

Process variable   What you are trying to control: temperature, pressure, flow, composition, pH, etc. Also called the *measurement*.

Program   A sequence of instructions to be executed by the processor.

Program scan   The PLC processor executes/resolves every network/rung of the ladder logic program during each program scan.

Programmable controller   An intensive I/O computer designed to handle real-time requirement and equipped with software tools to implement industrial control.

Proportional band   The area around the set point where the controller output is neither fully ON nor fully OFF for the entire time cycle.

Proportional gain   This is the "P" part of the PID control, Prop. gain = 100/PB.

R

Rack   A housing used to hold modules and associated interconnections.

Random access memory (RAM)   A memory system allowing random access of any location during read, write, or fetch cycles.

Range   The area between two limits in which a quantity or value is quantified, represented, or measured. Usually described in terms of lower and upper limits.

Read-only memory (ROM)   A memory system allowing random read only of any location. Information is written in the ROM using special device before use.

Real number   Numbers having both integer and fraction parts.

Relay   An electromechanical device that switches electric circuits.

Relay contacts   Relay contacts are either normally open or normally closed relative to their coil de-energized state.

Relay logic   Representation of logic in a format suitable for relays operation.

Remote I/O   An I/O system located at any distance from the processor, which exchanges information through a communication medium.

Remote set point   A feature that allows set point setting from a remote location.

Resolution   The smallest quantifiable increment in a range of values defined for a given signal or entity.

Response time   The time duration required for a process or a device to respond to a change or a request.

Retentive timer   A timer that counts when it receives power and maintains its count when power is lost.

RTD   Resistance temperature detector.

Run   The continuous execution of a program by the PLC processor.

Rung (network)   A group of PLC instructions that control an output, modify storage values, or perform other functions.

S

Sampling rate   The rate at which input data is polled for information.

SCADA   Acronym used to refer to data acquisition and supervisory control system.

Scan time   The time required to update/activate an output connected to a PLC. It includes the updating of I/O image tables, execution of the program, and all diagnostic tasks.

Sensor   A device used to generate measurements of a physical value typically as an electric signal.

Set point   The desired value of a controlled variable.

Set point   The set point is where you would like the process variable to be.

Significant digit   The digit that contributes the most to a represented value.

Steady state   The error in a process control system is zero or within the error dead band.

Step response   The controlled variable response to a sudden change in the controlling variable.

Storage bit   A memory bit that can be set or reset but is not associated with any input or output device.

Subroutine   A program that executes only when it is called on by the PLC logic, which helps implement structured programs by eliminating code repetition.

T

Tag   A unique text name assigned to variable or a constant data stored in a memory area.

Terminal address   The text address assigned to a particular input or output point connection.

Thermocouple   The junction of two dissimilar metals that has a voltage output proportional to the difference in temperature between the hot and the cold junction connected to the sensor.

Time base A time unit generated from the microprocessor clock and used by the PLC timer instructions.

Toggle switch An ON/OFF panel mounted switch with an extended movable lever.

Transducer A device that converts physical quantity as pressure or temperature to an electric signal.

Transitional contact A contact programmed to be on for one program scan for every selected transition from 0 to 1 or 1 to 0.

Transmitter A primary device that translates a process measurement into a current, voltage, or pneumatic signal.

TRIAC A solid-state device used to switch alternating currents.

True An on, 1, or enabled state.

Truth table A table listing the truth value of a given output as function of all input variables state combinations.

U

Undershoot The amount by which a controlled variable falls below the set point before reaching stead state.

Unlatch instruction The second half of a PLC pair instruction that emulates the unlatching function of a latching relay.

Up-counter A counter that counts up typically from 0 to the preset count value.

V

Validation The process of validating and correcting input measurements and user inputs based on user-defined rules.

Variable A process or system entity that can be measured, altered, or controlled.

Variable data Numerical information that can be altered during program execution.

Volatile memory A memory structure that loses its content once power is removed.

W

Watchdog timer A timer that is reset for every scan, and it monitors the processor operation.

Word A group of adjacent bits treated as one unit, typically measured in 2 or 4 bytes.

X

XOR A logic operation that produces true output only when one input or the other is true, but not both.

Z

Zone A section of PLC ladder logic program that can be enabled or disabled by a control function.

Index

Note: Page numbers followed by "f" indicate figures. Page numbers followed by "t" indicate tables.